Antibiotics and Antiviral Compounds

Edited by
K. Krohn, H. A. Kirst and H. Maag

© VCH Verlagsgesellschaft mbH, D-6940 Weinheim (Federal Republic of Germany), 1993

Distribution:

VCH, P. O. Box 10 11 61, D-6940 Weinheim (Federal Republic of Germany)

Switzerland: VCH, P. O. Box, CH-4020 Basel (Switzerland)

United Kingdom and Ireland: VCH (UK) Ltd., 8 Wellington Court, Cambridge CB1 1HZ (England)

USA and Canada: VCH, 220 East 23rd Street, New York, NY 10010-4606 (USA)

Japan: VCH, Eikow Building, 10-9 Hongo 1-chome, Bunkyo-ku, Tokyo 113 (Japan)

ISBN 3-527-29040-0 (VCH, Weinheim)   ISBN 1-56081-745-3 (VCH, New York)

Antibiotics and Antiviral Compounds

Chemical Synthesis and Modification

Edited by

Karsten Krohn, Herbert A. Kirst and Hans Maag

VCH   Weinheim · New York · Basel · Cambridge · Tokyo

Prof. Dr. Karsten Krohn
Universität-GH Paderborn
FB 13 – Angewandte Chemie
Fachgebiet Organische Chemie
Warburger Str. 100
W-4790 Paderborn
Germany

Dr. Herbert A. Kirst
Lilly Research Laboratories
Lilly Corporate Center
Indianapolis, Indiana 46285
USA

Dr. Hans Maag
Syntex Research
3401 Hilview Avenue
P.O. Box 10850
Palo Alto, California 943003
USA

Published jointly by
VCH Verlagsgesellschaft mbH, Weinheim (Federal Republic of Germany)
VCH Publishers Inc., New York, NY (USA)

Editorial Director: Dr. Thomas Mager
Production Manager: Max Denk

Library of Congress Card No. applied for.

British Library Cataloguing-in-Publication Data: A Catalogue record for this book is available from the British Library.

Deutsche Bibliothek Cataloguing-in-Publication Data:
Antibiotics and antiviral compounds : chemical synthesis and modification / ed. by Karsten Krohn ... – Weinheim ; New York ; Basel ; Cambridge ; Tokyo : VCH, 1993
ISBN 3-527-29040-0 (Weinheim ...)
ISBN 1-56081-745-3 (New York)
NE: Krohn, Karsten [Hrsg.]

Foreword

Modern antibiotics play an important part in the treatment of infectious diseases and human cancer. Most of them are derived from natural products. Very often, nature has passed on to us only some basic ideas about their biological activity and, therefore, extensive systematic structure-activity investigations as well as serendipity have been necessary for drug optimization. Improved drugs also demand more sophisticated chemistry. The development of stereoselective methods and new protecting group strategies for building up complex chiral molecules is a continuing challenge. We look forward to exciting new technologies in the field of antisense and triplex forming oligonucleotides which may be used in the future to suppress pathogenic processes by the downregulation of gene expression.

Our knowledge of structural and conformational parameters of drugs and biomolecules is still insufficient for a complete understanding of complex pathogenic mechanisms. Many problems in chemotherapy remain unsolved and new ones have appeared on the scene, e. g., in the fields of bacterial resistance, retroviral infections and immunocompromised patients. One difficulty lies in the extreme diversity of microorganisms with their enormous variability under selective pressure and their multifold defence strategies. Infectious diseases will always be with us and will threaten people in many ways. The quest for more selective and more potent drugs will, therefore, remain a permanent area of research. We will have to be vigilant if we are to maintain the high standards which we have already attained.

I hope that this book will foster and consolidate research into natural products, promote drug development, and give us a glimpse of what we can expect in the years to come.

Frankfurt am Main, February 1993   Walter Dürckheimer

Contents

New Methodology Applied to Antibiotic Synthesis

Macrolide Antibiotics

β-Lactames, Quinolones and Cyclopentanoid Antibiotics

Peptides and Glycopeptides

Enediyne Antibiotics

Carbohydrates in Antibiotic Synthesis

Antiviral Agents

Subject Index

List of Authors to whom Correspondence Should be Addressed

Prof. Hans-Josef Altenbach
Organische Chemie
Univ. – GH Wuppertal
Gaußstr. 20
D-5690 Wuppertal
Germany

Prof. Robert K. Boeckman
Department of Chemistry
University of Rochester
Rochester
New York 14627
USA

Dr. Daniel Bouzard
Cetre de Recherche et de Production
Bristol-Myers S.A.
F-77185 Lognes
Rue de la Maison Rouge
France

Prof. Jan Balzarini
Rega Instituut
Katholieke Univ. Leuven
Minderbroederstraat 10
B-3000 Leuven
Belgium

Dr. Yves Chapleur
Lab. de Chimie Organique 3
Université de Nancy I
PB 239
F-54506 Vandoevre-lès-Nancy Cédex
France

Prof. Alessandro Dondoni
Dipartimento di Chimica
Laboratorio di Chimica Organica
I-4410 Ferrara
Via L. Borsanri
Italy

Prof. George W. J. Fleet
Dyson Perris Laboratory
Oxford University
Oxford, OX1 3QY
South Parks Road
England

Dr. Hans Fliri
Präklinische Forschung
SANDOZ PHARMA AG
CH-4002 Basel
Schweiz

Prof. Giovanni Franceschi
Research Director
Farmitalia Carlo Erba
I-20146 Milano
Via dei Gracchi 35
Italy

Prof. Léon Ghosez
Lab. de Chimie Organique
Université Catholique de Louvain
B-1348 Louvain-La-Neuve
Place L. Pasteur 1
Belgium

Dr. David S. Grierson
Institut de Chimie
des Subst. Naturelles
CNRS
F-91198 Gif sur Yvette
France

Dr. James V. Heck
Antibiotic Department
Merck Sharp & Dohme
Rahway
New Yersey 07065-0900
USA

Prof. Masahiro Hirama
Department of Chemistry
Tohoku University
Sendai 980
Japan

Prof. Dr. Gergard Höfle
GBF
Mascheroder Weg
D-3300 Braunschweig
Germany

Prof. Dr. Reinhard W. Hoffmann
Fachbereich Chemie
Philipps-Universität Marburg
Hans-Meerwein-Straße
3550 Marburg
Germany

Dr. Antonin Holý
Institute of Organic Chemistry
Czechoslovak Academy of Sciences
166 10 Praha 6
Flemingovo namesti 2
Czechoslovakia

Prof. Minoru Isobe
Lab. of Organic Chemistry
Nagoya Univ., Fac. of Agriculture
Chichusa
Nagoja 464
Japan

Dr. Gerhard Jähne
SEG Antiinfektiva SGE 838
Hoechst AG
6230 Frankfurt am Main 80
Germany

Prof. Dr. Janusz Jurczak
Institute of Organic Chemistry
Polish Academy of Sciences
01-224 Warszawa
Kasprzaka 44/52
Poland

Prof. Dr. Horst Kessler
Organisch-Chemisches Institut
TU München
Lichtenbergstr. 4
D-8046 Garching
Germany

Dr. Werner Klaffke
Inst. f. Organische Chemie
Universität Hamburg
2000 Hamburg 11
Martin-Luther-King-Platz 6
Germany

Dr. Herbert A. Kirst
Lilly Research Laboratories
Lilly Corporate Center
Indianapolis
Indiana 46285
USA

Prof. Dr. Horst Kunz
Institut für Organische Chemie
der Univ. Mainz
J.-J. Becher-Weg 18
D-6500 Mainz
Germany

Dr. Hans Maag
Synthex Research
3401 Hilview Avenue, P. O. Box 10850
Palo Alto
California 943003
USA

Prof. Marian Mikolajczyk
Centre of Molecul. Macromol. Studies
Polish Akademy of Science
90-363 Lódz
Sienkiewicza 112
Poland

Prof. Dr. Johann Mulzer
Freie Univ. Berlin
Institut für Organische Chemie (WE 02)
Takustraße 3
W-1000 Berlin 33
Germany

Prof. A. V. Rama Rao
Regional Res. Laboratory
Indien Inst. Chem. Tech.
Hyderabad-500 007
India

Prof. Dieter Schinzer
Institut für Organische Chemie
Universität Braunschweig
Hagenring 30
D-3300 Braunschweig
Germany

Prof. Victor Snieckus
Dept. of Chemistry
Univ. of Waterloo
Waterloo
Ontario N2L 3G1
Canada

Prof. Pierre Sinaÿ
E.N.S., Laboratoire de Chimie
24 Rue Lhomond
F-75231 Paris Cedex 05
France

Prof. Barry M. Trost
Dept. of Chemistry
Stanford University
Stanford
Calfornia 94305-5080
USA

Dr. Eugen Uhlmann
Pharma-Synthese, SGE 838
Hoechst AG
D-6230 Frankfurt am Main 80
Germany

Prof. Peter Welzel
Abt. f. Chemie
Ruhr Univ. Bochum
Iniversitätsstr. 150
D-4630 Bochum
Germany

Prof. Dr. James D. White
Dept. of Chemistry
Oregon State University
Corvallis
Oregon 97331-4003
USA

Dr. Hanno Wild
PH-FE F CWL
Bayer AG
D-5600 Wuppertal
Germany

Prof. Dr. Axel Zeeck
Institut für Organische Chemie
Universität Göttingen
Tammannstr. 2
D-3400 Göttingen
Germany

Dr. Zoltán Zubovics
Drug Research Institute
H-1325 Budapest
PO. Box 82
Hungary

Editors

Prof. K. Krohn
Fachbereich Chemie und Chemietechnik
der Univ. – GH Paderborn
Postfach 1621
D-4790 Paderborn
Germany

Dr. H. A. Kirst
Lilly Research Laboratories
Lilly Corporate Center
Indianapolis
Indiana 46285
USA

Dr. Hans Maag
Syntex Research R6-E3
3401 Hilview Avenue
P. O. Box 10850
Palo Alto, CA 94303
U S A

Antibiotics. A Challenge for New Methodology

Barry M. Trost

Department of Chemistry, Stanford University
Stanford, Ca 94305-5080

Summary: Palladium catalyzed allylic substitutions provide chemo-, regio-, diastereo- and enantioselective strategies for the synthesis of carbocyclic analogues of carbohydrates and nucleosides. The invention of two particular reactions, vicinal hydroxyamination and hydroxycarbation from epoxides or their synthetic equivalents, provide particularly direct syntheses of valienamine, mannostatin A, allosamizoline, allosamidin, aristeromycin and carbovir. The design of a new class of modular chiral ligands provides, for the first time, a high asymmetric induction in palladium catalyzed allylic alkylations in a general way and converts the above syntheses into asymmetric ones.

Introduction

The myriad of biologically important molecules challenges the synthetic chemist to develop simple and practical synthetic strategies. Carbocyclic analogues of amino sugars and nucleosides represent classes of compounds of growing importance because of the tremendous promise for innovative therapy. Glycosidase inhibitors, some of the most potent being carbocyclic analogues of amino sugars, are aiding in developing an understanding of glycoprotein processing in which glycoconjugates present on the surface of mammalian cells constitute functional domains for carbohydrate protein interactions involved in recognition, adhesion, transport, etc.[1] Nucleoside antibiotics of which the carbanucleosides are some of the most promising aid the understanding of diverse cellular functions, especially as they may differ among diverse cell types, ranging from the fundamental source of cellular control, the transcription and translation of nuclei acids, to metabolic roles such as metabolite carriers, energy donors, secondary messengers and enzymatic cofactors.[2] Possible applications of these classes of compounds in immunology, virology, diabetes, systemic mycoses and cancer as well as their activities as herbicides and insecticides stimulates general interest into structures with specific biological function. Access to such compounds depends upon effective synthetic strategies which, in turn, evolve from the available synthetic reactions and reagents. Transition metals open the opportunity to create new types of reactivity from which more efficient strategies may evolve. Palladium catalyzed allylic substitution [3] represents one such "tool" which offers the molecular artisan the opportunity to think about synthetic strategies from a new perspective.

Regioselectivity

Vinyl epoxides and pronucleophiles undergo simple additions under the influence of a Pd(0) catalyst [4] with clean 1,4-regioselectivity (eq. 1).[5] For example, cyclopentadiene monoepoxide and adenine combine

(1)

with complete orientational control both with respect to the vinyl epoxide (1,4) and the purine (N-9). Variation of the heterocycle allows generation of a variety of including carbovir,[6] a clinical candidate for the treatment of AIDS.[7]

For many applications, attack of the pronucleophile in a 1,2 orientation with respect to the vinylepoxide is required. An approach envisions tethering the pronucleophile to the epoxide oxygen which should deliver the nucleophile to the adjacent carbon. For example, exposure of a vinyl epoxide in the presence of

(2)

1

carbon dioxide as the pronucleophile to a Pd(0) complex creates the oxygen nucleophile by formation of a zwitterionic carbonate **1** which can collapse only by 1,2-attack to give the 5-membered cyclic carbonate as the product.[8] The sequence of asymmetric epoxidation-carboxylation constitutes a convenient asymmetric *cis*-hydroxylation. The example illustrates its application in an asymmetric synthesis of the antiviral agent (+)-citreoviral.[9]

Replacing the carbon dioxide by other cumulative unsaturated moieties offers the opportunity to deliver nucleophiles other than oxygen. Among the most important is introduction of nitrogen which can be accomplished by employing an isocyanate as the pronucleophile. In this case, the zwitterionic intermediate

(3)

2

2 may undergo cyclization by either O or N attack.[10] The higher nucleophilicity of N towards π-ally palladium species favors this path to generate the oxazolidin-2-one **3** which proved to be a useful intermediate towards the aminosugar acosamine.[11] A most interesting feature of this process is the dependence of the geometry of the oxazolidin-2-one on the nature of the isocyanate. In contrast to the above example wherein the *trans*-epoxide generated the *trans*-oxazolidin-2-one, employment of 2-methoxy-1-naphthyl isocyanate generates the thermodynamically less stable *cis*-oxazolidin-2-one regardless of the stereochemistry of the starting material.[12] Because of the ease of conversion of these oxazolidin-2-ones to amino alcohols, the reaction of vinyl epoxides with isocyanates serves as a powerful approach to this important structural type.

Valienamine

The importance of valienamine (**4**) both because of its own biological activity as an α-glucosidase inhibitor[13] as well as an important structural unit of acarbose,[14] an antidiabetic, and validamycin A,[15] a fungicide, has stimulated many synthetic efforts-especially beginning from carbohydrates.[16]   The presence of a *cis*-vicinal amino alcohol immediately suggests the utility of an oxazolidin-2-one such as **6**

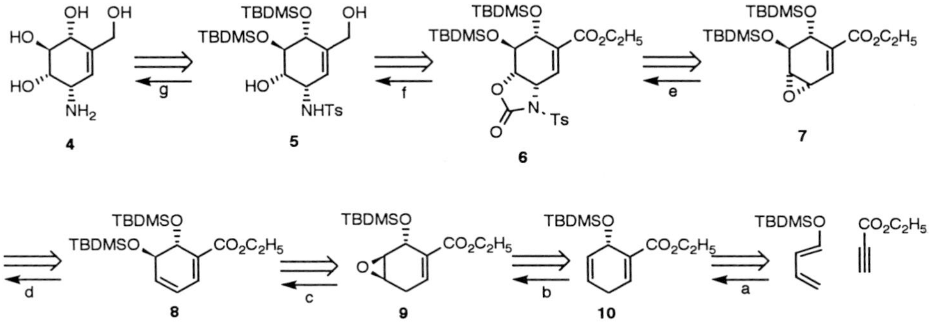

a) Neat, methylene blue, 80°, 91%. b) MCPBA, CH_2Cl_2, $NaHCO_3$, 78%. c) DBU, TBDMS-Cl, DMAP, PhH, rt, 76%. d) MCPBA, $NaHCO_3$, CH_2Cl_2, rt, 85%. e) **11**, 1.5% $Pd(OAc)_2$, nC_4H_9Li, TsNCO, $(CH_3)_3SnOAc$, THF, 54%. f) DIBAL-H, CH_2Cl_2, -78°, 72%. f) Na, NH_3, THF then TBAF, THF.

Scheme 1. Retrosynthetic Analysis and Synthesis of Valienamine

which obviously will derive from the Pd(0) catalyzed reaction of the vinyl epoxide **7** and p-toluenesulfonyl isocyanate (see Scheme 1). A straightforward retrosynthetic analysis of epoxide **7** suggests a siloxydiene and ethyl propiolate as the basic building blocks.[17]   While the Diels-Alder adduct **10** is only a breath away from aromatizing, its generation and diastereoselective epoxidation proceeds without incident to generate the monoepoxide **9**. Treatment with a tertiary amine base in the presence of a silylating agent creates a second cyclohexadiene **8** which also might have been anticipated to be sensitive towards aromatization but which is not.

Molecular mechanics calculations reveal the molecule preferentially adopts a half chair conformation in which both siloxy groups adopt a pseudo-axial orientation (see Figure 1). Thus, epoxidation proceeds

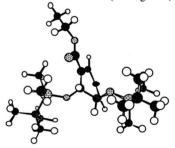

Figure 1. Conformation of Diene **8**

diastereoselectively to give a single monoepoxide **7**. Difficulties in the Pd(0) catalyzed reaction were anticipated because of electronic deactivation caused by the presence of an electron withdrawing group on the double bond and of steric deactivation caused by the presence of an axially oriented siloxy group blocking approach of the Pd(0) complex distal to the epoxide oxygen. It was therefore not surprising that treatment of the epoxide **7** with p-toluenesulfonyl isocyanate in the presence of tetrakis (triisopropyl

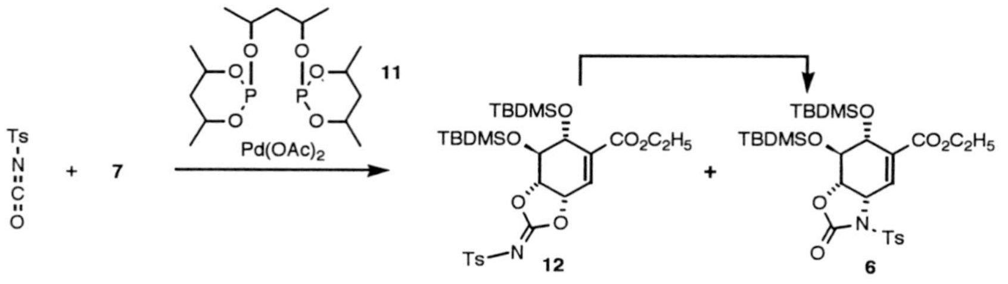

phosphite) palladium gave no reaction. To overcome the steric inhibition, a sterically less demanding surrogate for triisopropyl phosphite, i.e. the 1,3,2-dioxaphosphorinanyloxyl pentane **11**,[18] was employed as ligand (see eq. 4). To counteract the electronic deactivation, a co-catalyst to make the oxygen a better leaving group was employed. Using the simplest co-catalyst, a proton, the product became the iminocarbonate **12** that derived from O rather than N alkylation. While Pd(0) catalysis can isomerize **12** to the desired oxazolidin-2-one **6**, use of trimethyltin acetate as the co-catalyst[19] proved more effective by generating the desired product directly in 54% yield.

DIBAL-H reduction simultaneously generates the hydroxymethyl side chain and cleaves the oxazolidin-2-one. Removal of both the sulfonamide and silyl groups completes an eight step sequence which compares quite favorably with the 14-20 steps of all the previous syntheses.[16]

Mannostatin A

Mannose inhibitors have attracted special attention because of their potential as anti-AIDS agents. The isolation of mannostatins A (**13**) and B,[20] nanomolar inhibitors of high specificity, stimulate great

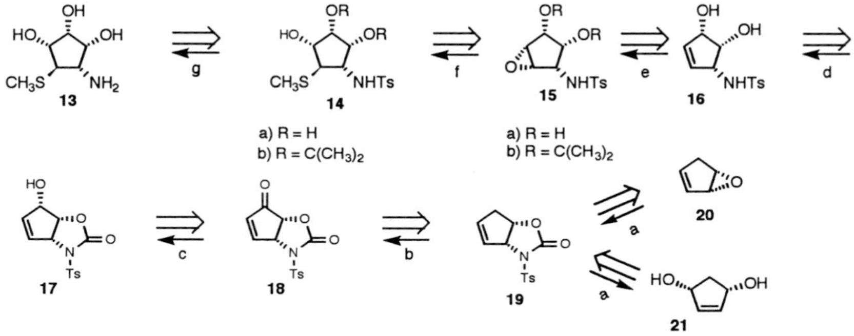

a) TsNCO, 9 mol% (dba)$_3$Pd$_2$ • CHCl$_3$, 7.2 mol% (iC$_3$H$_7$O)$_3$P, THF, reflux, 97%. b) SeO$_2$, diglyme, 170° then Dess-Martin periodinane, CH$_2$Cl$_2$, rt, 65%. c) NaBH$_4$, CH$_3$OH, C$_2$H$_5$OAc, CeCl$_3$ 83%. d) K$_2$CO$_3$, CH$_3$OH-H$_2$O, rt, 95%. e) i. CH$_3$C(OCH$_3$)$_2$CH$_3$, CH$_3$COCH$_3$, CSA, rt, 93%. ii. CF$_3$CO$_3$H, Na$_2$HPO$_4$, CH$_2$Cl$_2$, 90%. f) CH$_3$Li, CH$_3$SSCH$_3$, THF, -5°, 78%. g) i. Na, NH$_3$, -78°, 97%, ii) 60% aq. CF$_3$CO$_2$H, 60°, 86%.

Scheme 2. Retrosynthetic Analysis and Synthesis of Mannostatin A interest in synthetic approaches to such compounds.[21] As revealed by the retrosynthetic analysis of Scheme 2, the presence of the *trans*-thiohydrin suggests its creation via epoxide ring opening (**15→14**) with the epoxide naturally derived from the olefin **16**.[20,21] Allowing for the regioselective introduction of the allylic hydroxyl group by allylic oxidation of **19** makes the presence of the *cis* amino alcohol adjacent to the double bond a natural target for the Pd(0) catalyzed methodology.

Indeed, reaction of the vinyl epoxide **20** with p-toluenesulfonyl isocyanate in the presence of a catalyst generated from palladium acetate and triisopropyl phosphite gives a virtually quantitative yield of the oxazolidin-2-one.   Because of our interest in ultimately effecting an asymmetric synthesis using enantioselective Pd(0) reactions, an achiral synthon of the epoxide, the meso diol **21** was also examined as a precursor for the oxazolidin-2-one **19**. Indeed, treatment of the diol with 2 eq. of p-toluenesulfonyl isocyanate followed by addition of the same Pd(0) catalyst to the resulting solution gives a 97% yield of the same oxazolidin-2-one **19**.

Allylic oxidation with selenium dioxide followed by treatment with the Dess-Martin periodinane generated the enone **18** which was directly reduced with sodium borohydride in the presence of cerium chloride to give the all *cis* product **17**. The alcohol and/or amide functionality present adjacent to the double bond of **16** serves to hydrogen bond to a peracid and atom transfer the oxygen in a *cis* fashion to give the epoxide **15** as a single diastereomer. The most difficult step of the sequence proved to be the regioselective ring opening of this epoxide. All efforts led to 1:1 regioisomeric mixtures at best. To the extent that any selectivity was observed, it was in the wrong direction.

When attempts to affect the selectivity favorably by coordinating in some fashion to the existing functionality failed, we turned to the concept of stereoelectronic factors to affect regioselectivity. While such effects are well known in six membered rings (Fürst-Plattner rule),[23] extension of such concepts to five membered rings is lacking. Reasoning that a bicyclo[3.3.0]octyl system preferentially adopts a

butterfly type conformation, the ring opening of the acetonide **22**, easily derived from diol **15a,** would proceed regiopreferentially to thiohydrin **14b** (eq. 5). Indeed, the latter proved to be the major product (4:1 dr) upon treatment of the acetonide simply with lithium methylmercaptide. Desulfonylation and hydrolysis completes an efficient ten step synthesis of mannostatin A in 27% overall yield.

Allosamizoline and Allosamidin

Identification of a potent chitinase inhibitor, the pseudo trisaccharide allosamidin **23**, revealed it was composed of two molecules of N-acetylallosamine and a novel carbocyclic analogue of an amino sugar whose structure was initially assigned as **26** and thus was named allosamizoline to suggest its structural

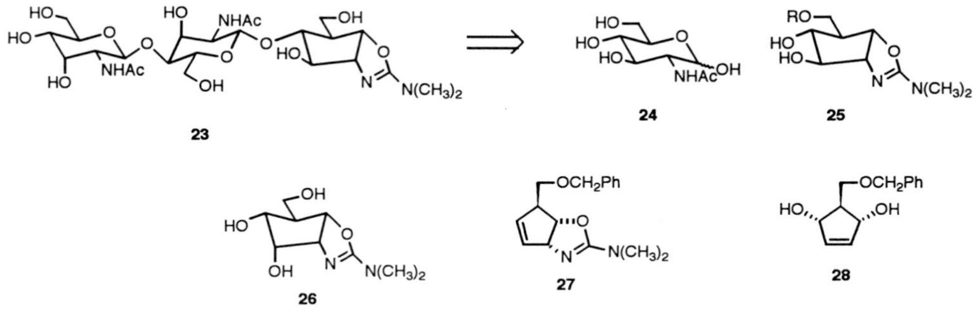

23

24   **25**

26   **27**   **28**

relationship to allosamine.[24]   Subsequent characterization studies led to the revised stereochemical assignment depicted in **25**.[25,26] The cyclopentene **27** is an excellent intermediate since it could serve not only to affirm this structural revision but also provide access to all four diastereomeric diols. This proposed intermediate contains the exact structural features easily assembled by our Pd(0) methodology - a *cis* amino alcohol adjacent to a double bond. As in the case of mannostatin, while the monoepoxide of a substituted cyclopentadiene would serve as the necessary substrate, considerations for asymmetric synthesis suggest the meso diol **28** to be more suitable. Scheme 3 depicts the synthesis starting from commerically available

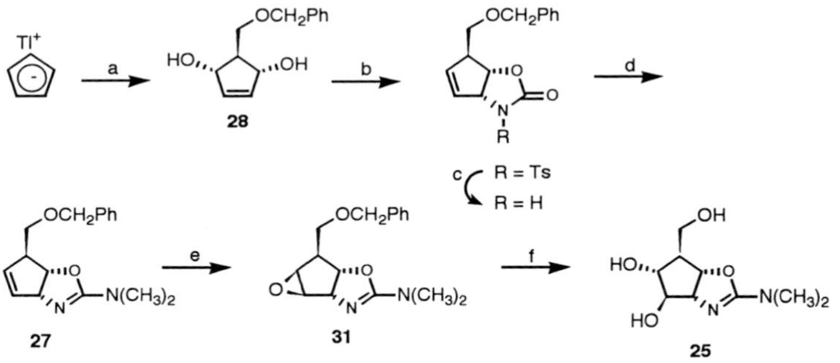

28

R = Ts
R = H

27   **31**   **25**

a) PhCH$_2$OCH$_2$Cl, ether, then CH$_3$OH, methylene blue, O$_2$, hv, 33-60%. b) 2 eq. TsNCO, 0.75 mol% (dba)$_3$Pd$_2$ • CHCl$_3$, 2 mol% (iC$_3$H$_7$O)$_3$P, PhCH$_3$, rt, 96%. c) Na, C$_{10}$H$_8$, DME, -78°, 92%. d) CH$_3$OSO$_2$CF$_3$, CH$_2$Cl$_2$, 0° then (CH$_3$)$_2$NH, rt, 100%. e) 5.4 M CF$_3$CO$_3$H, CF$_3$CO$_2$H, 0°. f) CF$_3$CO$_2$H, H$_2$O then 2 atm H$_2$, 10% Pd/C, CH$_3$OH, rt, 68% for e and f.

Scheme 3. A Total Synthesis of Allosamizoline

thallium cyclopentadienide.[27] Monoalkylation was followed directly by singlet oxygen cycloaddition with *in situ* reduction of the endoperoxide to give **28** directly to avoid regioequilibration of the 5-monoalkylcyclopentadiene either by thermal 1,5-hydrogen shift or by base. Formation of the bis-urethane was followed by direct addition of the Pd(0) catalyst to give the N-tosyloxazolidin-2-one nearly quantitatively. Activation of the detosylated oxazolidin-2-one with methyl triflate followed by direct addition of anhydrous dimethyl amine provided the pivotal cyclopentene **27** in only four steps. *cis*-Hydroxylation using catalytic osmium tetroxide provided the expected attack from the convex face which,

(6)

after catalytic hydrogenolysis, gave the triol **29** (eq. 6). *cis*-Hydroxylation from the concave face was accomplished by coordinating the substrate to cyclodextrin to shield the convex face to give the originally assigned structure, triol **30**, after catalytic hydrogenolysis. Neither structure was identical to the natural product, confirming the need for a structural revision to a *trans* isomer. On the other hand, epoxidation of **21** with 1.2 M trifluoroperacetic acid produced a mixture of the two monoepoxides **31** and **32**, the former undergoing much more rapid solvolysis so that a mixture of a monobenzyl ether of allosamizoline **33** and the α-monoepoxide **32** was isolated (eq. 7). More robust solvolysis (50% aq. trifluoroacetic acid, 65°) and catalytic hydrogenolysis completed the synthesis of the *trans*-diol **34**. On the other hand, epoxidation with 5.4 M trifluoroperacetic acid produced predominantly the β-epoxide **31** whose solvolysis and debenzylation was performed as one operation to give the *trans*-triol **25** directly (see Scheme 3) which was

(7)

identical to the natural allosamizoline, confirming its structural assignment. Since epoxidation and solvolysis was performed as a single step, the overall sequence requires just six steps from cyclopentadiene (36% overall yield).

The ease of availability of the monobenzyl ether of allosamizoline **33** led to the examination of its direct glycosylation with the disaccharide **35** to examine the intrinsic chemoselectivity with respect to the two secondary alcohols. Using racemic allosamizoline but scalemic disaccharide **35**, excellent chemoselectivity for reaction at the required secondary alcohol was observed (eq. 8) to produce allosamidin and one of its epimers after deprotection.[29]

(8)

Asymmetric Induction

Conversion of our syntheses of mannostatin and allosamizoline into asymmetric ones requires a chiral Pd(0) complex to differentiate the two enantiotopic leaving groups. Unlike other transition metal catalyzed reactions in which high asymmetric induction occurs with chiral ligands because bond making occurs between groups directly bound to the metal, allylic alkylation has the metal and its attendant ligands on the face of the substrate opposite from where bond breaking (eq. 9) and/or bond making occurs.

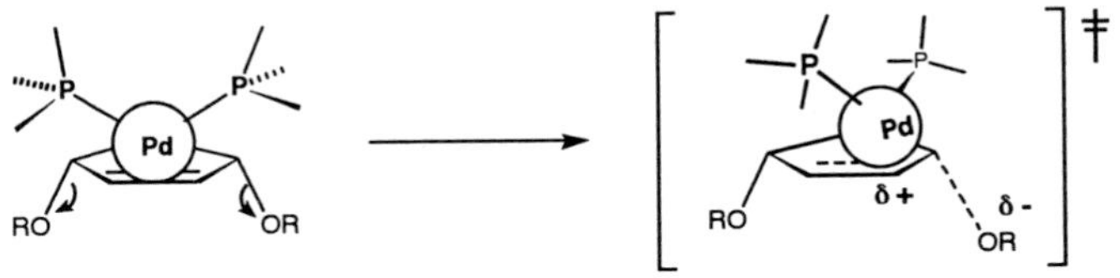

The concept we pursued was to design chiral ligands in a modular sense consisting of a chiral scaffold to which the phosphine metal binding posts are attached.[30] The buttressing interaction between the diphenylphosphino moieties and the scaffold would create a chiral pocket which may embrace the substrate as depicted in Figure 2.

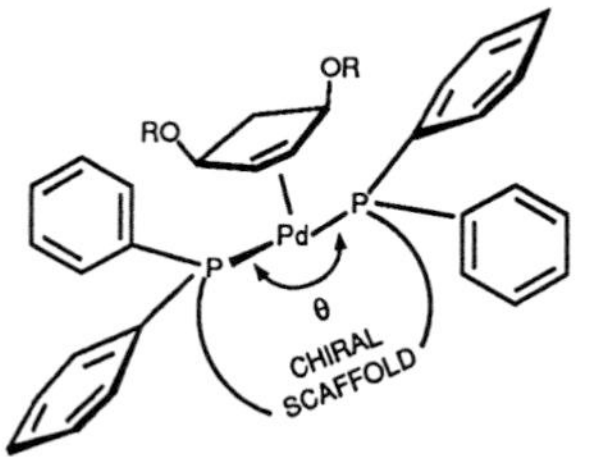

Figure 2. A Model for Ligand Effect

For this purpose, we designed ligands easily derived from C_2 symmetric diols or diamines. For example, acylation of the diol **36**, easily derived from tartaric acid by heating with benzyl amine, or the well known R,R-1,2-diaminocyclohexane (**37**) with 2-diphenylphosphino-benzoic acid provided the two C_2 symmetric ligands **38** (eq. 10) and **39** (eq. 11).

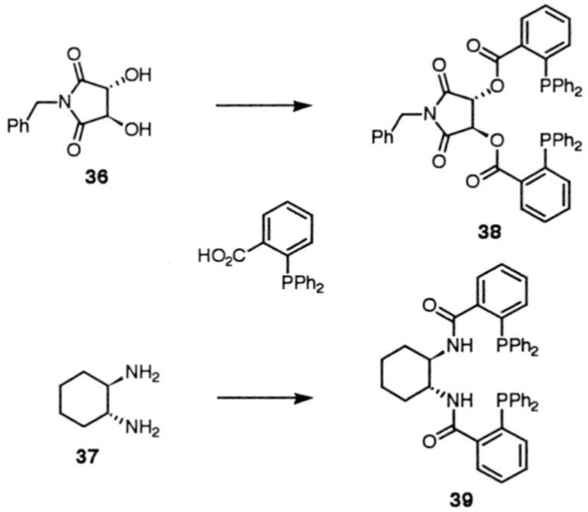

Using the cyclization of the bis-urethane **40** generated *in situ* from 3,5-dihydroxycyclopentene as our test reaction (eq. 12) with the diester ligand **38** at 0° gave (-)-**19** possessing an enantiomeric ratio (*er*) of

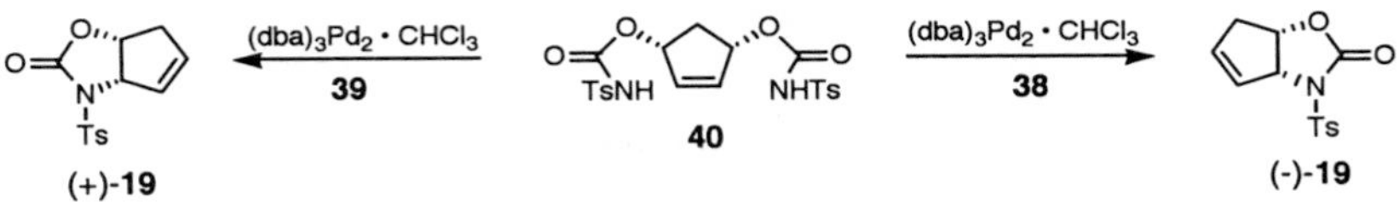

(12)

(+)-19   39   40   38   (-)-19

87.5:12.5; whereas the diamide ligand **39** at 0° gave (+)-**19** having an *er* of 89:11. Envisioning the depth of the chiral pocket would be related to the P-Pd-P angle (so called bite angle θ in fig. 2) which in turn would be related to the dihedral angle between the tethering points of the binding posts (i.e. the diol or

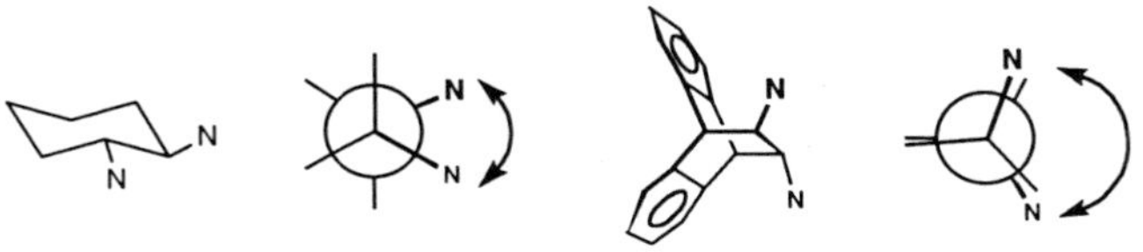

Fig. 3. Enhancement of Dihedral Angle

diamine), we sought to increase this angle by going from the cyclohexyl ring to a bicyclo [2.2.2] octyl system (see fig. 3 and eq. 13). Indeed, cyclization with ligand **40** produces (-)-**19** consisting of a 94:6 *er*.

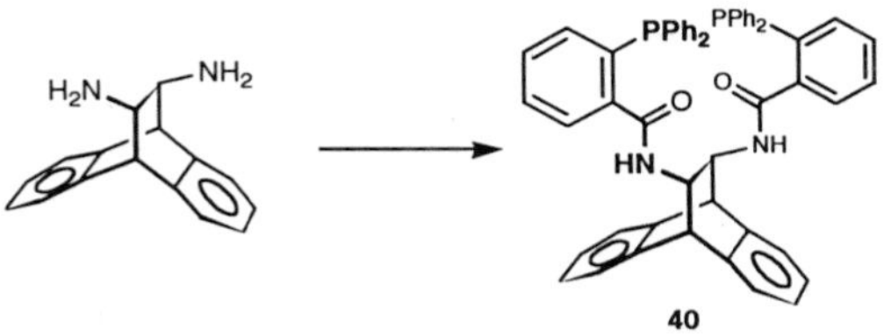

(13)

40

Furthermore, the stereochemical course of the reaction correlates with the sense of chirality of the ligand as outlined in Figures 4 and 5. Thus, a simple mnemonic emerges whereby clockwise oriented binding posts initiated preferential ionization of the pro-R leaving group and counterclockwise oriented binding posts initiated preferential ionization of the pro-S leaving group.

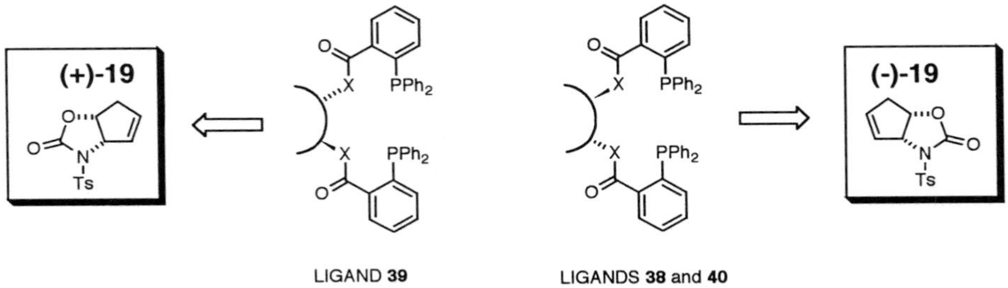

(+)-19   (-)-19

LIGAND **39**   LIGANDS **38** and **40**

Figure 4. Correlation of Absolute Stereochemistry of the Product with that of the Variable Chiral Linker

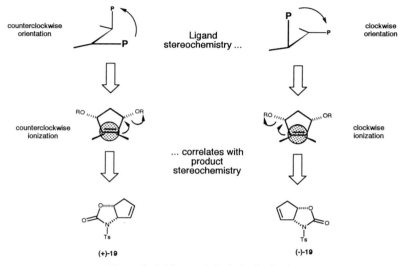

Figure 5. A Mnemonic Rule for Ionization

Access to (+)- or (-)-**19** then provides access to either enantiomer of mannostatin. Furthermore, extension of this concept to the diol precursor **29** of allosamizoline gave the (+)-oxazolidine depicted with an 85:15 *er* using the R,R-diamide ligand **41** (eq. 14). Thus, these ligands also provide access to either enantiomer of allosamizoline as well.

$$(14)$$

Carbanucleosides

The discovery of high enantiodiscrimination in the Pd(0) catalyzed ionization of prochiral leaving groups led to our exploration of this concept for an asymmetric synthesis of the carbocyclic analogues of nucleosides, several of which are being advanced as anti-AIDS agents.[2,31]   Subjecting the bis(diphenylacetate) or preferably dibenzoate of *cis*-3,5-dihydroxycyclopentene (**42**) to lithium phenylsulfonylmethylnitronate in the presence of chiral Pd(0) catalysts derived from ligands **38** or **39**

$$(15)$$

gave a novel cycloalkylation product **43** (eq. 15) in which the initial C-alkylated product undergoes enolization and a subsequent O-alkylation of the resultant nitronate to the observed product.[32] In accord with the earlier observations, switching from the diester ligand **38** to the diamide ligand **39** saw a dramatic increase in enantioselectivity from an *er* of 82:18 to 98:2.   Furthermore, the complementary sense of chirality of the ligands created products which were the opposite mirror image isomers. The diester ligand **38** gave (-)-**43** (86% yield), the product resulting from kinetic ionization of the pro-R leaving group; whereas, diamide ligand **39** gave the product from kinetic ionization of the pro-S leaving group, (+)-**43** in 94% yield.

The resultant isoxazoline-2-oxides are extremely versatile building blocks. Whereas, the sulfone moiety of **43** was relatively inert towards substitution, deoxygenation to the isoxazoline **44** with stannous chloride dihydrate (CH₃CN, rt, 94% yield) labilized this group so that a broad range of nucleophiles may

readily displace it.   For example, solvolysis in basic methanol produced a 91% yield of the methoxyisoxazoline **45**. Reduction of either **44** or **45** with molybdenum hexacarbonyl in moist acetonitrile gave the *cis*-hydroxynitrile **46** or *cis*-hydroxyester **47** respectively of very high *er*.

The diastereo- and enantioselective *cis*-hydroxyalkoxycarbonylation readily translates into a general strategy for the synthesis of carbanucleosides.   Reduction (LAH, ether, 95%) and acylation (nC₄H₉Li, THF, then ClCO₂CH₃, 98%) readily produces the scalemic dicarbonate **48**. This pivotal intermediate, available in 63% overall yield from **42a** in six steps, requires S_N2' substitution with retention of configuration with any desired purine or pyrimidine, a process that requires the selectivity imposed by Pd(0) catalysis, to complete the synthesis of whatever carbanucleoside may be desired. Exposing a neutral

solution of adenine and the bis-carbonate **48** to a Pd(0) catalyst in DMSO gave the cyclopentene **49** as the exclusive product in 96% yield. Its *cis*-hydroxylation completed a synthesis of (-)-aristeromycin **50**.[33] Alternatively, Pd(0) catalyzed substitution using 2-amino-6-chloropurine produced cyclopentene **51** exclusively in 77% yield. Simple hydrolysis completed a synthesis of (-)-carbovir.[34] The directness and flexibility of this strategy makes it a very attractive general approach to this increasingly important family of compounds.

Conclusion

Pd(0) catalyzed allylic alkylations provides opportunities for extraordinary chemo-, regio-, diastereo- and, for the first time in a general sense, enantioselectivity. By its ability to introduce and manipulate functionality on five and six membered rings, exciting, practical approaches to carbocyclic analogues of both monosaccharides and nucleosides have resulted contributing a powerful tool for investigation of the biological consequences of such structures.

Acknowledgment   Foremost, I am indebted to an especially talented group of co-workers, D.L. Van Vranken, T. Lübbers, L. Li, Y. Shi and S.D. Guile who have contributed so much to the intellectual development of this program as well as being responsible for all the experimental work.   The National Science Foundation generously supported the underlying methodological studies and the National Institutes of Health, General Medical Sciences, their applications to the specific targets.

References

1. Springer, T. A. *Nature* **1990**, *346*, 425; Stoolman, L. M. *Cell* **1989**, *56*, 907; Sharon, N.; Lis, H. *Science* **1989**, *246*, 227; Clark, A.E.; Wilson, I.A. Eds. "Carbohydrate-Protein Interactions" Springer Verlag, Heidelberg, 1988; Elbein, A.D. *Ann. Rev. Biochem.* **1987**, *56*, 497.

2. Deuholm, K. L.; Pedersen, E. B. *Synthesis* **1992**, 1; Isono, K. *J. Antibiot.* **1988**, *41*, 1711; Marquez, V. E.; Lim, M. *Med. Res. Rev.* **1986**, *6*, 1.

3. Trost, B. M. *Angew. Chem. Int. Ed. Engl.* **1989**, *28*, 1173; Trost, B.M. in "Stereochemistry of Organic and Bioorganic Transformations," Bartmann, W.; Sharpless, K.B. Eds.; VCH Verlagsgesellschaft mbH, Weinheim, 1987, 191; Trost, B. M. *J. Organomet. Chem.* **1986**, *300*, 263; Trost, B. M. *Pure Appl. Chem.* **1981**, *53*, 2357; Trost, B. M. *Accounts Chem. Res.* **1980**, *13*, 385. For a recent review, see Godleski, S.A. in "Comprehensive Organic Synthesis," Trost, B.M.; Fleming, I. Eds. Pergamon Press, Oxford, 1991, Vol. 4, p. 585.

4. Trost, B. M.; Molander, G. A. *J. Am. Chem. Soc.* **1981**, *103*, 5969.

5. Trost, B. M.; Kuo, G. H.; Benneche, T. *J. Am. Chem. Soc.* **1988**, *110*, 621.

6. Peel, M. R.; Sternbach, D. D.; Johnson, M. R. *J. Org. Chem.* **1991**, *56*, 4990.

7. Vince, R.; Brownell, J. *Biochem. Biophys. Res. Commun.* **1990**, *168*, 912.

8. Trost, B. M.; Angle, S. R. *J. Am. Chem. Soc.* **1985**, *107*, 6123.

9. Trost, B. M.; Lynch, J.; Angle, S. R. *Tetrahedron Lett.* **1987**, *28*, 375.

10. Trost, B. M.; Hurnaus, R. *Tetrahedron Lett.* **1989**, *30*, 3893.

11. Trost, B. M.; Sudhakar, A. R. *J. Am. Chem. Soc.* **1987**, *109*, 3792.

12. Trost, B. M.; Sudhakar, A. R. *J. Am. Chem. Soc.* **1988**, *110*, 7933.

13   Furumoto, T.; Yoshioka, T.; Kamata, K.; Kameda, Y.; Matsui, K. *J. Antibiot.* **1991**, *44*, 371.

14. Clissold, S. P.; Edwards, C. *Drugs* **1988**, *35*, 214; Ogawa, S.; Shibata, Y. *Chem. Commun.* **1988**, 605.

15. Suami, T.; Ogawa, S.; Chida, N. *J. Antibiot.* **1980**, *33*, 98; Kameda, Y.; Asano, N.; Teranishi, M.; Matsui, K. *J. Antibiot.* **1980**, *33*, 1573.

16. Knapp, S.; Naughton, A. B. J.; Murali Dhar, T. G. *Tetrahedron Lett.* **1992**, *33*, 1025; Nicotra, F.; Panza, L.; Ronchetti, F.; Russo, G. *Gazz. Chim. Ital.* **1989**, *119*, 577; Yoshikawa, M.; Cha, B. C.; Okaichi, Y.; Takinami, Y.; Yokokawa, Y.; Kitagawa, I. *Chem. Pharm. Bull.* **1988**, *36*, 4236; Schmidt, R. R.; Köhn, A. *Angew. Chem. Int. Ed. Engl.* **1987**, *26*, 482; Ogawa, S.; Chida, N.; Suami, T. *J. Org. Chem.* **1983**, *48*, 1203; Paulsen, H.; Heiker, F. R. *Annalen* **1981**, 2180; Ogawa, S.; Toyokuni, T.; Suami, T. *Chem. Lett.* **1980**, 713.

17. Lübbers, T.; Shi, Y. unpublished work in these laboratories.

18. Trost, B. M.; Vos, B. A.; Brzezowski, C. M.; Martina, D. P. *Tetrahedron Lett.* **1992**, *33*, 717.

19. Trost, B. M.; King, S. A. *J. Am. Chem. Soc.* **1990**, *112*, 408; Trost, B. M.; King, S. A.; Schmidt, T. *J. Am. Chem. Soc.* **1989**, *111*, 5902.

20. Aoyagi, T.; Yamamoto, T.; Kojiri, K.; Morishima, H.; Nagai, M.; Hamada, M.; Takeuchi, T.; Umezawa, H. *J. Antibiot.* **1989**, *42*, 883; Morishima, H.; Kojiri, K.; Yamamoto, T.; Aoyagi, T.; Nakamura, H.; Iitaka, Y. *J. Antibiot* **1989**, *42*, 1008.

21   Trost, B. M.; Van Vranken, D. L. *J. Am. Chem. Soc.* **1991**, *113*, 6317.

22. For contemporaneous syntheses see Knapp, S.; Dhar, T. G. M. *J. Org. Chem.* **1991**, *56*, 4096; King, S. B.; Ganem, B. *J. Am. Chem. Soc.* **1991**, *113*, 5089; Ogawa, S.; Yuming, Y. *Chem. Commun.* **1991**, 890.

23. Fürst, A.; Plattner, P. A. *Helv. Chim. Acta* **1949**, *32*, 275.

24. Sakuda, S.; Isogai, A.; Matsumoto, S.; Suzuki, A. *Tetrahedron Lett.* **1986**, *27*, 2475.

25. Sakuda, S.; Isogai, A.; Matsumoto, S.; Suzuki, A. *J. Antibiot* **1987**, *40*, 296.

26. Sakuda, S.; Isogai, A.; Matsumoto, S.; Koseki, K.; Kodema, H.; Yamada, Y. *Agr. Biol. Chem.* **1988**, *52*, 1615; Sakuda, S.; Isogai, A.; Makita, T.; Matsumoto, S.; Koseki, K.; Kodama, H.; Suzuki, A. *Agr. Biol. Chem.* **1987**, *51*, 3251; Nishimoto, Y.; Sakuda, S.; Takayama, S.; Yamada, Y. *J. Antibiot.* **1991**, *44*, 716.

27. Trost, B. M.; Van Vranken, D. L. *J. Am. Chem. Soc.* **1990**, *112*, 1261.

28   Also see Simpkins, N. S.; Stokes, S.; Whittle, A. J. *Tetrahedron Lett.* **1992**, *33*, 793; Nakata, M.; Akazawa, S.; Kitamura, S.; Tatsuta, K. *Tetrahedron Lett.* **1991**, *32*, 5363; Takahashi, S.; Terayama, H.; Kuzuhara, H. *Tetrahedron Lett.* **1991**, *32*, 5123; Griffith, D. A.; Danishefsky, S. J. *J. Am. Chem. Soc.* **1990**, *112*, 5811.

29. Maloisel, J.-L.; Vasella, A.; Trost, B. M.; Van Vranken, D. L. *Chem. Commun.* **1991**, 1099.

30. Trost, B. M.; Van Vranken, D. L. *Angew. Chem. Int. Ed. Engl.* **1992**, *31*, 228; Trost, B. M.; Van Vranken, D. L.; Bingel, C. *J. Am. Chem. Soc.* , in press.

31. Hobbs, J.B. in "Comprehensive Medicinal Chemistry," Hansch, C.; Sammes, P.G.; Taylor, J.B. Eds.; Pergamon; Oxford, 1990; Vol. 2, pp. 306-322; Hovi, T. in "Antiviral Agents: The Development and Assessment of Antiviral Chemotherapy," Field, H.J. Ed.; CRC Press, 1988; Chapter 1, pp. 1-12; Jones, M.F. "Chemistry in Britain," **1988**, 1122; Dolin, R. *Science* **1985**, *227*. 1296.

32. Trost, B. M.; Li, L.; Guile, S.D. *J. Am. Chem. Soc.* in press.

33. For very recent syntheses of aristeromycin see, Arai, Y.; Hayashi, K.; Matsui, M.; Koizumi, T.; Shiro, M.; Kuriyama, K. *J. Chem. Soc. Perkin I.* **1991**, 1709.; Katagiri, N.; Muto, M.; Nomura, M.; Higashikawa, T.; Kaneko, C. *Chem. Pharm. Bull.* **1991**, *39*, 1112; Saville-Stones, E. A.; Lindell, S. D.; Jennings, N. S.; Head, J. C.; Ford, M. J. *J. Chem. Soc. Perkin I.* **1991**, 2603; Wolfe, M. S.; Anderson, B. L.; Borcherding, D. R.; Borchardt, R. T. *J. Org. Chem.* **1990**, *55*, 4712; Maggini, M.; Prato, M.; Scorrano, G. *Tetrahedron Lett.* **1990**, *31*, 6243.

34. For very recent syntheses of carbovir, see Evans, C. T.; Roberts, S. M.; Shoberu, K. A.; Sutherland, A. G. *J. Chem. Soc. Perkin I* **1992**, 589; Gundersen, L.; Benneche, T.; Undheim, K. *Tetrahedron Lett.* **1992**, *33*, 1085; Peel, M. R.; Sternbach, D. D.; Johnson, M. R. *J. Org. Chem.* **1991**, *56*, 4990; Exall, A. L.; Jones, M. F.; Mo, C.-L.; Myers, P. L.; Paternoster, I. L.; Singh, H.; Storer, R.; Weingarten, G. G.; Williamson, C.; Brodie, A. C.; Cook, J.; Lake, D. E.; Meerholz, C. A.; Turnball, P. J.; Highcock, R. M. *J. Chem. Soc. Perkin I* **1991**, 2467; Jones, M. F.; Myers, P. L.; Robertson, C. A.; Storer, R.; Williamson, C. *J. Chem. Soc. Perkin I* **1991**, 2479; Taylor, S. J. C.; Sutherland, A. G.; Lee, C.; Wisdom, R.; Thomas, S.; Roberts, S. M.; Evans, L. *Chem. Commun.* **1990**, 1120; Vince, R.; Hua, M. *J. Med. Chem.* **1990**, *33*, 17.

New Methodology and Applications to the Synthesis of Antibiotics and Other Bioactive Complex Molecules

Robert K. Boeckman, Jr.

Department of Chemistry, University of Rochester

Rochester, New York 14627-0216

For the past 40 years or more, chemists have been actively engaged in the isolation, characterization, and biological evaluation of new naturally occurring materials from marine sources. Marine organisms have proven to be a particularly rich source of structurally novel materials possessing a wide range of biological activities ranging from antibiotic and antitumor activity to activity in the central nervous system. Quite apart from their biological/chemotherapeutic potential, marine sources often afford materials which are structurally unusual or unique when compared to terrestrial sources. In particular, a growing number of substances, possessing a medium ring (or several) containing a heteroatom or atoms often oxygen, have been isolated and structurally characterized. Typical of these substances are oxepins and oxacenes such as Isolaurepinnacin, Laurencin, and Laurenyne (1-3), isolated from the *Laurencia*, and the unique and biologically potent Brevitoxins such as Brevitox B (4).[1,2]

isolaurepinnacin (1)   laurencin (2)   laurenyne (3)

Brevitoxin B (4)

The challenges posed by the construction of these medium ring heterocycles has resulted in the development of a variety of approaches to these ring systems.[3,4] These approaches are typified by epoxide cleavage, transketalization/reduction, and cation-olefin cyclization strategies (Scheme 1).

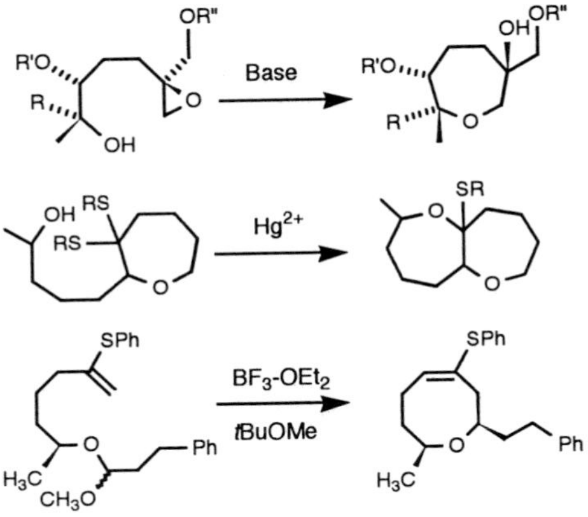

Scheme 1

Our interest in these systems can be traced, in part, to an interest in extending our work on cyclic vinyl ether metalation from 6-membered to larger rings.[5] Having available a ready source of these medium ring vinyl ethers was essential to the pursuit of these studies. It was clear that, in spite of the elegant chemistry cited previously, no truly general approach to these systems existed. Therefore, we set out to develop such methodology based upon the following design criteria. The route to 2,3,4,5-tetrahydrooxepins and 1,2,3,4,5,6-hexahydrooxacenes must be: 1) general with respect to substitution patterns, 2) permit the assembly of fused ring systems comprised of 2 medium rings, 3) permit ready stereospecific and/or stereoselective introduction of chiral centers, 4) proceed from readily available precursors, and 5) be amenable to the production of enantiomerically pure materials.

Our attention was drawn to the observations of Rhodes and others that vinyl cyclopropane carboxaldehyde is in equilibrium with the corresponding dihydrooxepin, even at room temperature.[6] The equilibrium constant was observed to strongly favor the cyclopropane isomer ($K_{eq} \approx 0.05$), and we have confirmed these observations in related substituted systems such as 5 (Eqn. 1). For this reason, the potential utility of this transformation has not been exploited for synthesis in the intervening nearly 25 years.[7] We had observed some years earlier, that it was possible to modulate the position of the Claisen-retroClaisen equilibrium by introduction of suitably positioned π-stabilizing functional groups which induce the equilibrium between a [2.2.2]bicyclic system to an oxadecalin to favor the retroClaisen

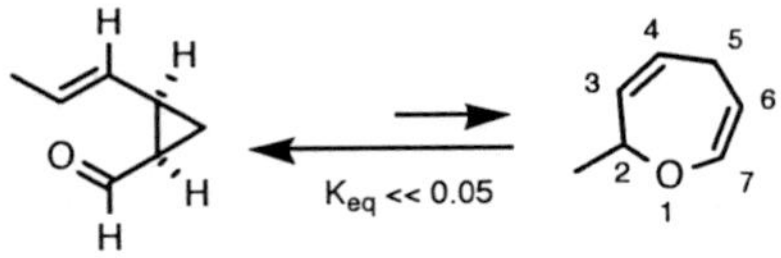

(1)

isomer exclusively.[8]   We, therefore, sought to exploit this fact in order to modify the Claisen-retroClaisen equilibrium in the dihydrooxepin system, and to determine if the concept could be readily extended to the homologous oxacene systems.

The key to shifting the equilibrium is to position a π stabilizing group such as a carbonyl group so as to stabilize the product oxepin. This can be accomplished by situating this group at C_6 (or C_7) as well as potentially C_3 or C_4.[9] Thus, the required substrates become 1,1-dicarbonyl vinylcyclopropanes. These materials are readily available via the general route shown in Scheme 2 involving seqential or tandem alkylation of bis allylic electrophiles with malonates or other doubly stabilized carbanions.[10] Subsequent reduction using either Dibal-H or LAH then affords the corresponding dicarbinol derivatives in yields typically in the 60-85% range.

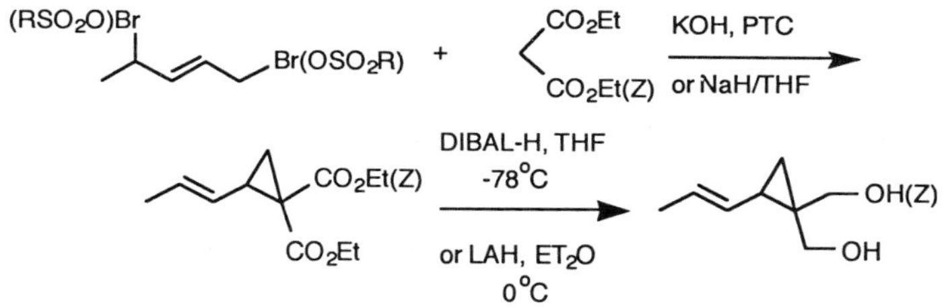

Scheme 2

Oxidation of these1,3-diols by chromium-based reagents and under Swern conditions   were complicated by fragmentation and over oxidation. We found the most suitable and general reagent to be the,[11] although for larger scale work, catalytic tetrapropylammonium perruthenate (TPAP) was found to be suitable in some cases although generally at the expense of a somewhat lower yield.[12] Somewhat surprisingly, the expected dicarbonyl systems are not observed, but these materials directly undergo the desired [3,3] rearrangement to afford the dihydrooxepins such as **6** in excellent overall yield (68-95%) as shown in Scheme 3.

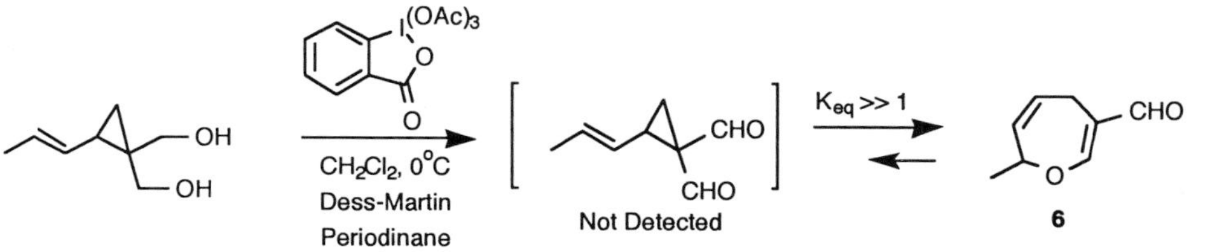

Scheme 3

A selection of examples of substituted cyclopropanes has been examined as shown in Scheme 4. The rearrangement seems to be largely insensitive to substitution, and the nature of the stabilizing group affording the desired dihydrooxepins in very good overall yields. A variety bifunctional electrophiles are suitable precursors, as are a variety of methods of obtaining the requisite vinyl cyclopropanes. Particularly noteworthy is the successful rearrangement of sulfone **7**. Although in this case, the approach to equilibrium may be somewhat slower in that a minor amount of an aldehyde is observed upon workup which is subseqeuntly converted to **8** on standing. Evidence that the rearrangement does indeed proceed *via* the expected mechanism involving a concerted [3,3] sigmatropic process can readily be gleaned from the disparate behavior of the cyano cyclopropymethanols **9** and **10** upon oxidation. The *cis* isomer **9** affords the expected dihydrooxepin **11** whereas the *trans* isomer **10**, which is geometrically incapable of rearrangement, affords the aldehyde **12**.

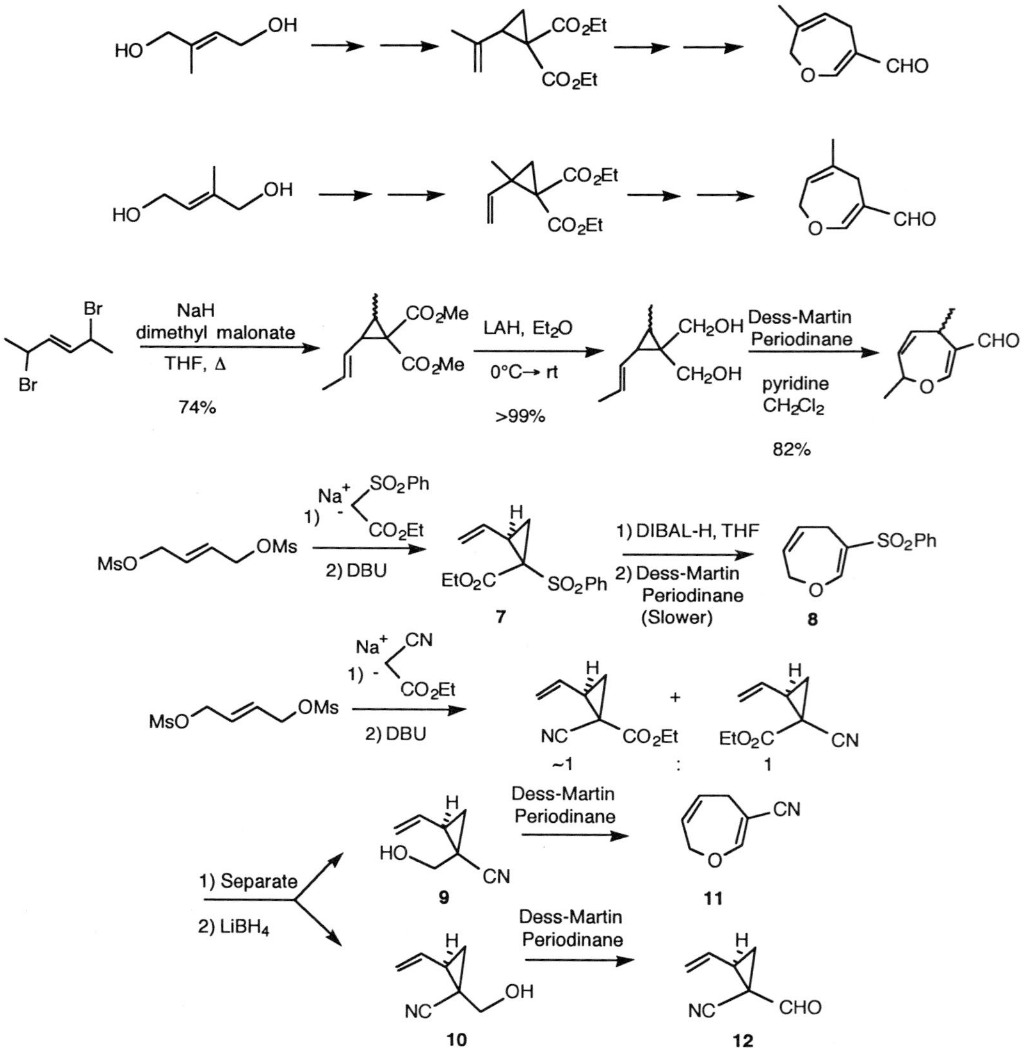

Scheme 4

An example of the variety of bifunctional electropliles which can be employed is depicted in Scheme 5. A cyclic carbonate derived from an α-hydroxy aldehyde derivative, serves as the bifunctional electrophile in an initial palladium catalyzed alkylation.

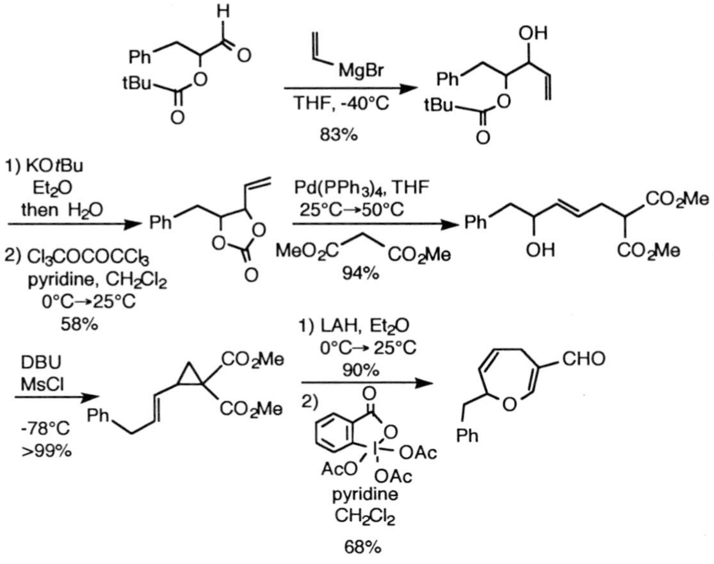

Scheme 5

The only exception to the generality of the rearrangement was observed in the attempted conversion of the cyclopropyl carbinol derived from **13** to dihydrooxepin **14** as shown in Scheme 6. In cases where a particularly stable cation is formed upon ionization of the allylic system, we speculate that a stepwise acid catalyzed rearrangement to the observed dihydrofuran **15** occurs subsequent to the expected [3,3] sigmatropic rearrangement. Although not yet in the case of **13**, formation of dihydrofuran products can be suppressed by effecting the oxidation in the presence of pyridine (3 fold excess relative to periodinane).

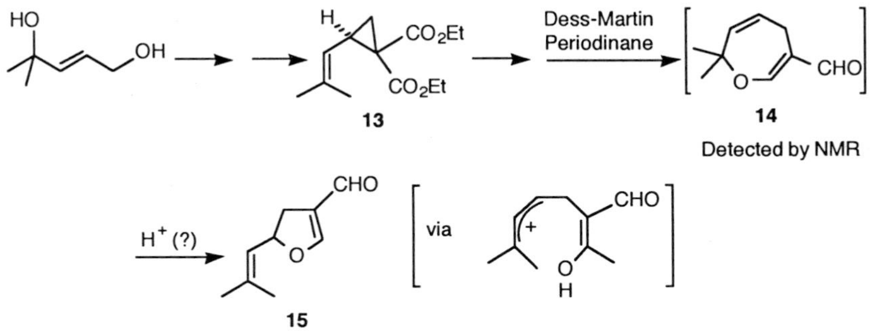

Scheme 6

Our interpretation is supported by the observation that the related carbinol **16** undergoes smooth conversion to the dihydrooxepin **17**, possibly owing to the inductive destabilization of the allylic cation resulting from ionization by the acetoxy group (Eqn 2).

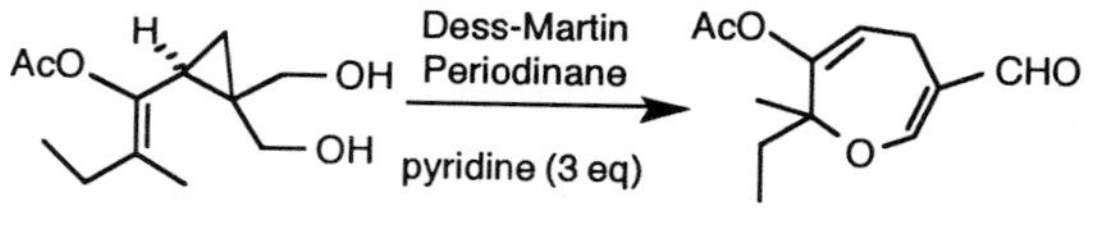

(2)

16   **17**

The versatility of this process arises from the ability to control the position of the Claisen /retroClaisen equilibrium to facilitate further functionalization of the oxepin isomer. For example, controlled reduction of **6** with NaBH$_4$ affords the related mono aldehyde **19** which exists in equilibrium with the dihydrooxepin **18** (~4:1). Advantage of this equilibrium can be taken to introduce nitrogen by reaction   for example, with benzyl amine to afford the sensitive imine **20**, followed by immediate oxidation to the dihydroazepine **21** in 85% yield overall from **6** (Scheme 7).

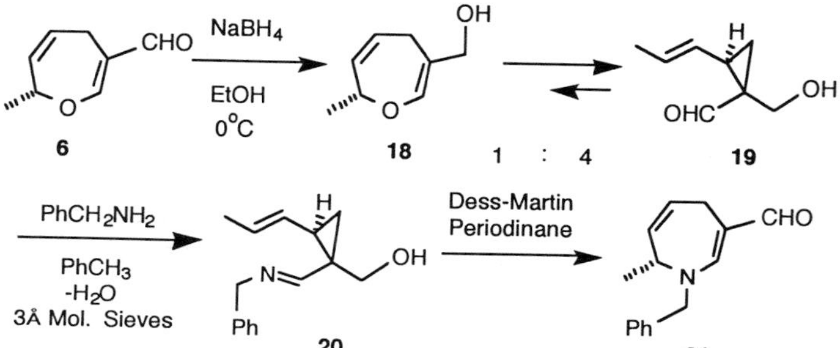

6   **18**   1 : 4   **19**

20   **21**

Scheme 7

Addition of organometallic reagents to **19** and its analogues is also feasible, affording primary, secondary diols which upon oxidation afford 7-substituted oxepins as outlined in Scheme 8 (unoptimized). Athough protection of the alcohol function was employed as shown in Scheme 8, it seems plausible that use of excess organometallic reagent, when feasible, would avoid the use of a protecting group in this sequence.

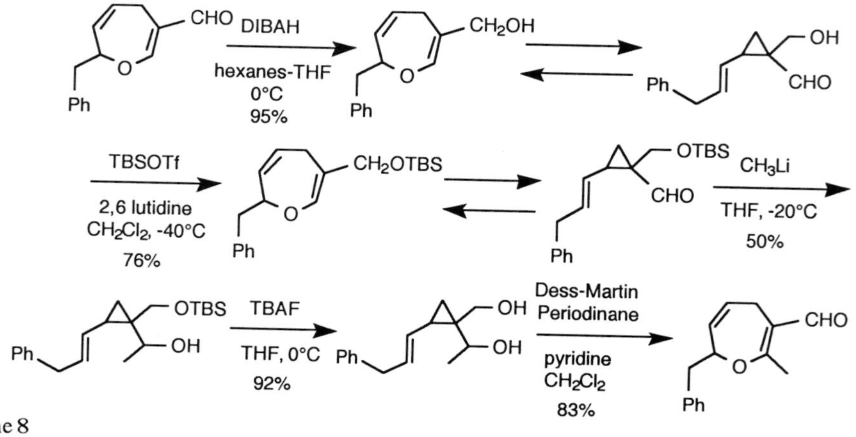

Scheme 8

A most interesting system for applications to both the marine natural products, and macrolide antibiotics resuts from the extension of the rearrangement to the homologous cyclobutanes, which upon rearrangement would afford the homologous dihydrooxacenes. An example of the required class of precursors, the cyclobutane dicarbinols derived by reduction of **22** and **23**, was obtained straightforwardly as outlined in Scheme 9.[13] Oxidation of the mixture of dicarbinols was accompanied by rearrangement *in situ* as before affording the oxacene derivative **24** in excellent yield. The barrier to rearrangement and the magnitude of the equilibrium constant appear qualitatively similar to the cyclopropane systems. It is, again, useful and significant that both olefin geometric isomers rearrange readily, and that the alternative transition state discussed below (Eqn. 3) does not participate although it might be expected to be significantly lower energy than the equivalent transition state for the cyclopropane systems.

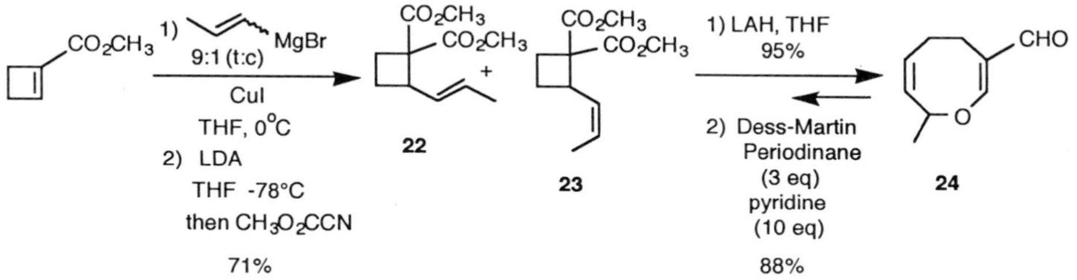

Scheme 9

Since many applications of this methodology to complex molecule synthesis which we envisioned require the preparation of intermediates in enantiomerically pure form, methods for preparation of the precursors were then investigated which would provide enantiomerically pure materials. The methodology itself, owing to the apparent concerted nature of the rearrangement, should provide excellent chirality transfer. However, it was not known whether rearrangement *via* non-concerted pathways was in competition with the concerted mechanism. Furthermore, as shown in Eqn. 3, for 7-membered and 8-membered rings, rearrangement occurs through only one of two possible diastereomeric transition states owing to the strain attendant the introduction of *trans* double bonds into the medium ring isomer. This is fortunate since the two diastereomeric transition states would lead to enantiomeric series with respect to the newly created chiral center (although the products would be diastereomeric as the result of the differing olefin geometry).

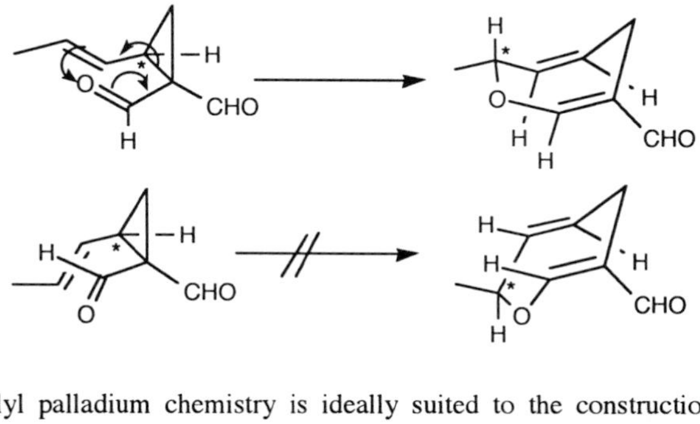

(3)

The use of π–allyl palladium chemistry is ideally suited to the construction of the required enantiomerically pure cyclopropanes.[14,15] As shown in Scheme 10, beginning with the known enan-

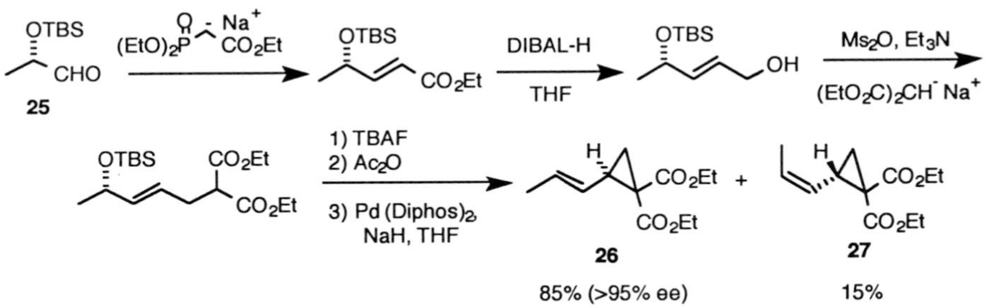

Scheme 10

tiomerically pure protected lactaldehyde **25** elaboration of the required enantiomerically pure cyclopropane diesters **26** and **27** was straightforward. As had been observed previously, a mixture (typically 7-10:1 E/Z) was obtained.[15] At the time, the major isomer was separated and transformed further, since the double bond isomers could have resulted in the formation of partially racemic products depending on the stereochemistry of the cyclopropane ring center in the minor Z isomer. The enantiomeric excess (ee) of the major isomer was determined by reduction to the diol and conversion to the bis Mosher ester.[16]

The dicarbinol resulting from reduction of **26** was oxidized to oxepin **6** (Eqn. 4). To determine the extent of chirality transfer, the oxepin **6** was reduced with Wilkinson's catalyst to tetrahydrooxepin **28**

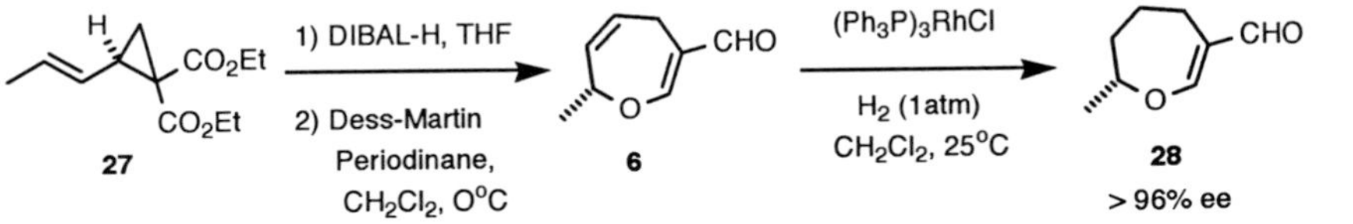

whose ee was ascertained to be >96% by NMR after conversion to the (-)SAMP hydrazone.[17] The extent of the chirality transfer during rearrangement was further tested by reduction of **6** back to the

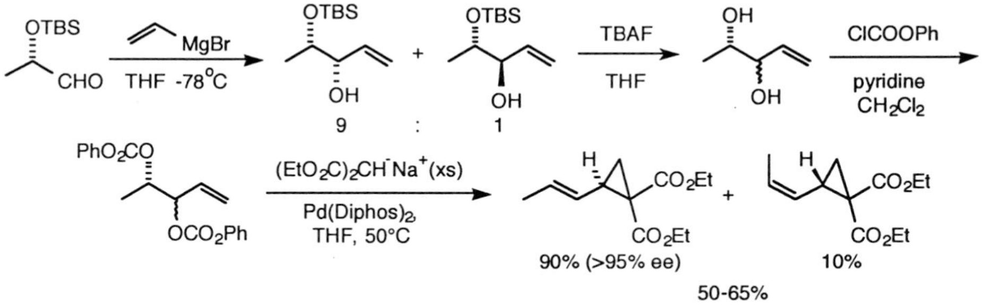

Scheme 11

dicarbinol and reoxidation. After 4 cycles, no loss of ee was noted by examination of the Mosher ester of the dicarbinol. Calculation shows that the lower limit of the fidelity of the chirality transfer must therefore be greater than 99%.

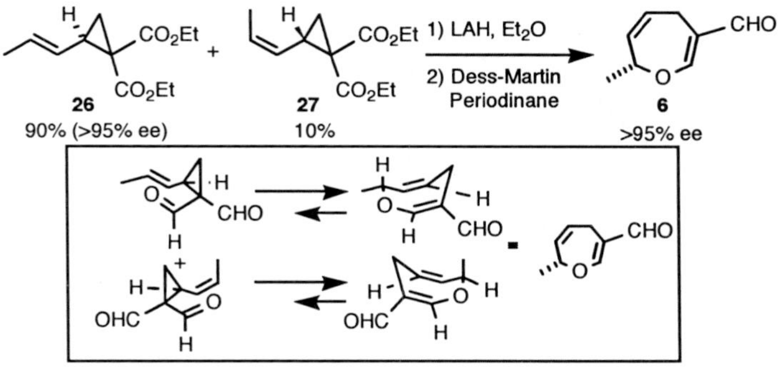

Scheme 12

The route above is somewhat long, thus a substantial effort has been made to increase the efficiency. As shown in Scheme 11, a tandem Pd catalyzed alkylation can be accomplished in reasonable yield (50-65%, unoptimized), thereby shortening the route significantly. At this point, after consideration of the mechanistic model for the π-allylpalladium alkylation, it became apparent that both elements of chirality should be stereochemically linked, thus the minor Z isomer should possess the opposite chirality at the ring center as shown in Scheme 11.[18] This stereochemical outcome had not been previously demonstrated unequivocally in a carbon alkylation although a similar stereochemical outcome has been seen in heteroatom substitution.[19] This fact permits the use of the mixture of diastereomers in the rearrangement since after rearrangement they afford the same dihydrooxepin enantiomer, as shown in Scheme 12. The derived rearrangement product shows no loss of enantiomeric purity thereby verifying the previous hypothesis and the absolute stereochemistry of the minor Z cyclopropyl diester **27.**

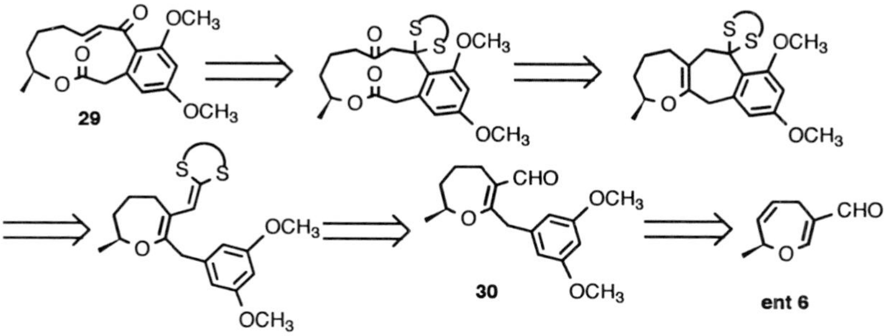

Scheme 13

With the methodology established, specific applications were now considered. The general strategy which seeks to construct 7,7-fused ring vinyl ethers followed by cleavage to a macrocyclic ketolactone can be seen in the retrosynthetic analysis of the dimethyl ether of the simple fungal metabolite dehydrocurvularin (**29**) shown in Scheme 13.

Construction of the required tetrahydrooxepin **30** proceeded from the known optically active dihydrooxepin **6**. The general route follows that in Scheme 8 above and is outlined in Scheme 14.

Addition of the Grignard reagent prepared from 3,5-dimethoxybenzyl bromide, followed by fluoride deblocking and oxidation affords the 3,5-dimethoxybenzyl dihydrooxepin **31** in very good overall yield.

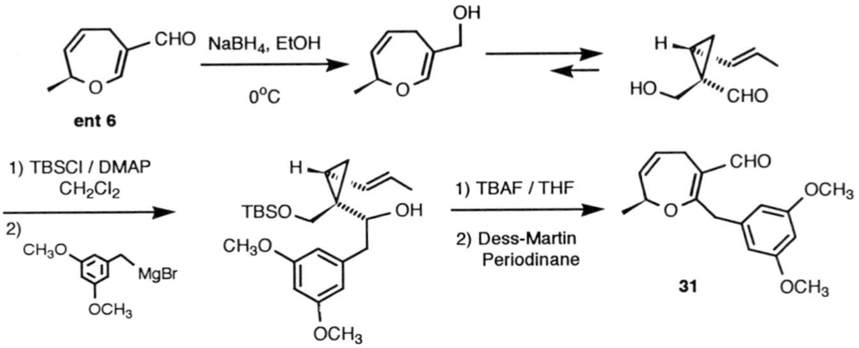

Scheme 14

Since dihydrooxepin **31** was still in equilibrium with the cyclopropane isomer, the double bond is removed by reduction with Wilkinson's catalyst to prevent unexpected interconversion affording **30**, as shown in Scheme 15. The crucial annulation of the remaining 7-membered ring was envisioned *via* a

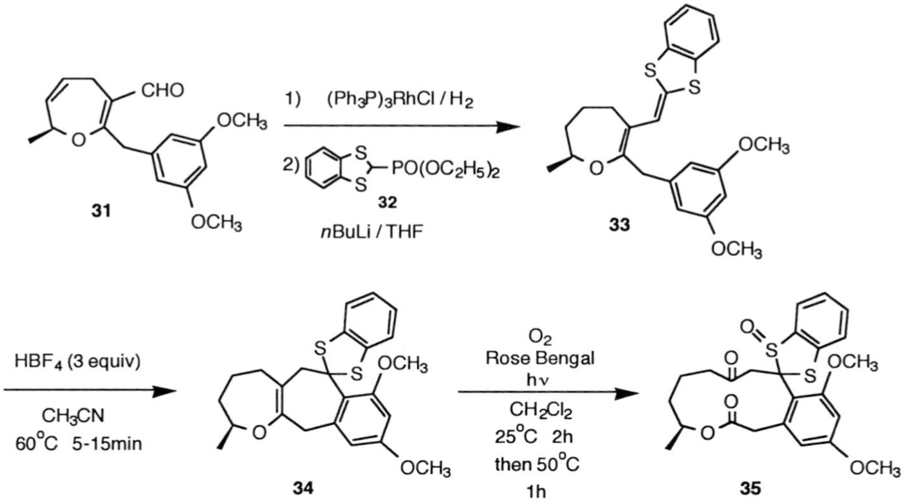

Scheme 15

Friedel-Crafts like cationic cyclization of a doubly vinylogous dithioortho ester (see Scheme 13). Our initial attempts employed acyclic thioethers in the construction of the required precursor. However, attempts to effect the required cyclization were unsuccessful leading only to hydrolysis and attendant cleavage of the 7-membered ring ether. We surmised that the difficulty arose as the result of protonation at the vinyl ether moity rather than the required protonation of the dithioketene acetal. As a result, we employed the recently described benzodithiolane phosphonate **32** for homologation to the required cyclization substrate **33**.[20] The cation derived from protonation of the dithioketene acetal double bond in this system benefits from aromatic-like stabilization and is known to be of stability intermediate between trityl cation and tropilium cation.[20] Thus, we hoped to reverse the regioselectivity in the protonation step under conditions of thermodynamic control.

Treatment of the tetrahydrooxepin derived from **31** with the lithium anion derived from **32** smoothly provided the required ketene acetal **33** in 75-80% yield. However, exposure to up to 2 equivalents of fluoroboric acid in Et_2O/CH_2Cl_2 still resulted in hydrolysis. Apparently, kinetic protonation still occurs at the vinyl ether moity, and hydrolysis is faster than equilibration and/or cyclization. However in the presence of excess HBF_4 (3 equiv), cyclization takes place affording the tetracyclic vinyl ether **34** in yields up to 60% (unoptimized). Our present interpretation of these results invokes double protonation and cyclization *via* a dicationic intermediate, although these mechanistic details must yet be firmly established. Vinyl ether **34** has been oxidatively cleaved with singlet oxygen as we have described earlier with concomitant oxidation at sulfur affording the macrocyclic ketolactone **35** in 70% yield.[21] Thus, the essential feasibility of this strategy to macrocyclic lactones has been verified. We are presently investigating the conversion of **35** to dehydrocurvularin and curvularin dimethyl ethers.

The second application is directed toward the macrolides, specifically methynolide (**36**) as shown in Scheme 16.[22] As can be seen, the key features are the same as the approach to the curvularins in that the construction and subsequent cleavage of a 7,7-fused bicyclic vinly ether is envisioned.[23] In this discussion, attention will be focused on our construction of the key enantiomerically pure tetrahydrooxepin **37** and its use in preliminary cyclization attempts.

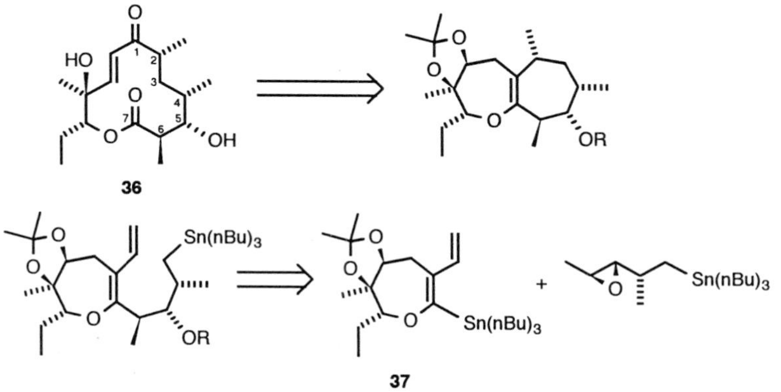

Scheme 16

The sequence to **37** was initated from commercially available allylic alcohol **38** as shown in Scheme 17. In this instance, the inital chiraltiy is obtained by Sharpless epoxidation of **38**.[24] The resulting epoxide **39** is obtained in 80-90% yield and 95% ee as judged by Mosher ester formation. Oxidation of **39** to the derived aldehyde, and Wittig olefination affords the sensitive and volatile epoxide **40** which is directly converted by sequential palladium catalyzed alkylations to the inconsequential mixture of enantiomerically pure (95% ee) vinyl cyclopropane diesters **41** and **42**.[14,15] Use of the phenyl carbonate for the second alkylation provided the best compromise between reactivity and preservation of chirality of any leaving group examined to date.[25] As described above, the mixture of

41 and **42** is converted to a single enantiomerically pure dihydrooxepin **43** in ~85% yield as a consequence of the mechanism of the π-allyl palladium catalyzed alkylation coupled with mechanism of the retroClaisen rearrangement.

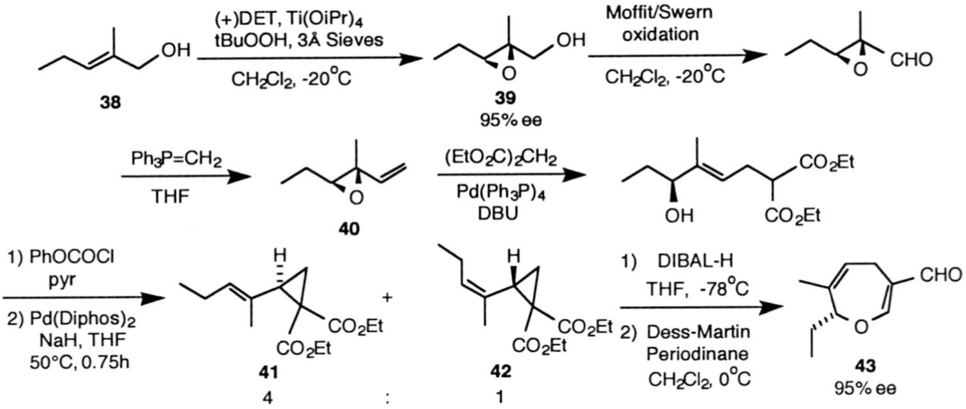

Scheme 17

The key oxepin **37** was then elaborated from oxepin **43** as outlined in Scheme 18. Oxepin **43** is cleanly regioselectively hydroxlyated with catalytic OsO$_4$/NMMO at -20°C providing the diol **44** as the

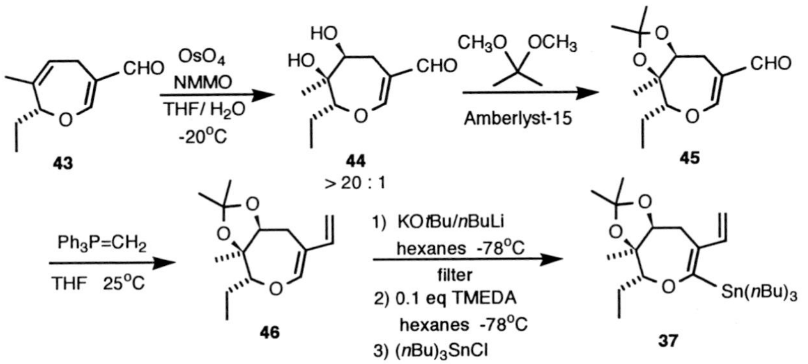

Scheme 18

only detectable diastereomer (NMR) in 70-75% yield.[26] The stereoselectivity of this hydroxylation was anticipated as shown in Eqn. 5. The bulky osmium reagent approaches the peripheral face of the olefin in the most stable conformation in which the ethyl group is in the equatorial-like minimizing A$_{1,2}$ strain arising from interaction with the adjacent vinyl methyl group and avoiding the prow interaction with the transannular methylene group.

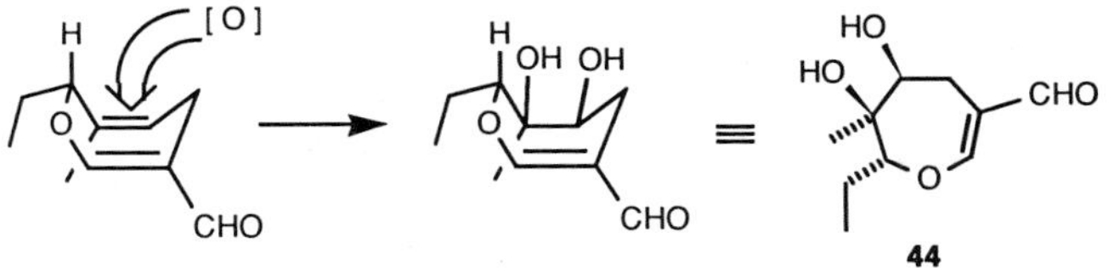

$$(5)$$

The acid sensitive diol **44** is then converted to the acetonide **45** in high yield under mild, nearly neutral conditions by exposure to 2,2-dimethoxypropane in the presence of acidic Amberlyst IR-15 resin. Wittig olefination of **45** proceeded smoothly affording the acid sensitive dienol ether **46** in 65-75% yield. Direct metalation of **46** did not prove possible, in fact, only very potent metalation conditions followed by trapping with nBu$_3$SnCl sufficed to convert **46** to the desired precursor for the lithium reagent, vinyl stannane **37** in ~75% yield.[27]

With the fully functionalized oxygen containing ring in hand in the form of **37**, attention was turned to the annulation methodology. Space and time do not permit a complete description of our ongoing work in the area. However, to test the feasibility of such an annulation, stannane **37** was converted to the related lithium reagent by treatment with nBuLi at low temperature. The lithium reagent was alkyated *in situ* with the known iodobutylstannane **47** affording the alkylated dienol stannane **48** in 50-60% yield, as outlined in Scheme 19.

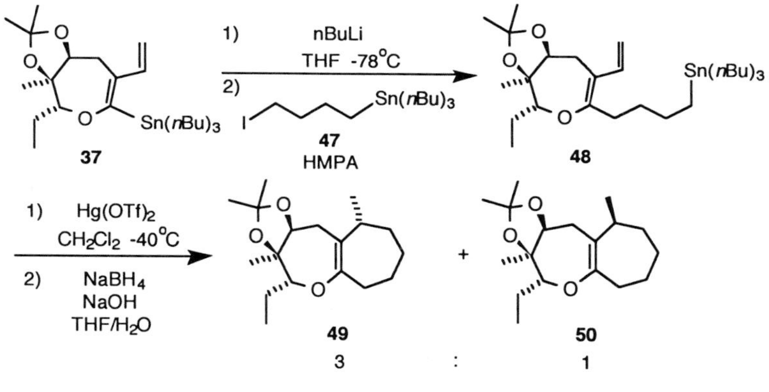

Scheme 19

The required annulation was then accomplished by treatment of dienol stannane **48** with freshly prepared Hg(OTf)$_2$ at -40°C affording, in a single experiment thus far, a mixture of organomercurials, which upon reductive demercuration, provided a mixture of the tricyclic tetrasubstituted enol ethers **49** and **50** (3:1) in ~50% yield.[28] The products can be clearly demonstrated to be epimeric only at the newly created secondary methyl group (by NMR). However, due to the complex array of similar conformations available to these butterfly-like molecules, it has not yet been possible to unequivocally assign their relative stereochemistry. The assignments given above for **49** and **50** are based upon the expected steric bias for ring closure on the face opposite that of the acetonide ring. Efforts are ongoing in our laboratories to apply this strategy to the preparation of methynolide (**36**), and other naturally occurring macrolide antibiotics.

Hopefully, the foregoing has provided a glimpse of the potential of this new methodology for the creation of unsaturated medium ring heterocycles containing both oxygen, nitrogen, and conceivably sulfur in enantiomerically pure form if required.   As intermediates, these classes of medium ring compounds provide access to a potentially powerful and thus far unexploited strategy for the creation of macrocyclic lactones, when coupled with the new annulation methodology for creation of 7,7-fused bicyclic systems under development in our laboratories.

Acknowledgment - I wish to acknowledge a dedicated group of coworkers: Dr. J. Ramon Vargas, Matthew Shair, Theodore Kameneka, Dr. Steven W. Andrews, Dr. Alain Commerçon, Dr. Lesley A. Stolz, and Dr. Kay M. Brummond who contributed substantially to both the conceptual development as well as experimental execution of the chemistry reported herein.   I also wish to thank the National Institutes of General Medical Sciences of the National Institutes of Health for a research grant (GM-29290) in support of this research.

References

1. Shimizu, Y.; Chou, H.-N.; Bando, H.; Van Duyen, G.; Clardy, J. *J. Am. Chem. Soc.* **1986**, *108*, 514. (b) Faulkner, D. J., *Nat. Prod. Rept.*, **1987**, *4*, 539; **1986**,*3*,1; **1984**, *1*, 251, 555. (c) Moore, R. E., *Marine Natural Products: Chemical and Biological Perspectives*; Scheuer, P. J., Ed.; Academic Press; New York; 1978, vol.1.

2. Lin, Y. Y.; Risk, M.; Ray, M. S.; Van Engen, D.; Clardy, J.; Golik, J.; James, J. C.; Nakanishi, K. *J. Am. Chem. Soc.* **1981**, *103*, 6773.

3. (a) Masamune, T.; Matsue, H.; Murase, H., *Bull. Chem. Soc. Jpn.*, **1979**, *52*, 127. (b) Masamune, T.; Murase, H.; Matsue, H.; Murai, A.; *Bull. Chem. Soc. Jpn.*, **1979**, *52*, 135. (c) Chen, R.; Rowland, D. A. *J. Am. Chem. Soc.* **1980**, *102*, 6609. (d) Nicolaou, K. C.; Claremon, D. A.; Barnett, W. E. *J. Am. Chem. Soc.* **1980**, *102*, 6611. (e) Kotsuki, H; Ushio, Y.; Kadota, I.; Ochi, M. *J. Org. Chem.* **1989**, *54*, 5153. (f) Schreiber, S. L.; Kelly, S. E. *Tetrahedron Lett.* **1984**, *25*, 1757.

4. (a) Kane, V. V.; Doyle, D. L.; Ostrowski, P. C. *Tetrahedron Lett.* **1980**, *21*, 2643. (b) Jackson, W. P.; Ley, S. V.; Morton, J. A. *Tetrahedron Lett.* **1981**, *22*, 2601. (c) Overman, L. E.; Thompson, A. S. *J. Am. Chem. Soc.* **1988**, *110*, 2248. (d) Nicolaou, K. C.; Prasad, C. V. C.; Hwang, C.-K.; Duggan, M. E.; Veale, C. A. *J. Am. Chem. Soc.* **1989**, *111*, 5321. (e) Nicolaou, K. C.; Prasad, C. V. C.; Somers, P. K.; Hwang, C.-K. *J. Am. Chem. Soc.* **1989**, *111*, 5335 . (f) Castaneda, A.; Kucera, D.J.; Overman, L. E. *J. Org. Chem.* **1989**, *54*, 5695. (g) Blumenkopf, T. A.; Bratz, M.; Castaneda, A; Look, G. C.; Overman, L. E.; Rodriguez, D.; Thompson, A. S. *J. Am. Chem. Soc.* **1990**, *112*, 4386. (h) Carling, R.W.; Holmes, A. B. *J.Chem. Soc., Chem. Commun.* **1986**, 565.

5. Boeckman, Jr., R. K.; Bruza, K. J. *Tetrahedron* **1981**, *37*, 3997.

6.   (a) Rhoads, S. J., Cockroft, R. D. *J. Am. Chem. Soc.* **1969**, *91*, 2815. (b) Rey, M.; Dreiding, A. *Helv. Chim. Acta*, **1965**, *48*, 1985. (c) Hughes, M. T.; Williams, R. O. *Chem. Commun.* **1968**, 587. (d) Rhoads, S. J.; Raulins, N. R. *Org. React.* **1975**, *22*, 1.

7. For the related Cope rearrangement of divinyl epoxides: Clark, D. L.; Chou, W.-N.; White, J. B. *J. Org. Chem.* **1990**, *55*, 3975.

8. Boeckman, R. K., Jr.; Flann, C.J.; Poss, K. M. *J. Am. Chem. Soc.* **1985**, *107*, 4359.

9. Wenkert, E.; Greenberg, R. S.; Kim, H-S. *Helv. Chim. Acta*, **1987**, *70*, 2159. (b) Alonso, M. E.; Jano, P.; Hernandez, M. I.; Greenberg, R. S.; Wenkert, E. *J. Org. Chem.* **1983**, *48*, 3047.

10. (a) Singh, R. K.; Danishefsky, S. *J. Org. Chem.* **1975**, *40*, 2969. (b) Näf, F.; Decorzant, R. *Helv. Chim. Acta* **1978**, *61*, 2524.

11. Dess, D. B.; Martin, J. C. *J. Org. Chem.* **1983**, *48*, 4155.

12. Griffith, W. P., Ley, S. V., Whitcombe, G. P., White, A. D. *J. Chem. Soc., Chem. Commun.* **1987**, 1625. (b) Griffith, W. P., Ley, S. V. *Aldrichimica Acta*, **1990**, *23*, 13.

13. (a) Clark, R.D. *Synth. Commun.*, **1979**, *9*, 325. (b) Corey, E.J.; Boaz, N.W. *Tetrahedron Lett.* **1985**, *26*, 6015, 6019. (c) Mander, L. N. ; Sethi, S. P. *Tetrahedron Lett.* **1983**, *24*, 5425.

14. (a) Trost, B. M.; Molander, G. A. *J. Am. Chem. Soc.* **1981**, *103*, 5969. (b) Trost, B. M.; Mignani, S. *J. Org. Chem.* **1986**, *51*, 3435.

15. (a) Genêt, J. P.; Balabane, M.; Charbonnier, F., *Tetrahedron Lett.* **1982**, *23*, 5027. (b) Genêt, J. P.; Piau, F.; Ficini, J. *Tetrahedron Lett.* **1980**, *21*, 3183. (c) Genêt, J. P.; Piau, F. *J. Org. Chem.* **1981**, *46*, 2414. (d) Bäckvall, J. -E.; Vågberg, J. O.; Zercher, C.; Genêt, J. P.; Denis, A. *J. Org. Chem.* **1987**, *52*, 5430. (e) Genêt, J. P.; Gaudin, J. M. *Tetrahedron* **1987**, *43*, 5315.

16. Dale, J. A. Dull, D.L., Mosher, H. S. *J. Org. Chem.* **1969**, *34*, 2543.

17. Enders, D.; Fey, P.; Kipphardt, H. *Org. Synth.* **1987**, *65*, 173.

18. (a) Bosnich, B.; MacKenzie, P. B. *Pure and Appl. Chem.* **1982**, *54*, 189. (b) Trost, B. M.; Verhoeven, T. R. *J. Am. Chem. Soc.* **1980**, *102*, 4730. (c) Bosnich, B. *Asymmetric Catalysis*: Martinus Nijhoff Publishers, Dortrecht, The Netherlands, 1986, Chap. 3, pp. 54-60. (d) Faller, J. W.; Thomsen, M. E.; Mattia, M. J. *J. Am. Chem. Soc.* **1971**, *93*, 2642.

19. Spears, G. W.; Nakanishi, K.; Ohfune, Y. *Tetrahedron Lett.* **1990**, *31*, 5339.

20. Rigby, J. H.; Kotnis, A.; Kramer, J. *J. Org. Chem.* **1990**, *55*, 5078.

21. Boeckman, Jr., R. K.; Bruza, K. J.; Heinrich, G. R. *J. Am. Chem. Soc.* **1978**, *100*, 7101.

22. Boeckman, Jr.; R.K.; Goldstein, S.W.; *Total Synthesis of Natural Products* **1988**, *7*, 1-140.

23. Borowitz, I.J.; Gonis, G.; Kelsey, R.; Rapp, R.; Williams, G.J. *J. Org. Chem.* **1966**, *31*, 3032

24. Hill, J. G.; Sharpless, K. B.; Exon, C. M.; Regenye, R. *Org. Synth.* **1990**, *Coll. Vol. 7*, 461.

25. (a) Granberg, K. L.; Bäckvall, J. -E. *J. Am. Chem. Soc.* **1992**, *114*, 6858. (b) Bäckvall, J.-E.; Granberg, K. L.; Heumann, A. *Isr. J. Chem.* **1991**, *31*, 17.

26. VanRheenen, V.; Cha, D. Y.; Hartley, W. M. *Org. Synth.* **1978**, *58*, 44.

27. Brown, H. C.; Zaidlewicz, M.; Bhat, K. S. *J. Org. Chem.* **1989**, *54*, 1764.

Integrated Metalation - Cross Coupling Approaches to Bioactive Natural Products

Victor Snieckus

Guelph-Waterloo Centre for Graduate Work in Chemistry
University of Waterloo, Waterloo, Ontario CANADA N2L 3G1

Summary: Three synthetic strategies, individually and combined, are developed for the construction of bioactive aromatic natural products. The Directed *ortho* Metalation (DoM) (**Scheme 1**) process serves as the fundamental strategy providing access to aromatics with regiospecific substitution patterns. The DoM tactic is combined with the Pd(0)-catalyzed cross coupling reaction (**Schemes 2, 3**) which allows new modes of aryl-aryl bond formation affording a variety of biaryls, heterobiaryls, and condensed analogues. The DoM and cross coupling methodologies are intertwined with new remote metalation possibilities (**Schemes 7, 13, 24, 31, 36**) which add new dimensions to synthetic aromatic chemistry.
The use of these strategies in the synthesis of specific targets within several diverse classes of natural products are delineated.

Introduction

During the last decade, fueled by discoveries and systematic studies in a number of laboratories, the directed *ortho* metalation (DoM) reaction (**Scheme 1**) has emerged as a prominent strategy for the regiospecific construction of aromatic and heteroaromatic substrates [1,2]. Although mechanistic rationalization of the DoM process is in its infancy [3], the synthetic exploration of a variety of directed metalation groups (DMGs) has provided sufficient empirical data to allow analysis of target structures with some assurance that the DoM process will offer at least partial solutions. Work in our group has been focussed on the use of both carbon-based ($CONEt_2$, $CON(Me)CH(TMS)_2$), CON-t-Bu) and heteroatom-based (NCOt-Bu, NCO_2t-Bu, $OCONEt_2$, $OCSNEt_2$, OMOM, OSEM, $S(O)_n$t-Bu, n = 1,2) DMGs in methodological and total synthesis endeavours [1,4].

The Directed ortho Metalation (DoM) Way to Polysubstituted Aromatics

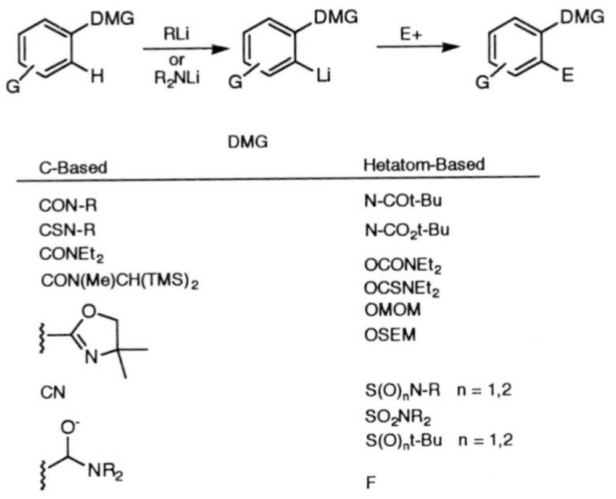

C-Based	Hetatom-Based
CON-R	N-COt-Bu
CSN-R	N-CO$_2$t-Bu
CONEt$_2$	
CON(Me)CH(TMS)$_2$	OCONEt$_2$
	OCSNEt$_2$
	OMOM
	OSEM
CN	S(O)$_n$N-R n = 1,2
	SO$_2$NR$_2$
	S(O)$_n$t-Bu n = 1,2
	F

Scheme 1

In 1985, stimulated by the salient observation by Suzuki that aryl boronic acids undergo Pd-catalyzed cross coupling with aryl halides [5], we initiated work [6] to connect the DoM process to this methodology by the simple expediency of *ortho* metalation-induced boronation of both carbon- and heteroatom-based DMG systems (**Scheme 2**) . In the interim, a variety of DMGs have been used in successful Pd(0)-catalyzed cross coupling reactions with aryl bromides and iodides [7] and, most recently, aryl triflates [8]. Although the *ortho*-DMG aryl boronic acids **4 (Scheme 3)** were initially derived directly from the corresponding *ortho*-lithiated species **1**, an alternate regimen which proceeds via the *ortho*-TMS derivatives **3**, usually obtained in high yields from **1**, may have merit in affording cleaner products [7d]. The boronic acids may also be "stored" as bromo derivatives, **1** → **2** to be revealed by the reverse (metal-halogen exchange) process at an appropriate time and systems **2** (DMG = other FG) are obligatory for preparation of aryl boronic acids with non-DMG groups, **2** → **1** [8]. A further interconnection in this grid, the *ipso* bromodesilylation, **3** → **2** may also have value [9].

The DoM-Pd(0)-Catalyzed Cross Coupling Connection

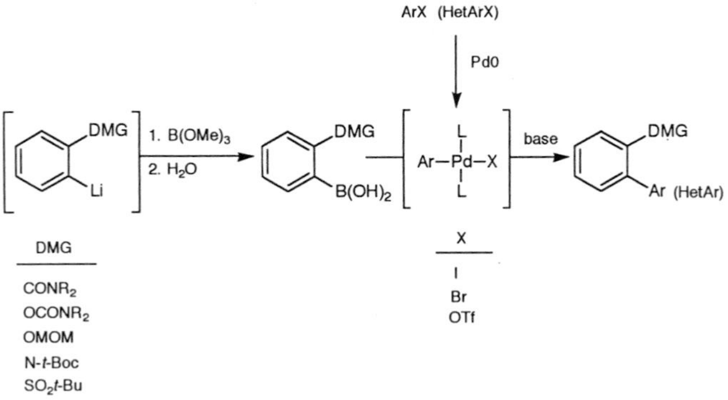

Scheme 2

A recent demonstration of the utility of the DoM-cross coupling link concerns the rapid construction of polyaryl systems by sequential coupling procedures (**Scheme 4**) [9]. In this route, mixed bromo-iodo aromatics **5** are allowed to undergo the standard cross coupling reaction in which the production of the intermediates **6** is dictated by the known greater reactivity of the Ar-I bond. This reaction completed, a second cross coupling may be carried out with different boronic acids, with or without the isolation of **6**, to give *m*- and *p*-terphenyl derivatives. Examples of such iterative processes afford highly substituted terphenyls (**Scheme 5**) while convergent 2:1 couplings furnish similarly functionalized quinquephenyls (**Scheme 6**) . In both series, the regiospecificity in the diverse substitution patterns of the products is assured from the DoM chemistry on the coupling partners [9].

Route to *ortho* - DMG Aryl Boronic Acids

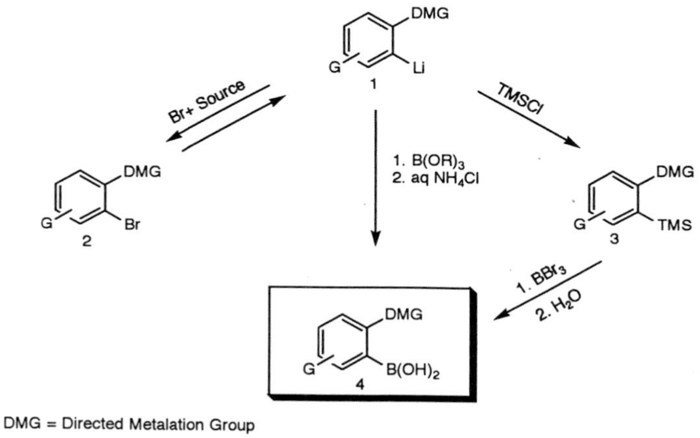

DMG = Directed Metalation Group

Scheme 3

Sequential Aryl Halide - Aryl Boronic Acid Cross Couplings

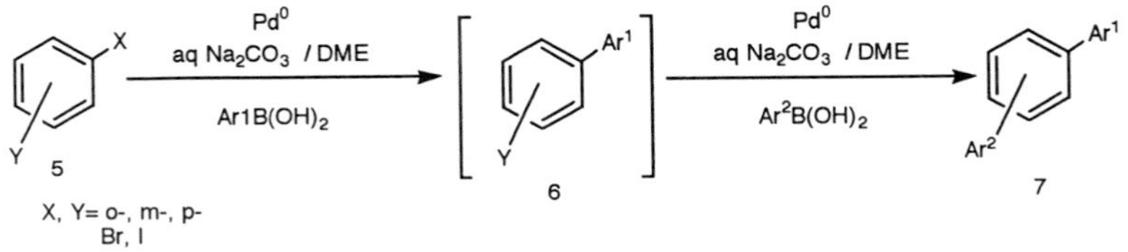

Scheme 4

Sequential 2:1 Cross-Coupling Illustrated

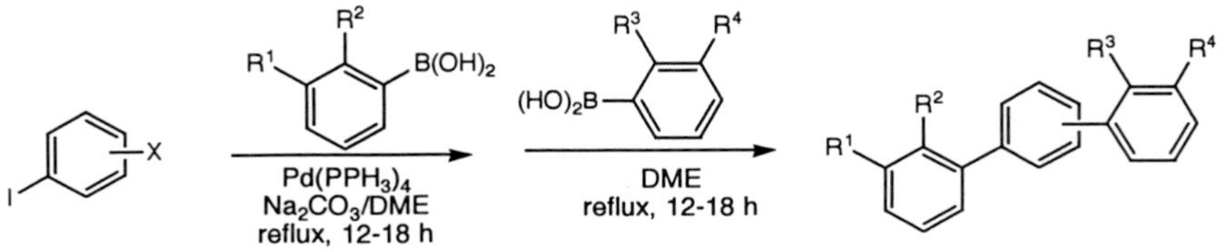

X	R¹	R²	R³	R⁴	Yield, %
p-Br	H	Me	H	H	79
m-Br	H	Me	H	H	77
p-Br	H	CON(iPr)₂	CON(iPr)₂	TMS	43
p-Br	TMS	CON(iPr)₂	CON(iPr)₂	H	47
p-Br	H	CON(iPr)₂	H	CON(iPr)₂	63
m-Br	H	CON(iPr)₂	H	CON(iPr)₂	61

Scheme 5

2:1 Cross Coupling Illustrated

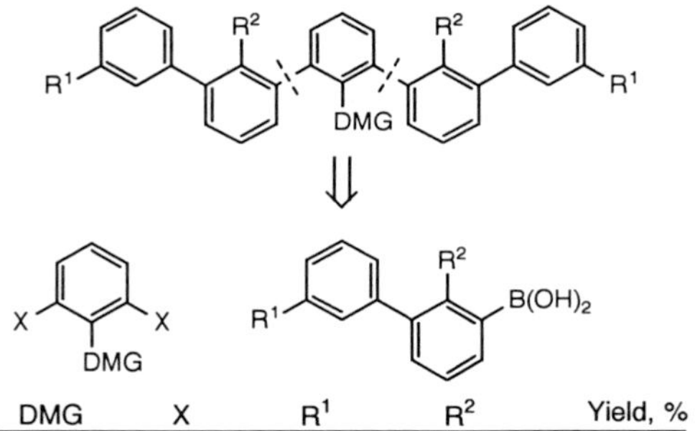

DMG	X	R¹	R²	Yield, %
CONEt₂	Br	H	H	44
CONEt₂	I	H	H	69
NH2	Br	H	H	67
CN	I	CH₉OMe	H	85
OMOM	Br	H	OMOM	86

Scheme 6

With the aim of further expanding the DoM-cross coupling connection, remote metalation routes to 9,10-phenanthraquinones have been developed (**Scheme 7**) . Thus products **9** of Suzuki cross coupling of **8** with *ortho*-boronic acid benzamides, when subjected to LDA treatment, undergo remote vinylogous "*ortho*-tolyl" metalation followed by cyclization to afford 9-phenanthrols **10** [7d], which can be easily oxidized to the corresponding quinones **11** by a variety of methods including salcomine-catalyzed oxygenation.[10] Alternatively, taking advantage of the excellent methodology for biaryl synthesis developed by Meyers, the Grignard reagents of **8**, when treated with *o*-methoxyphenyl oxazolines, furnish compounds **12** which, under the identical LDA conditions used for **9**, afford the phenanthrol ether amines **13**. Direct oxidation of **13** using, for example, Fremy's salt, also leads to the quinones **11** [10].

Cross Coupling - Remote " *ortho* - Toluyl" Metalation
Route to 9,10-Phenanthraquinones

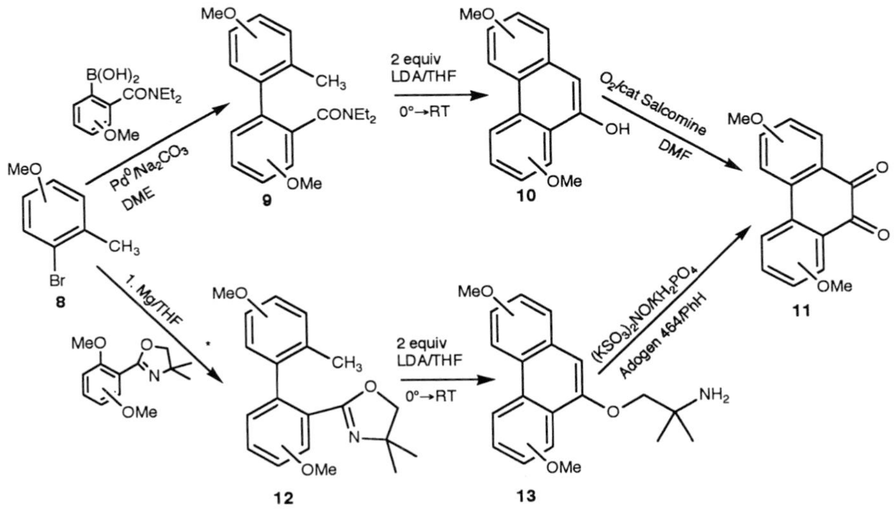

*Review: Reuman, M.; Meyers, A.I. Tetrahedron **1985**, *41*, 837

Scheme 7

Retene (**18**), a phenanthrene first obtained by degradation of abietic acid (**19**) and recently shown to be a useful marker for determination of extent of wood combustion in ambient air, was chosen as a target molecule to show the synthetic advantage of the DoM-cross coupling-remote metalation protocol [11]. Thus the aryl *O*-carbamate **14**, upon metalation-carbamoyl migration [1] and triflation provided compound **15**. Cross coupling with 2,3-dimethyl phenyl boronic acid, obtained from metal-halogen exchange-boronation of 2,3-dimethylbromo-benzene, smoothly led to the biaryl **16**. Deviating from the previous protocol (**Scheme 7**), **16** was treated with *n*-BuLi to afford the phenanthrol **17** in high yield. Reductive scission of the triflate of **17** under Pd-catalyzed conditions completed the synthesis of retene (**18**).

Synthesis of Retene, A Unique PAH Molecular Marker
of Wood Combustion in Ambient Air

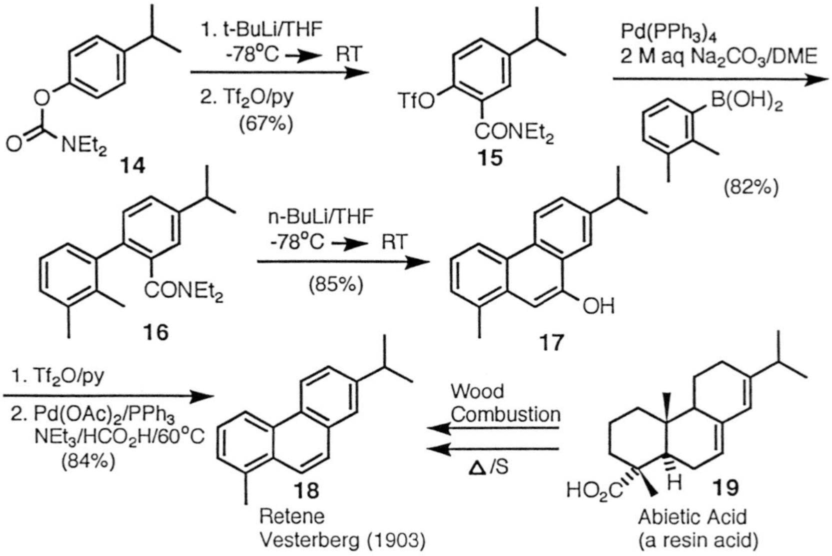

Scheme 8

The azaphenanthraquinone **24**, (**Scheme 9**), a key intermediate in an early demonstration by Weinreb of the intramolecular azadiene Diels-Alder reaction (**20**) in natural product synthesis [eupolauramine (**23**)] posed as an interesting target to demonstrate utility of the new phenanthraquinone synthesis (**Scheme 7**). According to this tactic, **24** would be derived by remote metalation-cyclization of intermediate **25** which, in turn, would be available by cross coupling of pyridine (**21**) and *ortho* toluyl (**22**) partners in which the origin of respective X,Y = Br, B(OH)$_2$ groups would be determined by the availability of starting materials. In the event [12], 2-bromonicotinic acid (**26**) (**Scheme 10**) was converted by a non-standard method into the corresponding amide **27** which upon metalation-carbamoylation afforded the diamide **28** in good yield. Cross coupling with a requisite aryl boronic acid afforded the azabiaryl **29** which was smoothly cyclized using LDA into the oxygenated azaphenanthrene **30**. Amide hydrolysis of **30** under refluxing acetic acid conditions was anchimerically assisted by the phenol group and resulted in the formation of lactone **31**. The demonstrably valuable Weinreb reagent, trimethylaluminum-methylammoniumchloride, allowed conversion into amide **32** which upon CAN oxidation led quantitatively to the azaphenanthraquinone **24** previously converted by Weinreb into eupolauramine (**23**).

Retrosynthetic Analyses for Azaphenanthrene
Alkaloid Eupolauramine

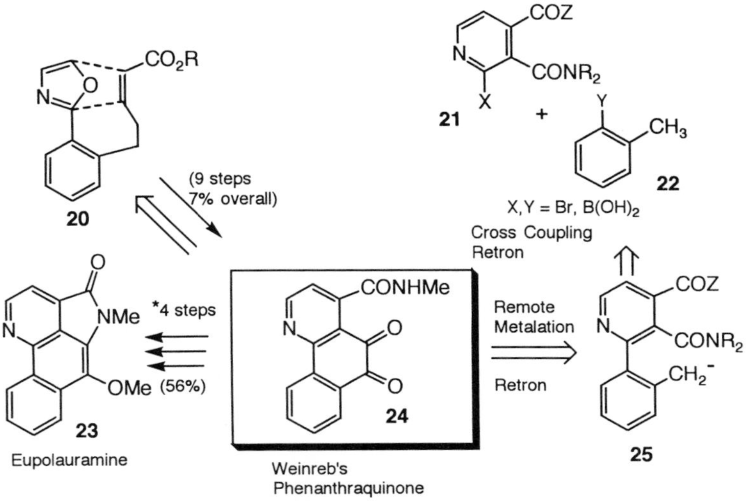

(9 steps
7% overall)

20

21   X   +   Y   **22**

X,Y = Br, B(OH)₂

X,Y = Br, $B(OH)_2$

Cross Coupling
Retron

23

Eupolauramine

*4 steps

(56%)

Remote
Metalation

Retron

24

Weinreb's
Phenanthraquinone

25

* Levin, J.I.; Weinreb, S.M. J. Org. Chem. **1984**, *49*, 4325

Scheme 9

Homomoschatoline (**33**) (**Scheme 11**) , a representative of the abundant oxoaporphine group of alkaloids, was considered as a heteroring-fused 9,10-phenanthraquinone and therefore a potential target to further test the remote metalation strategy. The envisaged intermediate **34** appears poised to undergo sequential methyl metalation-cyclization (step 1), oxidation (step 2), and side chain heteroring annelation (step 3) to give the target molecule **33**. With the proviso that the ethyl amine side chain could be introduced at the biaryl stage, compound **34** was envisaged to be derived by cross coupling of **35** and **36** partners with the X,Y substituents being determined as before (**Scheme 9**) . The synthesis was initiated [11] by the cross coupling of aryl iodide **37** with boronic acid **38**, a reaction with steric hindrance constraints which were not manifested (however, *vide infra*). The biaryl amide product **39** was subjected to a classical electrophilic formylation followed by nitromethane chain extension and reduction to afford the phenethylamine **40**. The LAH step was carefully controlled in order to avoid reduction of the amide function. The synthesis was initiated [11] by the cross coupling of aryl iodide **37** with boronic acid **38**, a reaction with steric hindrance constraints which were not manifested (however, *vide infra*). The biaryl amide product **39** was subjected to a classical electrophilic formylation followed by nitromethane chain extension and reduction to afford the phenethylamine **40**. The LAH step was carefully controlled in order to avoid reduction of the amide function.

Cross Coupling - Remote " *ortho*-Toluyl" Metalation
Approach to Eupolauramine

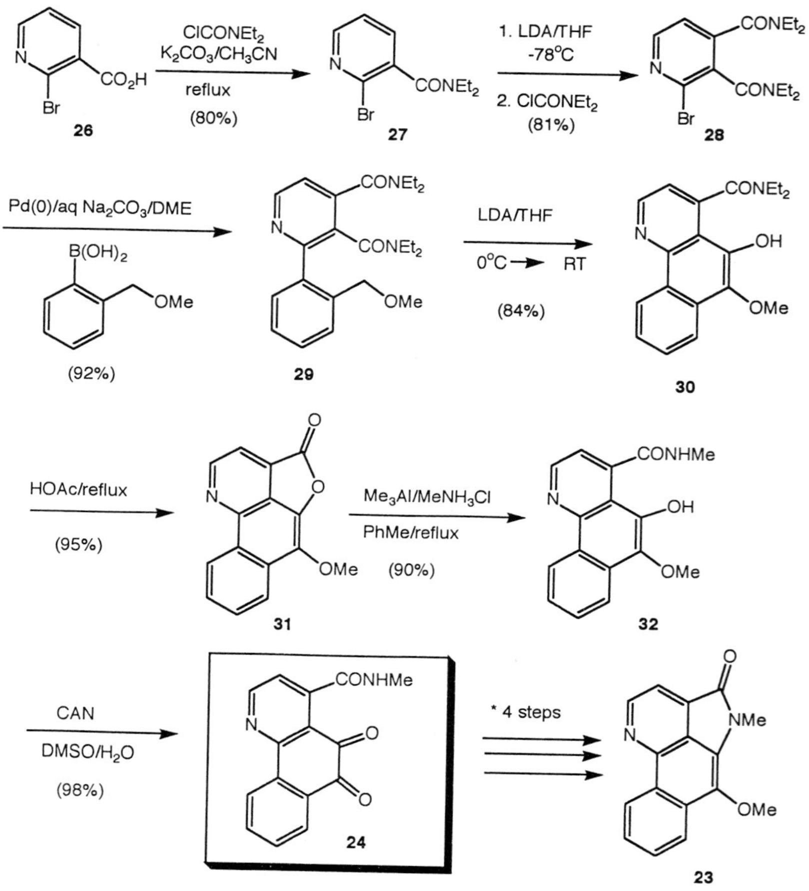

Scheme 10

The synthesis was initiated [11] by the cross coupling of aryl iodide **37** with boronic acid **38**, a reaction with steric hindrance constraints which were not manifested (however, *vide infra*). The biaryl amide product **39** was subjected to a classical electrophilic formylation followed by nitromethane chain extension and reduction to afford the phenethylamine **40**. The LAH step was carefully controlled in order to avoid reduction of the amide function.

Treatment of **40** with excess LDA resulted in the formation of the desired **41** which, owing to its instability, was immediately oxidized using the catalytic salcomine oxygenation method. Under these

* Levin, J.I.; Weinreb, S.M. J. Org. Chem. **1984**, *49*, 4325

conditions, cyclization and dehydrogenation ensued to give homomoschatoline (**33**) in 50% yield over the final two steps [11].

**Oxoaporphine Alkaloids: Retrosynthesis Based
on a Combined Metalation-Cross Coupling Strategy**

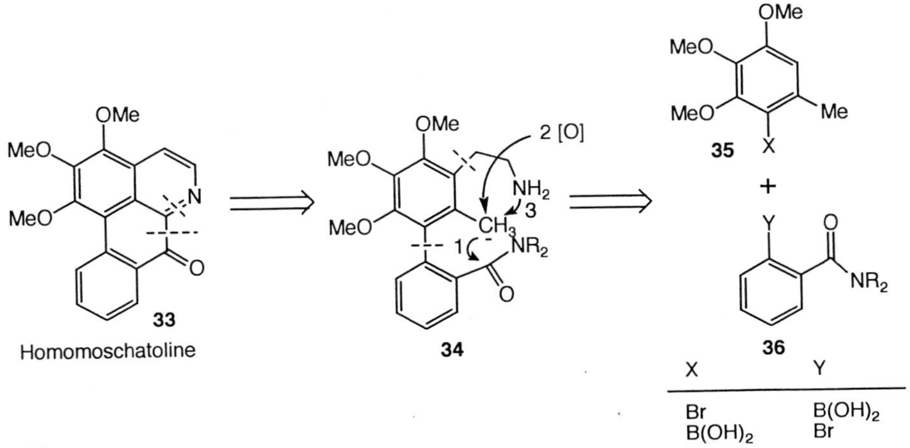

Scheme 11

Total Synthesis of Oxoaporphine Alkaloid Homomoschatoline

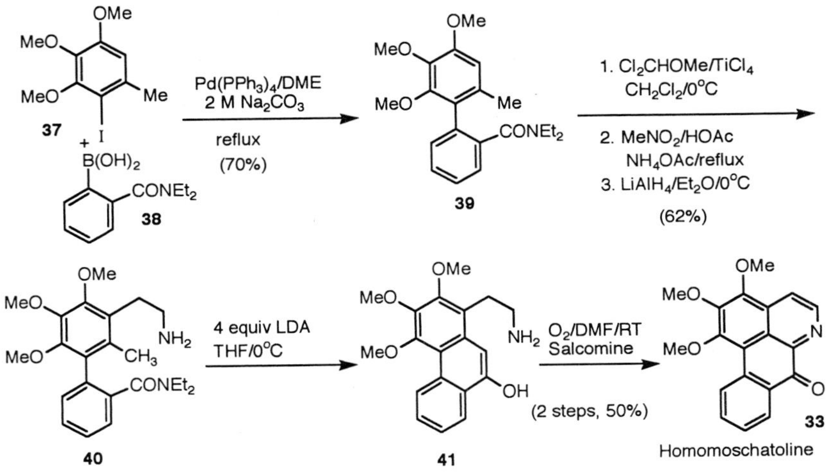

Scheme 12

A second remote metalation concept (**Scheme 13**) , whose roots may directly be traced to the complex-induced proximity effect (CIPE) enunciated by Beak and Meyers and by Klumpp [13], led to the development of a general regiospecific route to fluorenones and further expanded the combined metalation-cross coupling synthetic strategy. Using the CIPE hypothesis, it was envisaged that amide **42** - RLi complexation would induce remote deprotonation (arbitrarily an Ar_2 hydrogen) leading to intermediate **43** which would cyclize to fluorenone **44** in an intramolecular version of the well known intermolecular ArLi - tertiary benzamide condensation reaction. The successful demonstration and generalization of this remote metalation reaction for the regiospecific construction of condensed fluorenones and azafluorenones has been reported [14]; its potential to override Friedel-Crafts regiochemistry, dictated by normal electrophilic substitution rules, i.e. **45** → **47** (**Scheme 14**) , by DoM-induced effects, i.e. **46** → **47**, remains to be fully evaluated [14].

Remote Metalation: The Concept

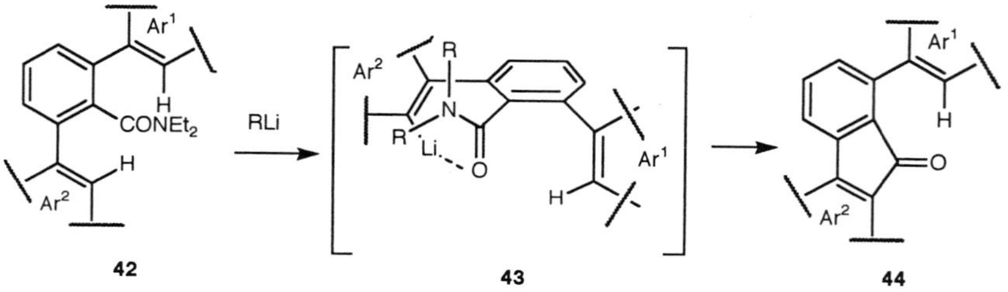

42   43   44

Scheme 13

Remote Metalation Route to Fluorenones.
Overriding the Friedel-Crafts Regiochemistry

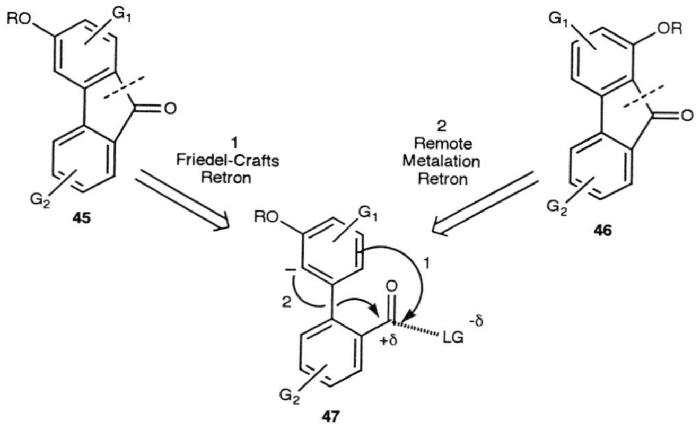

Scheme 14

The remote metalation route to fluorenones was "a method searching for an application" until the surprise discovery of naturally occurring fluorenones dengibsin (**48a**) (**Scheme 15**)   and dengibsinin (**48b**) in Indian orchids by Talapatra and coworkers [15]. Prior to the discovery of the biaryl amide remote metalation reaction [14] a retrosynthetic analysis, **48 → 49 → 50 + 51**, based on an ultimate Friedel-Crafts step was followed and gave unsatisfactory results [16]. The remote metalation tactic allowed expedient and unequivocal synthesis of **55a** and **55b** (**Scheme 16**) , the proposed structures for dengibsin dimethyl ether and dengibsinin dimethyl ether respectively [15]. The inconsistency in physical (**Scheme 16**) and spectroscopic properties [10] between synthetic materials **55a,b** and the dimethyl ethers of dengibsin and dengibsinin was corroborated by Talapatra [17] on the basis of further spectroscopic analysis and by Sargent [18] who synthesized both the putative (**48a,b**) and authentic (**56a,b**) (**Scheme 17**) natural products and several of their derivatives.

Fluorenones From *Dendrobium gibsonii*   (Orchidaceae)

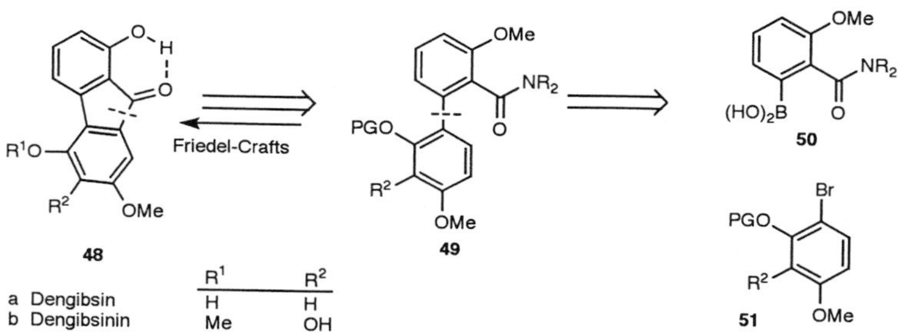

Scheme 15

**Synthesis of Dengibsin and Dengibsinin Dimethyl
Ethers (Putative).* Expediancy of Remote Metalation**

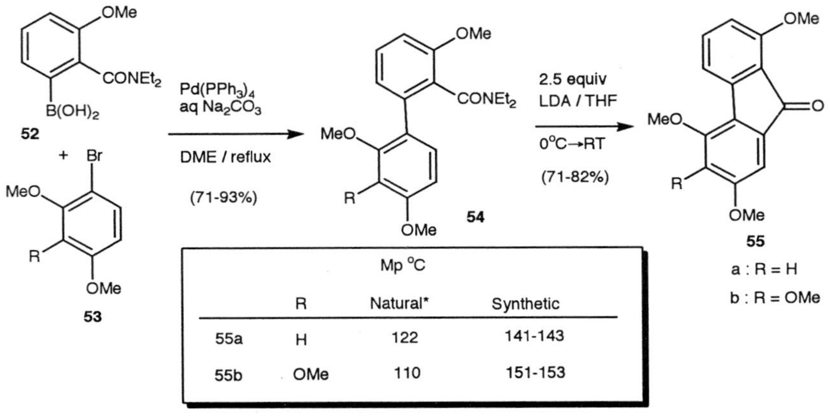

		Mp °C	
	R	Natural*	Synthetic
55a	H	122	141-143
55b	OMe	110	151-153

* Talapatra, S.K. et al Tetrahedron **1985**, *41*, 2765

Scheme 16

Fluorenones from *Dendrobium gibsonii* **(Indian Orchidaceae).**
Retrosynthetic Analysis based on Remote Metalation

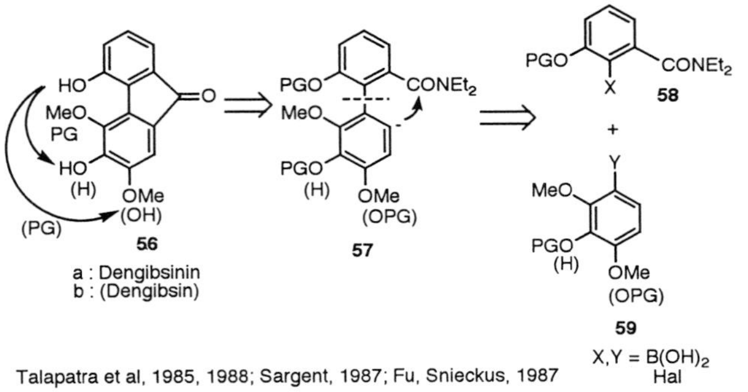

Talapatra et al, 1985, 1988; Sargent, 1987; Fu, Snieckus, 1987

Scheme 17

The revised structures of dengibsinin (**56a**) and dengibsin (**56b**) (**Scheme 17**) allowed a retrosynthetic analysis which, aside from being based on the remote metalation - cross coupling strategy, **56 → 57 → 58 + 59**, contained an exercise in differential phenol protection. The latter point was incorporated into the starting coupling partners, **60** and **61** (**Scheme 18**) according to the selective isopropyl ether cleavage protocol developed by Sargent [18]. In this coupling, noteworthy is also the use of boronate ester **60** rather than the corresponding boronic acid, an advantage established systematically [10] for cases which are subject to steric hindrance effects. The resulting biaryl amide **62** was subjected to the standard remote metalation protocol to give fluorenone **63** which upon chemoselective BCl₃ deprotection concluded the short synthesis of dengibsinin (**56a**) .

Total Synthesis of Dengibsinin

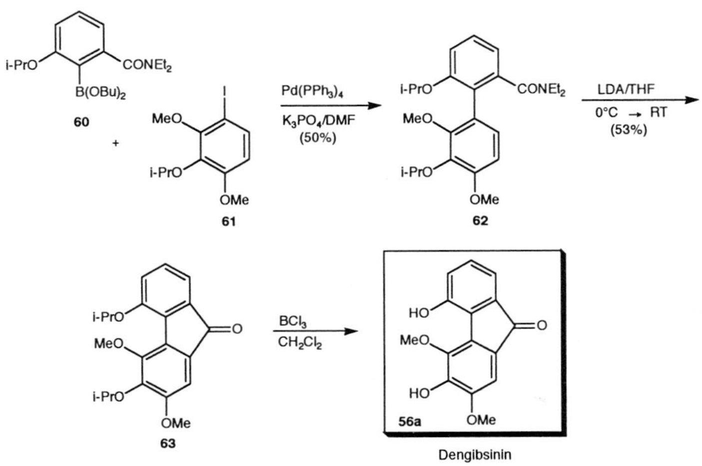

Scheme 18

The azafluoranthene class of alkaloids **64 (Scheme 19)** invited a retrosynthetic analysis conceptually reminiscent of that successfully executed for homomoschatoline **(Scheme 11)** . Thus the cyclopentane carbon in **64** (dotted lines) was viewed as a lynchpin, derived from a remote metalation-cyclization **(65)** and serving for heteroring annelation from the ethyl amine side-chain. Aside from the CIPE consideration [13], metalation as indicated in **65** was also envisaged to receive assistance from the weak DMG effect of the side chain N-atom. The key biaryl amide **65** would be derived unexceptionally from cross coupling of **66** and **67**, partners whose X,Y groups would again be decided upon availability of starting materials. To put these ideas to the experimental test, imeluteine was chosen as the target molecule [19]. Cross coupling of the bromobenzaldehyde **68 (Scheme 20)** with the o-boronic acid benzamide **69** under standard conditions afforded the biaryl **70** in high yield. Chain extension by condensation with nitromethane followed by LAH reduction, without the obligatory short reaction time and low temperature required for this reaction in an analogous diethyl amide case **(Scheme 12)** , gave the phenethyl amine **71**. Treatment of **71** with excess LDA resulted in the desired double cyclization to give **72** in low but as yet unoptimized yield. Dehydrogenation of **72**, already achieved in a previous preparation of imeluiteine, concluded this short and efficient synthesis of the natural product **(73)** .

Azafluoranthene Alkaloids: Retrosynthetic Analysis
Based on Integrated Metalation-Cross coupling Tactics

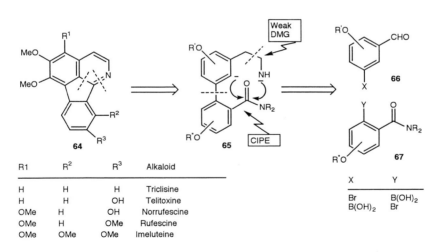

R1	R²	R³	Alkaloid
H	H	H	Triclisine
H	H	OH	Telitoxine
OMe	H	OH	Norrufescine
OMe	H	OMe	Rufescine
OMe	OMe	OMe	Imeluteine

X	Y
Br	B(OH)₂
B(OH)₂	Br

Menispermaceae (Abuta imene, Abuta rufescens,
Triclisia gilletti, Telitoxicum peruvianum)

Scheme 19

Total Synthesis of Azafluoranthene Alkaloid Imeluteine

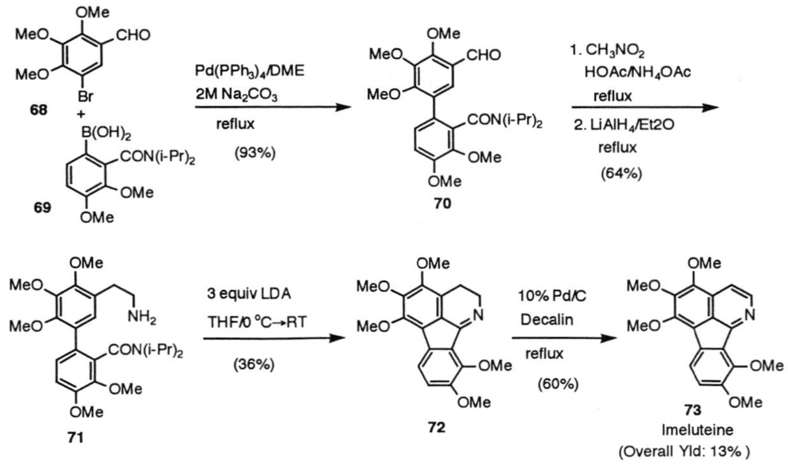

Scheme 20

The kinamycin antibiotics, constituting ring D reduced-oxygenated systems of prekinamycin (**79**, **Scheme 21**) discovered in 1970, have been of considerable interest owing to their potent and broad spectrum antitumor activity and intriguing biosynthetic origin [20]. While the polyketide origin of the kinamycins and the intermediacy of the benz[a]anthraquinone **74** has been established [20], a number of fascinating biosynthetic uncertainties remain.

Biosynthesis of the Kinamycin Antibiotics

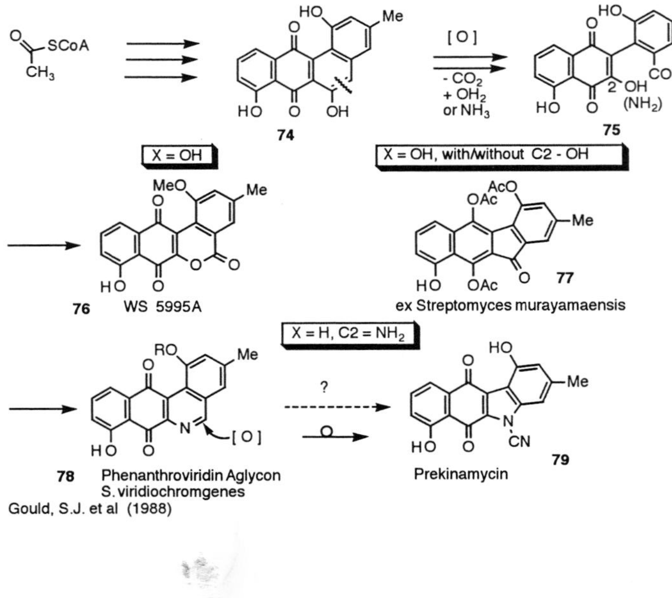

Scheme 21

The intermediates **75** have been proposed on the biosynthetic grid leading to the metabolites WS 5995A **(76)** and the recently discovered phenanthroviridin aglycon **(78)** and the precursor relationship of the latter with prekinamycin **(79)** has been suggested [20]. In the meantime, Gould has isolated the fluorenone **77** from a *Streptomyces* species [20].

As an expression of general interest in the synthesis of the diverse structural types associated with the kinamycins **(Scheme 21)** , we have undertaken the synthesis of fluoreone **77 (Scheme 22)** and WS 5995A **(76) (Scheme 23)** by combined metalation-cross coupling methods. In the attack on **77 (Scheme 22)** [21] coupling of readily available bromo naphthoquinone **80** with the stannylated benzamide **81** (prepared by DoM chemistry) using the excellent Stille-Eschavarren protocol [22] afforded **82** which upon sequential reduction and methylation gave the oxygenated naphthalene **83**. LDA-induced cyclization afforded the yellow fluorenone **84** in low yield. The conversion of **84** into **77** is in hand.

Biarylamide Remote Metalation: Synthesis of Fluorenone   *ex*
Streptomyces murayamaensis

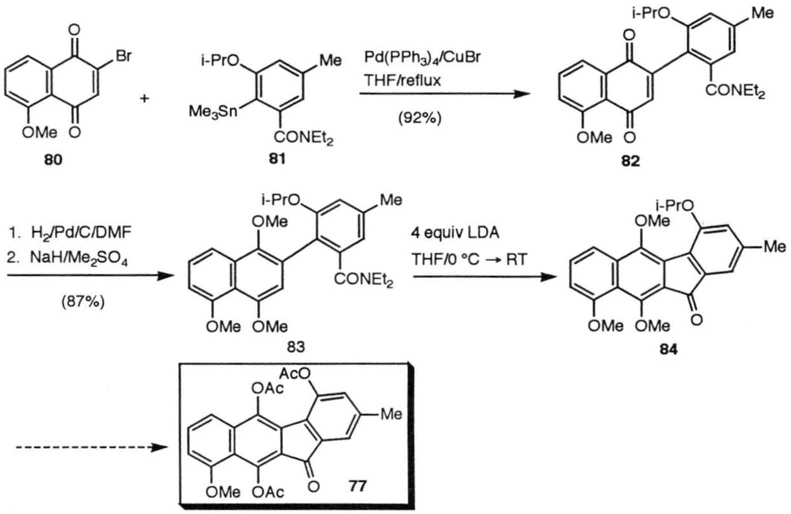

Scheme 22

The synthesis of WS 5995A **(76, Scheme 23)** commenced [21] with the naphthoquinone **85** which was prepared by a route analogous to that described for **82 (Scheme 22)** . Ammonolysis of **85** produced the aminonaphthoquinone **86** which upon treatment with heptafluorobutyric acid gave the lactone **87** in high yield. Since the conversion of **87** into the WS 5995A **(76)** has already been accomplished by Watanabe and coworkers [23], this concluded a formal synthesis of the natural product. Current efforts are aimed in adapting intermediates of both projects **(Schemes 22, 23)** for the synthesis of phenanthroviridin aglycon **(78, Scheme 21)** .

Synthesis of Antibiotic WS 5995A   *ex Streptomyces auranticolor*

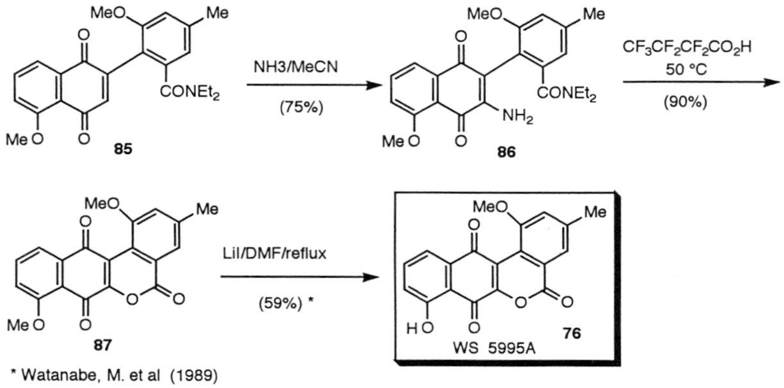

Scheme 23

In order to attempt diversification of the remote metalation concept, the reactivity of biaryl *O*-carbamates **88 (Scheme 24)** to strong base was tested. Previous experience had demonstrated kinetic deprotonation *ortho* to the carbamate (site 1) and rapid O to C carbamoyl transfer [1] even at low temperatures. Protection (PG_1) of this site was therefore envisaged prior to attempts at remote metalation (site 2) and ring to ring carbamoyl transfer leading to **89**. If successful, further useful synthetic ramifications of **89**, e.g. metalation *ortho* to amide subsequent to phenol protection (PG_2) was anticipated.

After some experimentation, the general strategy for the metalation-carbamoyl transfer, constituting an overall remote anionic Fries rearrangment, was defined [24]. Thus cross coupling of the standard aryl bromide and boronic acid partners, **90** and **91 (Scheme 25)** , followed by low temperature metalation-silylation (to avoid *ortho*-carbamoyl migration) led to the silylated biaryl *O*-carbamates **92**. To our delight, the proposed ring to ring carbamoyl transfer occurred upon treatment of **92** with 2.5 equiv of LDA in *refluxing* THF solution to give a variety of biaryl amide phenols **93** in good yields.

Remote Anionic Fries Rearrangement of   *O*-Biaryl Carbamates.
Ring to Ring Carbamoyl Transfer Concept

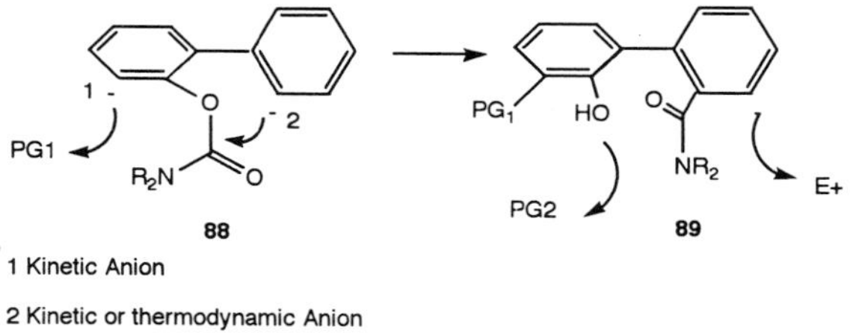

1 Kinetic Anion

2 Kinetic or thermodynamic Anion

Scheme 24

As illustrated in two selected cases (**94 → 95, 96 → 97, Scheme 26**) , the reaction proceeds under milder conditions and in high yield when the carbon to which carbamoyl migration occurs is *ortho* to a DMG (OMe, CONEt$_2$). Furthermore, a preliminary result (**98 → 99 + 100, Scheme 27**) indicates that the remote metalation-migration sequence is viable, with the expected regiochemical consequences, in various pyridine series.

Remote Anionic Fries Rearrangement via
Triethylsilyl (TES) Protection

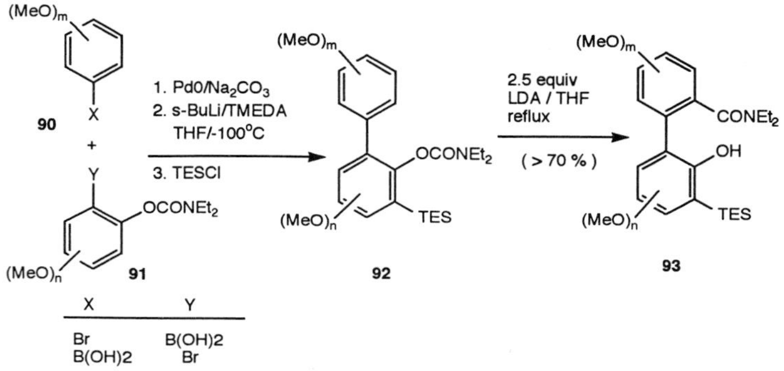

Scheme 25

DMG-Enhanced Remote Anionic Fries Rearrangement

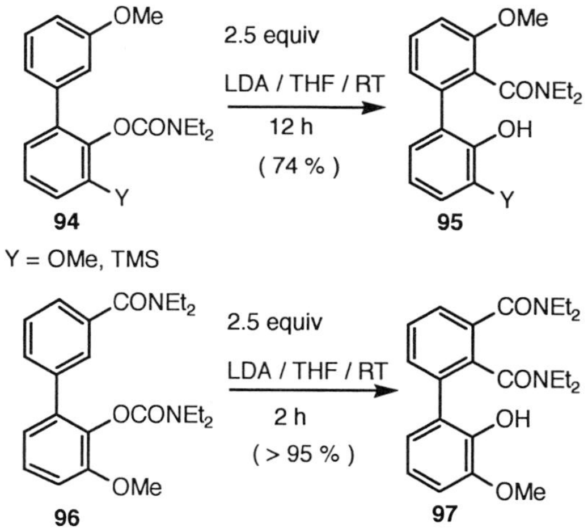

Scheme 26

Remote Anionic Fries Rearrangement of an
O-Heterobiaryl Carbamate

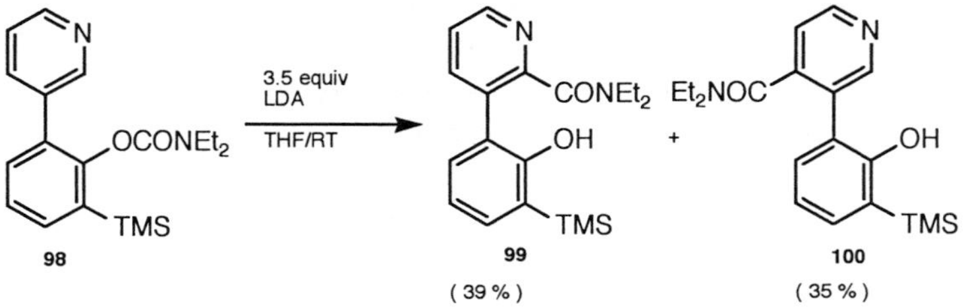

Scheme 27

The availability of 2-hydroxy-2'-carboxamido (**Schemes 25-27**) by the remote metalation tactic led to the development [24] of a new route to dibenzo[b,d]pyranones **101** → **102** (**Scheme 28**) and allowed the preparation of some highly sterically hindered systems (e.g. **103, 104**) and aza analogues **(105)** not attainable by the previously developed direct procedure [7h]. In two of these series, the synthesis of models **108** and **109** (**Scheme 29**) for the intensely studied gilvocarcin, ravidomycin, and chrysomycin classes of antibiotics **106** and the only recently conquered [25] isoschumanniophytine **107** group of alkaloids has been accomplished. The synthesis of the former **(108)** was initiated by Suzuki cross coupling of bromonaphthalene O-carbamate **110** (**Scheme 30**) with 2,4-dimethoxyphenyl boronic acid to give the biaryl **111**. Metalation-migration was achieved under the optimum LDA conditions to smoothly furnish **112** which under refluxing acetic acid conditions gave the gilvocarcin model **108**.

Remote Anionic Fries Rearrangement of   O-Biaryl Carbamates.
General Regiospecific Route to Dibenzo[b,d]pyran-6-ones

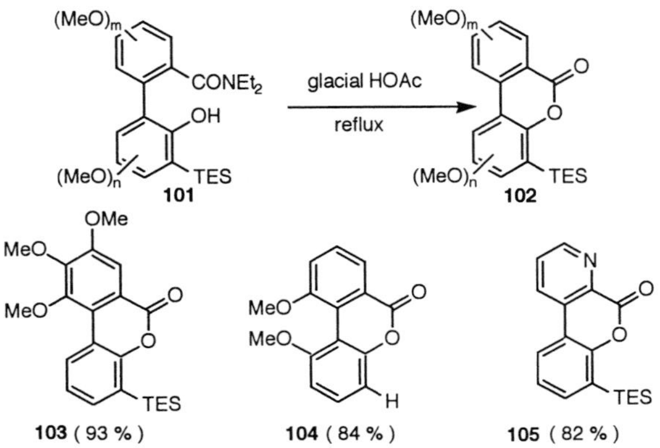

Scheme 28

The *ortho*-toluyl amide (**Scheme 7**) , and carbamate (**Scheme 24**) remote metalation discoveries nurtured ideas of combining some of these tactics in total synthesis endeavours.   Thus a sequence (**Scheme 31**)   was envisaged in which remote metalation-carbamoyl migration (**113**) would be followed by *ortho*-toluyl cyclization (**114**) (after suitable protection, PG$_2$+) to deliver concise and regiospecific

Synthesis of Ring Systems of Benzo[d]naphthopyran-
6-one Antibiotics and Chromone Alkaloids

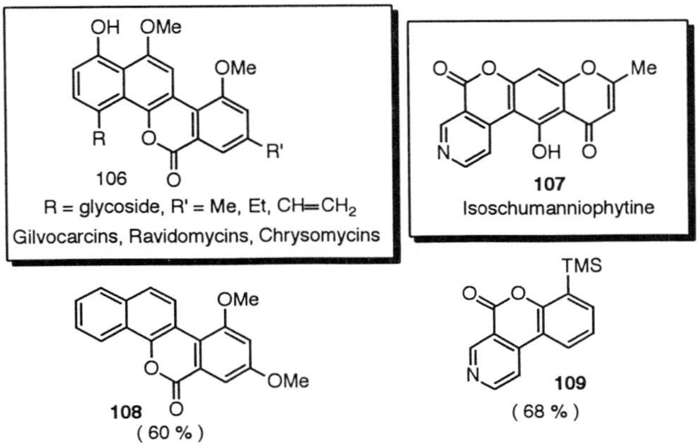

106
R = glycoside, R' = Me, Et, CH=CH$_2$
Gilvocarcins, Ravidomycins, Chrysomycins

107
Isoschumanniophytine

108
(60 %)

109
(68 %)

Scheme 29

routes to (by dissecting gymnopusin (**117, Scheme 32**)), a naturally occurring phenanthrene whose structure had undergone revision [26] partially on the basis of the total synthesis of the putative structure [10, 26b]. Thus C-9-C-10 scission in **117** allows consideration of 2-methyl-2'-carboxamidobiaryl **118** as the remote metalation precursor.

Biaryl *O*-Carbamate Remote Metalation. Ring to Ring Carbamoyl Transfer Route to a
Gilvocarcin Model

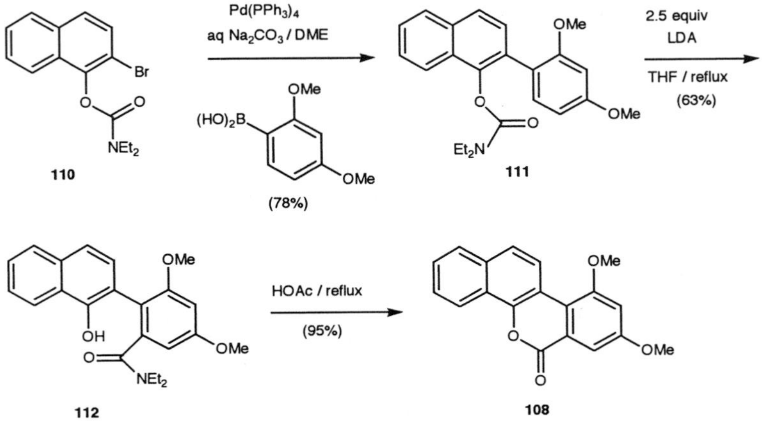

110

(HO)$_2$B

(78%)

Pd(PPh$_3$)$_4$
aq Na$_2$CO$_3$ / DME

111

2.5 equiv
LDA

THF / reflux
(63%)

112

HOAc / reflux

(95%)

108

Scheme 30

Intermediate **118**, in turn, would be derived from biaryl **119** by anionic carbamoyl transfer, a process which overcomes the problems associated with direct construction of this 2,2',6-trisubstituted biaryl. The biaryl carbamate **119** would be derived in the usual manner by cross coupling of **120** and **121** partners. To address the challenge of preparing the pentasubstituted benzene **120** with differential oxygen protection and activation, the blueprint outlined in **Scheme 33** was conceptualized. In this proposal, the real target **123** was viewed to result from a sequence (**122**) of metalation-OH⁺ synthon introduction-protection (step 1); a repeat performance with exception of carbamoyl group introduction (step 2); and finally, taking advantage of this potent metalation director, DoM reaction-electrophile (= Br⁺) quench.

Phenanthrols by Remote Metalation - Amide DMG Ring-to-Ring Translocation

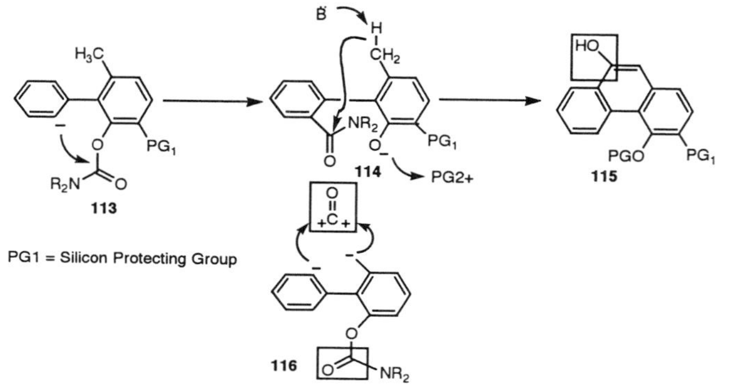

Scheme 31

Retrosynthetic Analysis of Orchid Natural Product Gymnopusin

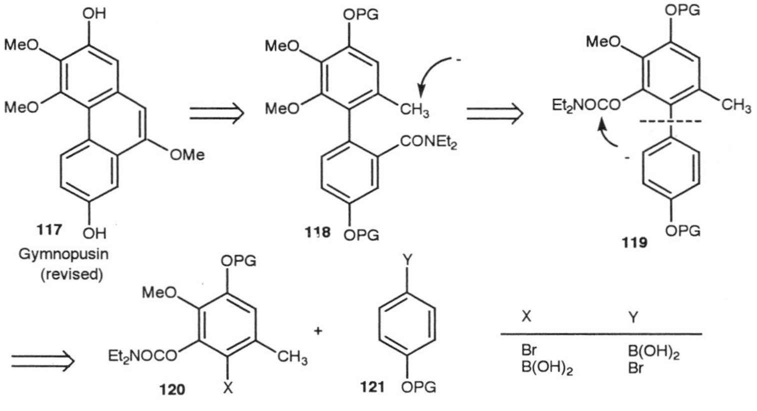

Scheme 32

Metalation-Cross Coupling Route to Gymnopusin.
The ArX Coupling Partner

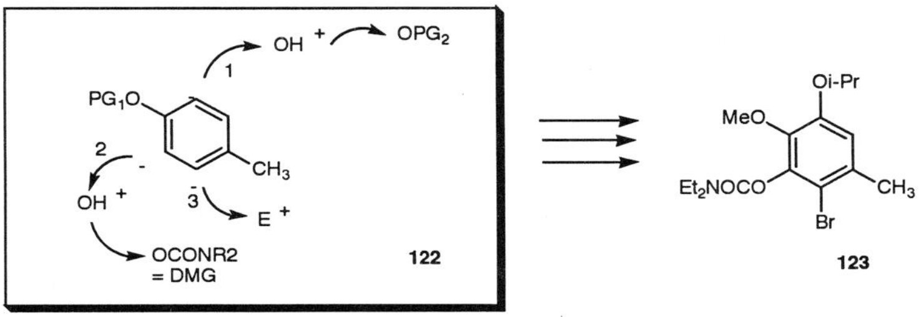

Scheme 33

The first step in the construction of gymnopusin [27] was dictated by the commercial availability of 2,6-dibromo-p-cresol (**124, Scheme 34**) . Conversion into the methyl ether **125** followed by sequential metal-halogen exchange, boronation, and oxidative workup afforded the phenol **126** in high yield. Isopropylation gave **127** which upon the identical metal-halogen exchange-boronation-oxidation steps furnished the phenol **128**. Carbamoylation led to **129** which upon metalation and treatment with a sparsely used Br+ source provided the pentasubstituted benzene **130**. This key intermediate displays a 1,2,3-trioxygenated pattern with each oxygen differentially protected. Standard cross coupling of **130** (**Scheme 35**) with the simple boronic acid **131** delivered the biaryl **132** which, upon treatment with LDA, was persuaded to rearrange to the isomeric biaryl **133**.

Metalation-Cross Coupling Route to Gymnopusin

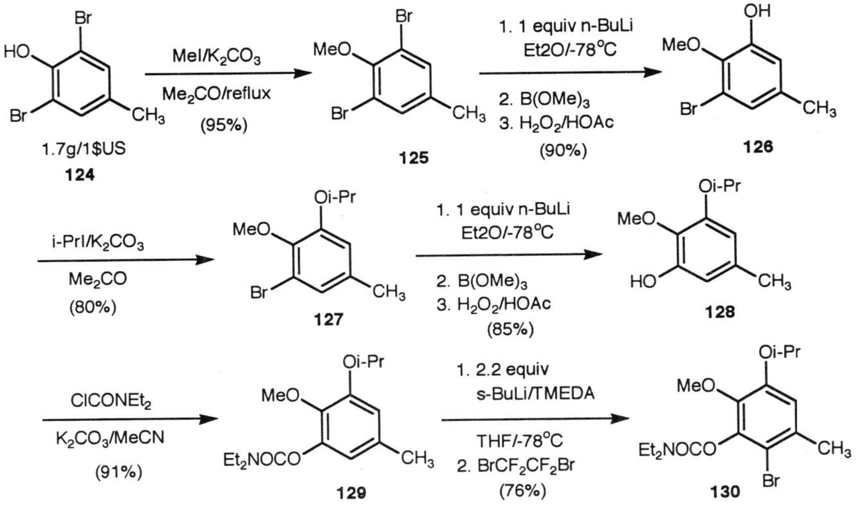

Scheme 34

The excess of LDA was apparently required to overcome the base "lost" in coordination to the multiply oxygenated system **132**. Phenol methylation followed by *ortho*-toluyl metalation-cyclization afforded the requisite phenanthrol which, owing to its instability, was immediately protected by a further methylation reaction to provide compound **134**. Selective deisopropylation according to Sargent [18] concluded the short, efficient synthesis of gymnopusin (**117**).

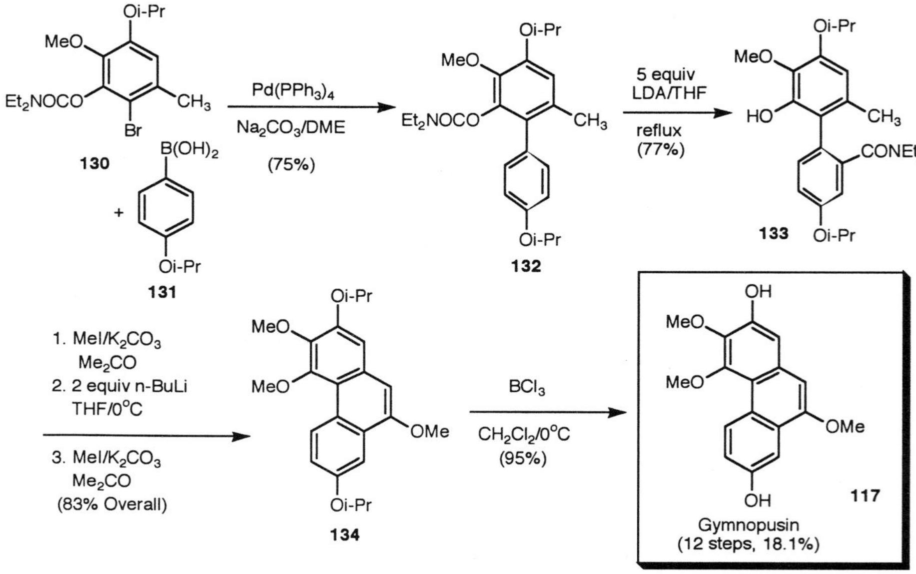

Scheme 35

As an alternate conceptual framework of combinational remote metalation reactions, the carbamoyl migration (**135** → **136, Scheme 36**) followed [after protection, PG2] by amide cyclization (**136** → **137**) was advanced. This sequence effectively bridges a 2,2'-biaryl dianion with a carbonyl 1,1-dication equivalent (**138**) which is derived from *internal* (compare with **Scheme 31**) electrophilic reagent. Dengibsin (**56b**) (**Scheme 37**) provided the playground in which to test this remote metalation-DMG translocation sequence (**Scheme 37**).

Thus retrosynthetic analysis posited that **56b** would be derived by remote metalation-cyclization of precursor **139** which in turn would be obtained, with a functional group interconversion (FGI from **140**) by the remote anionic Fries rearrangement. The cross coupling regimen, **141 + 142**, would be used to provide **140**.

Fluorenones by Remote Metalation - DMG Translocation

135   136   137

138

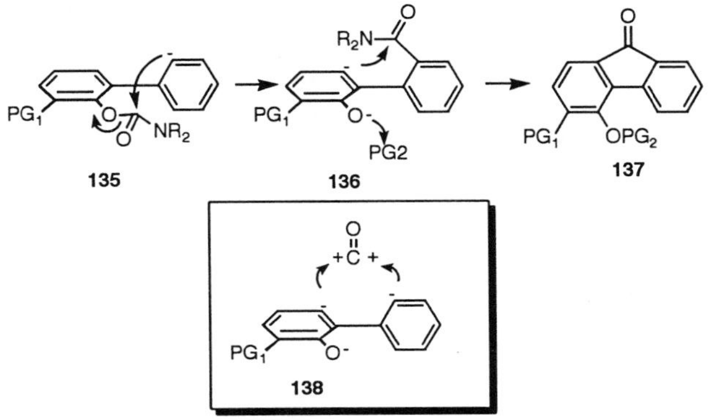

Scheme 36

Fluoreneones from *Dendrobium gibsonii* (Indian Orchidaceae).
Retrosynthetic Analysis based on Remote Metalation - DMG Translocation

56b
Dengibsin

139

140

141   142

Talapatra et al, 1985, 1988; Sargent, 1987;
Fu, Snieckus, 1987

X, Y = B(OH)$_2$, Hal

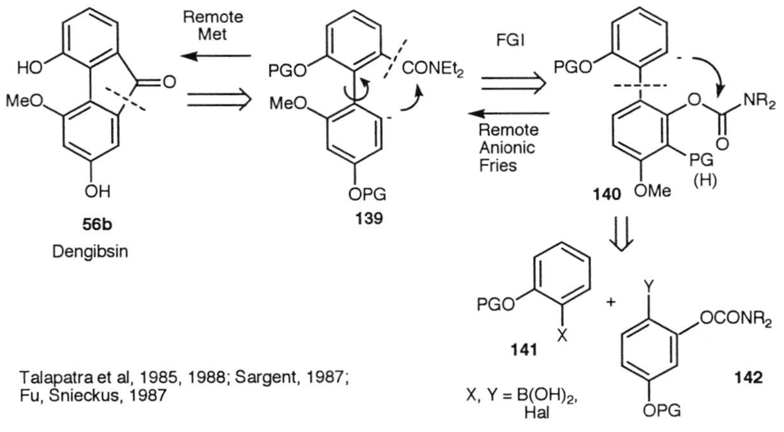

Scheme 37

In pursuit of the synthesis of dengibsin [24], the carbamate **143** (**Scheme 38**) was converted by two unexceptional steps into the *t*-butyl dimethylsilyl (TBS) derivative **144**. The proposal that the bulky TBS group would discourage metalation at the "in between" site in **144** proved correct when its metalation-bromination exclusively afforded the bromo derivative **145** in good yield. Cross coupling with the o-isopropoxyphenyl boronic acid led to biaryl carbamate **146**. Compound **146** was deprotected and reprotected and then, in a sequence dictated by previous general experience (cf **Scheme 25**), metalated and

54   V. Snieckus

triethylsilylated to furnish **147**. In the key experiment, LDA treatment of **147** resulted in the formation of two chromatographically separable products, **148** and **149**, in the ratio of 2:1. Although undesired, the major product (**148**) was recycled to provide additional material (**149**). Parenthetically, the regioselective deisopropylation of **147** is possibly driven by a CIPE and is of mechanistic interest. Methylation and desilylation of **149** yielded compound **151** which, upon LDA treatment, gave the fluorenone **150**. Boron trichloride deisopropylation completed the short and regiospecific construction of dengibsin (**56b**).

**Combined Directed Metalation-Cross Coupling Strategy
to Dengibsin**

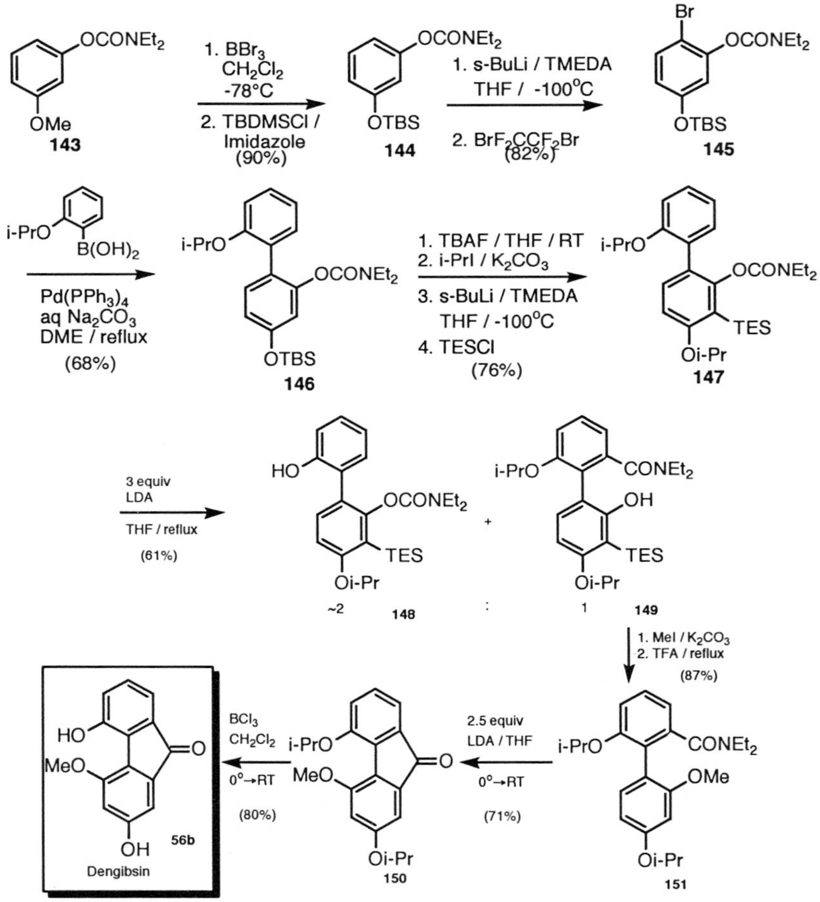

Scheme 38

Concluding Remarks

The Directed *ortho* Metalation (DoM) reaction [1] seeded by Gilman and Wittig in 1939 and systematically nurtured by C.R. Hauser in the late 1950s and by others in the ensuing years has reaped a rich harvest in modern synthetic practice during the last decade. A conceptual advance of similar magnitude, the transition metal catalyzed cross coupling reaction, originating perhaps with the work of Kharash but undoubtedly due to the fundamental studies of Corriu, Kumada, Negishi, Stille, Suzuki, among others, is bearing equally rich fruit today in synthetic laboratories worldwide.

The intertwining of synthetic methodologies provides strategies and processes which supercede the separate components, leads to bridging innovative proposals, and may influence areas far removed from the molecular prototype initially investigated. As illustrated in this article, the connections between DoM and cross coupling tactics invites new retrosynthetic thinking and new opportunities in aromatic chemistry.

The evolution of the DoM (*ortho* and remote) and the more recent cross coupling methodologies continues unabated. These fundamental reactions, separately or linked, will provide practicioners in synthetic aromatic chemistry with more ideas than they can handle. For any given target molecule, the task at hand is to set priority routes based on rationalization of efficiency, economy, and, with increasing concern, environmental issues. Undoubtedly, these roads will lead to the emergence of future generations of fundamentally different and practically useful synthetic methods.

References and Footnotes

1.   V. Snieckus, *Chem. Rev.* **90**, (1990) 879-933.

2.   G. Queguiner, F. Marsais, V. Snieckus, J. Epsztajn, *Advan. Heterocyclic Chem.* **52** (1992) 187-304.

3.   N. J. R. van Eikema Hommes, P. v. R. Schleyer, *Angew Chem. Internat. Ed. Engl.* **31** (1992) 755.

4.   Recent studies not reviewed in ref 1: OCONEt$_2$: M. Tsukazaki, V. Snieckus, *Heterocycles* **33** (1992) 533-536; OSEM: S. Sengupta, V. Snieckus, *Tetrahedron Lett.* **31** (1990) 4267-4270; CON-t-Bu: B. Zhao, V. Snieckus, *Tetrahedron Lett.* **32** (1991) 5277-5278; OCSNEt$_2$: F. Beaulieu, V. Snieckus, *Synthesis* **1** (1992) 112-118; SOt-Bu: C. Quesnelle, T. Iihama, T. Aubert, H. Perrier, V. Snieckus, *Tetrahedron Lett.*, **33** (1992) 2625-2628.

5.   N. Miyaura, T. Yanagi, A. Suzuki, *Synth. Commun.* **11** (1981) 513.

6.   M. J. Sharp, V. Snieckus, *Tetrahedron Lett.* **26** (1985) 5997-6000.

7.   a) M. J. Sharp, W. Cheng, V. Snieckus, *Tetrahedron Lett.* **28** (1987) 5093-5096; b) W. Cheng, V. Snieckus, *Tetrahedron Lett.* **28** (1987) 5097-5098; c) T. Alves, A.B. de Oliveira, V. Snieckus, *Tetrahedron Lett.* **29**, (1988) 2135-2136; d) J.-m. Fu, M. J. Sharp, V. Snieckus, *Tetrahedron Lett.* **29** (1988) 5459-5462; e) M. A. Siddiqui, V. Snieckus, *Tetrahedron Lett.* **29** (1988) 5463-5466; f) T. Iihama, J.-m. Fu, M. Bourguignon, V. Snieckus, *Synthesis* **3** (1989) 184-188; g) M.A. Siddiqui, V. Snieckus, *Tetrahedron Lett.* **31** (1990) 1523-1526; h) B. I. Alo, A. Kandil, P. A. Patil, M.J. Sharp, M. A. Siddiqui, P. D. Josephy, V. Snieckus, *J. Org. Chem.* **56** (1991) 3763-3768.

8.   J.-m. Fu, V. Snieckus, *Tetrahedron Lett.* **31** (1990) 1665-1668.

9.   C. Unrau, M.G. Campbell, V. Snieckus, *Tetrahedron Lett.* **33** (1992) 2773-2776.

10.   J.-m. Fu, Ph.D. Thesis, University of Waterloo, 1990.

11.   X. Wang, Ph.D. Thesis, University of Waterloo, 1992.

12.   X. Wang, V. Snieckus, *Tetrahedron Lett.* **32** (1991) 4883-4884.

13.   P. Beak; A. I. Meyers, *Acct. Chem. Res.* **19** (1986) 356; G.W. Klumpp, *Recl. Trav. Chim Pays-Bas*, **105** (1986) 1.

14.   J.-m. Fu, B.-p. Zhao, M. J. Sharp, V. Snieckus, *J. Org. Chem.* **56** (1991) 1683-1685.

15.   S.K. Talapatra, S. Bose, A.K. Mallik, B. Talapatra,*Tetrahedron* **41** (1985) 2765-2769.

16.   M. J. Sharp, M.Sc. Thesis, University of Waterloo, 1986.

17.   S. K. Talapatra, S. Chakraborty, B. Talapatra, *Indian J. Chem.* **27B** (1988) 250-252.   We are most grateful to Professor Talapatra for prior information of structural revision.

18.   M. V. Sargent, *J. Chem. Soc., Perking Trans. 1* (1987) 2553-2563.

19.   B.-p. Zhao, V. Snieckus, *Tetrahedron Lett.* **32** (1991) 5277-5278.

20.   For a leading reference with citations, see M. P. Gore, S. J.Gould, D. D.Weller, *J. Org. Chem.* **57** (1992) 2774-2783.

21.   B.-p. Zhao, V. Snieckus, unpublished results. *Abstracts of Papers*, 204th National Meeting of the American Chemical Society; American Chemical Society: Washington, D.C., 1992; Abstract ORG 422.

22.   N. Tamayo, A. M. Echavarren, M. C. Paredes, *J. Org. Chem.* **56** (1991), 6488-6491.

23.   M. Watanabe, M. Date, S. Furukawa, *Chem. Pharm. Bull. Jpn.* **37** (1989) 292-297 .

24.   W. Wang, V. Snieckus, *J. Org. Chem.* **57** (1992) 424-427.

25.   T. R. Kelly, M. H. Kim, *J.Org. Chem.* **57** (1992) 1593-1597.

26.   a) P. L. Majumder, S. Banerjee, *Indian J. Chem.* **28B** (1989) 1085.   b) A. B. Hughes, M. V. Sargent, *J. Chem. Soc. Perkin Trans. 1* (1989) 1787.

27.   X. Wang, V. Snieckus, *Tetrahedron Lett.* **32** (1991) 4879-4882.

New Approaches to the Stereoselective Synthesis of β-Amino-α-Hydroxy Acids, Dipeptide Isosteres, and Azasugars from Natural Amino Acids

Alessandro Dondoni

Laboratorio di Chimica Organica, Dipartimento di Chimica, Università, Ferrara, Italy

Summary: New synthetic approaches to the title bioactive compounds are centered on the use of the thiazole ring as a masked formyl group. The one-carbon homologation of α-amino acids via 2-thiazolyl amino ketones, followed by the stereocontrolled reduction of the carbonyl and thiazole-to-formyl deblocking, provides β-amino-α-alkoxy aldehydes which serve as advanced intermediates to the corresponding amino acids and hydroxyethylene and ketomethylene dipeptide isosteres. The two- and three-carbon homologation of α-amino acids via α-amino aldehydes, using 2-thiazolyl-armed phosphoranes or the lithium enolate of 2-acetyl thiazole, lead to precursors to nojirimycin, mannojirimycin, galactonojirimycin and their 3-deoxy derivatives.

The main theme of this lecture deals with the homologation of amino acids to (poly)hydroxy amino aldehydes and the conversion of these intermediates to the corresponding acids as well as to bioactive compounds of current interest such as azasugars and dipeptide isosteres (Scheme 1). Although different routes are followed to this goal, they all rely upon the use of the thiazole ring as a convenient protected form of the formyl group, a key fulcrum for organic synthesis. In fact this heterocycle appears to encompass numerous features that serve as an excellent masked functionality. It can be readily installed in a given substrate, it is resistant to oxidants and reductants as well as to bases and acids, and yet it can be easily cleaved to an aldehyde under essentially neutral conditions. The latter feature is crucial for maintaining the configurational integrity of stereogenic centers.

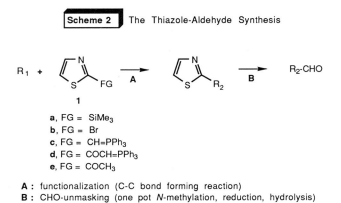

The *thiazole-aldehyde synthesis* (Scheme 2) involves two essential operations: **A**. the construction of a suitable carbon chain at C2 of the thiazole ring (*functionalization*); and **B**. the release of the formyl group from the thiazole ring (*unmasking*). While operation **B** is carried out by a general one-pot procedure consisting of three sequential reactions, i.e. *N*-methylation, reduction, and hydrolysis, operation **A** exploits different carbon-carbon bond forming reactions varying with the substrate and the thiazole-armed reagent employed. This flexibility enables one to control the extent of carbon-chain elongation of the substrate and the stereochemistry of newly formed stereocenters. Readily available and stable, yet rather unexpensive, 2-thiazolyl reagents **1** which offer the opportunity for various synthetic plans are listed in Scheme 2.

Scheme 2   The Thiazole-Aldehyde Synthesis

$$R_1 + \underset{\mathbf{1}}{\overset{N}{\underset{S}{\bigcirc}}} \text{-FG} \xrightarrow{A} \underset{S}{\overset{N}{\bigcirc}} \text{-}R_2 \xrightarrow{B} R_2\text{-CHO}$$

a, FG = SiMe₃
b, FG = Br
c, FG = CH=PPh₃
d, FG = COCH=PPh₃
e, FG = COCH₃

A: functionalization (C-C bond forming reaction)
B: CHO-unmasking (one pot *N*-methylation, reduction, hydrolysis)

As we have already demonstrated application of the thiazole-aldehyde synthesis in totally chemical routes to carbohydrates [1], we intend to employ this methodology in synthetic approaches to unusual amino acids [2] and various bioactive *N*- compounds .

Synthesis of β-Amino-α-Hydroxy Acids and Isosteric Dipeptides

The development of stereoselective approaches to β-amino-α-hydroxy acids [3,4] is of considerable interest because of their occurrence in many biologically active compounds and their use as precursors to modified peptides. Some molecules incorporating these amino acid units include the potent anticancer agent taxol [5], detoxin D$_1$, a selective antagonist of the antibiotic blastacidin S [6], and the aminopeptidase inhibitors bestatin [4,7] and amastatin [4,7], just to cite a few (Scheme 3).

Scheme 3

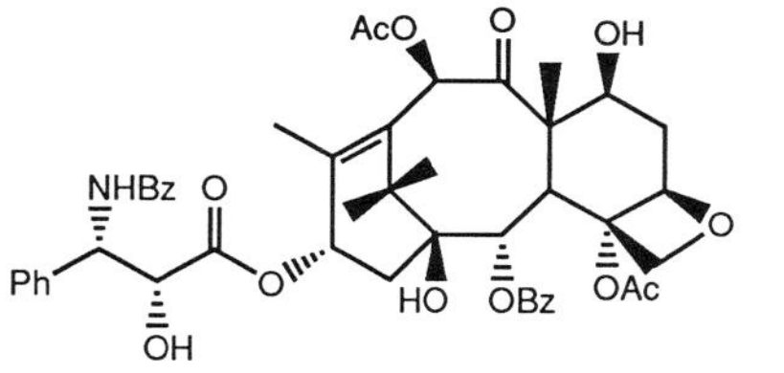

Taxol

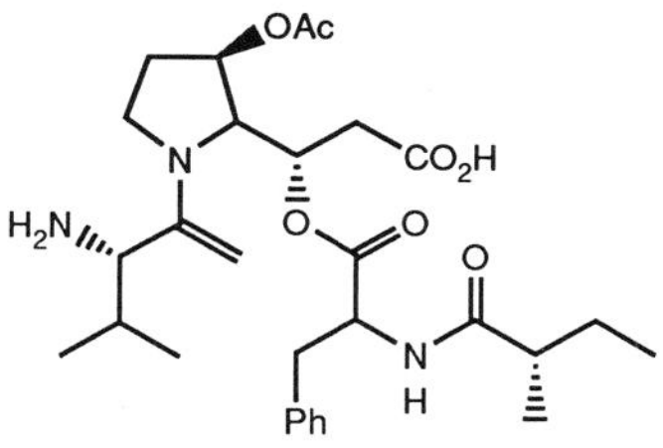

Detoxin D$_1$

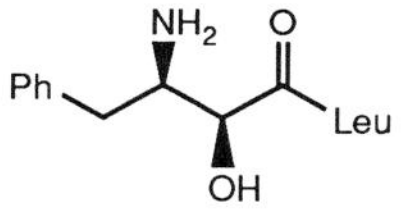

Bestatin

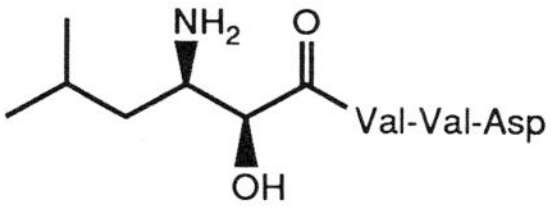

Amastatin

A direct entry to β-amino-α-hydroxy acids can be envisaged via one-carbon homologation of α-amino acids with concomitant formation of the new stereocenter bearing the hydroxy group (Scheme 4).

Scheme 4

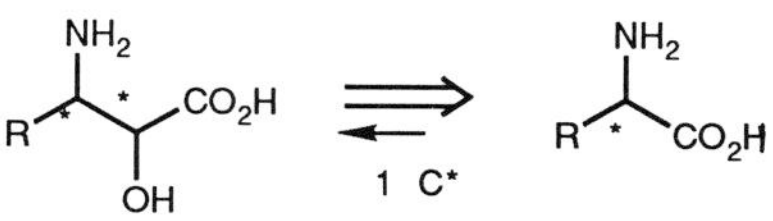

We have reported [8] the execution of this plan by stereoselective addition of 2-(trimethylsilyl)thiazole (2-TST, **1a**) to α-amino aldehydes, a class of reactive intermediates [9] readily available from α-amino acids. The subsequent thiazole-to-formyl unmasking converted the adducts to α-hydroxy-β-amino aldehydes, i.e. advanced precursors to the target amino acids. Quite interestingly, access to either epimer at C2 was achieved by stereocontrolled addition of 2-TST (**1a**) to differentially *N*-protected α-amino aldehydes. For example, *N*-Boc serinal acetonide (**2a**) (Scheme 5) afforded the amino alcohol *anti*-**3** (Felkin-Ahn adduct), whereas the monoprotected *N*-Boc *O*-benzyl analog (**2b**) produced the diastereomer *syn*-**3** (Cram chelate adduct).

Scheme 5

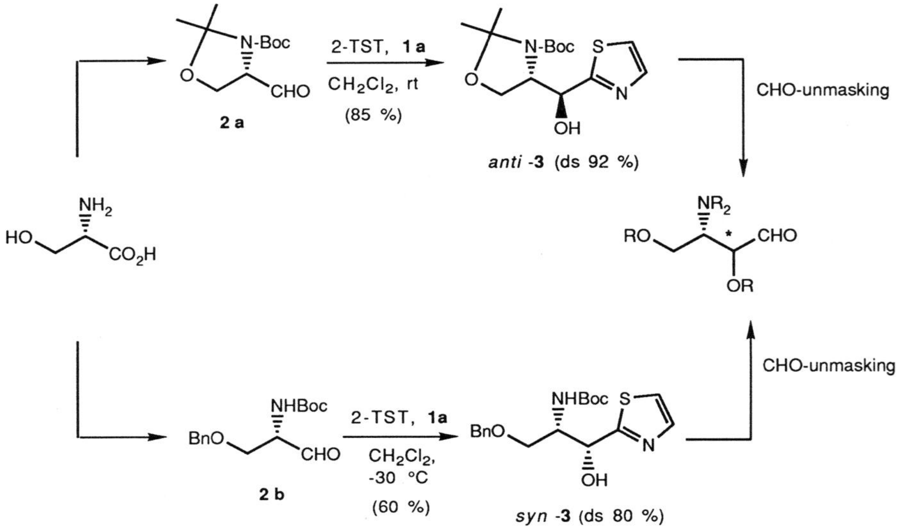

We devised an alternative route to *syn* and *anti* 2-thiazolyl amino alcohols by stereocontrolled reduction of thiazolyl amino ketone intermediates. This methodology was successfully applied [10] to L-threonine (Scheme 6). The amino acid was transformed into the protected methyl ester **4** which, by substitution with 2-lithiothiazole, 2-LTH, (from 2-bromothiazole (**1b**) and BuLi at -78 °C), afforded the key intermediate amino ketone **5**. Tunable diastereoselectivity (ds ≥ 95 %) of the carbonyl reduction was associated also in this case with the different protection of nitrogen, thus providing access to either amino alcohol epimer *anti*-**6** and *syn*-**6**. The synthesis was completed by the liberation of the aldehydes and their oxidation to β-amino-α-hydroxy acids **7**. The stereochemistry of **7a** and **7b** and consequently that of their precursor amino alcohols *syn*-**6** and *anti*-**6**, was supported by their NMR spectra. ^{1}H and ^{13}C NMR data were consistent with a chair conformation of the 1,3-dioxane ring of **7a** and a twist boat conformation of **7b**. Finally, the efficiency of this synthetic sequence was substantiated by the good yields and high levels of stereoselectivity.

Scheme 6

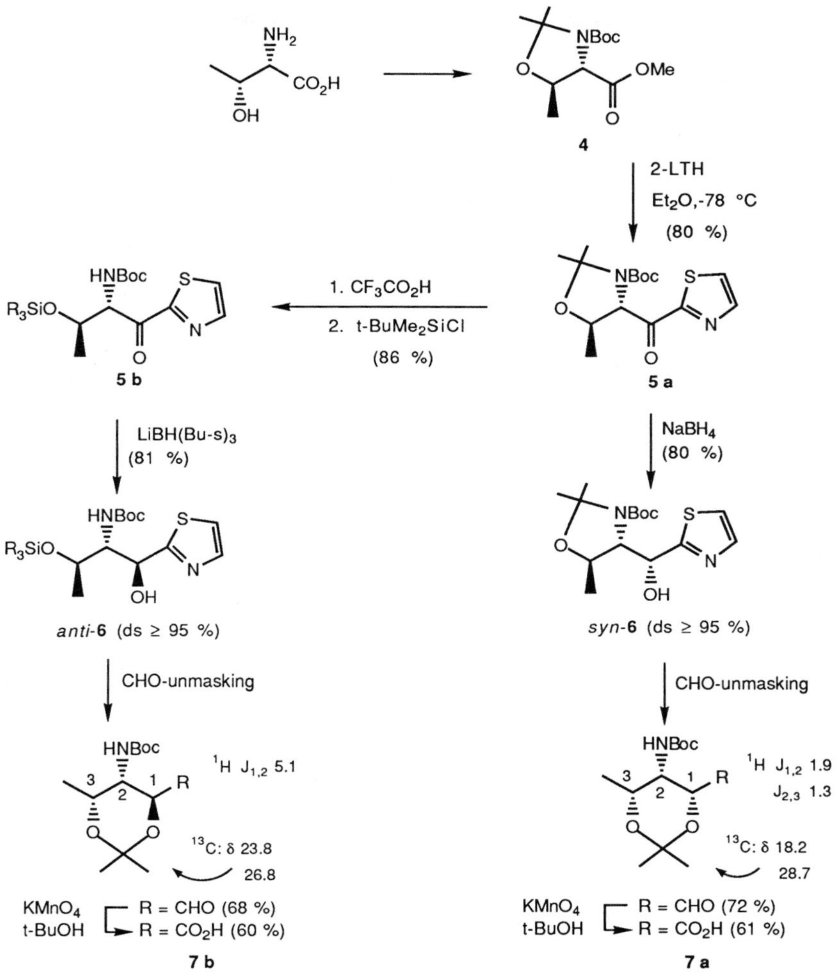

The same reaction sequence was employed with high efficiency for the stereocontrolled homologation of serine, phenylalanine, and leucine to the corresponding pairs of β-amino α-epimer α-hydroxy aldehydes and acids (Scheme 7). Hence, the thiazole-mediated synthesis of unusual amino acids employing amino ketones as reactive intermediates appears to be a convenient complementary route to that via amino aldehydes.

Scheme 7   Synthesis of β-Amino-α-hydroxy Aldehydes (X = CHO) and Acids (X = CO₂H)

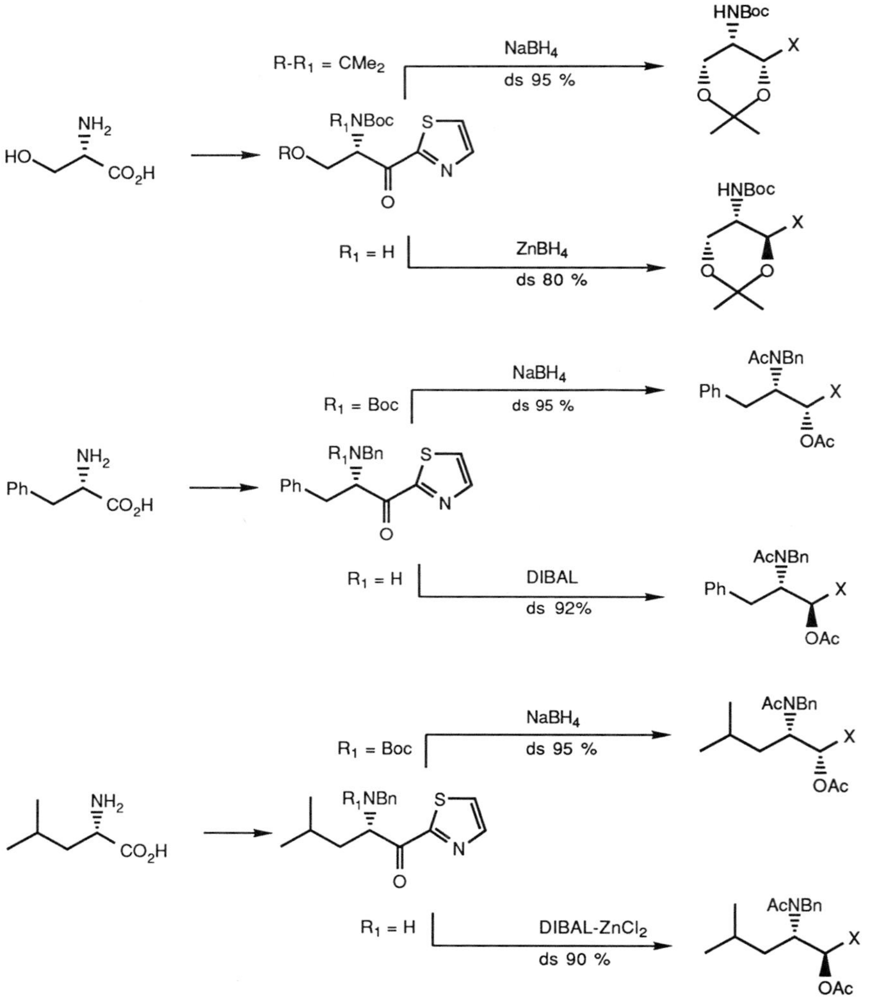

The β-amino-α-hydroxy aldehydes obtained by the above methodology turned out to be useful intermediates in synthetic approaches to isosteric dipeptides. These compounds, designed on the transition-state mimetic concept as renin and protease inhibitors, are modified peptides in that they possess a carbon-carbon bond bearing a suitable functionality in place of the scissile amide linkage [11]. Dipeptide isosteres are gaining even greater attention for their incorporation into potential drugs against HIV-1, the virus responsible for the AIDS disease. Consequently, enormous efforts have been recently stimulated in the synthesis of various types of these compounds [12]. A report [12p] by researchers of Merck Sharp and Dohme describes the synthesis of potent and selective inhibitors of HIV-1 protease incorporating the hydroxyethylene Phe-Phe isostere. This modified peptide was prepared in about 20 % yield from D-mannose via a key intermediate lactone (Scheme 8).

Scheme 8

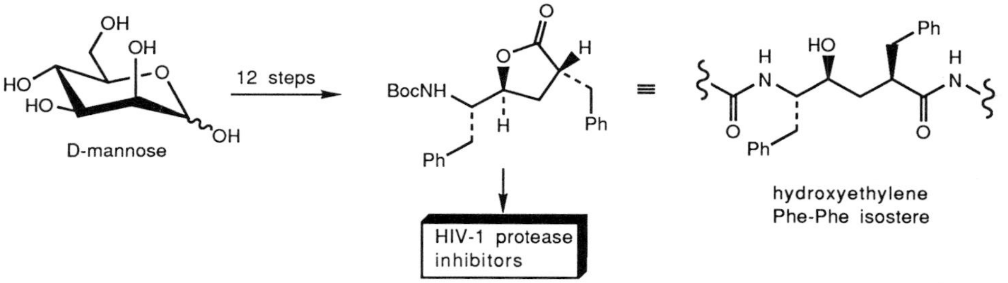

Our synthesis [13] of the lactone intermediate starts from L-phenylalanine methyl ester hydrochloride (**8**) (Scheme 9) which is converted via the 2-thiazolyl amino ketone into the protected β-amino-α-hydroxy aldehyde **9** (six steps, 56 %). A two carbon chain elongation of this aldehyde by Wittig olefination with a carboxylate armed phosphorane afforded the *E*-enoate **10** which upon carbon-carbon double bond reduction and desilylation produced the γ-lactone **11**. Finally, the target benzylated lactone **12** equivalent to hydroxyethylene L-Phe-L-Phe isostere was obtained by stereoselective benzylation (ds 92 %) of the anion of **11** with benzyl iodide. The overall unoptimized yield of **12** from the commercially available phenylalanine methyl ester (**8**) was 40 %.

Scheme 9

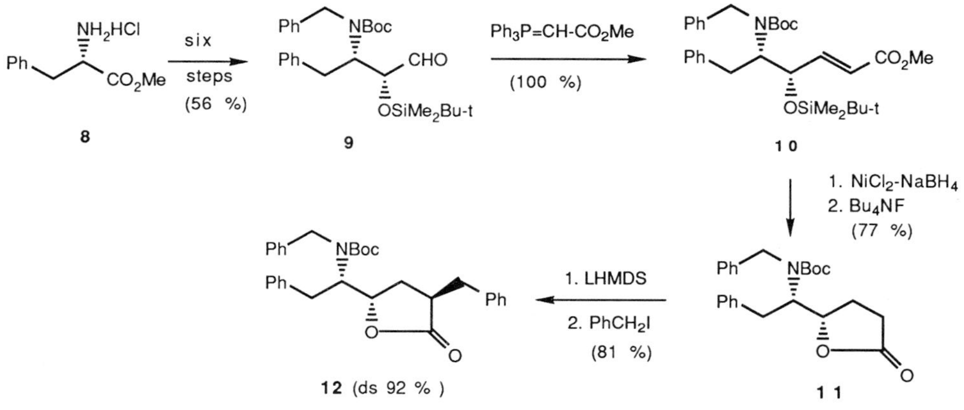

A similar sequence was followed [14] for the synthesis of the hydroxyethylene dipeptide isostere corresponding to L-Leu-L-Leu [15] in its lactone form **13**. Work is in progress toward the well known renin inhibitors hydroxyethylene isosteres L-Leu-L-Val **14** [16] and its cyclohexyl derivative **15** [17] (Scheme 10).

13   **14**   **15**

The *E*-enoate **10** appeared to be a useful intermediate toward the ketomethylene L-Phe-Gly isostere **16** . This pseudopeptide was readily obtained in about 70 % yield with the sacrifice of the stereocenter of **10** bearing the hydroxymethyl group by oxidation of the hydroxyl to a carbonyl followed by reduction of the ethylenic double bond (Scheme 11). The resulting dipeptide **16** was previously reported [18] as a component of angiotensin converting enzyme inhibitors.

Synthesis of Azasugars

Naturally occurring and synthetic polyhydroxylated piperidines (azahexoses) and pyrrolidines (azafuranoses) and their derivatives are well known potent and specific inhibitors of glycosidase enzymes [19]. For example, nojirimycin, related to glucose by replacing the ring oxygen by the NH group, is a powerful glucosidase inhibitor. Similarly, mannojirimycin and galactonojirimycin are active against mannosidase and galactosidase respectively (Scheme 12). In general, azasugars are very effective in carbohydrate associated metabolic disorders. As so, they have been recognized to possess potent antiviral properties because of the inhibition of glycoprotein processing necessary for viruses replication. This activity is exerted also against HIV-1.

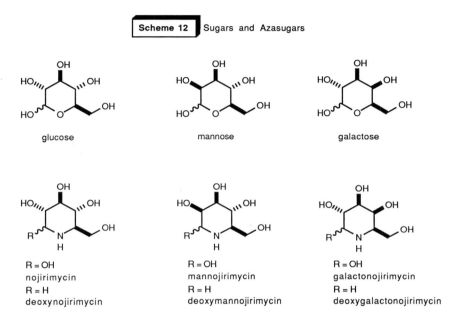

Scheme 12 Sugars and Azasugars

The numerous existing syntheses of azasugars employ sugar-based chiral starting materials in a variety of different strategies [20]. We have developed new approaches to these compounds from amino acids by the thiazole-aldehyde synthesis.

Galactonojirimycin [21]

Retrosynthesis (Scheme 13) points to serine (L-antipode is shown) as a precursor to this azasugar. The α–amino acid has to be subjected to two homologation processes, the first one extending the chain by one carbon atom, and the second by two carbon atoms. Overall, three contiguous stereocenters have to be formed.

Scheme 13

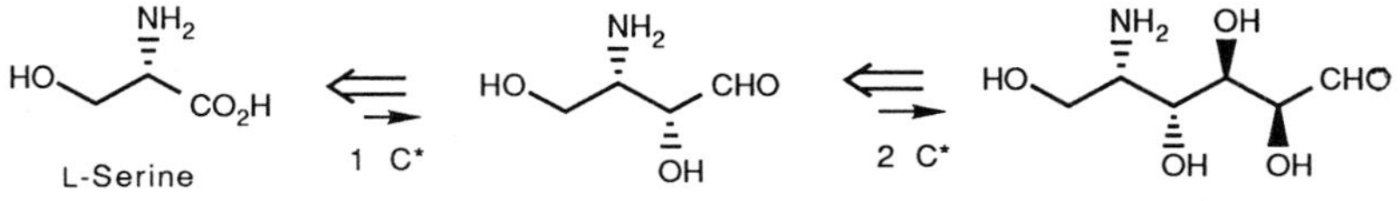

The first operation was effectively carried [22] out via the thiazolyl amino ketone intermediate **18** obtained from the serine derived amino ester **17** by the substitution with 2-lithiothiazole, 2-LTH (Scheme 14). The stereoselective reduction of **18** (ds 98 %) to the amino alcohol *syn*-**19** followed by thiazole to formyl deblocking afforded the suitably protected β-amino-α-hydroxy aldehyde (3-deoxy-3-amino threose) (**20**). The additional two carbon atoms were installed by Wittig olefination of the aldehyde **20** with 2-thiazolylmethylenetriphenylphosphorane (2-TMP, **1c**) to give the *E*-alkenylthiazole **21**. Two hydroxy groups were then readily added by *cis*-hydroxylation of **21** with a catalytic amount of osmium tetroxide with 2 equiv of *N*-methylmorpholine *N*-oxide as reoxidant (OsO₄-NMO). This reaction occurred with a modest level of stereoselectivity (ds 75 %) in favor of the required *cis*-diol *anti*-**22**. The aldehyde was then revealed from the thiazole ring and *N*- and *O*-protecting groups were removed to give (-)-galactonojirimycin in 22 % overal yield from the serine derived methyl ester **17**. Obviously, the natural (+)-antipode can be prepared by an identical route from D-serine.

Scheme 14

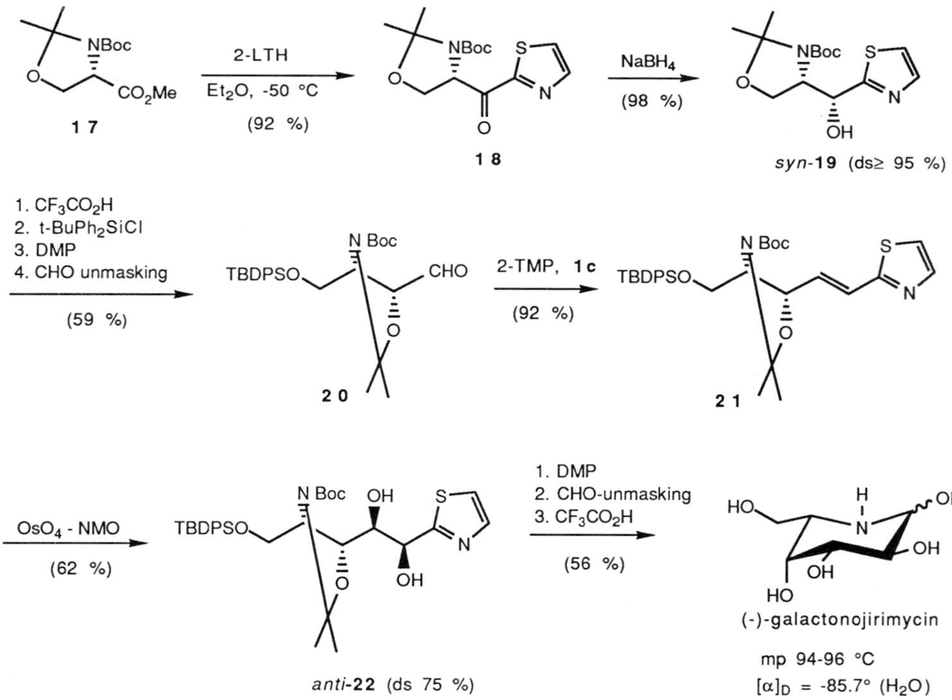

Nojirimycin [23] and Mannojirimycin [24]

Since these azasugars are epimers at C-2, retrosynthetic analysis indicates a route to both compounds via stereocontrolled reduction of the ketone carbonyl of a chiral α-keto aldehyde intermediate. The access to this intermediate from serine (L-antipode is shown) requires a chain elongation by three carbon atoms (Scheme 15).

Scheme 15

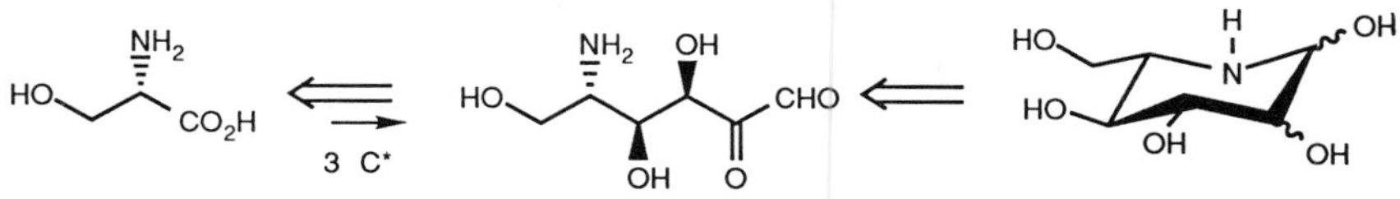

Serine was transformed [25] into a thiazole-masked α-keto aldehyde **28** by two complementary routes (Scheme 16), one involving the Wittig olefination of the protected L-serinal **2a** with 2-thiazolylcarbonylmethylene phosphorane (2-TCMP, **1d**), the other with the carbethoxymethylene phosphorane **23**. The *cis*-hydroxylation with *anti*-diastereoselectivity of the resulting olefins **24** and **25** was effectively carried out with the OsO₄-NMO mixture and the thiazole ring was installed in the ester *anti*-**27** by substitution with 2-lithiothiazole, 2-LTH.

Scheme 16

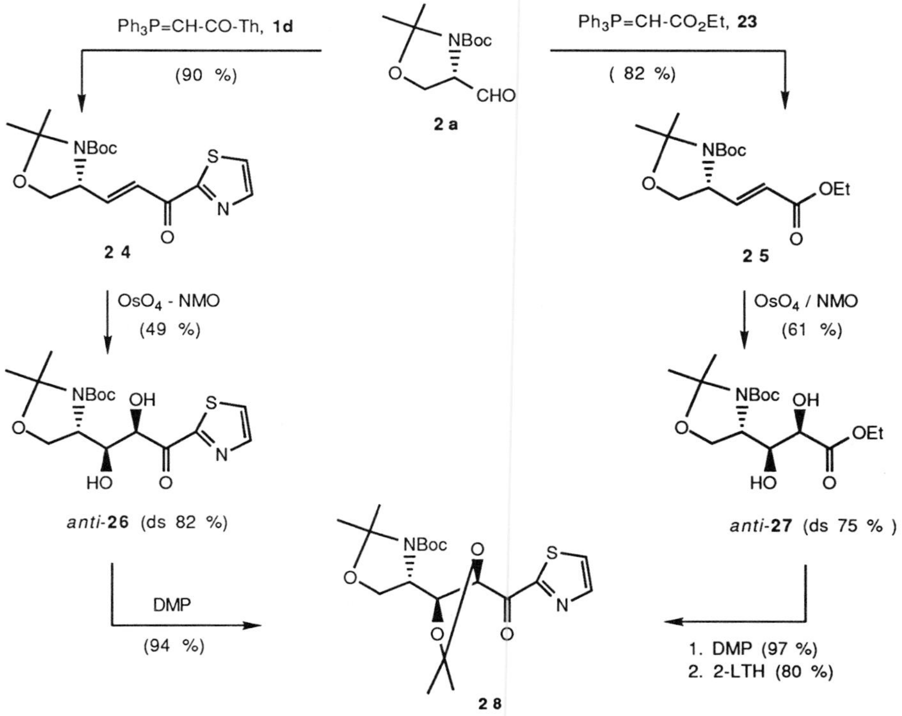

Having the chiral 2-thiazolyl ketone **28** in hand, i. e. a protected equivalent of the required α-keto aldehyde, the key reduction of the carbonyl with opposite diastereofacial selectivity was achieved using NaBH₄ to give the alcohol *syn*-**29** (ds 94 %) and Red-Al [Vitride, NaAlH₂(OCH₂CH₂OMe)] to give the epimer *anti*-**29** (ds ≥ 95 %) (Scheme 17). These intermediates were transformed into the target azasugars (about 20 % overall yield) by the now familiar thiazole-to-formyl deblocking protocol and removal of *N*- and *O*-protecting groups in the aldehydes *anti*-**30** and *syn*-**30**. It is also evident in this case that the total synthesis of the natural (+)-antipodes can be carried out by the same methodology starting from D-serine.

Scheme 17

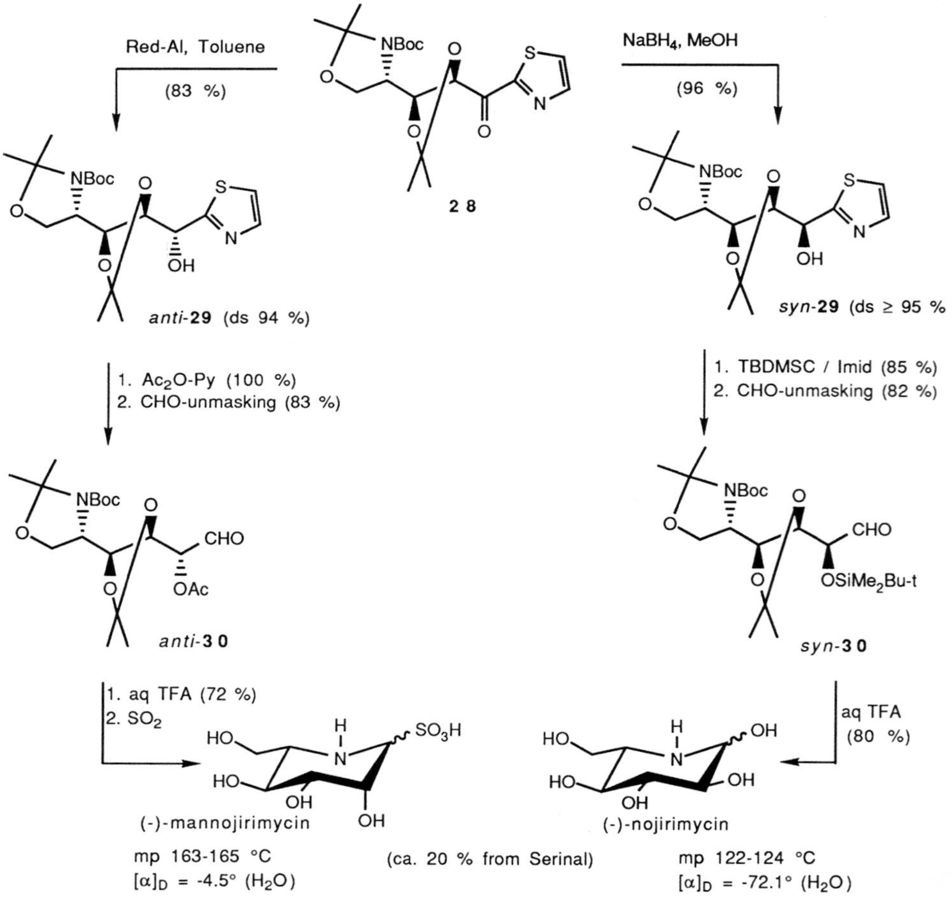

3-Deoxynojirimycin and 3-Deoxymannojirimycin

1-Deoxy derivatives of nojirimycin and mannojirimycin have been found to be equal to or more active than their parent compounds [19]. Moreover, given the interest in synthetic modifications of natural azasugars for biological activity explorations and structure-activity relationships, we decided to approach the synthesis of 3-deoxy derivatives by the thiazole-mediated methodology. In this case, [25, 26] the three-carbon chain elongation of the protected L-serinal (**2a**) to the thiazolyl ketone intermediate relied upon the aldol condensation with the lithium enolate of 2-acetylthiazole (2-ATT, **1e**) (Scheme 18). The resulting β-hydroxy ketone **31** was reduced to either epimeric 1,3-diol by using suitable borohydride reducing reagents. Thus, tetramethylammonium triacetoxy borohydride, Me₄NBH(OAc)₃, afforded the 1,3-diol *anti*-**32** (ds≥ 95 %) very likely via internal hydride delivery in a six-membered chair-like intermediate involving the β-hydroxy group [27]. On the other hand, the reduction with sodium borohydride in the presence of diethylmethoxyborane (NaBH₄-Et₂BOMe) afforded the expected 1,3-diol *syn*-**32** (ds ≥95%) presumably via external hydride delivery on a boron chelate half-chair intermediate [28]. The diastereomeric α,γ-dihydroxyalkylthiazoles *syn*-**32** and *anti*-**32** protected as the O-benzyl derivatives **33** were then elaborated by the usual techniques into the target 3-deoxy azasugars in very good overall yields (Scheme 19).

Scheme 18

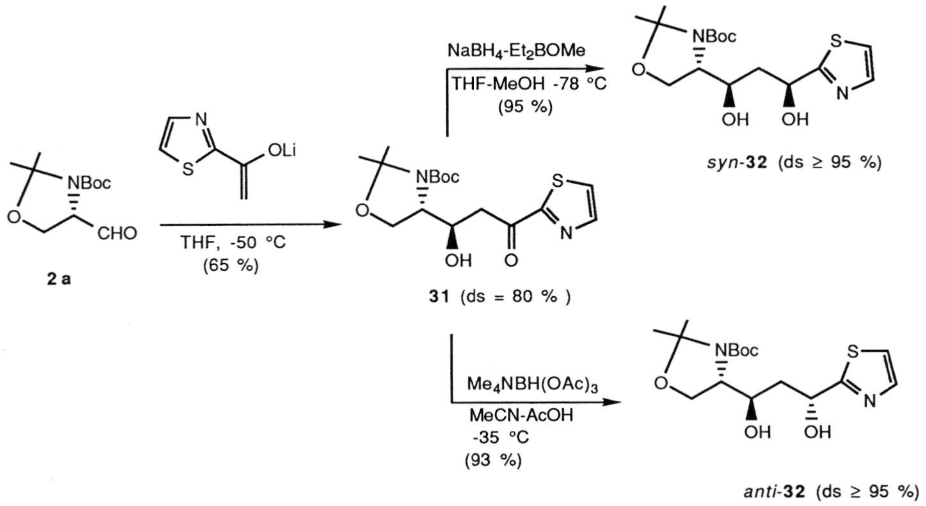

Scheme 19

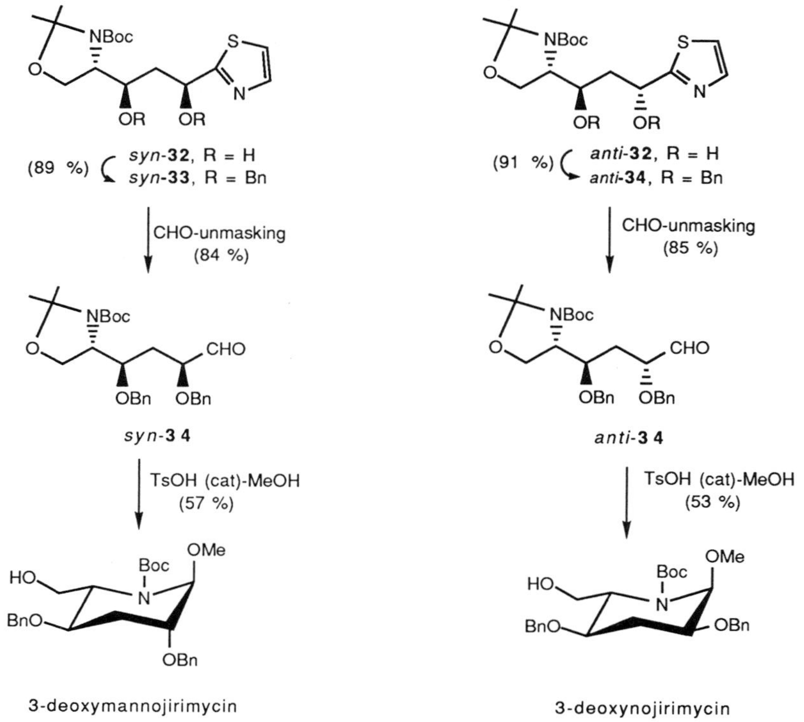

(89 %)
- syn-**32**, R = H
- syn-**33**, R = Bn

CHO-unmasking
(84 %)

syn-**34**

TsOH (cat)-MeOH
(57 %)

3-deoxymannojirimycin

(91 %)
- anti-**32**, R = H
- anti-**34**, R = Bn

CHO-unmasking
(85 %)

anti-**34**

TsOH (cat)-MeOH
(53 %)

3-deoxynojirimycin

Conclusions

The convenient synthetic equivalence of the thiazole ring with the formyl group has been demonstrated to apply quite well in synthetic approaches to two classes of interesting antibiotics, i. e. dipeptide isosteres and azasugars, from α-amino acids. The various strategies employed are based on the chain elongation of these chiral starting materials by the use of suitable 2-thiazolyl-armed reagents to give thiazole-masked aldehydes which are then elaborated to the final products. In all cases the initial new stereocenter is effectively constructed by exploiting the configuration at C-2 of the amino acid. The subsequent stereocenters are formed by substrate induced stereoselectivity as well. Quite interestingly, the presence of the thiazole ring did not interfere with any of the required synthetic elaborations nor did the application of the protocol for the aldehyde liberation from thiazole intermediates induce significant racemization. The exploration of other strategies based on these concepts for the synthesis of other classes of bioactive molecules becomes now even more attractive.

Acknowledgment

I am pleased to express my gratitude to Pedro Merino and Daniela Perrone who have developed with great diligence the work described here. Thanks are due to Ministero della Ricerca Scientifica e Tecnologica (MURST, Rome) and Progetto Finalizzato Chimica Fine e Secondaria n. 2, Consiglio Nazionale delle Ricerche (CNR, Rome) for finantial support.

References and Notes

1. Thiazole route to carbohydrates: synthesis of building blocks or precursors to carbohydrates and related compounds using C2 substituted thiazoles as intermediates. For a recent account see : Dondoni, A. *Carbohydrate Synthesis via Thiazoles* In *Modern Synthetic Methods*, Scheffold, R. (Ed.), Verlag Helvetica Chimica Acta, Basel, **1992**, p. 377.

2. For the synthesis and use of amino acids, see: a) Martens, J. *Top. Curr. Chem.* **1984**, *125*, 165. b) *Chemistry and Biochemistry of the Amino Acids*, Barrett, G. C. (Ed.), Chapman and Hall, London, **1983**. c) Williams, R. M. *Synthesis of Optically Active a-Amino Acids*, Pergamon Press, Oxford, **1989**. d) Coppola, G. M.; Schuster, H. F. *Asymmetric Synthesis. Construction of Chiral Molecules Using Amino Acids*, Wiley, New York, **1987**.

3. Y. Kobayashi, Y. Takemoto, Y. Ito, and S. Terashima, *Tetrahedron Lett.* **1990**, *31*, 3031; T. Matsumoto, Y. Kobayashi, Y. Takemoto, Y. Ito, T. Kamijo, H. Harada, and S. Terashima, *Tetrahedron Lett.* **1990**, *31*, 4175; C. Palomo, A. Arrieta, F. P. Cossio, J. M. Aizpurua, A. Mielgo, and N. Aurrekoetxea, *Tetrahedron Lett.* **1990**, *31*, 6429.

4. Drey, C. N. C. in ref. 2b, ch. 3.

5. Isolation: Wani, M. C.; Taylor, H. L.; Wall, M. E.; Coggon, P.; McPhail, A. I. *J. Am. Chem. Soc.* **1971**, *93*, 2325. Synthesis: Wender, P. A.; Mucciaro, T. P. *J. Am. Chem. Soc.* **1992**, *114*, 5878 and references cited therein.

6   Kakinuma, K.; Otake, N.; Yonehara, H. *Tetrahedron Lett.* **1972**, 2509. Synthesis of the precursor amino acid detoxinine : Ohfuna, Y.; Nishio, H. *Tetrahedron Lett.* **1984**, *25*, 4133.

7. Synthesis of precursor 3-amino 2-hydroxy amino acids : Herranz, R.; Castro-Pichel, J.; Vinuesa, S.; Garcia-Lopez, M. T. *J. Org. Chem.* **1990**, *55*, 2232.

8. Dondoni, A.; Fantin, G.; Fogagnolo, M.; Pedrini, P. *J. Org. Chem.* **1990**, *55*, 1439.

9. Jurczak, J.; Golebiowski, A. *From a-Amino Acids to Amino Sugars* In *Studies in Natural Products Chemistry*, Rahman, A. (Ed.), Elsevier, Amsterdam, **1989**, p. 111. Jurczak, J.; Golebiowski, A. *Chem. Rev.* **1989**, *89*, 149.

10. Dondoni, A.; Perrone, D.; Merino, P. *J. Chem. Soc. Chem. Commun.* **1991**, 1313.

11. a) Spatola, A. In *Chemistry and Biochemistry of Amino Acids, Peptides and Proteins,* Weinstein, B. (Ed.), Marcel Dekker, New York, **1983**, Vol. 27, p. 267. b) Rich, D. H. *Proteinase Inhibitors*, Barret, A. J.; Salveson, G. (Eds.), Elsevier, New York, **1986**, p. 179. c) Tourwé, D. *Janssen Chimica Acta* **1985**, *3*, 3.

12. Selected recent papers: a) Kano, S.; Yokomatsu, T.; Shibuya, S. *Tetrahedron Lett.* **1991**, *32*, 233. b) Moree, W. J.; van der Marel, G. A. *Tetrahedron Lett.* **1991**, *32*, 409. c) DeCamp , A. E.; Kawaguchi, A. T.; Volante, R. P.; Shinkai, I. *Tetrahedron Lett.* **1991**, *32*, 1867. d) Chakraborty, T.

K.; Gangakhedkar, K. K. *Tetrahedron Lett.* **1991**, *32*, 1897. e) Plata, D. J.; Leanna, M. R.; Morton, H. E. *Tetrahedron Lett.* **1991**, *32*, 3623. f) Kotsuki, H.; Miyazaki, A.; Ochi, M. *Tetrahedron Lett.* **1991**, *32*, 4503. g) Rivero, R. A.; Greenlee, W. J.; Patchett, A. A. *Tetrahedron Lett.* **1991**, *32*, 5263. h) Bol, K. M.; Liskamp, R. M. J. *Tetrahedron Lett.* **1991**, *32*, 5401. i) Ghosh, A. K.; McKee, S. P.; Thompson, W. J. *Tetrahedron Lett.* **1991**, *32*, 5729. j) Vara Prasad, J. V. N.; Rich, D. H. *Tetrahedron Lett.* **1991**, *32*, 5857. k) Rosenberg, S. H.; Boyd, S. A.; Mantei, R. A. *Tetrahedron Lett.* **1991**, *32*, 6507. l) Thompson, W. J.; Tucker, T. J.; Schwering, J. E.; Barnes, J. L. *Tetrahedron Lett.* **1991**, *32*, 6819. m) Hanko, R.; Rabe, K.; Dally, R. Hoppe, D. *Angew. Chem. Int. Ed. Engl.* **1991**, *30*, 1690. n) Cushman, M.; Oh, Y.; Copeland, T. D.; Oroszlan, S.; Snyder, S. W. *J. Org. Chem.* **1991**, *56*, 4161. o) Ibuka, T.; Habashita, H.; Otaka, A. Fujii, N.; Oguchi, Y.; Uyehara, T.; Yamamoto, Y. *J. Org. Chem.* **1991**, *56*, 4370. p) Ghosh, A. K.; McKee, S. P.; Thompson, W. J. *J. Org. Chem.* **1991**, *56*, 6500. q) Garret, G. S.; Emge, T.J.; Lee, S. C.; Fischer, E. M.; Dyehouse, K; McIver, J. M. *J. Org. Chem.* **1991**, *56*, 4823. r) Alewood, P. F.; Brinkworth, R. I.; Dancer, R. J.; Garnham, B.; Jones, A.; Kent, S. B. H. *Tetrahedron Lett.* **1992**, *33, 977.* s) Baker, W. R.; Condon, S. L. *Tetrahedron Lett.* **1992**, *33,* 1581.

13. Dondoni, A.; Perrone, D. *Tetrahedron Lett.* **1992**, *33,* 7259.

14. Dondoni, A.; Perrone, D., unpublished results

15. Fray, A. H.; Kaye, R. L.; Kleinman, E. F. *J. Org. Chem.* **1986**, *51*, 4828.

16. Szelke, M.; Jones, D. M.; Atrash, B.; Hallett, A.; Leckie, B. J. *Peptides, Structure and Function* In *Proc. Am. Pept. Symp. 8th*, Hruby, V. J. and Rich, D. J. (Eds.), Pierce Chemical Co., Rockford, Ill. **1983**, p. 579.

17. Bühlmayer, P.; Caselli, A.; Fuhrer, W.; Göschke, R.; Rasetti, V.; Rüeger, H.; Stanton, J. L.; Criscione, L.; Wood, J. M. *J. Med. Chem.* **1988**, *31*, 1839.

18. Almquist, R. G.; Chao, W.-R.; Ellis, M. E.; Johnson, H. L. *J. Med. Chem.* **1980**, *23*, 1392.

19. Fellows, L. E. *Chem. Br.* **1987**, *23*, 843. Fleet, G. W. J. *Chem. Br.* **1989**, *25*, 287. Fleet, G.W.J. *Tetrahedron Lett.* **1985**, *26*, 5073. Fleet, G.W.J.; Namgoong, S.K.; Berker, C.; Baines, S.; Jacob, G.S.; Winchester, B. *Tetrahedron Lett.* **1989**, *30*, 4439. Sinnott, M. L. *Chem. Rev.* **1990**, *90*, 1171.

20. For selected recent papers with leading references, see: a) Straub, A.; Effenberger, F.; Fischer, P. *J. Org. Chem.* **1990**, *55,* 3296. b) Aoyagi, S.; Fujimaki, S.; Kibayashi, C. *J. Chem. Soc. Chem. Commun.* **1990**, 1457. c) Anzeveno, P. B.; Creemer, L. J. *Tetrahedron Lett.* **1990**, *31*, 2085. d) Bernotas, R. C.; Papandreou, G.; Urbach, J.; Ganem, B. *Tetrahedron Lett.* **1990**, *31*, 3393. e) Reitz, A.B.; Baxter, E.W.; *Tetrahedron Lett.* **1990**, *31*, 6777. f) Bernotas, R. C. *Tetrahedron Lett.* **1990**, *31, 469.* g) Chen, S.-H.; Danishefsky, S. J. *Tetrahedron Lett.* **1990**, *31,* 2229. h) Wehner, V.; Jäger, V. *Angew. Chem. Int. Ed. Engl.* **1990**, *29*, 1169. i) Jäger, V.; Hümmer, *Angew. Chem. Int. Ed. Engl.* **1990**, *29*, 1171. j) W. Witte, J. F.; McClard, R. W. *Tetrahedron Lett.* **1991**, *32*, 3927. k) Hardick, D. J.; Hutchinson, D. W.; Trw, S. J. Wellington, E. M. H. *J. Chem. Soc. Chem. Commun.* **1991**, 729. l) Liu, P. S.; Rogers, R. S.; Kang, M. S.; Sunkara, P. S. *Tetrahedron Lett.* **1991**, *32,* 5853. m) Wagner, J.; Vogel, P. *Tetrahedron Lett.* **1991**, *32,* 3169. n) Ina, H.; Kibayashi, C. *Tetrahedron Lett.* **1991**, *32,* 4147. o) Kajimoto, T.; Liu, K. K.-C.; Pederson, R. L.; Zhong, Z.; Ichikawa, Y.; Porco, Jr., J. A.; Wong, C.-H. *J. Am. Chem. Soc.* **1991**, *113*, 6187. p) Kajimoto, T.; Chen, L.; Liu, K. K.-C.; Wong, C.-H. *J. Am. Chem. Soc.* **1991**, *113*, 6678. q) Liu, K. K.-C.;

Kajimoto, T. Chen, L.; Zhong, Z.; Ichikawa, Y.; Wong, C.-H. *J. Org. Chem.* **1991**, *56*, 6280. r) Hung, R. R.; Straub, J. A.; Whitesides, G. M. *J. Org. Chem.* **1991**, *56*, 3849. s) Vasella, A.; Ermert, P. *Helv. Chim. Acta*, **1991**, *74*, 2043. t) Duréault, A.; Portal, M.; Depezay, J. C. *Synlett*, **1991**, 225. u) Gersspacher, M.; Rapoport, H. *J. Org. Chem.* **1991**, *56*, 3700. v) Takahata, H.; Banba, Y.; Tajima, M.; Momose, T. *J. Org. Chem.* **1991**, *56*, 240. w) Rassu, G.; Pinna, L.; Spanu, P.; Culeddu, N.; Casiraghi, G.; Gasparri Fava, G.; Belicchi Ferrari, M.; Pelosi, G. *Tetrahedron* , **1992**, *48,* 727. x) Burgess, K.; Chaplin, D. A.; Henderson, I.; Pan, Y. T.; Elbein, A. D. *J. Org. Chem..* **1992**, *57*, 1103. y) Ballini, R.; Marcantoni, E.; Petrini, M. *Org. Chem.* **1992**, *57*, 1316. z) de Raadt, A.; Stütz, A. E. *Tetrahedron Lett.* **1992**, *33*, 189. z') Hardick, D. J.; Hutchinson, D. W.; Trew, S. J.; Wellington, E. M. H. *Tetrahedron* **1992**, *48*, 6285.

21. For a synthesis by others, see: Aoyagi, S.; Fujimaki, S.; Yamazaki, N.; Kibayashi, C. *J. Org. Chem.* **1991**, *56*, 815.

22. Dondoni, A.; Merino, P.; Perrone, D. *J. Chem. Soc., Chem. Commun.* **1991**, 1576.

23. For previous syntheses , see: a) Vasella, A.; Voeffray, R. *Helv. Chim. Acta.* **1982**, *65*, 1134. b) Ida, H.; Yamazaki, N.; Kibayashi, C. *J. Org. Chem.* **1987**, *52*, 3337. c) Tsuda, Y.; Okuno, Y.; Kanemitsu, K. *Heterocycles*, **1988**, *27*, 63. d) Chida, N.; Furuno, Y.; Ogawa, S. *J. Chem. Soc., Chem. Commun.* **1989**, 1230. e) Rajanikanth, B.; Seshadri, R. *Tetrahedron Lett.* **1989**, *30*, 755.

24. For a previous synthesis, see: Legler, G.; Jülich, E. *Carbohydrate Res.* **1984**, *128*, 61.

25. Dondoni, A.; Merino; P.; Perrone, D. *Tetrahedron*, submitted for publication.

26. Dondoni, A.; Fantin, G.; Fogagnolo, M.; Merino, P. *J. Chem. Soc., Chem. Commun.* **1990**, 854.

27. Evans, D. A.; Chapman, K. T.; Carreira, E. M. *J. Am. Chem. Soc.* **1988**, *110*, 3560. Evans, D. A.; Gauchet-Prunet, J. A.; Carreira, E. M.; Charette, A. B. *J. Org. Chem.* **1991**, *56*, 741.

28. Chen, K. M.; Hardtmann, G. E.; Prasad, K.; Repic, O.; Saphiro, M. J. *Tetrahedron Lett.* **1987**, *28*, 155.

Directed Biosynthesis as an Alternative to Synthetic Modifications of Antibiotics

Axel Zeeck and Isabel Sattler

Institut für Organische Chemie, Universität Göttingen,
Tammannstraße 2, D-3400 Göttingen

Summary: The strain *Streptomyces parvulus* produces mainly manumycin A (**1**). Manumycin B (**2**), C (**3**) and D (**4**) were isolated as minor components. Precursor-directed biosynthesis and conducting the fermentation under increased oxygen concentration, respectively, led to a variation in the metabolite pattern. Different aminobenzoic acids used as precursors in the high concentration method of precursor-directed biosynthesis resulted in new aromatic manumycin analogues (e.g. **8**, **9**). Increased oxygen concentration gave rise to the new metabolites 64p-A (**11**), 64p-B (**12**), and 64p-C (**13**). The combined application of both methods with 3-aminobenzoic acid as precursor resulted in the formation of 64-mABA (**8**) and 64p-mABA (**14**), showing that increased oxygen concentration specifically affects the polyketide synthase of the "eastern" C_{13} side chain. Both methods have been further extended to strains producing other manumycin-related antibiotics, e. g., asukamycin (**5**) and colabomycin A (**6**). The modified directed biosynthesis allowed variation in distinct structural details of complex polyene-like compounds. This can be used as a metabolic design for creating new natural product derivatives with selected activities.

Introduction

A common procedure for discovering natural products from plants and microorganisms follows their biological activities. In many cases the activity and pharmacological properties may be improved by suitable chemically prepared derivatives. Moreover, the derivatization of natural products can be carried out by biological methods, such as biotransformation, mutasynthesis, genetic recombinant techniques and precursor-directed biosynthesis. The biological derivatization not only permits changes in the periphery, but also in the skeleton of complex molecules. In the latter case, generation of new metabolites takes place because of a reduced specificity of the involved enzymes. This biosynthetic method of creating new molecules is an alternative to synthetic approaches in natural products chemistry (Fig. 1). In the following chapter this concept is exemplified within the manumycin group of antibiotics by feeding unnatural precursors or by increasing the oxygen concentration during the cultivation of *Streptomycetes species*.

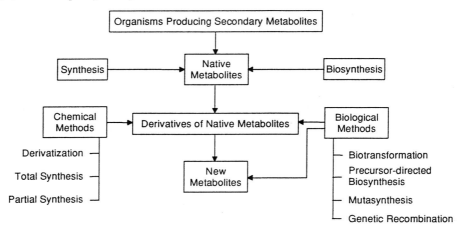

Fig. 1: Biological and Chemical Methods for Modification of Secondary Metabolites

The Manumycin Group of Antibiotics

The manumycin group is a small and discrete class of antibiotics, which includes about a dozen secondary metabolites of microbial origin. The first reported member, manumycin A (**1**), bears two unsaturated carbon chains linked in a meta relationship on a multifunctional six-membered ring (m-C_7N unit), and a 2-amino-3-hydroxycyclopent-2-enone (C_5N unit) positioned at the end of the triene chain.[1] Similar structural moieties are present in all members of the manumycin group, indicating their close structural relationship. Most of these antibiotics exhibit activity against Gram-positive bacteria as well as antifungal and cytotoxic activities. In addition, interesting insecticidal effects and inhibition of polymorphonuclear leucocyte elastase were described in the case of manumycin A (**1**).

Two homologues of manumycin A (**1**) are produced by *Streptomyces parvulus* (strain Tü 64) in minor yields (1%) compared to **1**, manumycin B (**2**) and manumycin C (**3**).[2,3] Their constitution differs within the amide-bound "eastern" side chain. In manumycin B (**2**), this chain only contains ten carbon atoms; compared with **1**, the last propionate building block is missing. In manumycin C (**3**), the final methyl branch at C-2´ is missing because the last propionate has been replaced by an acetate unit.

The absolute stereochemistry of C-4 within the central m-C$_7$N unit was deduced with the help of circular dichroism (CD) spectroscopy, making use of the exciton chirality method as in other manumycin group antibiotics (Fig. 2). In manumycin B (**2**), the center of chirality at C-4 was unchanged compared to **1** (4*R*). Most strikingly, manumycin C (**3**) exhibited the opposite (*S*)-configuration at C-4. The configuration of the oxirane ring in **3** (5*R*, 6*S*; as in **1**) was determined by NMR spectroscopy, making use of a nuclear Overhauser enhancement (nOe) effect between the oxirane proton at C-5 and 7-H of the triene chain. Consequently, manumycin A (**1**) and manumycin C

	3-H[1] (Type I)	3-H[1] (Type II)
CDCl$_3$	7.39	7.42
Pyridine-d$_5$	7.94	8.21
Δδ[2]	0.55	0.79

[1] δ in ppm, [2] Δδ = δ $^{Pyridin-d_5}$ - δ ^{CDCl_3}

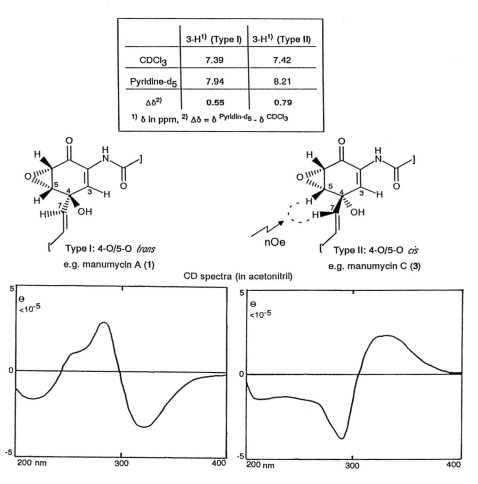

Fig. 2: Spectroscopic Determination of the Stereochemistry of the m-C$_7$N Unit

(3) are representatives of different dia-stereomeric types of the m-C$_7$N unit. Type I in manumycin A (1) is characterized by a *trans*-configuration of the two oxygen functionalities at C-4 and C-5, while they are *cis* orientated in the case of manumycin C (3) (type II). The comparison of ^{1}H-NMR chemical shift data of both compounds in chloroform and pyridine indicated a characteristic distinction. The Aromatic Solvent Induced Shift (ASIS) effect only showed a significant difference for the olefinic 3-H, $\Delta\delta = 0.55$ ppm for manumycin A (1) and $\Delta\delta = 0.79$ ppm for manumycin C (3). Other manumycin-type compounds could also be assigned by their ^{1}H-NMR spectroscopic data to one of the two stereochemical types. This method allows a facile determination of the relative stereochemistry in the central m-C$_7$N unit of the manumycin group.

Manumycin D (4) was discovered as a minor component of *Streptomyces parvulus* (strain Tü 64), but addition of 55 mM 3-aminobenzoic acid to the fermentation broth stimulates its formation. While the carbon skeleton of 4 coincided with that of manumycin A (1), structural differences were found in the substitution pattern of the central m-C$_7$N unit. The oxirane ring is replaced by a hydroxy-ethylene moiety. As in 1, C-4 shows an (R)-configuration, and conformational analysis led to the (S)-configuration at C-5. This metabolite demonstrates that the oxirane unit is not a distinctive structural feature of the manumycin-group.

Asukamycin (5), a metabolite of *Streptomyces nodosus* ssp. *asukaensis* (strain ATCC 29757), is another member of the manumycin group.[4] A prominent building block of the "eastern" triene side chain is a cyclohexane residue, a rare structural element in natural products. Revising the structure given in literature, the "southern" triene chain ranging from C-7 to the carboxamide bearing the C$_5$N unit exhibits all-(E) configuration as do the manumycins. In addition to the previously assigned (4S)-configuration, the absolute stereochemistry of the oxirane has been deduced by the above described NMR method (ASIS effect) to be 5R,6S.

The rather unstable colabomycin A (6), a secondary metabolite of *Streptomyces griseoflavus* (strain Tü 2880), is the only member of the manumycin group bearing a tetraene chain. [5] In addition, colabomycin D (7), similar to manumycin D (4) regarding the m-C$_7$N unit, has been isolated and structurally elucidated. [3,6] **6** and **7** correspond in their absolute stereochemistry of the m-C$_7$N unit (4*S*, 5*R*) with the adjacent oxygen functionalities on the same side of the ring. For the "eastern" tetraene side chain, *E/Z*-isomers of the main configuration are known. It is likely that they are due to an extreme photosensitivity of the olefinic system.

Implication of the Biogenetic Origin of Manumycin Group Antibiotics

The biogenetic origin of manumycin A (**1**), produced by *Streptomyces parvulus,* was studied using radioactive and stable isotope tracer techniques.[7] In a complex biosynthesis, precursors from the three main metabolic sources, the carbohydrate, the carboxylic acid, and the amino acid pool, are involved.[8] Both polyene chains are generated *via* two separate polyketide pathways. Thus, the methyl branches of the "eastern" C$_{13}$ side chain are introduced by incorporation of propionate units. The 2-amino-3-hydroxycyclopent-2-enone moiety (C$_5$N unit) is biosynthesized from succinate and glycine passing through 5-aminolevulinic acid as an intermediate.

The crucial question of manumycin biosynthesis concerns the central m-C$_7$N unit. This structural unit is a widespread, variably functionalized building block in natural products. In several cases the m-C$_7$N unit derives from aromatic precursors related to the shikimate pathway, e.g. 3-amino-5-hydroxybenzoic acid (ansamycins, mitomycins) or 3-aminobenzoic acid (pactamycin). These and other benzoic acids were not incorporated into manumycin A (**1**). Here, the skeleton of the m-C$_7$N unit is built from glycerol as a C$_3$- (C-1 to C-3) and succinate as a C$_4$-segment (C-6 to the exocyclic C-7) (Fig. 3). The nitrogen atom attached to C-2 is transferred in a usual manner from glutamine. The unique assembly from intermediates of the TCA cycle and the triose pool (carbohydrate metabolism) established a new biosynthetic pathway to the m-C$_7$N unit, which is therefore a characteristic feature for the manumycin group antibiotics.

The oxirane and the hydroxy functionality of the m-C$_7$N unit are formed by incorporation of aerial oxygen. This oxygenase activity of the producing strain gave rise to the idea that increased oxygen concentration may influence product formation in *Streptomyces parvulus*.

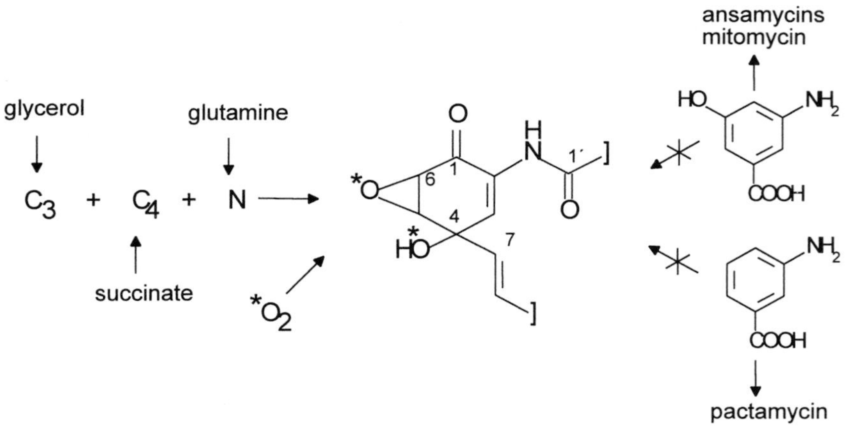

Fig. 3: Biogenetic Origin of the m-C$_7$N Unit in Manumycin A (**1**)

Metabolites from *Streptomyces parvulus* by Precursor-directed Biosynthesis

The precursor-directed biosynthesis is based on feeding of artificial precursors into the fermentation broth of a producing organism. The lack of specificity of the biosynthetic enzymes allows the replacement of a natural precursor by an artificial one, resulting in analogues of the parent metabolite. Because manumycin A (**1**) is assembled by different building blocks provided from independent biosynthetic pathways, it became possible to affect its biosynthesis by displacing the natural precursor of the m-C$_7$N unit with different benzoic acids. Such precursors are forced into the natural pathway by feeding unnaturally high amounts (about 50 mM).[9] This new approach, the so-called high concentration method of precursor-directed biosynthesis, was extended to a whole series of artificial precursors. Thus, three classes of manumycin A (**1**) analogues with a broad variety of structures were obtained.[10] This method turned out to be a versatile and attractive approach because it works without any requirement of specific enzyme inhibitors or blocked mutants.

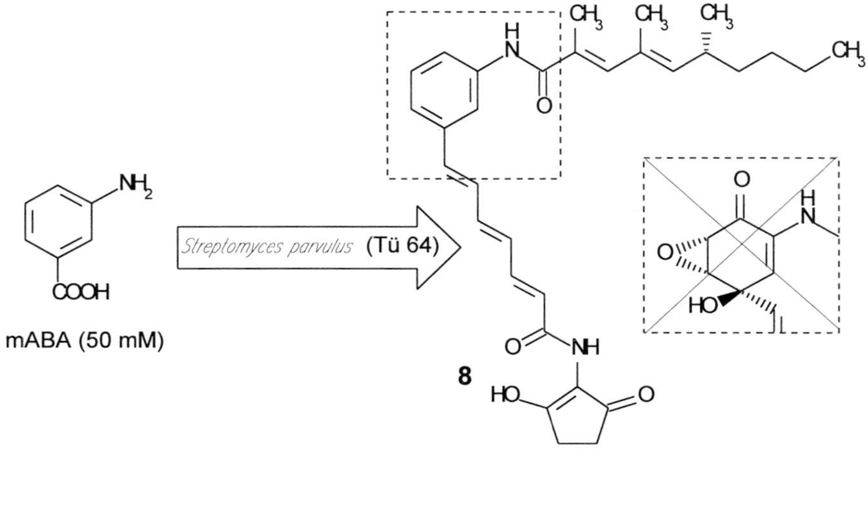

The early studies were based on a hypothesis that the m-C7N unit might originate from 3-aminobenzoic acid (mABA). But, upon feeding in physiological amounts (2-4 mM mABA), production of manumycin A (**1**) is suppressed. A 55 mM feeding concentration led to the new metabolite 64-mABA (**8**), in which the natural multifunctional m-C7N unit is replaced by mABA. The assembly of the molecule follows the original procedure.

In contrast to 64-mABA (**8**, class 3) other analogues carry only one of the two side chains, either the methyl-branched "eastern" chain (class 2), or the "southern" triene chain with the C5N unit (class 1) (Fig. 4). For example, the hydrazide of 3-aminobenzoic acid carries only the branched C13 chain (class 2) and 4-methyl-3-aminobenzoic acid is elongated only at the carboxy group (class 1).

The transacylase, which connects the m-C7N unit with the "eastern" C13 side chain works even when the amino group at C-3 is closely adjacent to amino or hydroxy substituents. For example, the class 3 compounds 64-3,4DABA and 64-4HmABA-2 were built upon feeding of 3,4-diaminobenzoic acid and 4-hydroxy-3-aminobenzoic acid, respectively.

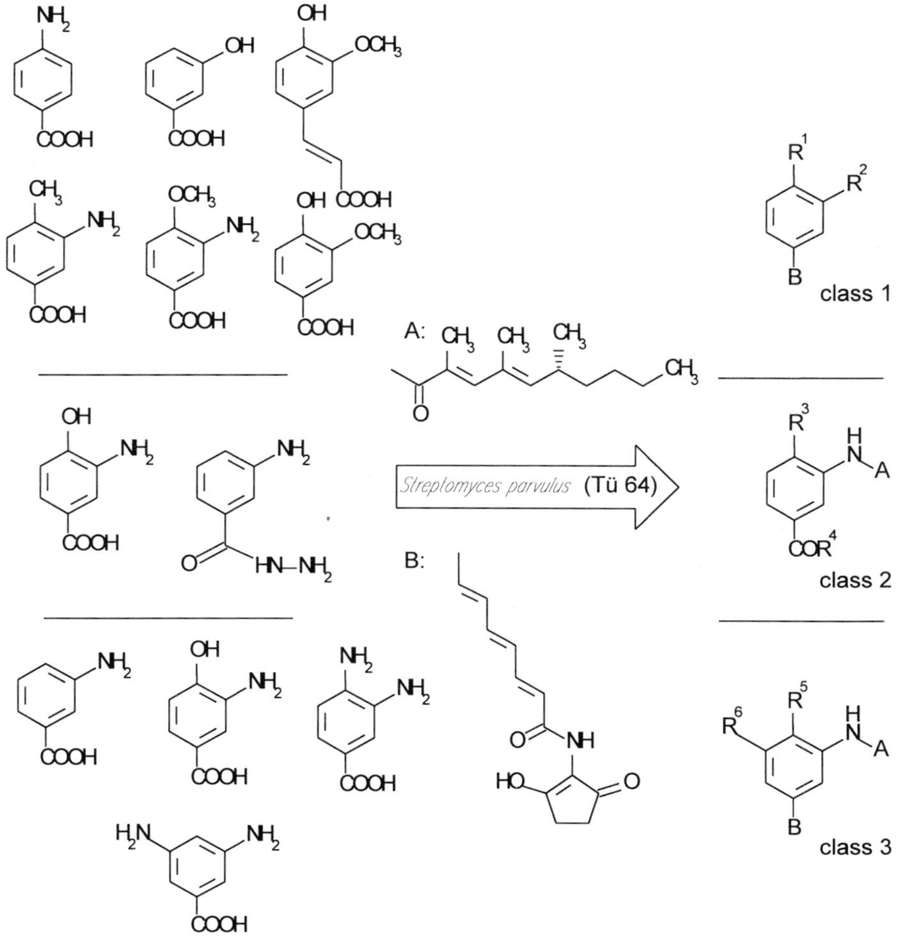

Fig. 4: Metabolites from *Streptomyces parvulus* by Precursor-directed Biosynthesis

Feeding of 4-aminobenzoic acid (pABA) mainly resulted in the already known 64-pABA-1 by elongation at the carboxy group. Moreover, 64-pABA-2 (**9**) turned up as a minor component, in which additionally, the C_{13} side chain is linked to the amino group. Remarkably, the enzymes work even when the reactive functional groups are different from the original substrate in spatially arrangement.

In the case of 4-hydroxy-3-aminobenzoic acid, the phenoxazinone **10** turned up as an additional compound. It is probably a product of biotransformation, since phenoxazinones are widespread as secondary metabolites of microorganisms (e.g. actinomycins). However, **10** was chemically prepared by oxidative dimerization of 4-hydroxy-3-aminobenzoic acid with manganese dioxide.[11]

10

New Metabolites from *Streptomyces parvulus* under the Influence of Increased Oxygen Concentration

Increased oxygen concentration has important regulatory and toxic effects on microbial metabolism and growth. Besides distinct influences on primary metabolism, like reversible reduction of cell growth up to 1200 mbar $p(O_2)$ or enhanced production of organic acids [12], the secondary metabolism can also be manipulated. For example, the tetracycline/oxytetracycline mixture of different *Streptomyces* strains is shifted to the higher oxygenated product oxytetracycline as a result of increased oxygen concentration.[13]

Regarding the involvement of molecular oxygen described above in the biosynthesis of manumycin A (**1**), attempts were made to change the metabolite pattern of the producing organism by fermentation under increased oxygen partial pressure. This was accomplished in a 10-liter scale under standardized fermentation conditions by increasing the atmospheric pressure (1 to 10 bar) or by variation of the oxygen content in the aeration gas (20 % to 100 %).[14]

The growth of *Streptomyces parvulus* (strain Tü 64) as well as manumycin-formation decreases continuously with increasing oxygen concentration. Above a $p(O_2)$ level of 1800 mbar, secondary metabolite formation is completely ceased. High amounts ($\sim$ 0.2 M) of 2-ketoglutarate and (R)-lactate are excreted into the culture broth after 40 hours of fermentation.

Several new products, which were not detectable in fermentations under atmospheric pressure, turned up in differing amounts between 630 and 1680 mbar p (O_2). The new metabolites 64p-A (11), 64p-B (12) and 64p-C (13) represent different parts of the parent metabolite manumycin A (1). 64p-B (12) is the amide of the original "eastern" C_{13}-side chain carboxylic acid already mentioned above. The non-chiral 64p-A (11) exhibits a novel polyketide structure as a result of an altered pathway. 64p-C (13) is the first natural aromatic derivative of 1 that was built without feeding of an aromatic precursor. Apparently, the polyketide pathway of the branched "eastern" side chain can be decoupled from the biosynthetic machinery, while the biosynthesis of the other parts of manumycin A (1) is closely connected.

The combined application of increased oxygen concentration and precursor-directed biosynthesis using 3-aminobenzoic acid (50 mM) in *Streptomyces parvulus* (strain Tü 64) led to a dramatic 6-fold increase of the already described 64-mABA (8). Additionally, the novel analogue 64p-mABA (14), in which the "eastern" side chain corresponds to 64p-A (11), was isolated in comparable yields.

The elevated yield upon feeding the artificial m-C_7N unit under increased oxygen concentration is quite remarkable in the light of the decreasing manumycin A (1) production under the same conditions. This might be explained by an inhibition of the m-C_7N unit biosynthesis, which is compensated by feeding of the precursor. The increase in precursor-directed biosynthesis might be due to different effects, e. g., a less hindered uptake of the precursor caused by membrane defects or a stimulation by increased oxygen concentration on the biosynthetic pathways.

64p-A (11) and 64p-mABA (14) revealed a specific effect of the increased oxygen concentration on the polyketide pathway of the "eastern" branched side chain (Fig. 5). The influence on the synthetase might be located in the processing after the introduction of the first propionate unit. According to the "module hypothesis" of the polyketide assembling of macrolides [15], the enoyl reductase that usually hydrogenates the double bond of the first propionate is inhibited. After the correct elongation with the second propionate the chain is detached, since the enzyme recognizes the usual final sequence of two olefinic propionate units.

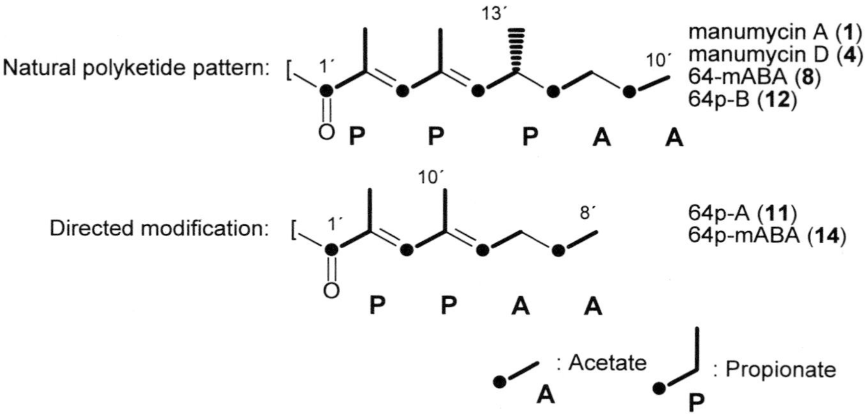

Fig. 5: Directed Modification of the Polyketide Metabolism in *Streptomyces parvulus*

In normal aerobic conditions, the precursor-directed biosynthesis with ferulic acid leads to a class 1 analogue of manumycin A (**1**). Under increased oxygen concentration, the precursor was decarboxylated yielding 4-vinylguaiacol (**15**). This intermediate dimerized by an oxidative phenol coupling to 64p-Fer-2 (**16**), which is optically active. Thus, this biotransformation is an enzymatic process, possibly catalyzed by the action of a cytochrome-P_{450} enzyme in a similar manner as the above described dimerization of 3-amino-4-hydroxybenzoic acid to the phenoxazinone **11**.

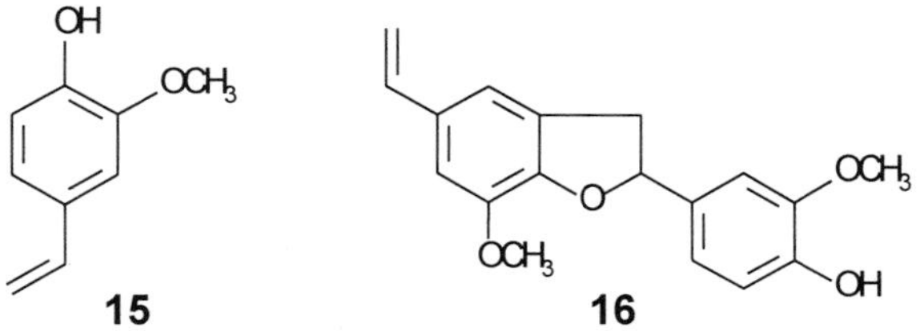

New Metabolites from *S. nodosus* and *S. griseoflavus* by Precursor-directed Biosynthesis

With the asukamycin (**5**) producing *Streptomyces nodosus* (strain ATCC 29757), the previously described precursor-directed biosynthesis with 3-aminobenzoic acid led to asuka-mABA, which structurally resembles 64p-C (**13**). In contrast to 64-mABA (**8**), the amide-bound side chain is missing in asuka-mABA. Perhaps, 3-aminobenzoic acid interferes as a related compound with the biosynthesis of the cyclohexyl carboxylic acid that is the starter unit of the "eastern" polyketide chain of **5** and is derived from the shikimate pathway.

Streptomyces griseoflavus (strain Tü 2880), the producer of colabomycins A (**6**) and D (**7**), can be forced with 3-aminobenzoic acid into precursor-directed biosynthesis to produce 2880-mABA (**17**). Although final evidence is left, the structure of **17** was deduced in analogy to 64-mABA (**8**) as the aromatic analogue of the parent metabolite colabomycin A (**6**).

The production of asuka-mABA (**13**) and 2880-mABA (**17**) shows that the high concentration method of precursor-directed biosynthesis can be applied to other producing *Streptomycetes* sp. of manumycin group antibiotics. Preliminary results indicated a similar enhancement in production of 2880-mABA (**18**) by *Streptomyces griseoflavus* under increased oxygen concentration as for 64-mABA (**8**) in *Streptomyces parvulus*.[6]

Discussion

The precursor-directed biosynthesis allows incorporation of unnatural precursors into the frame of manumycin-group antibiotics. Thus, it becomes possible to introduce different functionalities into the central part of the molecules. The cultivation under increased oxygen concentration provides a new method for achieving different secondary metabolites. It is a device to shift the product pattern in a qualitative as well as in a quantitative way. The actual yields of the new metabolites range from 1 to 100 mg/l without optimization of the productivity of the microorganisms. Thus, cultivation under increased oxygen concentration is a new approach to biological derivatization of secondary metabolites. It can also be successfully combined with the high concentration method of the precursor-directed biosynthesis.

Additionally, the studies described above offer further insight into the enzymes involved in manumycin biosynthesis. Increased oxygen concentration interferes mainly with the pathway leading to the m-C7N unit. Possibly the oxygenases that catalyze the introduction of oxygen atoms at C-4 and the oxirane are inhibited. Regarding the different structures of the "eastern" side chain, the polyketide syn-thase of this moiety seems to be affected in a specific manner. This resulted in variations of the original polyketide elongation steps concerning the substrates as well as their processing. The strong increase in the yield of 64-mABA (**8**) gives evidence that the assembly of the molecule remains unaffected.

Considering the rather unstable and complex structures, especially the polyene systems, the biological methods are more efficient than chemical synthesis or derivatization. Within some limitations, we are able to create new metabolites with a defined structure, thereby developing the precursor-directed biosynthesis into a metabolic design. The principles are schematically summarized in Fig. 6.

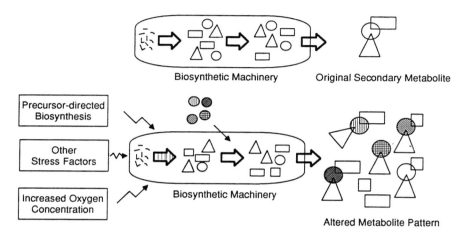

Fig. 6: Metabolic Design by Modification of the Biosynthetic Pathways

Treatment with unnatural precursors, increased oxygen concentration, and other stress factors resulted in the formation of new metabolites. The central and controlled alteration of a parent compound is conditional for the metabolic design. The further extension of this concept to other producers of manumycin-group antibiotics and also to other classes of microbial metabolites will be subject of future efforts.

Acknowledgements: We thank Prof. Dr. H. Zähner (Universität Tübingen) for providing strains Tü 64 and Tü 2880. Finally, we appreciate the collaboration with Prof. Dr. U. Onken and co-workers (Universität Dortmund).

References

1. A. Zeeck, K. Schröder, K. Frobel, R. Grote, R. Thiericke, *J. Antibiot.* **40** (1987) 1530-1540.

2. C. Gröne, Diploma thesis, Universität Göttingen (1988).

3. I. Sattler, Ph.D. thesis, Universität Göttingen (1992).

4. K. Kakinuma, N. Ikekawa, A. Nakagawa, S. Omura, *J. Am. Chem. Soc.* **101** (1979) 3402-3404.

5. R. Grote, A. Zeeck, H. Drautz, H. Zähner, *J. Antibiot.* **41** (1988) 1178-1185.

6. O. Dick, Ph.D. thesis, Universität Dortmund (1993).

7. R. Thiericke, A. Zeeck, A. Nakagawa, S. Omura, R. E. Herrold, S. T. Wu, J. M. Beale, H. G. Floss, *J. Am. Chem. Soc.* **112** (1990) 3979-3987.

8. J. Rohr, A. Zeeck, in *Jahrbuch der Biotechnologie* Bd.2 (P. Präve, M. Schlingmann, W. Crueger, K. Esser, R. Thauer, F. Wagner, eds.) pp. 263-295, Carl Hanser Verlag, München 1988.

9. R. Thiericke, A. Zeeck, *J. Chem. Soc. Perkin I* 1988 2123-2127.

10. R. Thiericke, H.-J. Langer, A. Zeeck, *J. Chem. Soc. Perkin I* 1989, 851-854.

11. C. Boddien, Diploma thesis, Universität Göttingen (1992).

12. U. Onken, E. Liefke, *Adv. Biochem. Eng. Biotechnol.* **40** (1989) 137-169.

13. E. Liefke, D. Kaiser, U. Onken, *Appl. Microbiol. Biotechnol.* **32** (1990) 674-679.

14. D. Kaiser, Ph.D. thesis, Universität Dortmund (1992).

15. S. Donadio, M. J. Staver, J. B. McAlpine, S. J. Swanson, L. Katz, *Science* **252** (1991) 675-679.

Total Synthesis of Macrolide Antibiotics.
A Route to Rutamycin B

James D. White

Department of Chemistry, Oregon State University
Corvallis, Oregon 97331-4003 USA

Summary. A new method for the synthesis of macrocyclic lactones based on the thermal decomposition of ω-hydroxy-S-acyl xanthates is presented. The requirement for 2,6-di-*tert*-butyl-4-methylpyridine for optimization of the lactone:diolide ratio indicates a unique role for this pyridine which is formulated in a proposed mechanism involving an acylpyridinium radical. An approach to the synthesis of the macrolide antibiotic rutamycin B is outlined in which four principal subunits are constructed. Stereocontrol is exercised in these synthetic sequences through the application of crotylboronate and crotylstannane addition to aldehydes, aldol condensation with chiral ketone enolates, and directed hydride reduction.

Macrocyclic lactones ("macrolides") are ubiquitous in nature, occurring in a very wide variety of organisms including marine species. Many members of the family possess antibiotic activity and a few, such as erythromycin, are highly valued for their medicinal properties. As a structural class, macrolides surpass even the steroids in their spectrum of biological activities. Not surprisingly, there has been a great deal of interest in the synthesis of macrolide antibiotics, with the result that many innovative strategies have been developed for assembling the complex architecture of these substances [1].

Two issues of paramount importance must be faced when considering the synthesis of a complex macrolide. These are, first, the systematic creation of multiple chiral centers, some of which may be subject to stereomutation en route to the target and, second, the closure of a seco acid to a highly functionalized, often labile lactone at or near the final stage of the synthesis. Our studies of lactonization methodology in general and of synthetic routes to the antibiotic rutamycin B (**1**) in particular have addressed these dual problems with interesting, and sometimes unexpected results.

Of the two most frequently employed techniques for macrolactone synthesis from ω-hydroxy-carboxylic acids (scheme I), that invoking carbonyl activation is by far the most common [2]. Hydroxyl activation, as in the Mitsunobu lactone synthesis, is a useful but more limited alternative. Methods based on radical intermediates, either acyl or alkoxyl, have not been explored although, in principle, this approach could be equally effective for cyclization of ω-hydroxycarboxylic acid derivatives.

Methods for Macrolactonization

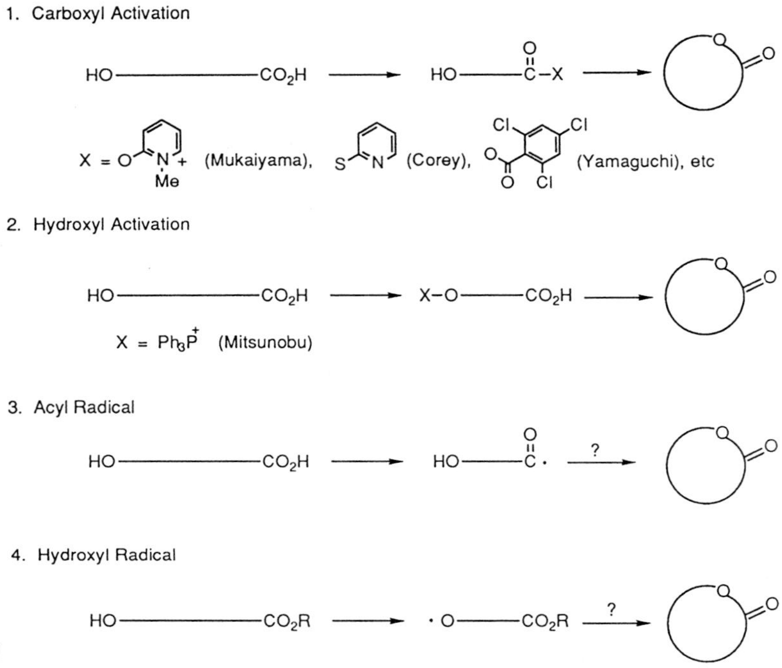

1. Carboxyl Activation

$$HO\text{------------}CO_2H \longrightarrow HO\text{------------}\overset{\overset{\displaystyle O}{\|}}{C}-X \longrightarrow$$

$$X = O\overset{+}{\underset{\underset{Me}{}}{N}} \text{ (Mukaiyama),} \quad S\overset{\cdot}{N} \text{ (Corey),} \quad O \text{ (Yamaguchi), etc}$$

2. Hydroxyl Activation

$$HO\text{------------}CO_2H \longrightarrow X\text{-}O\text{------------}CO_2H \longrightarrow$$

$$X = Ph_3\overset{+}{P} \quad \text{(Mitsunobu)}$$

3. Acyl Radical

$$HO\text{------------}CO_2H \longrightarrow HO\text{------------}\overset{\overset{\displaystyle O}{\|}}{C}\cdot \quad \overset{?}{\longrightarrow}$$

4. Hydroxyl Radical

$$HO\text{------------}CO_2R \longrightarrow \cdot O\text{------------}CO_2R \quad \overset{?}{\longrightarrow}$$

Scheme I

ω-Hydroxyl-S-acyl xanthates, easily prepared by treatment of the parent acid with diethyl chlorophosphate followed by potassium xanthate, undergo thermal decomposition at *ca* 120 °C and, in the presence of 2,6-di-*tert*-butyl-4-methylpyridine, afford lactones in fair to good yield (scheme II). In the absence of the pyridine or if the reaction is initiated by tri-*n*-butyltin chloride and AIBN, the major product is the alkanol resulting from decarbonylation and hydrogen abstraction from solvent (octane).

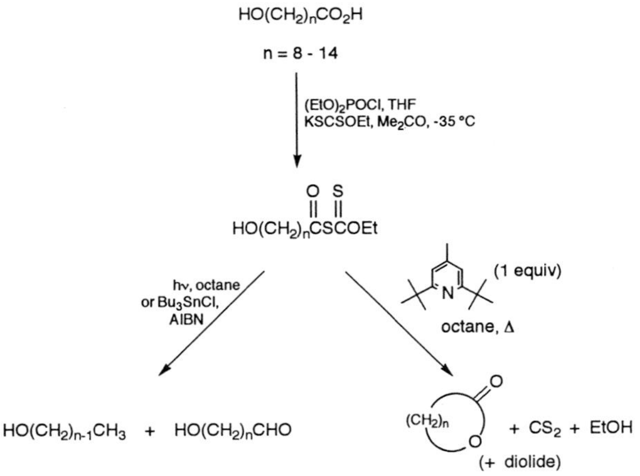

$$HO(CH_2)_nCO_2H$$

$$n = 8 - 14$$

(EtO)$_2$POCl, THF
KSCSOEt, Me$_2$CO, -35 °C

$$HO(CH_2)_n\overset{\overset{\displaystyle O}{\|}}{C}\overset{\overset{\displaystyle S}{\|}}{S}COEt$$

hν, octane
or Bu$_3$SnCl,
AIBN

(1 equiv)

octane, Δ

$$HO(CH_2)_{n-1}CH_3 \ + \ HO(CH_2)_nCHO$$

$(CH_2)_n$ lactone + CS$_2$ + EtOH

(+ diolide)

Scheme II

Yields (isolated) of lactone from hydroxy acyl xanthates of varying chain length are shown in Table I. For shorter chains, eg 10-hydroxydecanoyl, the diolide predominates over lactone, as is found with other methods of lactonization, but as the chain length is extended the lactone:diolide ratio steadily increases until, at 16-hydroxyhexadecanoyl and beyond, macrolactone is the exclusive product.

Table I. Lactonization of ω-Hydroxy-O-alkyl-S-acyl Xanthates

Xanthate		Yield (%)	
$\text{RCH(CH}_2)_n\text{CSCOEt}$ $\quad$ $\overset{|}{\text{OH}}$ $\quad$ $\overset{\|}{\text{O}}\overset{\|}{\text{S}}$		Lactone	Diolide
R	n		
H	8	16	42
H	10	54	28
H	11	63	21
H	13	75	18
H	14	81	0
$n\text{-C}_6\text{H}_{13}$	10	34	17

A tentative mechanism for this process is shown in scheme III. Initial thermolysis of the xanthate leads to an acyl radical, for which there is good precedent [3], and before this species has an opportunity to decarbonylate, it is trapped by the pyridine. The exact nature of the acyl-pyridine complex formed is unclear — it is arbitrarily represented as an acylpyridinium radical — but a variety of experiments designed to probe the mechanism has indicated a unique role for this pyridine which is not associated with ketene generation from the acyl xanthate nor with an ionic cyclization initiated by deprotonation of the terminal hydroxyl group.

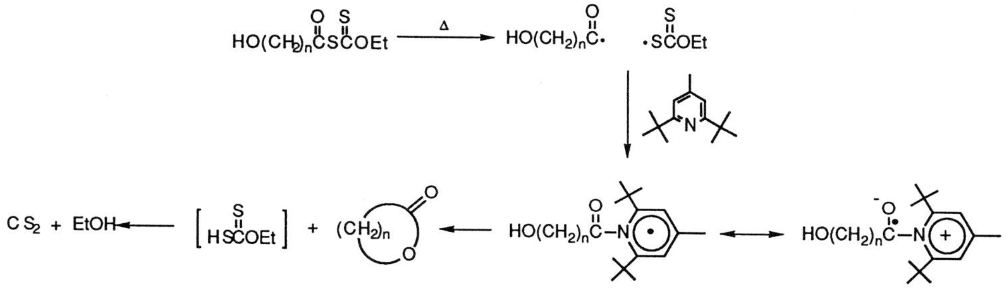

Scheme III

Ongoing studies are examining the scope and mechanism of this interesting lactonization. As an application of acyl radical chemistry it opens new doors to cyclic structures providing, of course, decarbonylation can be suppressed. The true test of this methodology lies in the future with its application to the closure of complex seco acids such as that leading to rutamycin B (**1**).

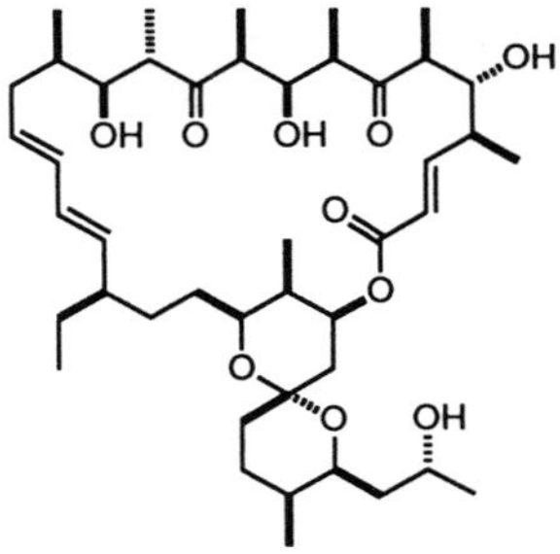

Rutamycin B (**1**)

Rutamycin B (**1**), a metabolite of *Streptomyces aureofaciens* is a member of the oligomycin family of antibiotics and has potent antifungal activity [4]. It has been shown to bind with high affinity to ATPase [5]. Rutamycin B was initially assigned the absolute configuration enantiomeric with **1** on the basis of a correlation with rutamycin A, whose constitution was established by X-ray crystallography [6], but this has been corrected by asymmetric synthesis of a known degradation product of **1** [7].

A plausible synthetic plan for **1** would first segregate the stereogenic centers of the molecule into two distinct sets defined as the upper and lower halves shown in scheme IV. These sections would be joined later through the trans,trans diene linker unit at their left-hand ends. Spiroketalization and final lactonization would yield **1**.

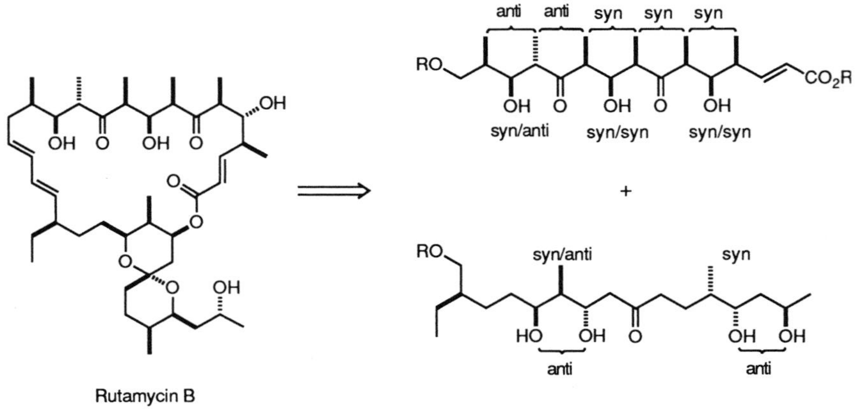

Rutamycin B

Scheme IV

The upper segment can be visualized as the aldol product of **2** and **3**, with the syn,syn stereochemistry resulting from the transition state shown in scheme V.

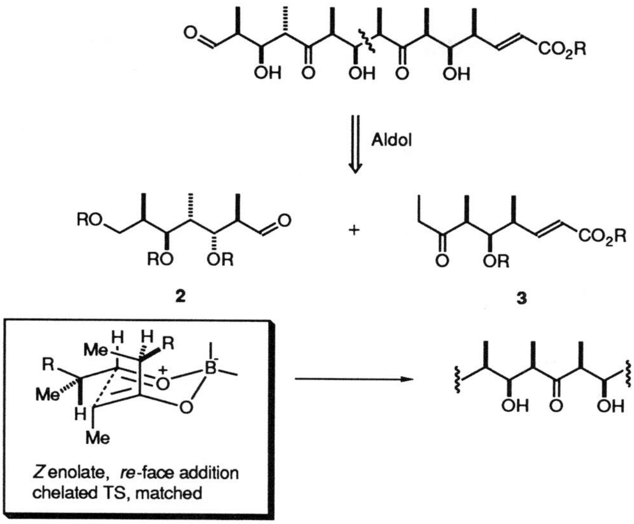

Scheme V

Further dissection of **2** and **3** reduces the stereochemical issue to reactions which can generate a syn,anti relationship (scheme VI) and a syn,syn orientation (scheme VII) in the crotylation of an appropriate aldehyde. Precedent for anticipating good stereocontrol in this process through the use of crotylboronates exists in the work of Roush [8].

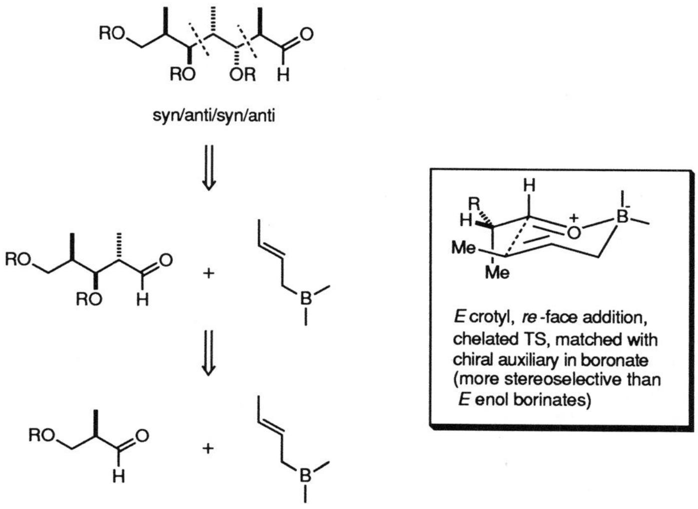

Scheme VI

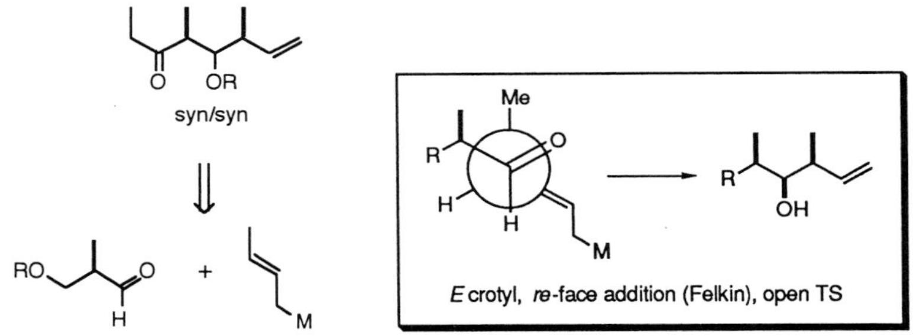

Scheme VII

The synthesis of **1** begins from (2*R*)-2-methyl-3-hydroxypropionate and proceeds through aldehyde **4** which is treated with the *E*-crotylboronate derived from (*S,S*)-tartrate. This alkylation gives **5** with good diastereoselectivity, and the latter is then protected and ozonized to aldehyde **6** (scheme VIII).

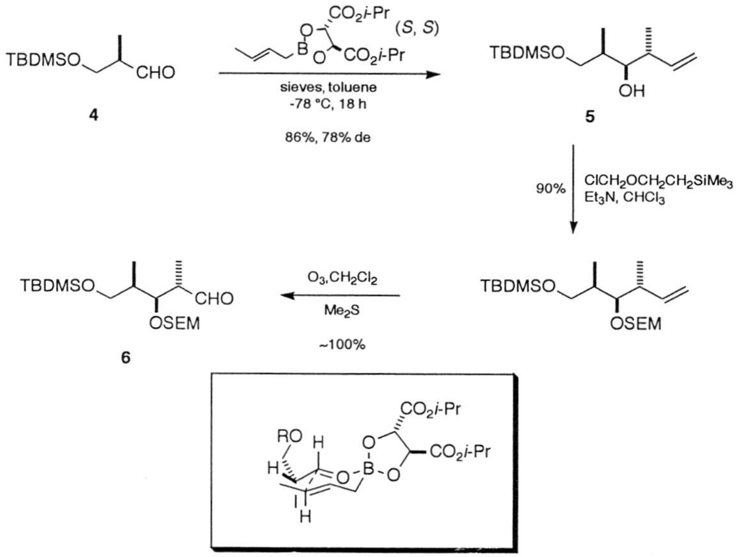

Scheme VIII

A second crotylation, now with the boronate prepared from (*R,R*)-tartrate, yields **7** (scheme IX). This confirms that *re*-face addition, controlled by the chirality of the tartrate, and anti configuration at the newly formed C-C bond, dictated by the E geometry of the crotyl group, results from a highly stereoselective alkylation. The preparation and use of these crotylboronates is greatly facilitated by a modification which enables the initially obtained boronate to be crystallized as its diethanolamine complex, thereby permitting removal of minor quantities of the geometrical isomer. After protection as its trimethylsilyl ether and ozonolysis, aldehyde **8** is obtained.

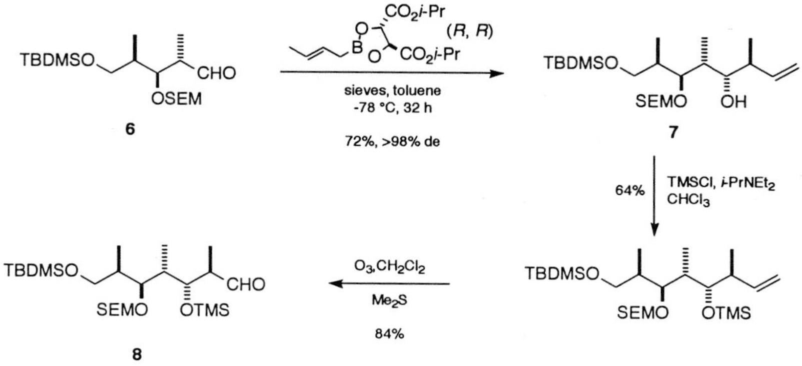

TBDMSO ⟶ CHO   OSEM   **6**

$\overset{(R, R)}{\underset{\text{sieves, toluene} \atop -78\,°C,\,32\,h}{\xrightarrow{\hspace{2cm}}}}$ reagent with CO_2i-Pr, B, CO_2i-Pr

72%, >98% de

TBDMSO ⟶   SEMO   OH   **7**

64% | TMSCl, i-PrNEt$_2$ CHCl$_3$

TBDMSO ⟶ CHO   SEMO   OTMS   **8**

$\xleftarrow[\text{Me}_2\text{S} \atop 84\%]{O_3, CH_2Cl_2}$

TBDMSO ⟶   SEMO   OTMS

Scheme IX

The right-hand segment required for the upper half of **1** embodies a syn,syn relationship of vicinal substituents, an array not easily accessible via crotylboronate methodology. For this, a crotylstannane alkylation of aldehyde **4** proved to be the solution [9], affording **9** in good yield and with fair stereoselectivity (scheme X). In contrast, to the Zimmerman-Traxler transition state invoked for boronate addition (see scheme VIII), crotylstannylation of **4** is not under chelation control; product geometry is therefore decided by the Felkin-Anh rule. In this circumstance, the configuration of the crotyl group is inconsequential, both E and Z isomers furnishing the same product **9**. This is fortunate since our method of preparing crotyltri-n-butylstannane produces a 1:1 mixture of geometrical isomers.

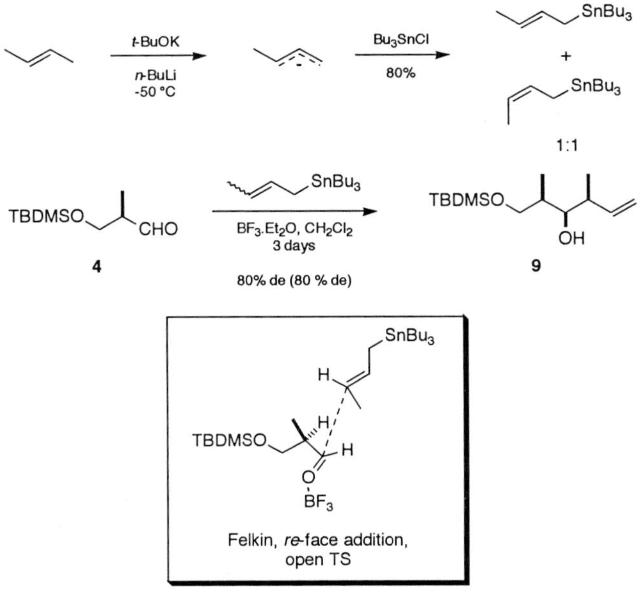

$\xrightarrow[\substack{n\text{-BuLi} \\ -50\,°C}]{t\text{-BuOK}}$ $\xrightarrow[80\%]{\text{Bu}_3\text{SnCl}}$

SnBu$_3$
+
SnBu$_3$

1:1

TBDMSO ⟶ CHO   **4**

$\xrightarrow[\substack{\text{BF}_3\cdot\text{Et}_2\text{O, CH}_2\text{Cl}_2 \\ 3\ \text{days}}]{\text{SnBu}_3}$

80% de (80 % de)

TBDMSO ⟶   OH   **9**

Felkin, re-face addition, open TS

Scheme X

Alcohol **9** is processed by first protecting the secondary hydroxyl and then selectively removing the primary blocking group (scheme XI). The resulting alcohol **10** is oxidized to aldehyde **11**, subjected to a Grignard reaction and oxidized again to ketone **12**. The vinyl group is transformed to the unsaturated ester **13** by ozonolysis, followed by a Wittig reaction. The two segments **8** and **13** now stand ready for the aldol condensation which will unite them and complete the upper half of **1**.

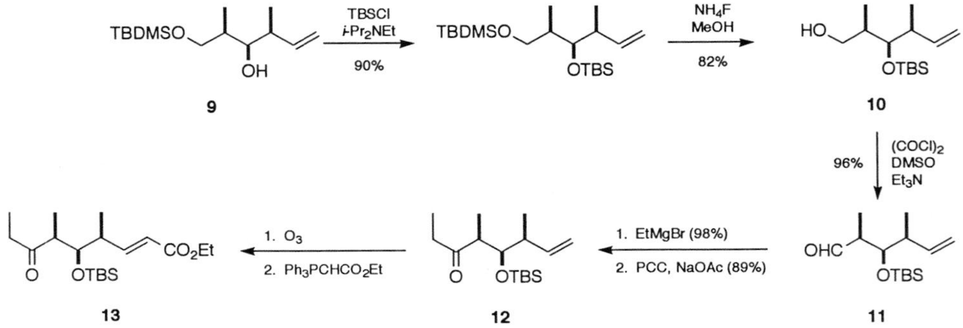

Scheme XI

Uncoiling of the lower half of rutamycin B (**1**) leads to an acyclic segment represented as **14** (scheme XII). Our expectation is that the polyhydroxy ketone will close under mild acid catalysis to generate the spiroketal moiety of **1** in the desired configuration [7]. The rationale for this outcome is that the spirocenter with natural configuration places the larger alkyl groups in an equatorial orientation while conferring maximum anomeric stabilization. The spiroketal precursor **14** logically arises from aldol condensation of aldehyde **15** with the enolate of ketone **16**. For this condensation to be under rigorous stereocontrol it will be necessary to prepare a chiral enolate from **16**. The (+)-isopinocampheylboronate of **16** is envisioned for this purpose since this auxiliary can be expected to dominate over the stereogenic center α to the aldehyde in influencing addition to the *si* face of **15** [10].

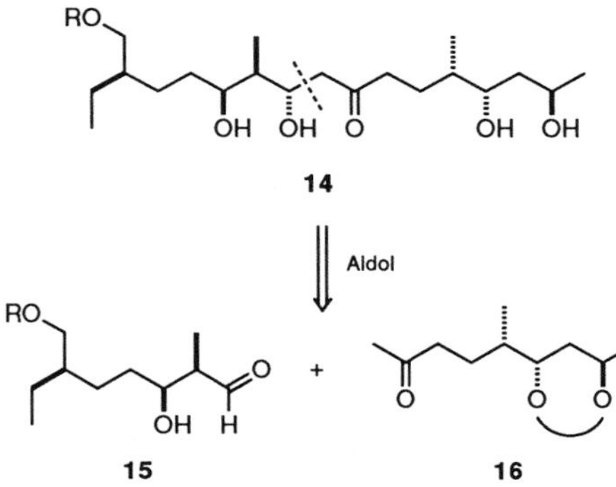

Scheme XII

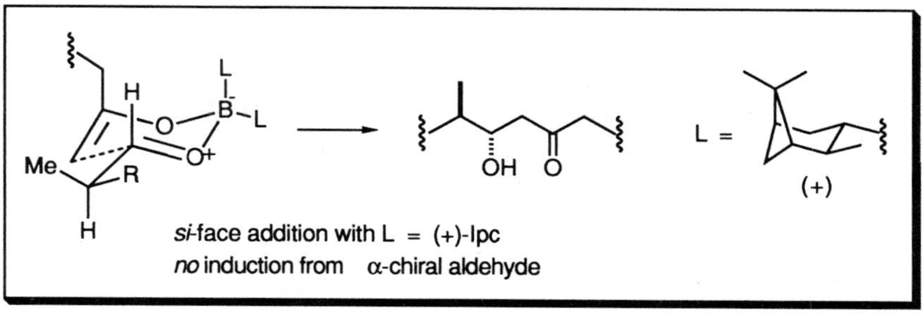

si-face addition with L = (+)-Ipc
no induction from α-chiral aldehyde

L =

(+)

Aldehyde **15** can again be visualized as the crotylation product of a precursor aldehyde **17**, the syn relationship in **15** now requiring Z configuration of the boronate. Installation of the ethyl substituent in **17** with (*R*) configuration is accomplished by an asymmetric Michael addition to the enolate of 3-butyroyl-4-benzyloxazolidin-2-one (**18**), prepared from (*S*)-phenylalaninol (scheme XIII).

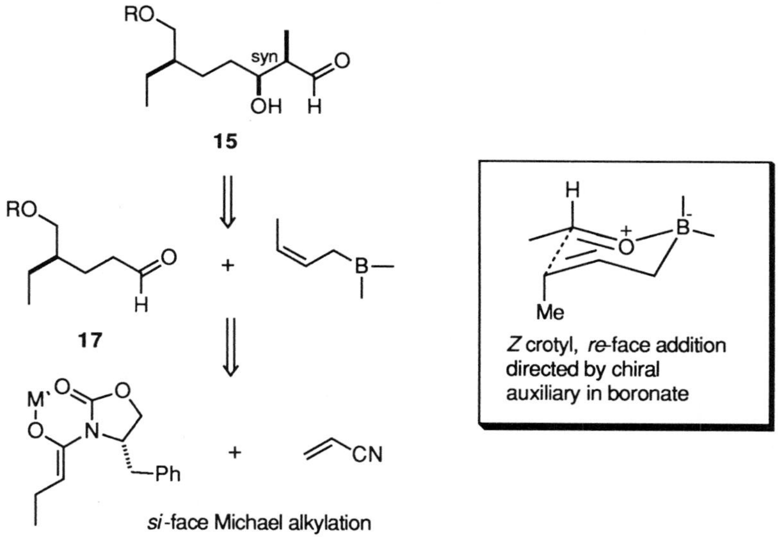

15

17

Z crotyl, *re*-face addition
directed by chiral
auxiliary in boronate

si-face Michael alkylation

Scheme XIII

The Michael addition of **18** as its titanium enolate to acrylonitrile takes place with very high stereoselectivity under Evans' conditions [11] yielding **19**. The auxiliary is then cleaved by reduction and the resulting hydroxy nitrile **20** is transformed to **21** by standard protocols (scheme XIV).

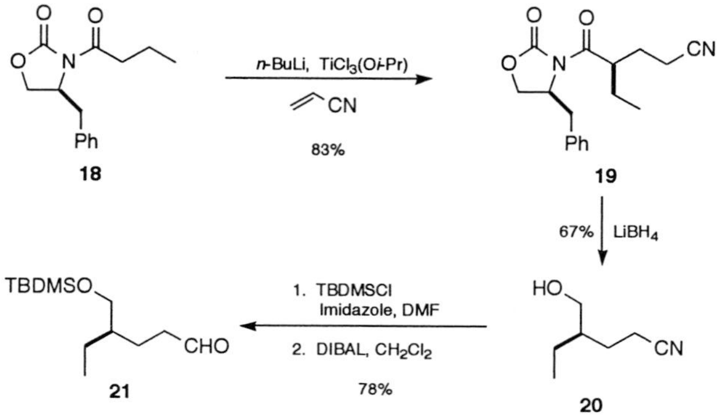

Scheme XIV

Alkylation of aldehyde **21** with the Z crotylboronate prepared from (S,S)-tartrate affords **22** with good stereoselectivity but in only modest yield, a result which appears to be associated with the instability of this particular boronate. Nevertheless, the alcohol **22** can be protected and converted to **23** by ozonolysis in a satisfactory manner (scheme XV).

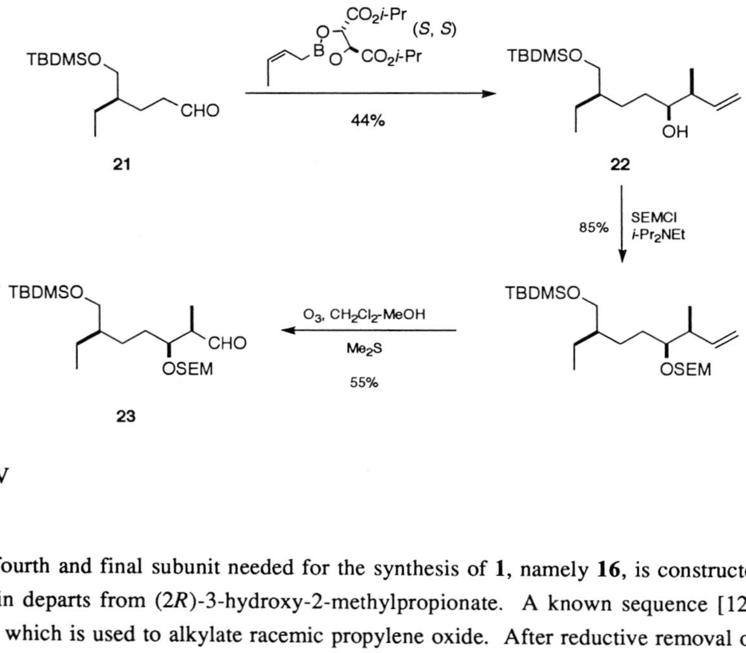

Scheme XV

The fourth and final subunit needed for the synthesis of **1**, namely **16**, is constructed by a route which again departs from (2R)-3-hydroxy-2-methylpropionate. A known sequence [12] takes us to sulfone **24** which is used to alkylate racemic propylene oxide. After reductive removal of the sulfone from **25** and oxidation of the alcohol, ketone **26** is obtained (scheme XVI).

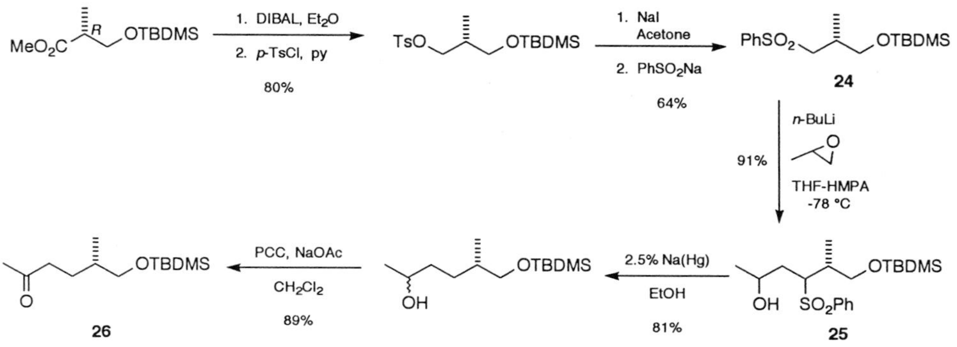

Scheme XVI

Elaboration of **26** to β-hydroxy ketone **28** was effected after adjustment of protecting groups and oxidation level by addition of a chiral acetone enolate to aldehyde **27**. Here again the (+)-isopinocampheylboronate[10] served to furnish syn stereochemistry by addition to the *si* face of **27** (scheme XVII). The steric bias imparted by the Ipc ligands on boron is due principally to a protruding secondary methyl substituent of one of the Ipc groups, the other ligand providing a buttressing effect.

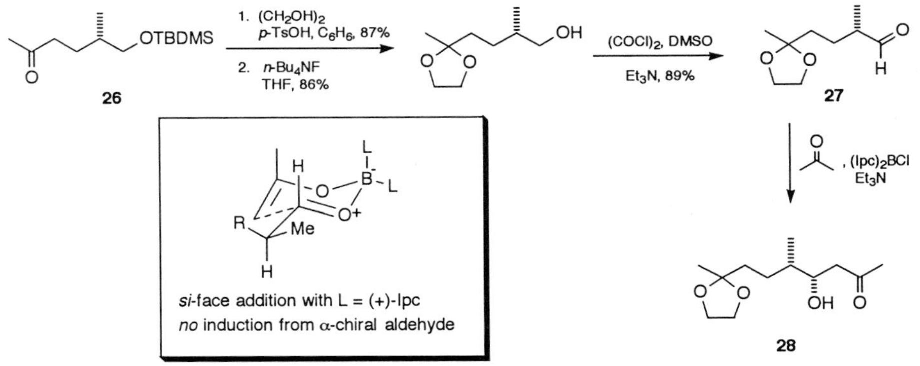

si-face addition with L = (+)-Ipc
no induction from α-chiral aldehyde

Scheme XVII

The anti-1,3-diol moiety present in **16** is now set in place by a directed reduction [13] of **28** in which a coordinated borohydride delivers hydride to the *re* face of the ketone. The diol is protected, and the ketal deprotected to yield **29** (scheme XVIII).

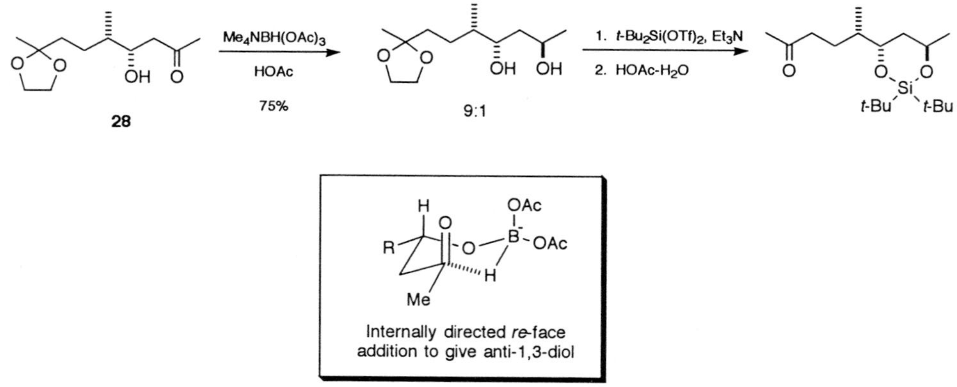

Internally directed *re*-face
addition to give anti-1,3-diol

Scheme XVIII

The sequences described above complete syntheses of the four principal subunits of **1** with a high degree of stereocontrol. It will next be necessary to converge **8** with **13** and **23** with **29**, again in a stereocontrolled fashion, to assemble the upper and lower halves of the macrolide. The lower half will then be connected via an *E,E*-diene linker unit to the upper half and a final spiroketalization and lactonization should yield rutamycin B (**1**).

It is with much pleasure that I acknowledge the contributions of my coworkers to this project: Yoshihiro Ohba, Thomas Tiller, Warren Porter, Shan Wang, Neal Green, Annapoorna Akella, and Fraser Fleming. Financial support was provided by the National Science Foundation and the National Institutes of Health. Dr. Tiller was the recipient of a fellowship from the Fonds der Chemischen Industrie and Neal Green is the recipient of a National Institutes of Health Postdoctoral Fellowship.

References

1.   K. Tatsuta: Total Synthesis of Macrolide Antibiotics. In *Recent Progress in the Chemical Synthesis of Antibiotics* (G. Lukacs, M. Ohno, Eds.), Springer-Verlag, Berlin, Germany, 1990, p. 1-38.

2.   T. G. Back: The Synthesis of Macrocyclic Lactones: Approaches to Complex Macrolide Antibiotics. *Tetrahedron* **33** (1977) 3041-3059.

3.   D. H. R. Barton, M. V. George, M. V. Tomoeda: Photochemical Transformations. Part XIII. A New Method for the Production of Acyl Radicals. *J. Chem Soc.* (1962) 1967-1974.

4.   D. Wulthier, W. Keller-Schierlein, B. Wahl: Stoffwechselprodukte von Mikroorganismen: Isolierung und Strukturaufklärung von Rutamycin B. *Helv. Chim. Acta* **67** (1984) 1208-1216.

5.   R. J. Fisher, A. M. Liang, G. C. Sundstrom: Selective Disaggregation of the H^+-translocating ATPase. *J. Biol Chem.* **256** (1981) 707-715.

6. B. Arnoux, M. C. Garcia-Alvarez, C. Marazano, B. C. Das, C. Pascard: X-Ray Structure of the Antibiotic Rutamycin. *J. Chem. Soc., Chem. Commun.* (1978) 318-319.

7. D. A. Evans, D. L. Rieger, T. K. Jones, S. W. Kalder: Assignment of Stereochemistry in the Oligomycin/Rutamycin/Cytovaricin Family of Antibiotics. Asymmetric Synthesis of the Rutamycin Spiroketal Synthon. *J. Org. Chem.* **55** (1990) 6260-6268.

8. (a) W. R. Roush, K. Ando, D. B. Powers, A. D. Palkowitz, R. L. Halterman: Asymmetric Synthesis Using Diisopropyl Tartrate Modified (*E*)- and (*Z*)-Crotylboronates: Preparation of the Chiral Crotylboronates and Reactions with Achiral Aldehydes. *J. Am. Chem. Soc.* **112** (1990) 6339-6348. (b) W. R. Roush, A. D. Palkowitz, K. Ando: Acyclic Diastereoselective Synthesis Using Tartrate Ester Modified Crotylboronates. Double Asymmetric Reactions with α-Methyl Chiral Aldehydes and Synthesis of the C(19)-C(29) Segment of Rifamycin S. *J. Am. Chem. Soc.* **112** (1990) 6348-6359.

9. G. E. Keck, D. E. Abbott: Stereochemical Consequences for the Lewis Acid Mediated Additions of Allyl and Crotyltri-*n*-butylstannane to Chiral β-Hydroxyaldehyde Derivatives. *Tetrahedron Lett.* **25** (1984) 1883-1886.

10. I. Paterson, J. M. Goodman, M. A. Lister, R. C. Schumann, C. K. McClure, R. D. Norcross: Enantio- and Diastereoselective Aldol Reactions of Achiral Ethyl and Methyl Ketones with Aldehydes: The Use of Enol Diisopinocampheylborinates. *Tetrahedron* **46** (1990) 4663-4684.

11. D. A. Evans, D. L. Rieger, M. T. Bilodeau, F. Urpí: Stereoselective Aldol Reactions of Chlorotitanium Enolates. An Efficient Method for the Assemblage of Polypropionate-Related Synthons. *J. Am. Chem. Soc.* **113** (1991) 1047-1049.

12. J. D. White, J. M. Kawasaki: Total Synthesis of (+)-Latrunculin A. *J. Am. Chem. Soc.* **112** (1990) 4991-4993.

13. D. A. Evans, K. T. Chapman, E. M. Carreira: Directed Reduction of β-Hydroxy Ketones Employing Tetramethylammonium Triacetoxyborohydride. *J. Am. Chem. Soc.* **110** (1988) 3560-3578.

Towards Erythronolides, Efficient Synthesis of Contiguous Stereocenters

Reinhard W. Hoffmann and Rainer Stürmer

Fachbereich Chemie der Philipps-Universität Marburg, D - W 3550 Marburg / Lahn

Summary: Stereoselective crotylboration reactions have been applied to the synthesis of two building blocks **2** and **3** conceived for a convergent synthesis of erythronolide A. The building block **2** comprising the 6 stereocenters from C9 to C13 was obtained in ten steps involving two crotylboration reactions. Building block **3** containing the 4 stereocenters at C2 to C5 of erythronolide was synthesized in seven steps relying again on two crotylboration reactions.

Introduction

Synthesis of erythronolides [1] is a continuing challenge [2] of our abilities in stereoselective synthesis. Syntheses of erythronolides therefore serve to demonstrate the reliability and efficiency of strategies and methods in stereoselective synthesis.

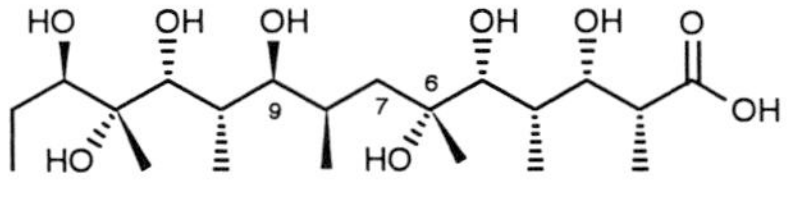

(9S)-Dihydro-Erythronolide A
seco-Acid

1

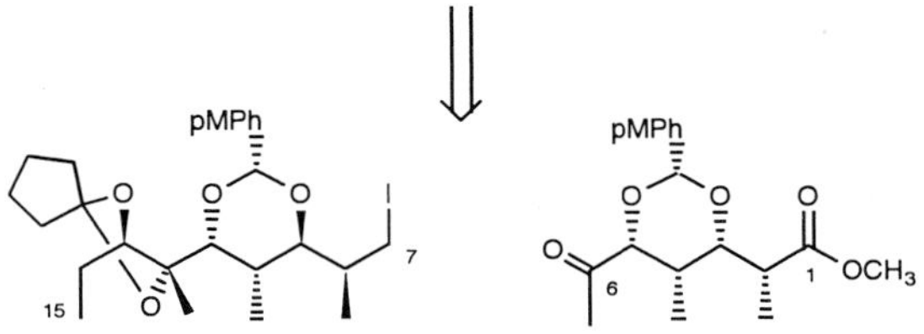

2   **3**

Of the seven completed or formal syntheses of erythronolide A reported up untill now, the majority are convergent ones: four of these utilize a retrosynthetic disconnection at the C6/C7 bond, [3,4,5] cf. **1**. Thus, obvious synthetic precursors for erythronolide A are compounds such as the C1/C6 building block **3** with 4 stereocenters and the C7/C15 building block **2** with 6 stereocenters. We describe here efficient stereoselective syntheses of these building blocks, syntheses which rely solely on reagent control of stereoselectivity.

The Tool: Stereoselective Crotylboration Reactions

Synthesis always involves steps which generate the molecular backbone. Efficient syntheses cannot cut down on those steps, but limit the number of refunctionalization steps. This requires that stereocenters be generated in the same step that was used to form the backbone of the target molecule, and not by subsequent stereoselective refunctionalizations. When the contiguous stereocenters in **2** or **3** are considered, an aldol- [6] or allylmetal-addition to an aldehyde might constitute the most efficient inroad.

For instance, the aldehyde **4** may be homologated to an aldol **5** or to the homoallyl alcohol **6**, which after protection and oxidative cleavage of the double bond furnishes another aldehyde **7**. The latter could then be placed at the starting point of a new chain extension sequence. Of course, the main problem is stereoselectivity, because starting from a given chiral aldehyde **8,** crotyl-metal addition may lead to four stereoisomeric triads **A** - **D**. [7]

It is compulsory to have methods at one`s disposal, which allow the selective transformation of a given aldehyde **8** at will into each of the four stereotriads **A** - **D**. One type of reaction by which this may be accomplished is the crotylboration reaction. This reaction shows high simple diastereoselectivity, [8] such that Z-crotylboronates **10** generate the stereotriads **A** and **B**, with a *syn*-arrangement at the two new stereocenters, whereas E-crotylboronates **9** lead to the stereotriads **C** and **D**, with an *anti*-arrangement at the two new stereocenters.

To direct the synthesis specifically to any individual stereotriad, chiral reagents are needed which enforce the selective formation of the desired stereotriad either by double stereodifferentiation or by reagent control of diastereoselectivity. [9] In the case of the crotylboration reaction, highest asymmetric induction can be attained by chirality transfer using α-substituted chiral crotylboronates. [7] The underlying principles can be delineated with reference to the pentenylboronate 11 .[10]

Stereocontrol in the crotylboration reaction relies on the fact that these additions to aldehydes proceed via cyclic six-membered transition states. A given enantiomer of 11 may add via two diastereomeric transition states to an aldehyde, each of which has the aldehyde residue in an equatorial position. The two transition states 12 and 14 lead to products of opposite configuration at both new stereocenters. The two transition states 12 and 14 differ in the arrangement of the methyl group at the stereogenic center of the reagent. In transition state 12 it is equatorial, whereas in transition state 14 it is axial and suffers substantial allylic 1,3-strain by interaction with the Z-positioned methyl group of the double bond. [11] As a consequence, the reaction utilizes only the reaction path via transition state 12.

Another aspect is worth of comment: the product 13 resulting from transition state 12 has an E-double bond, whereas that being formed (15) via transition state 14 has a Z-double bond. The geometry of the double bond in the products 13 and 15 and the configuration at the newly formed stereocenters in 13 and 15 are thus mechanistically linked. As the geometry of the double bond in the products is easy to analyse, it may thus serve as an indicator for the configuration of the newly formed stereocenters, vide supra.

	R	=	C_6H_5-	71 %	99.5 % e.e.
		=	$CH_2=C(CH_3)$-	86 %	99.0 % e.e.
		=	$CH_3\text{-}CH_2$-	79 %	98.5 % e.e.

In practice, the Z-pentenylboronate 16 with a dicyclohexylethanediol auxiliary is used. Reaction of the enantiomerically and diastereomerically pure Z-pentenylboronate 16 with achiral aldehydes generates the homoallylic alcohols 17 with high asymmetric induction. [10] This reaction is the key to the efficient synthesis of the erythronolide building blocks **2** and **3** described below.

The C1 - C6 Building Block 2

The first two stereocenters of the building block **2,** were indeed generated with high asymmetric induction by crotylboration of 2-methyl-pentenal (**18**) by the reagent **16**. The alcohol function in **19** was then protected as the p-methoxybenzyl ether.

For the further elaboration of the desired building block **2** a selective cleavage of the disubstituted double bond was required in the presence of a trisubstituted double bond, cf. **19**. This was initiated by careful bishydroxylation using N-methyl-morpholine-N-oxide and catalytic amounts of osmium tetroxide. Subsequent periodate cleavage furnished the aldehyde **20**, which was immediately allowed to react again with the reagent **16**. Inherent to a chiral aldehyde such as **20** is a high asymmetric induction favoring the formation of the stereotriad **B**, whereas the formation of the stereotriad **A**, and therefore reagent control of diastereoselectivity, is required in the synthesis of **3**. In this case, the reaction of the aldehyde **20** with the reagent **16** furnished two products; one (**21**) had an *E*-double bond, and the other (**22**) a *Z*-double bond. This made it clear at once which of the products was the one of reagent control of diastereoselectivity (**21**) and which was formed as a consequence of substrate induced asymmetric induction (**22**). Fortunately, the asymmetric induction from the reagent **16** was high enough to override the asymmetric induction of the aldehyde **20**, such that the desired product **21** was formed with substantial selectivity.

The synthesis of the desired building block **3** was then completed in a straightforward way. This synthesis of **3** in seven steps and 20 % overall yield compares quite favorably with other syntheses of similar C1 - C6 building blocks of erythronolide. [5,12]

The C7-C15 Building Block 2

The asymmetric crotylboration reaction, having served so well, was also envisioned to be the decisive tool for creating the C8, C9, and the C10, C11 stereocenters of the erythronolide building blocks. However, the C12/C13 stereocenters of erythronolide are at present not within the reach of the crotylboration reaction. Hence, other inroads were tested towards the key starting material **24**.

Our first synthesis relied on the "chiral pool" and used the glucosaccharinic acid lactone **23** as starting point. [13] Next, a more efficient route based on Seebach's [14] lactic acid synthons **25** was utilized. [15] A major improvement based on the Sharpless epoxidation was realized by Colvin, [16] who published a five step conversion of the allylic alcohol **26** into a compound related to **24**. Yet further improvement was possible: Sharpless epoxidation of **26** was followed by oxidation to the aldehyde **27**. Subsequent Lewis acid catalyzed acetalization with inversion at C13 resulted in a three step (67%) conversion of **26** into **24**. [17]

At the stage of **24**, the crotylboration reactions could again be used to advantage: Chain extension with the (*S,S,S*)-reagent **16** furnished an 81% yield of the homoallyl alcohol **28** as a single stereoisomer. [10] The *E*-geometry of the double bond in **28** made us confident about the proper configuration at the two new stereocenters. The stage was set for the next chain extension by protection of the hydroxyl group in **28** as a p-methoxybenzyl ether, followed by ozonolytic cleavage of the double bond.

For setting up the stereocenters at C8/C9, the enantiomeric (R,R,R)-reagent *ent*-**16** had to be applied. Chain extension resulted in a single stereoisomer **29** in 79% yield. Again, the fact that the product **29** contained an *E*-double bond left no doubt that the product obtained is that of reagent control of diastereoselectivity. The correctness of this statement was supported by an X-ray analysis at a later stage of the synthesis. [15] With **29** we had reached a compound with six stereocenters in seven steps from a simple starting material.

The next step is the protection of the C9-hydroxyl function as the p-methoxybenzylidene-acetal **30**. The usefulness of the building block **2** depends critically on the fact that the p-methoxyphenyl group has to be in the α-orientation on the dioxane ring. [4, 18] Unfortunately, this orientation is the thermodynamically less stable one. Thus, the p-methoxybenzylidene-acetal has to be formed under kinetic control. This can be realized [19] in a DDQ oxidation of **29**, which forms the oxonium ion **32** with a *syn*-arrangement of the two hydrogens as a consequence of allylic 1,3-strain.[11] Thus, the proper stereoisomer **30** can be reached in 88% yield, provided acidic conditions are strictly avoided.

Ozonolysis of the alkene **30** followed by borohydride reduction gave the alcohol **31**. The alcohol function was converted into the alkyl iodide **2** by mesylation and iodide displacement. This synthesis furnished **2** in 10 steps from **26** with 37% overall yield, demonstrating the versatility and efficiency in stereocontrol germane to the Sharpless epoxidation as well as the asymmetric crotylboration reactions.

Towards Erythronolide A?

The two building blocks **2** and **3** are very similar to **33** and **34**, which Stork and Rychnovsky had successfully joined in their synthesis of (9S)-dihydroerythronolide A. [4]

We have been able to convert **2** to the lithium derivative or the Grignard reagent **35**, which could be added to acetone, but which refused to add to the ketone **3**. Other transmetalation reactions or addition of various Lewis acids was to no avail.

At this stage we could have modified the ester function in **3** to a building block resembling more closely the one (**34**) used by Stork. [4] We are convinced that a synthesis of (9S)-dihydroerythronolide A could have been completed in this way. However, while this modification of the building block **3** would have required at most three additional steps, this would have rendered such a convergent synthesis of (9S)-dihydroerythronolide A less efficient than the linear synthesis that we just completed [20] by further elaboration of the alkene **30**. For this reason, the "arrows" **2** and **3** will continue to remain in the quiver.

Acknowledgement. We thank the Deutsche Forschungsgemeinschaft and the Fonds der Chemischen Industrie for support of this study. R.S. acknowledges gratefully a fellowship from the Graduierten-Kolleg "Metallorganische Chemie" at the Philipps-Universität Marburg.

References

1. I. Paterson, M.M. Mansuri, *Tetrahedron* **41** (1985) 3569.
2. J. Mulzer, *Angew. Chem.* **103** (1991) 1484; *Angew. Chem. Int. Ed. Engl.* **30** (1991) 1452.
3. M. Nakata, M. Arai, K. Tomooka, N. Ohsawa, M. Kinoshita, *Bull. Chem. Soc. Jpn* **62** (1989) 2618; I. Paterson, D. D. P. Laffan, D. J. Rawson, *Tetrahedron Lett.* **29** (1988) 1461.
4. G. Stork, S. D. Rychnovsky, *J. Am. Chem. Soc.* **109** (1987) 1565.
5. N. K. Kochetkov, A. F. Sviridov, M. S. Ermolenko, D. V. Yashunsky, V. S. Borodkin, *Tetrahedron* **45** (1989) 5109.
6. C. H. Heathcock, J. P. Hagen, S. D. Young, R. Pilli, D.- L. Bai, H.-P. Märki, K. Kees, U. Badertscher, *Chem. Scripta* **25** (1985) 39.
7. R. W. Hoffmann, *Angew. Chem.* **99** (1987) 503; *Angew. Chem., Int. Ed. Engl.* **26** (1987) 489.
8. R. W. Hoffmann, H.-J. Zeiß, *J. Org. Chem.* **46** (1981) 1309.
9. S. Masamune, W. Choy, J. S. Petersen, L. R. Sita, *Angew. Chem.* **97** (1985) 1 - 78; *Angew. Chem. Int. Ed. Engl.* **24** (1985) 1.
10. R. W. Hoffmann, K. Ditrich, G. Köster, R. Stürmer, *Chem. Ber.* **122** (1989) 1783.
11. J. L. Broeker, R. W. Hoffmann, K. N. Houk, *J. Am. Chem. Soc.* **113** (1991) 5006.
12. J. Mulzer, H. M. Kirstein, J. Buschmann, C. Lehmann, P. Luger, *J. Am. Chem. Soc.* **113** (1991) 910; G. Stork, I. Paterson, F. K. C. Lee, *J. Am. Chem. Soc.* **104** (1982) 4686.
13. R. W. Hoffmann, W. Ladner, *Chem. Ber.* **116** (1983) 1631.
14. D. Seebach, R. Naef, G. Calderari, *Tetrahedron* **40** (1984) 1313; D. Seebach, EPC Syntheses with C,C Bond Formation via Acetals and Enamines, in *Modern Synthetic Methods* (R. Scheffold, Edit.), Vol. 4, p. 125 ff, Springer, Berlin, 1986.
15. R. Stürmer, *Dissertation, Univ. Marburg* **1992**.
16. E. W. Colvin, A. D. Robertson, S. Wakharkar, *J. Chem. Soc., Chem. Commun.* **1983** 312; cf. also R. K. Boeckman Jr., J. R. Pruitt, *J. Am. Chem. Soc.* **111** (1989) 8286.
17. R. Stürmer, *Liebigs Ann. Chem.* **1991** 311.
18. R. B. Woodward et al., *J. Am. Chem. Soc.* **103** (1981) 3210,3213,3215; M. Hikota, H. Tone, K. Horita, O. Yone mitsu, *Tetrahedron* **46** (1990) 4613.
19. Y. Oikawa, T. Nishi, O. Yonemitsu, *Tetrahedron Lett.* **24** (1983) 4037.
20. R. Stürmer, K. Ritter, R. W. Hoffmann, *Angew. Chem.* submitted.

Total Synthesis of Erythronolide B Derivatives

Johann Mulzer*, Holger Kirstein, Peter Mareski

Institut für Organische Chemie der Freien Universität Berlin,
Takustraße 3, W-1000 Berlin 33, FRG

Summary : Efficient total syntheses of stereochemically pure erythronolide B (**12**) and several derivatives thereof are described. 2,3-Isopropylidene glyceraldehyde (**11**) is used as the only source of chirality. Key reactions are the coupling of two major fragments (Eastern and Western Zone) by two different methods (allyl sulfide and acetylide anion aldehyde additions) and conformationally controlled high yield macrolactonization.

Erythromycins are important macrolide antibiotics produced by via a polyketide pathway resulting in the formation of 6-deoxyerythronolide B (**1**) as the first isolable metabolite. **1** is hydroxylated in the 6-position to give erythronolide B (**2**) which is then glycosylated to erythromycin B (**3**) and subsequently hydroxylated at C-13 to erythromycin A (**4**), the most active member in the family [1]. Erythronolide A (**5**) is not a natural product, but has to be prepared from **4** by deglycosylation. The total synthesis of **1**, **2** and **4** is one of the largest projects in the history of organic chemistry, involving altogether about twenty research groups all over the world. So far several syntheses of **5**, one of **4** and two of **2** have been reported [2]. We describe our approach to **2** and some 9-dihydro- and 6- and 8-epimers thereof [3,4].

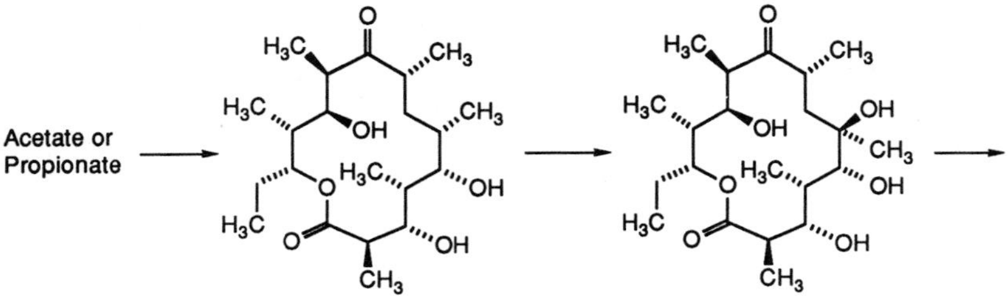

Acetate or Propionate →

6-Deoxyerythronolide B
(1)

Erythronolide B
(2)

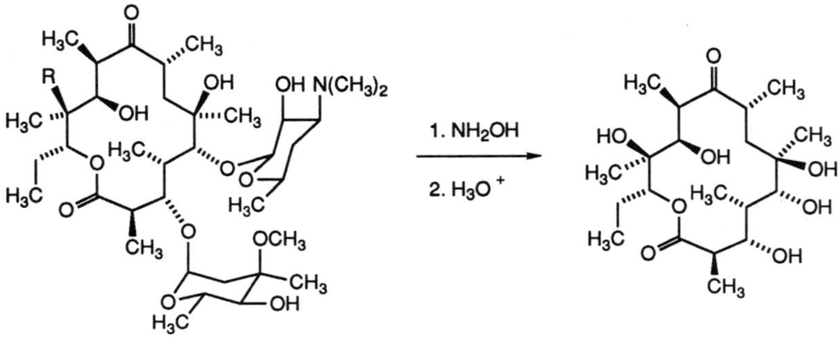

(3) R = H   Erythromycin B
(4) R = OH Erythromycin A

1. NH₂OH

2. H₃O⁺ →

Erythronolide A
(5)

Our retrosynthetic analysis of **2** leads to three fragments, namely the "western zone" (C-9 to 13) (**6**), the "eastern zone" (C-1 to 6) (**8**) and the connecting link, a formal di-carbanion (C-7 and 8) **7**, which combines **6** and **8** via two successive carbonyl additions. Both **6** and **7** contain similar structural subunits, the propionate triads A (**9**) and B (**10**), which have one stereogenic center (C-10 and C-2) in common and differ in the configurations of the two remaining stereogenic centers.

Both **9** and **10** are derived from 2,3-isopropylidene glyceraldehyde **11** [5], so that the following overall strategy emerges :

1. *Triads 9 / 10 from 11*
2. *East- and West-Fragments 6 and 8 from 9 /10*
3. *East-West-Coupling of 6 and 8*
4. *Macrolactonization*
5. *Final Functionalization*
6. *Removal of Protective Groups*

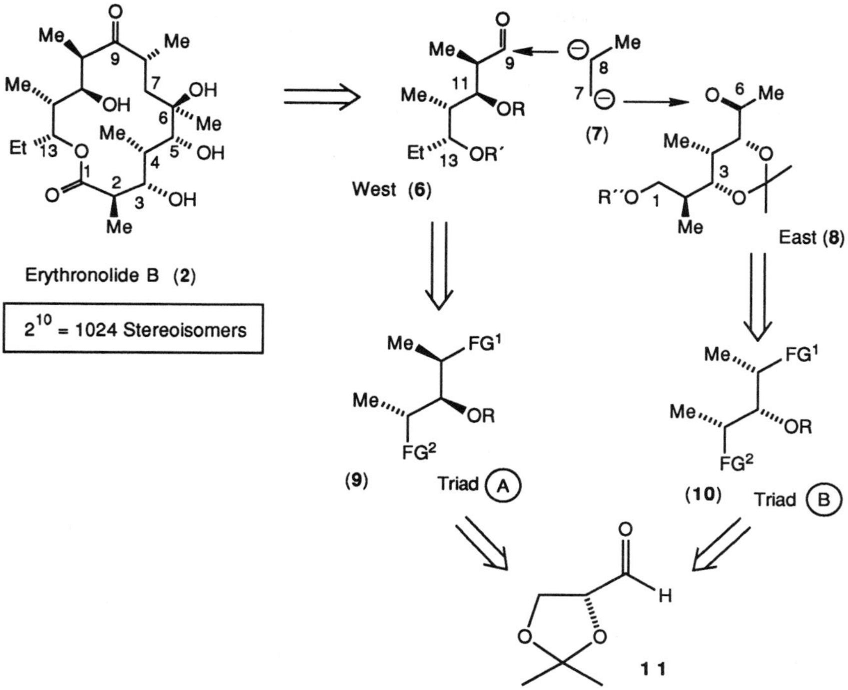

Me. O Me
      9
Me,,, 7   OH
      OH   6   Me
Et '''13 O   4 5   OH
      Me
O   1   2   '''OH
      3
Me

Erythronolide B (2)

2^{10} = 1024 Stereoisomers

⟹

Me. O
      9   ⟵   ⊖ Me
Me,,, 11   8
      OR   7 ⊖   ⟶   O   6   Me
Et '''13 OR'   (7)   Me,,,   3   O
                              O
West (6)   R''O   1   O
                        Me
                              East (8)

⟱   ⟱

Me.   FG¹   Me,,,   FG¹
Me,,,   OR   Me,,,   OR
FG²   FG²
(9)   Triad (A)   (10)   Triad (B)

⟱   ⟱

O
        H
1 1

The synthesis of **9** and **10** starts with the Hiyama-Nozaki reaction [6] of crotyl bromide with **11** furnishing a 60 : 40-mixture of the adducts **12** and **13** in 70 % overall yield [7]. This process has pros and cons, but has proven its high efficiency despite its lack of stereoselectivity. Therefore we did not switch to one of the well known stereocontrolled processes to convert **11** in either **12** or **13** [8].

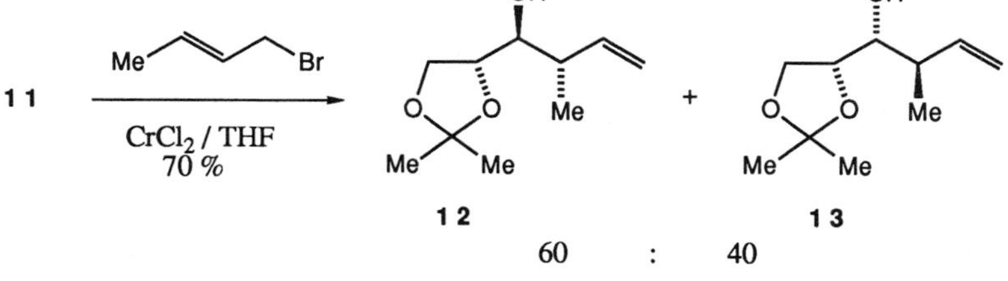

11

Me ⟍⟍⟋ Br
⟶
CrCl₂ / THF
70 %

OH
O   O   Me
Me   Me
1 2

+

OH
O   O   Me
Me   Me
1 3

60 : 40

Contra : low stereoselection, chromium

Pro : large scale, good separability, 10 out of 13 C - atoms and 6 out of 10 stereocenters from one reaction

Synthesis of Triads A (16) and B (18)

Adduct **12** was tosylated and deketalized to form diol **14**, which was cyclized to epoxide **15** exclusively without any formation of the Payne rearranged product or the oxetane. With dimethyl lithium cuprate, cleanly regiocontrolled ring opening to **16** was observed. The primary and secondary alcohol functions were easily differentiated by tritylation / benzylation to form derivative **17**. By an analogous sequence, adduct **13** was converted into triad B (**18**).

To generalize our synthetic approach with respect to the formation of as many diastereomers as possible, the problem of triad synthesis was analyzed more deeply. Clearly two types (meso and C2) may be distinguished. Given the possibility of easily converting the terminal functional groups of **19** and **20** into each other (e.g. vinyl into CH_2OH and vice versa), triad A represents structures **21** and **22**, and triad B **25** and **26**, accordingly. Type D is the enantiomer of B and can therefore easily be prepared from (S)-**11**. Only type C is not accessible by this procedure.

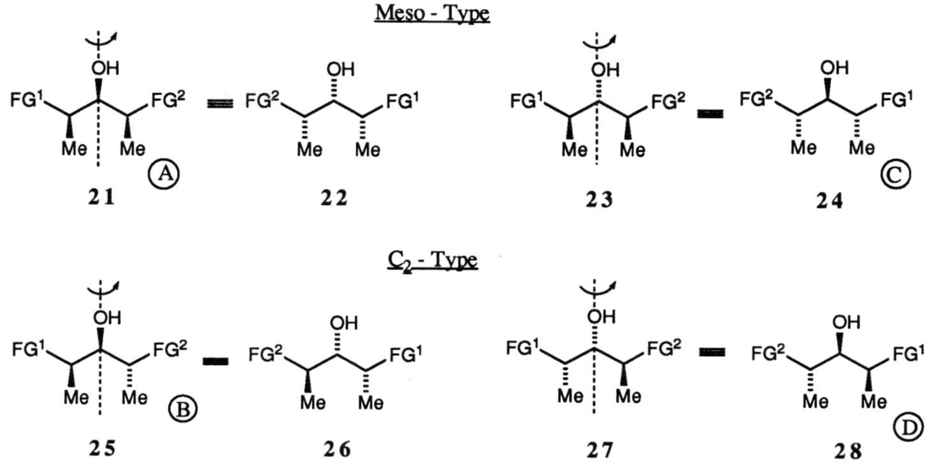

Meso - Type

21   22   ⒶⒸ   23   24

C₂ - Type

25   26   ⒷⒹ   27   28

It was therefore advisable to develop an alternative method starting from D-mannitol directly, which was converted into the di-tosylate **29**. Selective monoprotection of one of the homotopic diol units led to **30** which cyclized to epoxide **31** on treatment with weak base. Reaction with Lipshutz´cuprate resulted in a tandem process, generating triad C in the form of structure **33** via intermediate **32** . In a related tandem sequence, **37** was prepared via **34, 35** and **36** as intermediates. The regiocontrol in this case crucially depended on a bulky protective group for the terminal OH-functions [9].

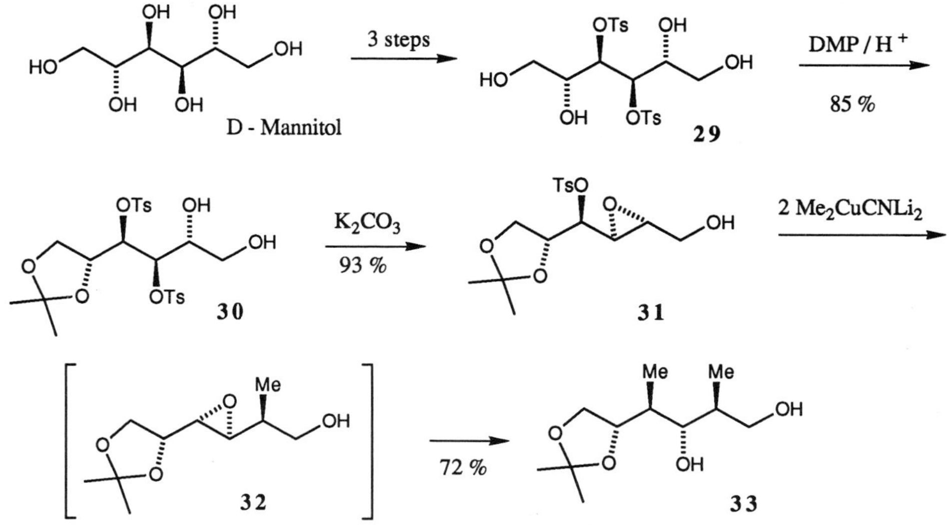

Synthesis of Western and Eastern Zones (39 and 42)

Triad A (17) was detritylated and oxidized to form the 13-aldehyde which added ethylmagnesium bromide with 4:1 Felkin-Anh selectivity. Silylation and ozonolysis furnished 39 without problems.

Similarly triad 18 was monobenzylated (mono : di-benzylether = 8 : 1) and silylated to give 40, which after ozonolysis was treated with either vinyl or isopropenyl magnesium bromide to furnish 41a/b with low stereocontrol. However, conversion into 42a/b and base catalyzed equilibration led to almost exclusive formation of the desired equatorial (5R)-epimers without significant loss of material.

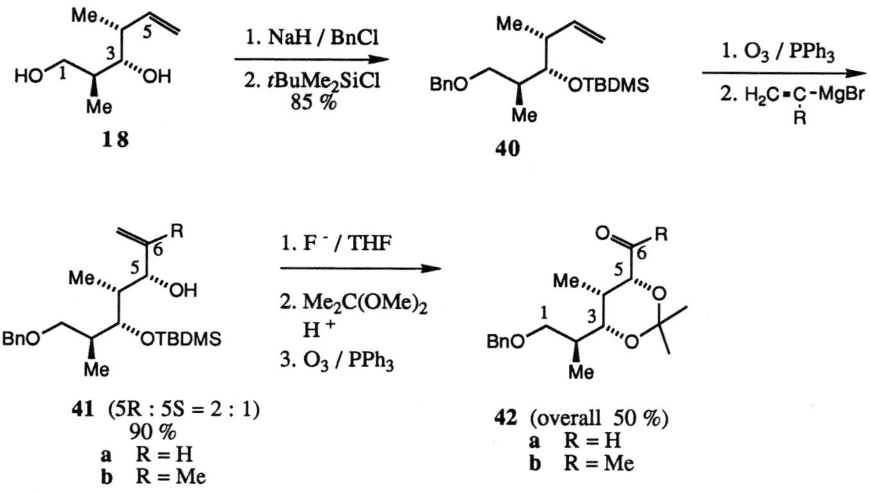

18 → 1. NaH / BnCl 2. tBuMe₂SiCl 85 % → 40

40 → 1. O₃ / PPh₃ 2. H₂C=C-MgBr (R) →

41 (5R : 5S = 2 : 1) 90 %
  a R = H
  b R = Me

1. F⁻ / THF
2. Me₂C(OMe)₂ H⁺
3. O₃ / PPh₃

→ 42 (overall 50 %)
  a R = H
  b R = Me

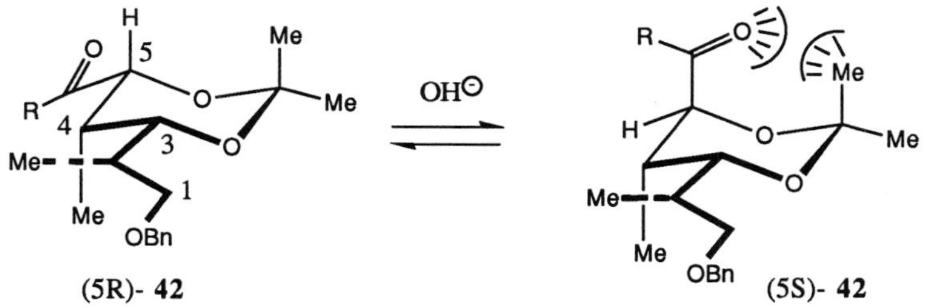

(5R)- 42   OH⊖ ⇌   (5S)- 42

before :   2 : 1
after equilibration :   > 99 : 1 !

Coupling of Western and Eastern Zone

Most of the earlier approaches to erythronolide derivatives suffered from serious problems on coupling major fragments. For instance in Kochetkov's synthesis, sulfoxide **43** was deprotonated at C-7 and added to ketone **45** to give adduct **46**; however, **43** could only be prepared along with its S(O)-epimer **44**, which was unreactive and had therefore to be converted into **43**. This led to a low overall yield of the coupling process [10].

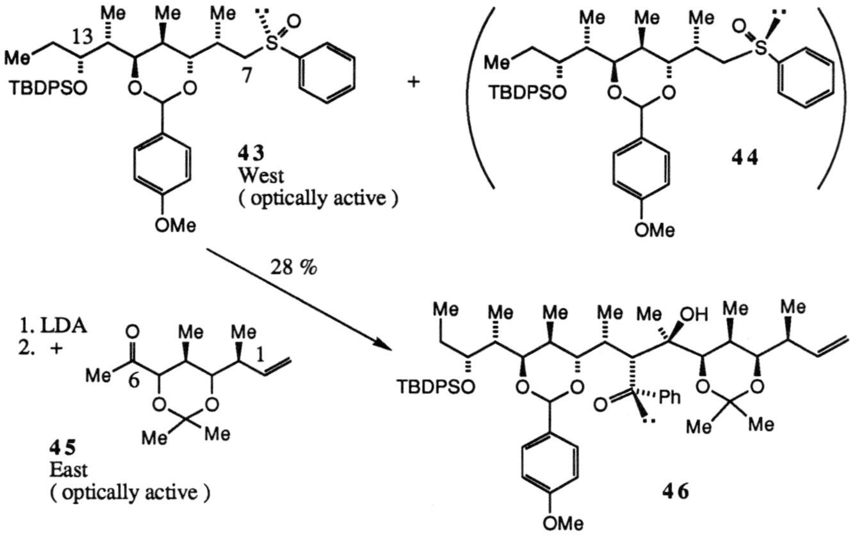

43
West
(optically active)

+

44

28 %

1. LDA
2. +

45
East
(optically active)

46

We therefore envisaged an alternative coupling which also makes use of the anion stabilizing effect of sulfur, but avoids the formation of too stable and hence unreactive anions. Western fragment **39** was converted into the allyl sulfide **47**, deprotonated with n-butyllithium and added to the eastern fragment **42b**.

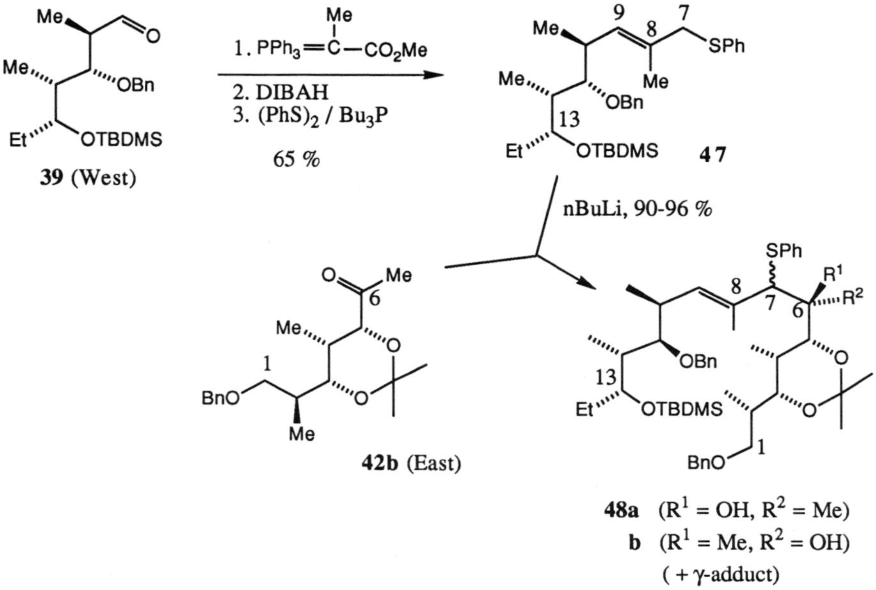

39 (West)

1. PPh$_3$=C—CO$_2$Me
2. DIBAH
3. (PhS)$_2$ / Bu$_3$P

65 %

47

nBuLi, 90-96 %

42b (East)

48a (R^1 = OH, R^2 = Me)
  b (R^1 = Me, R^2 = OH)
    (+ γ-adduct)

It turned out that in addition to the desired α-adduct **49,** the undesired γ-product **50** was also formed, in a ratio of 1:1.2 ! Obviously, the ambident anion **49** exhibited a higher nucleophilic activity in the γ- and not in the α-position as had been expected. However, after changing the solvent from tetramethyl ethylenediamine/hexane (low basicity) to tetramethyl ethylenediamine/THF/ hexamethyl-phosphoric triamide (HMPA) (higher basicity), the ratio of the regioisomers **48:50** switched to >99 : 1!

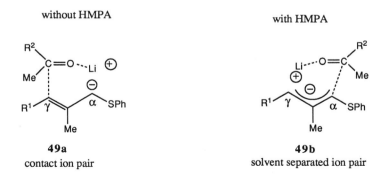

| TMEDA / Hexane | 1 | : | 1.2 | kinetic control ! |
| TMEDA / THF / HMPA | >99 | : | 1 | kinetic control ! |

We interpret these findings by assuming that in the less basic mediu,m **49** exists as a contact ion pair **49a**. In the presence of the strongly donating solvent HMPA, it is disrupted into a solvent separated ion pair **49b**. In both **49a** and **b,** the α- position carries the major part of the negative charge. Hence, lithium blocks the α-position in **49a**, and directs the ketone mainly into the γ-position, whereas in **49b** the reactive α-position is free for the attack of the ketone.

Stereoselection is also a major issue in this reaction. NMR studies demonstrated that the ketone **42b** exists in a chair conformation with an equatorial ketone function. The 5-methyl group has to adopt an axial position, and shields the bottom face of the chair against the attack of anion **49**. The stereochemical result of the addition depends on the relative participation of conformers A (undesired) and B (desired) respectively. Conformation A leads to Felkin-Anh and B to chelate-Cram-type addition [11]. Clearly in the medium chosen by us for optimum regiocontrol, chelation cannot occur and thus it is no surprise that conformation A vastly dominates and **48** is generated in the unnatural 6R-conformation **48b**. After some experimentation it turned out that the stereochemical course of the addition could be largely reversed by precomplexing ketone **42b** with boron trifluoride before adding it to **49**. In this case the ratio of **48a** : **48b** was 85:15.

(A) (undesired) (6R)   (B) (desired) (6S)

HMPA / TMEDA / THF   **48b** : **48a** (A : B) = 95 : 5

(A) (undesired)   (B) (desired)

BF₃ - Precomplexation   **48b** : **48a** (A : B) = 15 : 85

An alternative strategy of combining eastern and western fragment smade use of the stability and extreme nucleophilicity of acetylide anions. **39** was treated with monolithium acetylide to give **51** after THP-protection and a second deprotonation of the acetylide moiety. Addition of **51** to aldehyde **42a** afforded the coupling product **52** as an epimeric mixture with respect to C-6, C-9 and the anomeric center of the THP-group (C-1´) in practically quantitative chemical yield.

The stereochemical situation could easily be improved by chromatographca separation of the epimeric adduct **53**. Alternatively, the mixture was oxidized to the ketone and then reduced stereoselectively with Alpine-borane either to the 9R- or the 9S-alcohol, depending on the pinene enantiomer applied [12]. The THP-anomers **54** were separable by chromatography, so that the undesired anomer could be recycled by alcoholysis and re-ketalization. In effect, any one out of the four stereoisomers of **54** became available in pure form.

The stereoambiguity with respect to C-6 was eliminated by oxidation of alcohol **53** to ketone **55** under Swern conditions (90 % yield). The next problem was the stereocontrolled introduction of the two missing methyl groups at C-8 and C-6, respectively. This implied a conjugate addition to C-8 of the ynone system in **55** followed by a Grignard type addition to C-6.

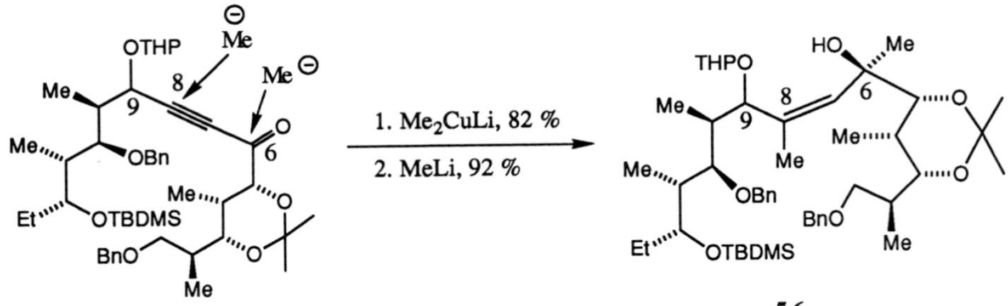

1. Me$_2$CuLi, 82 %

2. MeLi, 92 %

55   **56**

1. E/Z-selectivity with cuprates?

The E/Z-selectivity of the conjugate addition to **55** strongly depends on the configurations of both C-9 and C-1′. Only an unlike arrangement at these two stereocenters leads to high stereocontrol: the 9R,1′S combination induces Z- and the 9S,1′R-combination induces E-configuration around the newly formed trisubstituted 7,8-double bond. So far it is unclear how the mechanism of this addition works; quite obviously the addition of the cuprate and the subsequent protonation both are responsible for the stereochemical outcome and must therefore be controlled by the configurations of C-9 and C-1′!

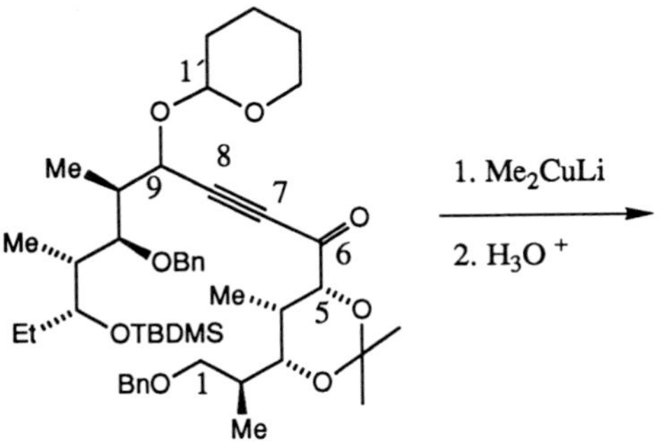

1. Me$_2$CuLi

2. H$_3$O$^+$

cis : trans	**56**	
8- Me / 7-H	9	1′
95 : 5	R	S
51 : 49	R	R
3 : 97	S	R
48 : 52	S	S

55

2. Stereochemistry of 6-MeLi-addition?

The addition of methyllithium to the 6-ketone function in **56** could be stereocontrolled by the solvent. Similar to the allyl sulfide anion addition to ketone **42b** Felkin-Anh and chelate-Cram transition states (**56a** and **b**) compete, the former being predominant in THF and the latter one in ether.

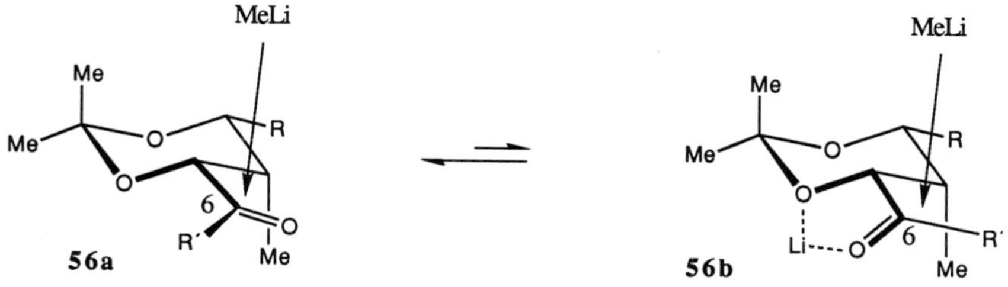

	6 - si		6 - re
Et$_2$O	< 3	:	> 97
THF	81	:	19

In consequence, the tertiary alcohol **57** was accessible in both the 6R (natural) and 6S (unnatural) configurations via **56a** and **56b**, respectively. To proceed to the seco acids, **57** was debenzylated to give the diol **58** and oxidized selectively at the primary position to give, after desilylation, seco acid **59** in the form of four stereoisomers **a-d**. In a similar fashion, intermediates **48a** and **b** were converted into seco acids **61a, b**.

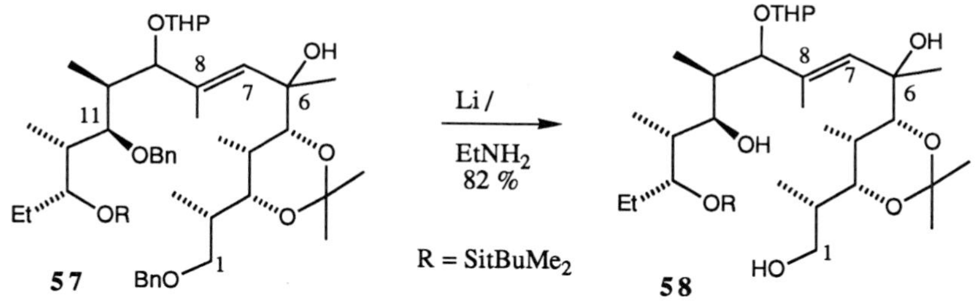

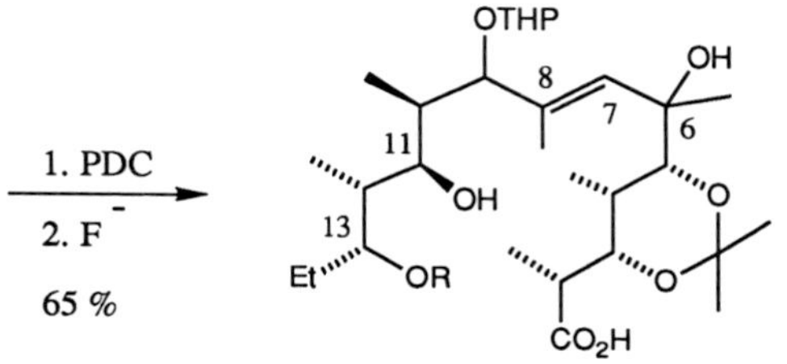

59
a   9S,7E,6R
b   9S,7E,6S
c   9R,7Z,6R
d   9R,7Z,6S

48a,b → (Li / EtNH$_2$, 82 %) → **60a,b**

(PDC, 65 %) → **61a,b**

a R^1 = OH, R^2 = Me
b R^1 = Me, R^2 = OH

Macrolactonization

Next the factors influencing the macrolactonization of seco acids **59** were investigated. For optimal ring closure, the conformation of **59** should involve three crucial sections: 1. an all anti backbone having OH-functions on both flanks of the chain. 2. Attached to this all anti chain there should be a hook ending in the (activated) carboxyl group. The curvature of the hook should be such that, together with a rigid spacer section, the carboxy-carbon should approach the appropriate OH-function along a Bürgi-Dunitz trajectory. 3. To do this, the requisite OH-function must be in a "weatherside" position with respect to the all anti backbone. "Leeward" OH-functions are not accessible for lactonization.

In fact the crystal structure of acid **63b** shows all the structural features just mentioned. The C-10 to 13 section forms an all-anti chain with the 11-OAc in the leeward and the 13-OTBDMS in the weatherside position. Due to allylic 1,3-strain [13], the C-6,7 bond is oriented perpendicularly on the plane of the C-8,9-olefin and thus forms the hook for the C-1-6-chain, which is held rigidly in an almost linear arrangement by the 3,5-acetonide chair.

Simple rotation around the C-5,6- and C-6,7-bond axes brings the 13-OH and the carboxyl carbon into an almost perfect arrangement for lactonization, whereas the 11-OH cannot react, although it may activate the carboxyl-oxygen via an intramolecular hydrogen bridge ("double activation").

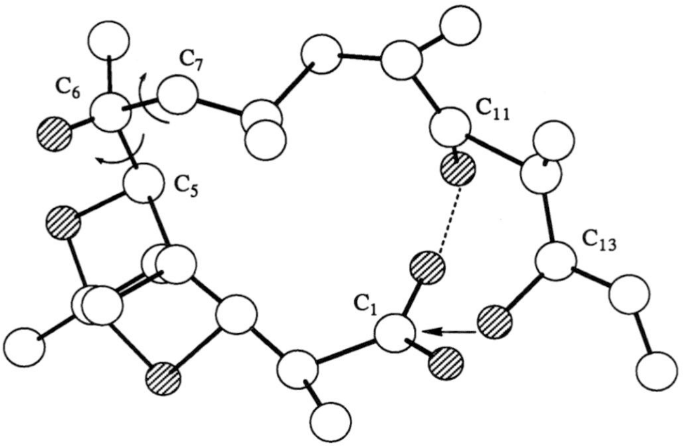

In view of these considerations ,it is no surprise that all our seco acid derivatives **59** smoothly lactonize under Yamaguchi´s conditions [14] to give macrolactones **62**, without any concomitant formation of 12-membered ring systems. Similarly, seco acids **61a,b** were converted via **63a,b** into dihydroxy compounds **64a,b**, which also cyclized chemospecifically to the 14-membered lactones **65a,b**.

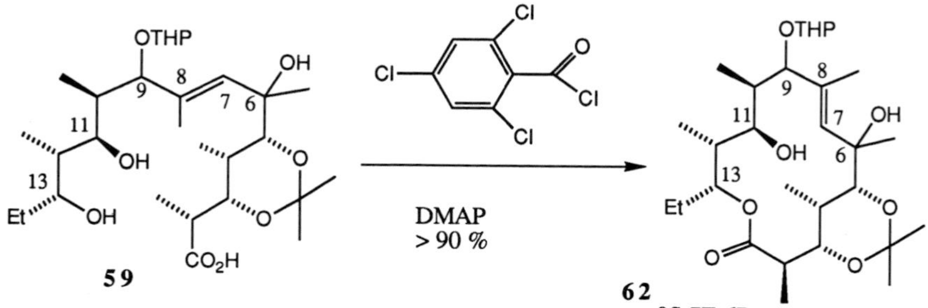

62
a   9S,7E,6R
b   9S,7E,6S
c   9R,7Z,6R
d   9R,7Z,6S

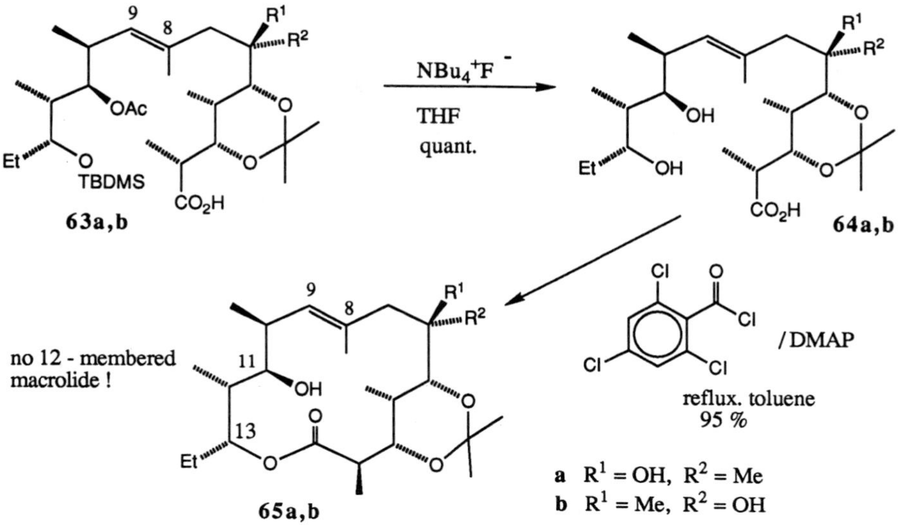

63a,b → NBu$_4^+$F$^-$ / THF / quant. → **64a,b**

no 12 - membered macrolide ! → **65a,b**

/ DMAP
reflux. toluene
95 %

a R^1 = OH, R^2 = Me
b R^1 = Me, R^2 = OH

Final Transformations

The final transformations of the lactones **62a-d** to erythronolide derivatives involve removal of the protective groups and hydrogenation of the 7,8-double bonds. Acid catalyzed deketalization led to complete destruction of the molecule, probably due to the formation of a tertiary carbenium ion at C-6 and subsequent cation-olefin polymerization.

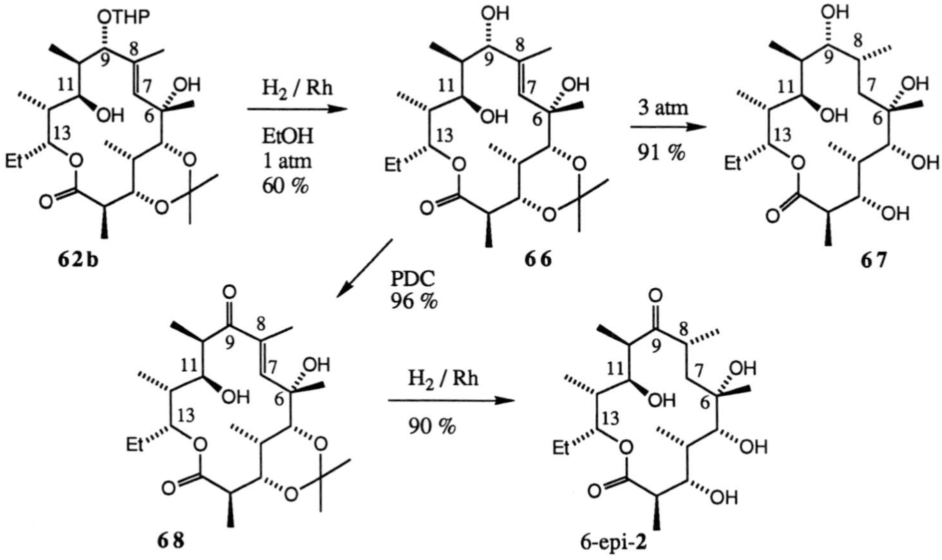

62b → H$_2$ / Rh / EtOH / 1 atm / 60 % → **66** → 3 atm / 91 % → **67**

66 → PDC 96 % → **68** → H$_2$ / Rh / 90 % → **6-epi-2**

However, catalytic hydrogenation over Rh/C (5 %) in ethanol under normal pressure cleanly removed the THP-group to give **66**. The unprotected 9-OH-group was oxidized to ketone **68**, which was hydrogenated to 6-epi-erythronolide B. Alternatively, **65** under forced hydrogenation conditions was transformed into the 9-dihydro-6-epi-erythronolide **67**.

Macrolide **65a**, on the other hand, was hydroborated to give **69** along with its 8,9-epimer (ratio 6:1). After HPLC separation **69** was oxidized and deketalized to form erythronolide B (**2**) in 70 % overall yield.

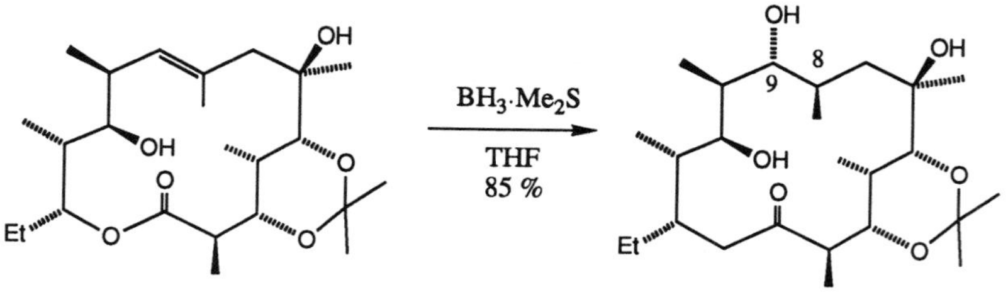

To confirm the configurational assignments, a number of crystal structure analyses were performed, two of which are shown below. Independently of the configuration of the 7,8-double, bond the macrolides show a rigid conformation, having two polypropionate chains (C-2 to C-6) and (C-9 to C-28) parallel to each other, crosslinked by essential planar three-atom moieties (C-6 to C-9 and C-1, O-8 and C-13).

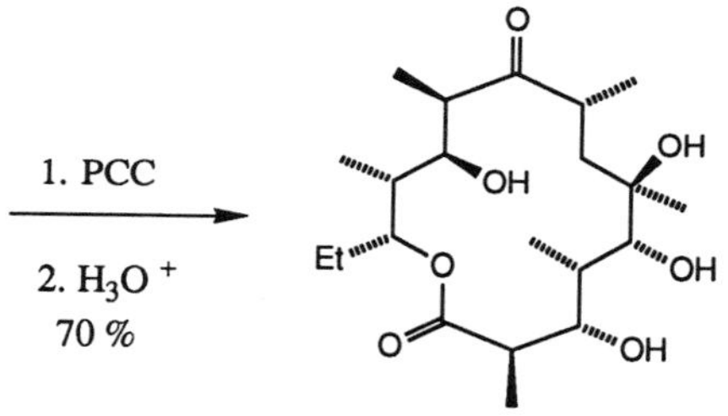

Acknowledgement

The X-ray crystal structure analyses were kindly performed by Prof. P. Luger and Dr. J. Buschmann, Institut für Kristallographie der Freien Universität Berlin. Financial support was provided by the Deutsche Forschungsgemeinschaft and the Schering AG, Berlin.

(1´S)- **62a**

(1′R)-**62d**

References

1. J. Staunton, *Angew. Chem. Int. Ed. Engl.* **30** (1991) 1302.

2. Review: J. Mulzer, *Angew. Chem. Int. Ed. Engl.* **30** (1991) 1452; I. Paterson, M. M. Mansuri, *Tetrahedron* **41** (1985) 3569.

3. J. Mulzer, H. M. Kirstein, J. Buschmann, C. Lehmann, P. Luger, *J. Am. Chem. Soc.* **113** (1991) 910.

4. J. Mulzer, P. A. Mareski, J. Buschmann, P. Luger, *Synthesis* (1992) 215.

5. J. Mulzer, L. Autenrieth-Ansorge, H. Kirstein, T. Matsuoka, *J. Org. Chem.* **52** (1987) 3784.

6. Y. Okude, S. Hirano, T. Hiyama, H. Nozaki, *J. Am. Chem. Soc.* **99** (1977) 3179.

7. J. Mulzer, P. deLasalle, A. Freißler, *Liebigs Ann. Chem.* (1986) 1152.

8. W. R. Roush, R. L. Halterman, *J. Am. Chem. Soc.* **108** (1986) 294.

9. J. Mulzer, B. Schöllhorn, *Angew. Chem. Int. Ed. Engl.* **29** (1990) 1529.

10. N. K. Kochetkov et at., *Tetrahedron* **45** (1989) 5109.

11. M. Cherest, H. Felkin, N. Prudent, *Tetraderon Lett.* (1968) 199 ; N.T. Anh, *Top. Curr. Chem.* **88** (1988) 145; E .L. Eliel in "Asymmetric Synthesis", Vol. 2, J. D. Morrison, Ed., Academic Press, N.Y. (1983) 125; M. T. Reetz, *Angew. Chem. Int. Ed. Engl.* 23 (1984) 556; J. Mulzer in J. Mulzer, H. J. Altenbach, M. Braun, K. Krohn, H. U. Reißig, Organic Synthesis Highlights, VCH, Weinheim-New York 1991, 3.

12. M. M. Midland, S. Greer, A. Tromantano, S. A. Zderic, *J. Am. Chem. Soc.* **101** (1979) 2352.

13. R. W. Hoffmann, *Chem. Rev.* **89** (1989) 1841.

14. M. Yamaguchi et al., *Bull. Chem. Soc. Jpn.* **52** (1979) 1989.

Chemical Modification of the Antifungal Macrolide Soraphen A1α: Cleavage and Reconstruction of the Lactone Bond

Dietmar Schummer, Bettina Böhlendorf, Michael Kiffe and Gerhard Höfle

GBF, Gesellschaft für Biotechnologische Forschung mbH, Mascheroder Weg 1,
3300 Braunschweig, FRG

Summary: Attempts to hydrolyse the lactone bond of soraphen $A_{1\alpha}$ (**1**) under basic or acidic conditions led to the rearranged δ- and γ-lactones **7** and **9**. After protection of the 3- and 5-hydroxyl groups as methyl and p-methoxybenzyl ethers to give **16**, reductive cleavage of the lactone with LAH gave the 1,17-diol **18**. With suitable protection, the diol **18** was oxidized and transformed into the 17-bromo-soraphen acid **22**, which was cyclized as its cesium salt to soraphen $A_{1\alpha}$. Analogously, borohydride reduction of the 17-keto-soraphen acid **24** gave a 1:1 mixture of the 17-epimeric soraphens **26** from which pure 17-epi-soraphen $A_{1\alpha}$ (**27**) was isolated after deprotection and chromatography.

Introduction

The *soraphens* were first isolated from the cellulose degrading soil bacterium *Sorangium cellulosum* in 1986 [2]. They have powerful broad spectrum antifungal properties including activity against a number of plant pathogens [2,3] which may lead eventually to a practical application. The basic compound soraphen $A_{1\alpha}$ as well as its 32 structurally related analogs [4] produced by *Sorangium cellulosum* are novel 18-membered macrolides. They are characterized by a ß-keto-lactone which is in the form of a hemi-acetal, a δ-hydroxyl group and an unsubstituted phenyl group. The soraphens also exhibit a novel mode of action. They are highly efficient inhibitors of the acetyl-CoA carboxylase and thus interfere with fungal lipid synthesis [5]. In addition to the isolated compounds, a large number of chemical derivatives at the allyl ether moiety C-9 to C-12 [6] and the 5-hydroxyl group [7,8] has been prepared to investigate the structure-activity relationship in more detail.

Soraphen A$_{1\alpha}$ (1)   2

Selective hydrogenation or electrophilic substitution of the phenyl group was not successful due to the sensitivity of the other functional groups. Also in nature, the only compound substituted in this part of the molecule has a m-hydroxyphenyl group [4] which is not of practical use because of further modifications of the molecule. We decided therefore to synthesize phenyl-modified soraphens by excision of the phenyl-C-17 ring segment and introduction of a synthetic building block with a variety of substituents R. In addition, this route would allow variation of the ring size of the macrocycle and exchange of the 18-oxygen for nitrogen, sulphur, and carbon substituents. A prerequisite for these transformations is the cleavage and reconstruction of the lactone group which is described here.

Reaction of soraphen A$_{1\alpha}$ (1) with base [7]

In an attempt to hydrolyze the lactone, a methanolic solution of soraphen A$_{1\alpha}$ (1) was treated with aqueous sodium hydroxide. At pH 13, an almost instantaneous quantitative formation of the ß-keto-enolate 4 was observed, which is characterized by a strong UV absorption at 286 nm (ε = 4.29). At lower pH, 1 and 4 co-exist in a pH-dependent equilibrium from which a pK_a = 12.4 for 4 can be derived in good agreement with that of ethyl 2-methylacetoacetate (pK_a = 12.3 [9]). When the enolate 4 was quenched at pH 3, soraphen A$_{1\alpha}$ (1) with the original stereochemistry at C-2 was isolated in almost quantitative yield. However, when the enolate was quenched at pH 7, the keto-form 3 (seco-soraphen A$_{1\alpha}$) was isolated in high yield and stereochemical purity. At room temperature, 3 cyclizes slowly to the initial hemi-acetal 1. This process takes place rapidly on addition of a catalytic amount of acid. The occurence of enol forms during this ring closure can be excluded since in CH$_3$OD solution, no deuterium was incorporated into 1 or recovered 2. In most organic solvents, soraphen A$_{1\alpha}$ (1) exists only in the hemi-acetal form 1. Only in very polar solvents, e.g. ethylene glycol, 5% of the keto-form 3 was detected in equilibrium by proton NMR spectroscopy.

When the enolate 4 was kept at room temperature, the UV maximum was slowly shifted to λ = 283 nm (ε = 4.19). After protonation, a new compound was isolated and identified as the ß-keto-δ-lactone 6 which is completely enolized. For steric reasons, a concerted process for this transformation is not possible. We rather assume that a 17-alkoxide is eliminated from the enolate 4 to give a highly reactive acyl ketene intermediate **5** which is subsequently trapped by the 5-hydroxyl group. Surprisingly no external nucleophile like the solvents methanol and water can compete with this reaction. The δ-lactone **6** is stable towards further hydrolysis and of no use for the desired purpose.

Reaction of soraphen A$_{1\alpha}$ (1) with acids [7]

In acidic aqueous methanolic solution, soraphen A$_{1\alpha}$ (**1**) is very stable even in the presence of 50% sulphuric acid. However, in the absence of water, a catalytic amount of sulphuric or another strong acid instigates a complex reaction sequence resulting in the formation of the γ-lactone **9a** in 18% yield.

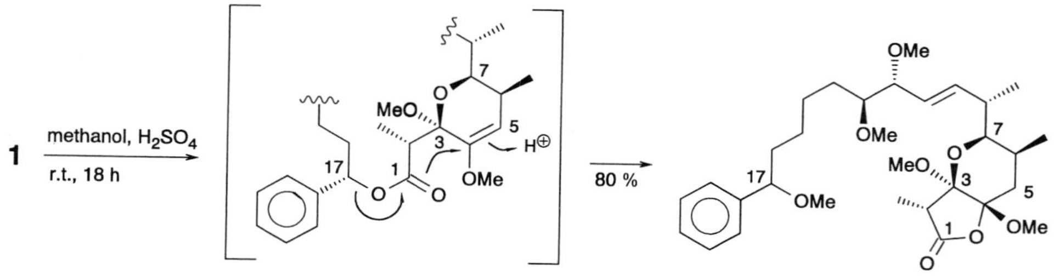

8   **9a**

Apparently after elimination of water from the 5-hydroxyl group and formation of the 3-methyl acetal, protonation of the intermediate enol ether **8** induces the rearrangement to a γ-lactone, leaving a benzyl cation which is subsequently trapped by methanol. The stereochemistry of **9a** was deduced from extensive NOE measurements.

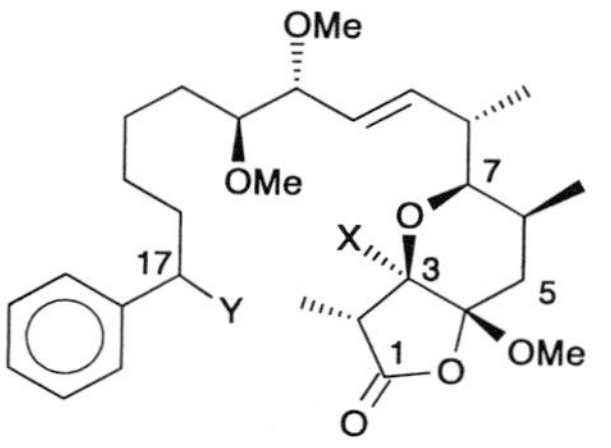

	reagent, solvent	X	Y	Yield
9a	H$_2$SO$_4$, methanol	OMe	OMe	80 %
b	H$_2$SO$_4$, THF	OH	OH	34 %
c		OH	$\Delta^{16,17}$	10 %
d	SOCl$_2$, dichloromethane	Cl	Cl	33 %
e	PBr$_3$, chloroform	OH	Br	5 %
		Br	Br	7 %
f	AlI$_3$, acetonitrile	H	I	29 %

Later we found that a wide range of Lewis acids can induce this reaction leading to related compounds **9b-e** with hydroxyl or halogen substituents at C-3 and C-17. The solvent has a surprisingly strong influence on the outcome of these reactions. Thus in contrast to the production of **9e** in chloroform (see Table), phosphorus tribromide in diethyl ether gave the enol ether **10**, with inverted configuration at C-2 [10]. Phosphorus pentachloride in dichloromethane yielded the 5-chloro derivative **11** with retention of configuration at C-5 through neighbor group participation of the 4-methoxy group. Both compounds are useful intermediates for further modification at C-4 and C-5.

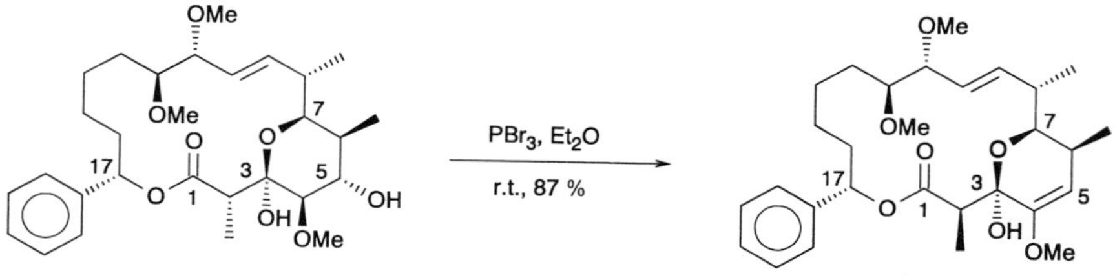

1   **10**

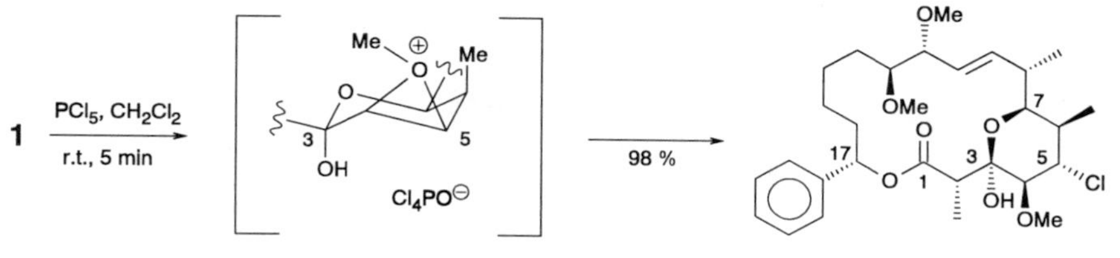

11

Protection of the 3,5-hydroxyl groups

At this stage of our work, we realized that protection of both hydroxyl groups of soraphen $A_{1\alpha}$ was necessary and that this had to be achieved under basic conditions. Of the various mono- and bifunctional silyl groups, only the 3,1-bis-t-butylsilylene could be introduced. Careful control of the reaction temperature was necessary. Although the cyclic ether **12** was sterically hindered at the silicon,

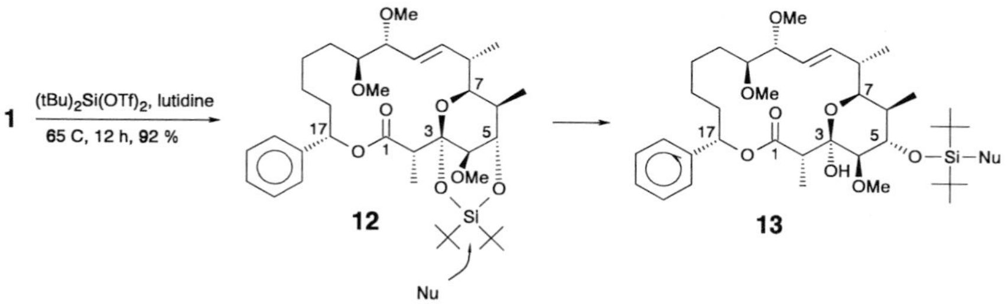

12   **13**

it was readily cleaved by nucleophiles before any reaction at the lactone group occurred. After some experimentation, ether protecting groups were finally chosen. Consecutive alkylation by p-methoxy-benzyl bromide in DMF and methyl iodide in HMPA gave **16** in high yield. Considerably

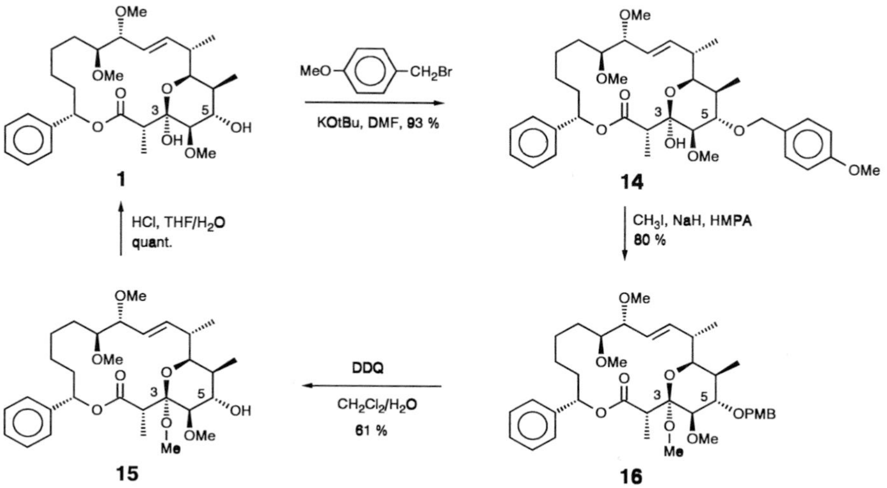

1

14

HCl, THF/H₂O
quant.

CH₃I, NaH, HMPA
80 %

15

DDQ
CH₂Cl₂/H₂O
61 %

16

lower yields were obtained in the second step with DMF or the less toxic tetramethylurea as solvents.

Stepwise deprotection with DDQin moist dichloromethane followed by 1 N HCl/THF returned soraphen $A_{1\alpha}$ (1) in good yield. Alternatively the reverse order of reagents worked equally well with 14 as intermediate.

The doubly protected soraphen 16 turned out to be extremely stable towards nucleophilic attack. It resisted heating in concentrated sodium hydroxide solution, and even n-butyl lithium or the super base n-butyl lithium/potassium tert.-butanolate [11] did not add to the carbonyl group nor abstract the α-proton, as judged by quenching the reaction mixture with D_2O. However, with the complex base sodium amide/sodium tert.-butanolate [12], a rapid elimination of methanol occurred and the α,β-unsaturated lactone 17 was isolated in 90% yield. 17 proved to be stable towards acidic or basic hydrolysis.

Cleavage and reconstruction of the lactone bond

The lactone bond in 16 was eventually cleaved by reduction with LAH in refluxing THF to the diol 18. Protection of the 17-hydroxyl group as the TMS ether 19 was achieved in two steps only in moderate yield due to low selectivity of the hydrolysis step. Consecutive oxidation of 19 with perruthenate/N-methyl morpholine-N-oxide and sodium chlorite afforded soraphen acid (21) via the aldehyde 20. It should be mentioned that the aldehyde 20 resisted oxidation by several common oxidants such as Cr(VI)-reagents, silver oxide, hydrogen peroxide and potassium permanganate. After protection of soraphen acid (21) as its TBDMS-ester, the bromoacid 22 was formed smoothly by reaction with 1-bromo-1-dimethylaminoisobutene followed by hydrolysis of the silyl ester. Finally cyclization of 22 to protected soraphen 16 with cesium carbonate in DMF (*Kellogg's* method [13]) was achieved in 45% yield over four steps.

According to the ^{1}H-NMR spectrum of the crude reaction product, the conversion of the 17-hydroxyl to the lactone proceeded with high stereospecificity by double inversion. Only small amounts of the 17-epimer 27 (*vide infra*) could be detected. After deprotection and preparative HPLC, soraphen $A_{1\alpha}$ (1) was obtained which was identical to the natural compound by chromatographic and spectroscopic methods and biological activity [14].

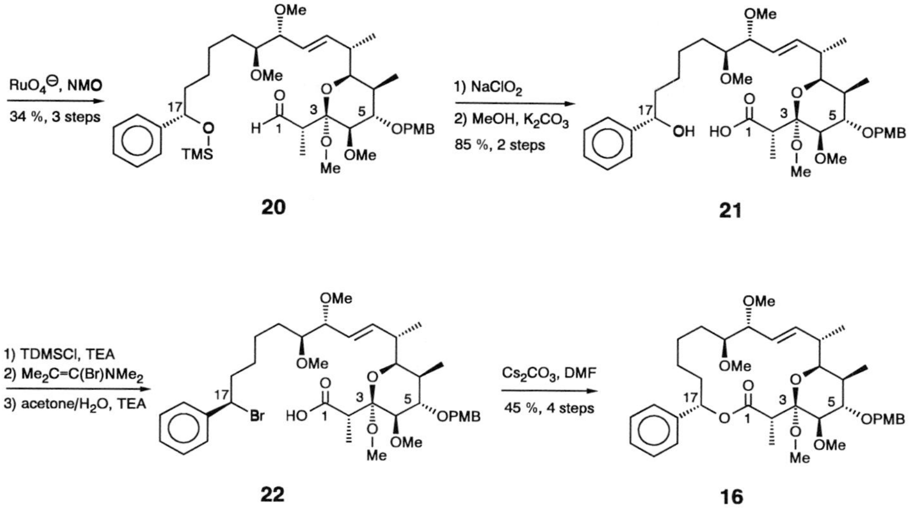

20

21

22

16

Alternatively cyclization of hydroxy acid **21** by the *Corey* [15] or *Yamaguchi* [16] method failed completely. After prolonged reaction time in the presence of excess DMAP, only activation of the carboxyl group and slow independent reaction at the 17-hydroxyl were observed. Highly activated carboxyl derivatives formed by reaction with mesyl chloride or thionyl chloride made an electrophilic attack at the 4-methoxyl group. The intermediate acyloxonium ion was demethylated to give the γ-lactone (**23**). Depending on an excess of reagents, the 17-hydroxyl group was sometimes modified.

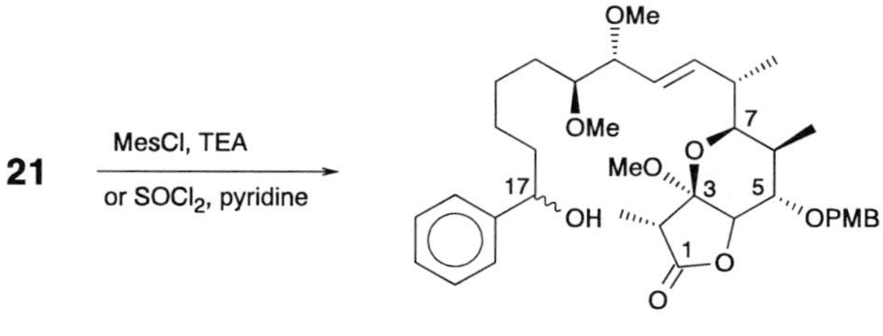

23

A common feature of protected soraphen derivatives is the very low reactivity of the C-1 carbonyl group as observed in derivatives of **21**, the aldehyde **20** and the lactone **16**. This is obviously caused by the highly substituted ß-carbon with a number of lone pair electrons on oxygen rather than the α-methyl group.

Preparation of 17-epi-soraphen A$_{1\alpha}$ (27)

When the diol **18** was oxidized without protection of the 17-hydroxyl group, the 17-keto acid **24** was obtained in high yield. For large scale preparation, oxidation with the pyridine·SO$_3$ complex according to the procedure described by Parikh and Doering [17] was more economical than perruthenate/NMO. Sodium borohydride reduction of the keto acid **24** gave a quantitative yield of soraphen acid (**25**) as a 1:1 mixture of diastereomers. However, since separation of the mixture by chromatography was not successful, it was converted as described above to a mixture of the lactones **26** in four steps. Deprotection yielded 17-epi-soraphen A$_{1\alpha}$ (**27**) and soraphen A$_{1\alpha}$ (**1**) which were easily separated by HPLC. The 17-epimer **27** proved to be slightly less polar on TLC, and significant changes in the 1H-NMR spectrum were observed for H-5 ($\Delta\delta$ -0,36), H-7 ($\Delta\delta$ +0,62), H-9 ($\Delta\delta$ -0,39) and H-17 ($\Delta\delta$ +0,08). Whereas synthetic **1** showed full antifungal activity, 17-epi-soraphen A$_{1\alpha}$ (**27**) was more than 10^3 less active against *Candida albicans* and the acetyl CoA carboxylase from *Ustilago maydis*.

The preparation of 17-keto-soraphen acid (**24**) opens the way to a variety of semi-synthetic soraphens with substituents R replacing the phenyl group. It is hoped that the knowledge of the basic chemistry of soraphens and the successful cyclization of soraphen acid to soraphen A$_{1\alpha}$ will assist work on the total synthesis of this fascinating group of natural fungicides.

Acknowledgement: We thank Dr. K. Gerth, Dr. T. Stammermann, T. Gehrs for biological and enzymatic tests and Dr. A.G. O'Sullivan for helpful discussions. This work was generously supported by Ciba-Geigy AG.

References

1. Article No. 51 on antibiotics from gliding bacteria. Article No. 50: D. Schummer, H. Irschik, G. Höfle, *Liebigs Ann. Chem.* in print.

2. H. Reichenbach, G. Höfle, H. Augustiniak, N. Bedorf, E. Forche, K. Gerth, H. Irschik, R. Jansen, B. Kunze, F. Sasse, H. Steinmetz, W. Trowitzsch-Kienast, EP 282455; *C.A.* **111**, 132597 v (1988).

3. K. Gerth, N. Bedorf, G. Höfle, H. Reichenbach, *J. Antibiot.*, in preparation.

4. N. Bedorf, B. Böhlendorf, E. Forche, K. Gerth, G. Höfle, H. Irschik, R. Jansen, B. Kunze, H. Reichenbach, F. Sasse, H. Steinmetz, W. Trowitzsch-Kienast, EP 358606; *C.A.* **115**, 112809 v (1990).

5. L. Pridzun, doctoral thesis, Technical University Braunschweig (1991).

6. M. Sutter, B. Böhlendorf, N. Bedorf, G. Höfle, EP 359706; *C.A.* **113**, 131887 u (1990).

7. B. Böhlendorf, doctoral thesis, Technical University Braunschweig (1991).

8. B. Böhlendorf, N. Bedorf, G. Höfle, D. Schummer, M. Sutter, EP 358608; *C.A.* **113**, 78893 q (1990). B. Böhlendorf, N. Bedorf, G. Höfle, M. Sutter, EP 358607; *C.A.* **113**, 41213 s (1990).

9. R. Brouillard, J.-E. Dubois, *J. Org. Chem.* **1974**, *39*, 1137.

10. After short reaction time the enol ether with the original configuration at C-2 can be isolated as intermediate in 20% yield.

11. M. Schlosser and S. Strunk, *Tetrahedron Lett.* **1984,** *25*, 741.

12. M. C. Carre, G. Ndebeka, A. Riondel, P. Bourgasser and P. Caubère, *Tetrahedron Lett.* **1984**, *25*, 1551.

13. W. N. Kruizinga, R. M. Kellogg, *J. Am. Chem. Soc.* **19981**, *103*, 5183.

14. In addition to **1** and other not identified byproducts of the cyclisation reaction 7% to 12% of the 17-epimer **27** was estimated by analytical HPLC after deprotection. Presumably some razemisation may occurr in the transformation of the benzylic alcohol to the bromide depending on slightly different reaction conditions.

15. E. J. Corey, K. C. Nicolaou, *J. Am. Chem. Soc.* **1974**, *96*, 5614.

16. I. Inanaga, K. Hirata, H. Saeki, T. Katsuki, M. Yamaguchi, *Bull. Chem. Soc. Jpn.* **1979**, *52*, 1989.

17. J. R. Parikh, W. von E. Doering, *J. Am. Chem. Soc.* **1967**, *89*, 5505.

Expanding the Therapeutic Potential of Macrolide Compounds

Herbert A. Kirst

Natural Products Research Division, Eli Lilly and Company,
Lilly Corporate Center, Indianapolis, Indiana, U.S.A. 46285

Summary. The macrolide class continues to provide new opportunities for the discovery of potentially useful therapeutic agents in human and veterinary medicine. Subsequent to the discovery of tilmicosin and its expanded antibiotic spectrum, new derivatives of tylosin and erythromycin have been prepared in an effort to further broaden the *in vitro* biological activities and/or to improve the pharmacokinetic features of the parent compounds. Certain derivatives of tylosin have been found that exhibit *in vitro* antifungal as well as antibacterial activity, and some derivatives of erythromycin have been prepared for non-antimicrobial applications. During these synthetic studies on macrolides, a selective N-demethylation of tilmicosin was achieved using $K_3Fe(CN)_6$ and a stereospecific reductive cyclization of erythromycin was accomplished with $NaBH_3CN + ZnCl_2$.

Introduction

A variety of semi-synthetic derivatives of macrolide antibiotics have now been discovered which expand the antimicrobial spectrum and/or modify the pharmacokinetic and pharmacological properties of the parent compounds [1-3]. Several of these newer semi-synthetic macrolides with improved features have been successfully developed as therapeutic agents in either human or veterinary medicine. Macrolides constitute a very large class of fermentation-derived products and exhibit an extensive array of structural variations and biological activities [4]. As part of a comprehensive program to discover additional macrolides with potential therapeutic utility, the synthesis and broad biological evaluation of new derivatives of tylosin and erythromycin have been pursued.

Selective 3'-N-Demethylation of Tilmicosin

Tilmicosin (**1**) is a semi-synthetic derivative of tylosin that has significantly greater activity against gram-negative bacteria such as *Pasteurella hemolytica* and *P. multocida* [5, 6]. Due to its good *in vitro* antimicrobial activity and long *in vivo* half-life, tilmicosin effectively treats bovine respiratory disease (BRD) in cattle with a single injection [7, 8]. The principal metabolite of tilmicosin is its 3'-N-demethyl derivative (**2**) [9]. Synthesis of an authentic sample of **2** was undertaken to confirm its structure and to provide larger quantities of material.

1 (Tilmicosin)	$\longrightarrow$	2 (Metabolite)
R = CH$_3$	K$_3$Fe(CN)$_6$	R = H
	KOH - *i*-PrOH - H$_2$O	
	5° to r.t., 6 hr.	

The most efficient synthesis of **2** would be via direct conversion from tilmicosin itself. However, initial attempts to N-demethylate tilmicosin employing methodolgy used for erythromycin (I$_2$, NaOAc, pH 9, MeOH, 47°) [10] were unsuccessful. No reaction occurred when 1 equiv. of iodine was used, whereas complex mixtures were obtained when excess iodine was added, presumably due to reactivity at sites in addition to 3'-N-demethylation. The desired metabolite was obtained via a multi-step route from desmycosin which involved ketalization (EtOH, TsOH), 3'-N-demethylation (I$_2$, NaOAc, pH 8, aq. MeOH, 50°, 2 hr), deketalization (0.1N HCl, MeOH), and reductive amination (3,5-dimethylpiperidine, then NaBH$_3$CN, MeOH). In this case, the process employing iodine with the diethyl ketal of desmycosin was successful. In the final reaction, a second product was simultaneously produced and identified as a dimeric moiety arising from reductive amination of N-demethyldesmycosin and N-demethyltilmicosin. A careful search of the literature uncovered a 1951 publication reporting N-demethylation of water-soluble secondary amines with K$_3$Fe(CN)$_6$ [11]. Although this procedure could not be directly applied to tilmicosin due to its relative insolubility in water at high pH, the procedure was modified by addition of a water-miscible organic cosolvent (such as *i*-PrOH) that was compatible with the other reagents. Thus, selective N-demethylation of tilmicosin was achieved in 42% yield (not optimized) on a 2 gm scale. The product was identical in all respects to the material isolated from prior metabolism studies [9].

Synthetic Strategies for 16-Membered Macrolides

Derivatives of tylosin and related macrolides in which the aldehyde function has been modified exhibit enhanced oral bioavailability and higher and more prolonged concentrations of antibiotic activity *in vivo* [12, 13]. It has also been shown that derivatives of the demycarosyl compounds (desmycosin and lactenocin) exhibit greater activity against gram-negative bacteria than do the analogous derivatives of tylosin or macrocin [12, 13]. Consequently, 20-modified derivatives of desmycosin and lactenocin represent good starting materials for further chemical modification in the search for new 16-membered macrolides possessing potentially useful biological activities.

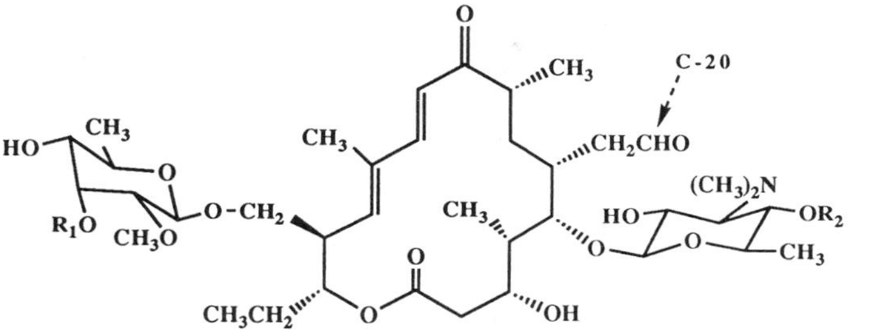

Tylosin: $R_1 = CH_3$; $R_2 = \alpha$-L-mycarosyl

Macrocin: $R_1 = H$; $R_2 = \alpha$-L-mycarosyl

Desmycosin $R_1 = CH_3$; $R_2 = H$

Lactenocin $R_1 = H$; $R_2 = H$

Three separate but interrelated strategies are envisioned for modifying the "left side" of 20-modified derivatives of desmycosin and lactenocin: 1. Conduct functional group manipulations on the existing saccharides; 2. Replace the naturally occurring sugars with other saccharide moieties; or 3. Replace the natural sugars by non-saccharide substituents. An initial survey of the literature indicates that modifications of this "left side" of tylosin-related macrolides can substantially affect *in vitro* activity [2]. Thus, the creation of an appropriately modified 20,23-bis-modified derivative might yield a compound in which both pharmacokinetics and antimicrobial potency have been optimized.

A wide variety of derivatives of the naturally occurring sugars (mycinose and 3-O-demethylmycinose) have previously been synthesized (strategy #1) and shown to exhibit good antibacterial activity both *in vitro* and *in vivo* [14, 15]. One of the most interesting compounds within this series was one with a 3'',4''-unsaturated sugar. As depicted below, the diol in 3-O-demethylmycinose was converted into a thionocarbonate group under standard conditions and was subsequently eliminated using either 1,3-dimethyl-2-phenyl-1,3,2-diazaphospholidine [16] or, in higher yield, bis(cyclooctadiene)nickel [17]. This conversion was readily applied to the synthesis of a pair of derivatives in which the C-20 substituent was either the 20-dihydro-20-O-phenyl group, obtained via Mitsunobu chemistry [12], or the 20-deoxo-20-(3,5-dimethylpiperidin-1-yl) group, prepared via reductive amination procedures [5].

Broad screening of these two compounds along with other macrolide derivatives unexpectedly uncovered *in vitro* antifungal activity. MIC values were determined following incubation for 48 hr at 35° by microtiter dilution tests using RPMI medium containing 10% fetal calf serum. For compound **4**, MICs were 5.0 and 2.5 µg/ml vs. *Candida albicans* and *Trichophyton mentagrophytes*, respectively. In contrast, compound **3** showed much weaker activity of 50 µg/ml vs each organism. Both compounds showed only weak activity against *Aspergillus flavus*, with each compound giving an MIC of 50 μg/ml. For comparison, amphotericin B gave MICs of 0.15 μg/ml against *C. albicans* and *A. flavus*, while tolnaftate had an MIC of 0.04 μg/ml against *T. mentagrophytes*.

The moderate antifungal activity of **4** was completely unanticipated. Although conventional 16-membered macrolides such as tylosin are potent antibacterial agents, they have not previously been reported to exhibit activity against fungi. However, a wide variety of other polyketide-derived macrocyclic compounds do exert antifungal activity, such as the polyenes (which include amphotericin B and nystatin), rhizoxin, rustmicin, and Sch 38516 [18-20].

Antifungal SAR for Macrolide Derivatives

Evaluation of a wide variety of derivatives of desmycosin and lactenocin failed to uncover any compound exhibiting better antifungal activity than the initial lead compound (**4**). Consequently, the strategy of replacing mycinose by non-saccharide substituents (strategy #3) was investigated. Using the readily available 5-O-mycaminosyltylonolide (OMT) as starting material [21], a large number of derivatives modified at both C-20 and C-23 were synthesized and tested. *In vitro* activity appeared to be greatest when the aldehyde at C-20 was modified by reductive amination. The hydroxyl group of OMT was converted into O-acyl, O-sulfonyl, O-tetrahydropyranyl, or O-aryl derivatives. Other modifications at C-23 included the replacement of oxygen as heteroatom by sulfur (S-aryl or S-alkyl), nitrogen (tertiary amines) or halogen. These compounds were synthesized by well known methods of Sn2 displacement reactions (via 23-O-triflate or 23-iodo-23-deoxy intermediates) and Mitsunobu reactions [22].

C-23 Substituent (X)	In Vitro Antifungal Activity: MIC (µg/ml)		
	C. albicans	*T. mentagrophytes*	*A. flavus*
O-(Tetrahydropyranyl)	1.25	5	20
O-Phenylsulfonyl	1.25	5	20
O-Phenyl	1.25	1.25	10
S-Phenyl	0.312	0.625	5
S-(2-Pyridyl)	2.5	2.5	2.5
S-(2-Pyrimidinyl)	10	10	40
S-(Cyclohexyl)	5	2.5	5
N-(3,5-Dimethylpiperidinyl)	0.312	2.5	2.5
N-(Octahydroindolyl)	0.625	1.25	5
Iodo	2.5	5	20

Despite good *in vitro* activity, all of the compounds failed to treat infections in mice caused by *C. albicans*. Numerous variations were made in the route of administration (s.c., i.p., p.o.), dosage schedule (number of doses, times of administration pre- and post-infection), formulation (solvent, pH), etc., but no evidence of efficacy was observed up to 100 mg/kg. In contrast, amphotericin B readily cured the infections at approximately 4 mg/kg x 3 i.p. At the high doses, evidence of compound-related toxicity was noted; subsequent testing in uninfected mice confirmed that several derivatives possessed an LD_{50} in mice near 100 mg/kg i.p. The lack of *in vivo* antifungal activity was probably not due to unfavorable pharmacokinetic properties, since many of the derivatives possessed good *in vitro* antibacterial activity and successfully treated standard bacterial infections in mice. Consequently, it was concluded that this series was another example of compounds possessing only *in vitro* antimicrobial activity. The mechanism by which these compounds exert their antifungal activity has not been determined.

Stereospecific Reductions of Erythromycin Derivatives

The intramolecular cyclization of erythromycin to its 8,9-anhydro-6,9-hemiketal derivative **5** has long been known to result in substantially diminished antimicrobial activity [23]. More recently, attention has focused on **5** and related compounds due to their potent gastrointestinal prokinetic activity [24]. Trans-lactonization of **5** to the interesting ring-contracted analog **6** has previously been reported [25, 26].

Hydrogenation of the 8,9-double bond within the dihydrofuran ring system of the 14-membered macrolide **5** proceeded from the back (α) face, yielding the 8,9-α-*cis*-dihydro derivative **7** [27]. Although another group had initially reported β-stereochemistry for this product [28], that assignment has subsequently been changed and is now in agreement with ours [29]. In contrast, hydrogenation of **6** under identical conditions unexpectedly produced addition of hydrogen to the upper (β) face, yielding the 8,9-β-*cis*-dihydro derivative **9**. The addition of hydrogen to opposite faces of **5** and **6** was confirmed by translactonization of **7** to **8**; the latter ring-contracted derivative was different than **9**, the product of direct hydrogenation of **6**.

$S_1 = β$-D-mycaminosyl; $S_2 = α$-L-cladinosyl

The structures of these products were established by detailed NMR studies [27]. Their proton and ^{13}C spectra were fully assigned and their relative stereochemistries were then determined from measurements of coupling constants and results from NOE experiments, which were compared with probable orientational conformations. The resultant stereochemical conclusions were in good agreement with results from molecular modeling experiments in which favorable conformations were derived by energy minimizations, using MacroModel 3D (version 3.5X) on a Cray 2 supercomputer [30].

The fourth compound in this series was the 8,9-β-*cis*-dihydro derivative **10**. It was synthesized by a novel reductive cyclization of erythromycin A with $NaBH_3CN$ and ZnI_2, as depicted below. The exact mechanism of deoxygenation following the likely acid-catalyzed intramolecular cyclization has not yet been determined. Zinc salts and borohydrides under these conditions have been used to deoxygenate activated groups such as aryl ketones and benzylic, allylic or tertiary alcohols [31]. Furthermore, certain glycosides have been converted to their 1-deoxy derivatives by the combination of borohydrides and acids [32].

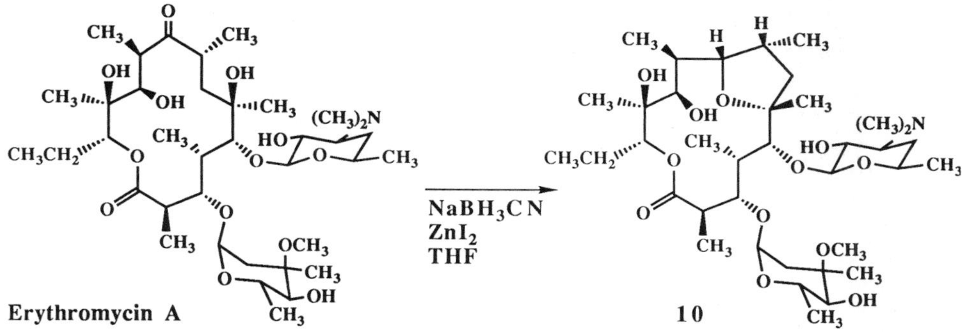

Erythromycin A

The antibacterial activity of these dihydro derivatives separated into two groups [27]. The 14-membered bicyclic derivatives **5, 7**, and **10** were comparable in inhibiting gram-positive bacteria such as staphylococci and streptococci, although substantially less active than erythromycin A itself. The activity of 12-membered bicyclic derivatives **6, 8**, and **9** were also similar to each other, but much less than for their 14-membered analogs. Consequently, neither the presence nor absence of the 8,9-double bond nor the stereochemistry of the 8,9-dihydro derivatives was critical for the relative degree of antibacterial activity. The most important factor was the type of macrolide ring: a monocyclic ring (erythromycin) was superior to all 14-membered bicyclic moieties, which were superior to all 12-membered bicyclic systems.

Acknowledgements

The author is indebted to his numerous colleagues at the Lilly Research Laboratories who have contributed to these projects, especially Mr. L. C. Creemer and Dr. A. Donoho on the tilmicosin metabolite work; Mr. J. P. Leeds, K. E. Willard, and D. J. Zeckner and Drs. M. Debono and R. S. Gordee on the antifungal project; and Mr. J. P. Leeds and J. W. Paschal and Drs. J. Martynow, F. T. Counter, and J. S. Gidda on the study of 8,9-dihydro derivatives of erythromycin.

References

1. H. A. Kirst, Annu. Rep. Med. Chem. **25** (1990) 119-128.
2. H. A. Kirst in G. Lukacs and M. Ohno (eds.), Recent Progress in the Chemical Synthesis of Antibiotics, Springer-Verlag, Heidelberg, 1990, pp. 39-63.
3. H. A. Kirst, Prog. Med. Chem. **30** (1993), manuscript accepted for publication.
4. S. Omura (ed.), Macrolide Antibiotics: Chemistry, Biology, and Practice, Academic Press, Orlando, Fla., 1984.
5. M. Debono, K. E. Willard, H. A. Kirst, J. A. Wind, G. D. Crouse, E. V. Tao, J. T. Vicenzi, F. T. Counter, J. L. Ott, E. E. Ose, and S. Omura, J. Antibiot. **42** (1989) 1253-1267.
6. E. E. Ose, J. Antibiot. **40** (1987) 190-194.
7. R. Laven and A. H. Andrews, Vet. Record **129** (1991) 109-111.
8. T. Picavet, E. Muylle, L. A. Devriese, and J. Geryl, Vet. Record **129** (1991) 400-403.
9. A. L. Donoho, T. D. Thomson, and D. D. Giera in D. H. Hutson, D. R. Hawkins, G. D. Paulson, and C. B. Struble (eds.), Xenobiotics and Food-Producing Animals, ACS Symposium Series #503, American Chemical Society, Washington, D. C., 1992, pp. 158-167.
10. L. A. Freiberg, U. S. Patent 3,725,385 (April 3, 1973).
11. T. D. Perrine, J. Org. Chem. **16** (1951) 1303-1307.
12. H. A. Kirst, J. E. Toth, M. Debono, K. E. Willard, B. A. Truedell, J. L. Ott, F. T. Counter, A. M. Felty-Duckworth, and R. S. Pekarek, J. Med. Chem. **31** (1988) 1631-1641.
13. H. A. Kirst, K. E. Willard, M. Debono, J. E. Toth, B. A. Truedell, J. P. Leeds, J. L. Ott, A. M. Felty-Duckworth, and F. T. Counter, J. Antibiot. **42** (1989) 1673-1683.
14. H. A. Kirst, M. Debono, G. D. Crouse, E. V. Tao, E. E. Ose, T. D. Thomson, T. Matsuoka, F. T. Counter, J. F. Quay, and S. Omura, 2nd Internat. Symp. on Chem. Synth. of Antibiotics and Related Microb. Prod., Oiso, Japan, Sept. 4-7, 1990, abstr. I-19.
15. H. A. Kirst, J. E. Toth, J. E. Wind, and J. P. Leeds, 6th European Symp. on Carbohyd. Chem., Edinburgh, Scotland, Sept. 8-13, 1991, abstr. B.103.
16. M. F. Semmelhack and R. D. Stauffer, Tetrahed. Lett. (1973) 2667-2670.
17. E. J. Corey and P. B. Hopkins, Tetrahed. Lett. **23** (1982) 1979-1982.
18. S. Omura and H. Tanaka in reference 4, pp. 351-404.
19. S. Omura in reference 4, pp. 509-552.
20. V. R. Hegde, M. G. Patel, V. P. Gullo, A. K. Ganguly, O. Sarre, M. S. Puar, and A. T. McPhail, J. Amer. Chem. Soc. **112** (1990) 6403-6405.
21. R. H. Baltz and E. T. Seno, Antimicrob. Agents Chemother. **20** (1981) 214-225.
22. H. A. Kirst, J. E. Toth, J. A. Wind, M. Debono, K. E. Willard, R. M. Molloy, J. W. Paschal, J. L. Ott, A. M. Felty-Duckworth, and F. T. Counter, J. Antibiot. **40** (1987) 823-842.
23. P. Kurath, P. H. Jones, R. S. Egan, and T. J. Perun, Experientia **27** (1971) 362.
24. S. Omura, K. Tsuzuki, T. Sunazuka, S. Marui, H. Toyoda, N. Inatomi, and Z. Itoh, J. Med. Chem. **30** (1987) 1943-1948.
25. H. A. Kirst, J. A. Wind, and J. W. Paschal, J. Org. Chem. **52** (1987) 4359-4362.

26. I. O. Kibwage, R. Busson, G. Janssen, J. Hoogmartens, H. Vanderhaeghe, and J. Bracke, J. Org. Chem. **52** (1987) 990-996.

27. H. A. Kirst in P. H. Bentley (ed.), Proc. 1st Intl. Symp. on Recent Advances in Chem. of Anti-Infective Agents, Cambridge, U.K., July 5-8, 1992.

28. K. Tsuzuki, T. Sunazuka, S. Marui, H. Toyoda, S. Omura, N. Inatomi, and Z. Itoh, Chem. Pharm. Bull. **37** (1989) 2687-2700.

29. Dr. T. Sunazuka, Sept., 1992, personal communication.

30. F. Mohamadi, N. G. J. Richards, W. C. Guida, R. Liskamp, M. Lipton, C. Caufield, G. Chang, T. Hendrickson, and W. C. Still, J. Compu\. Chem. **11** (1990), 440-467.

31. C. K. Lau, C. Dufresne, P. C. Belanger, S. Pietre, and J. Scheigetz, J. Org. Chem. **51** (1986) 3038-3043.

32. Y. Chapleur, P. Boquel, and F. Chretien, J. Chem. Soc. Perkin Trans. I (1989) 703-705.

Synthesis and Structure-Activity Relationships of Echinocandin Antifungal Agents

James V. Heck*, James M. Balkovec,
Regina Black, Milton L. Hammond and Robert Zambias,
Merck Research Laboratories
Rahway, NJ 07065

Summary: The echinocandins are naturally-occurring cyclic hexapeptides which exhibit antifungal activity. A series of analogs were prepared, both by derivatization of the natural product and by total synthesis, to define the minimum structure necessary for antifungal activity.

The echinocandins, of which echinocandin B **1** is a representative member, were independently isolated and characterized by a number of groups in the mid-1970's [1]. Subsequent research has established that these compounds owe their fungicidal activity to the inhibition of the synthesis of 1,3-β-glucan, a structural component of the fungal cell wall [2]. A limited number of structure-activity studies have been reported, primarily by the Lilly group [3]. These workers focused their attention on the lipophilic side chain, and prepared analogs by enzymatic deacylation and reacylation of echinocandin B. This effort led to the synthesis of cilofungin **2**, which exhibited reduced toxicity relative to echinocandin B and was selected for clinical study. Unfortunately, clinical trials of this compound were halted due to vehicle-related toxicity, and no antifungal agents of this type have reached the market.

In contrast to the extensive investigation of SAR in lipophilic side chain analogs, relatively little has been disclosed about structural analogs which differ in the cyclic hexapeptide nucleus. Our group became interested in these compounds when we discovered that a related lipopeptide isolated in our laboratories, L-688,786 **3**, exhibited potent activity in animal models of *Pneumocystis carinii* pneumonia (PCP), in addition to the activity against several species of *Candida* which is characteristic of this class [4]. Both PCP and *Candida* infections are problematic in immunocompromised patients, particularly those infected with HIV, so an agent with activity against both of these pathogens would be of considerable clinical interest. We therefore sought to explore structure-activity relationships in this class in more detail, with particular attention to modifications of the cyclic hexapeptide nucleus. The diversity of structure of the biologically active echinocandin natural products established that significant variation was permissible in the cyclic hexapeptide. For example, echinocandin C (**4**) and D (**5**), which lack some of the peripheral hydroxyl groups, both exhibit antifungal activity comparable to echinocandin B, and L-688,786 demonstrates that modification of one of the threonine residues is also consistent with biological activity. On the other hand, disruption of the cyclic hexapeptide, as in the base-catalyzed rearrangement product **6**, results in the complete loss of antifungal activity. The first goal of our studies was therefore the definition of the minimal hexapeptide pharmacophore in this class.

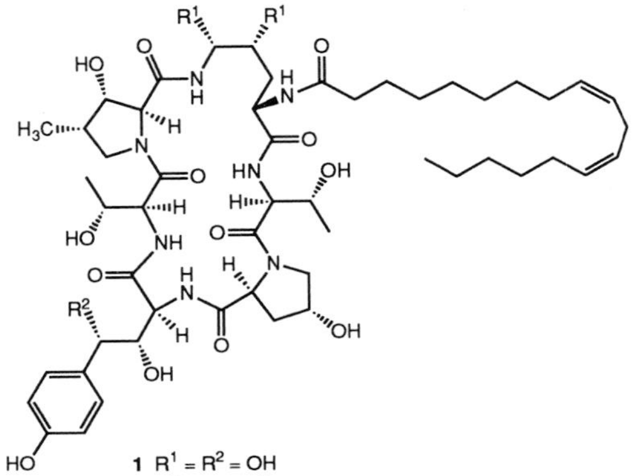

1 R^1 = R^2 = OH
4 R^1 = OH, R^2 = H
5 R^1 = R^2 = H

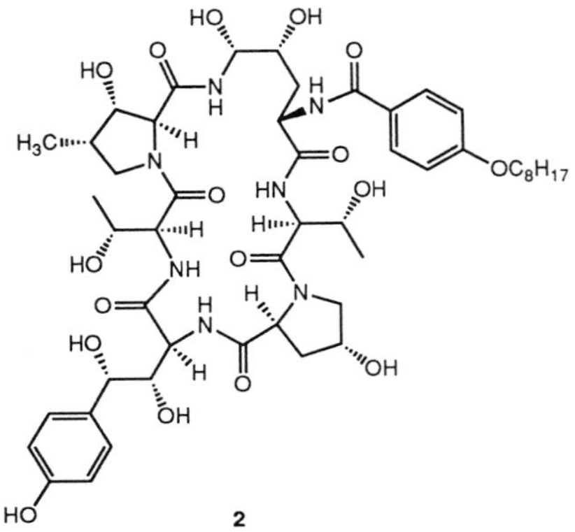

2

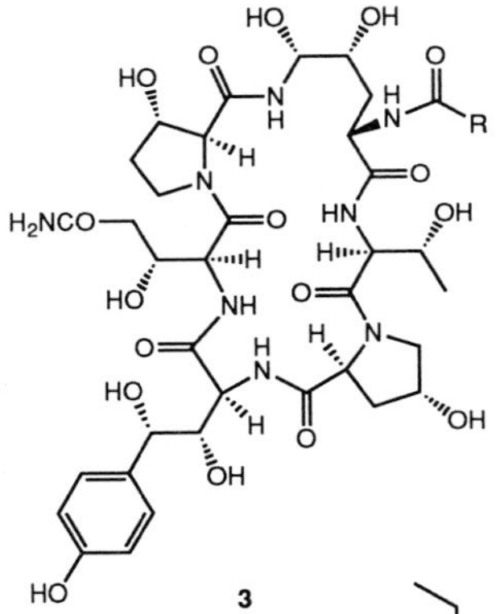

3   R =

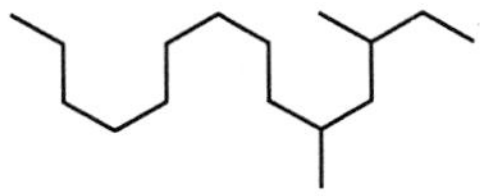

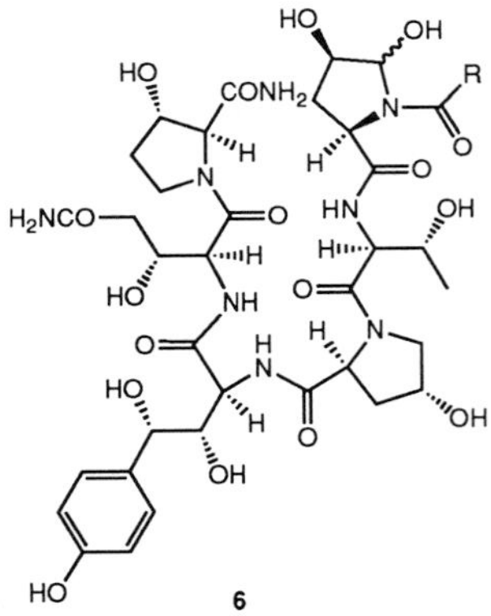

6

Total syntheses of the simplest member of this series, **5**, by Ohfune and Evans provided encouragement for the synthesis of simplified analogs by macrocyclization of hexapeptide precursors [5]. Two general strategies were employed for the assembly of the acyclic hexapeptide intermediates. The first, depicted in Scheme I, begins with resin-bound proline or proline analog and proceeds in a counterclockwise direction to the ornithine analog residue using BOC/carbodiimide chemistry. The terminal ornithine bears an α-BOC and a δ-ClCbz, which allows the lipophilic side chain to be introduced in the last resin-bound step, and deprotection of the δ-amine, secondary hydroxyl and phenol groups

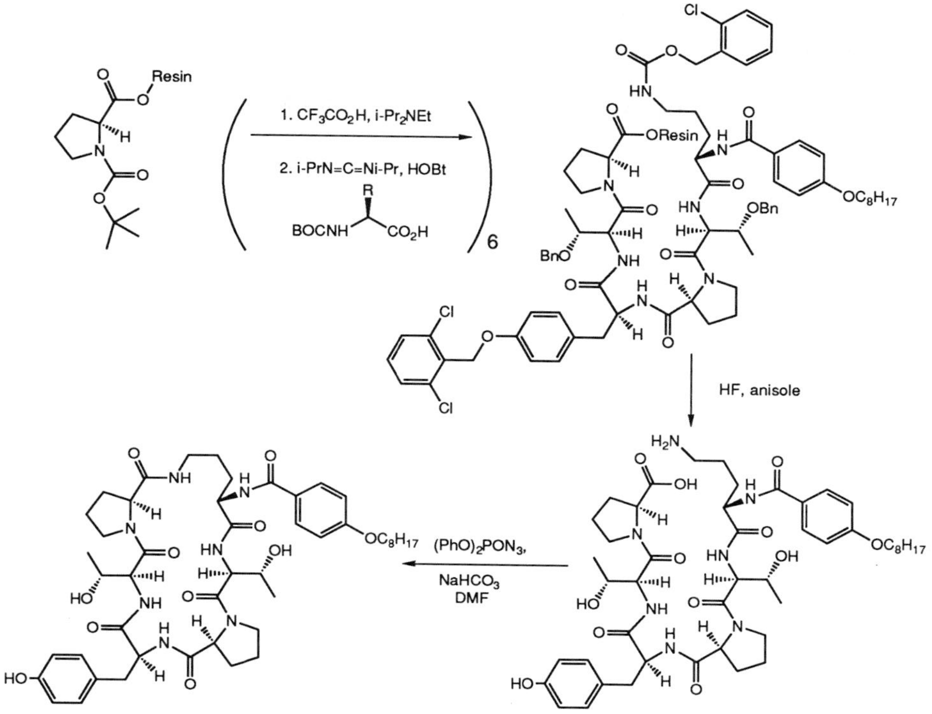

Scheme 1

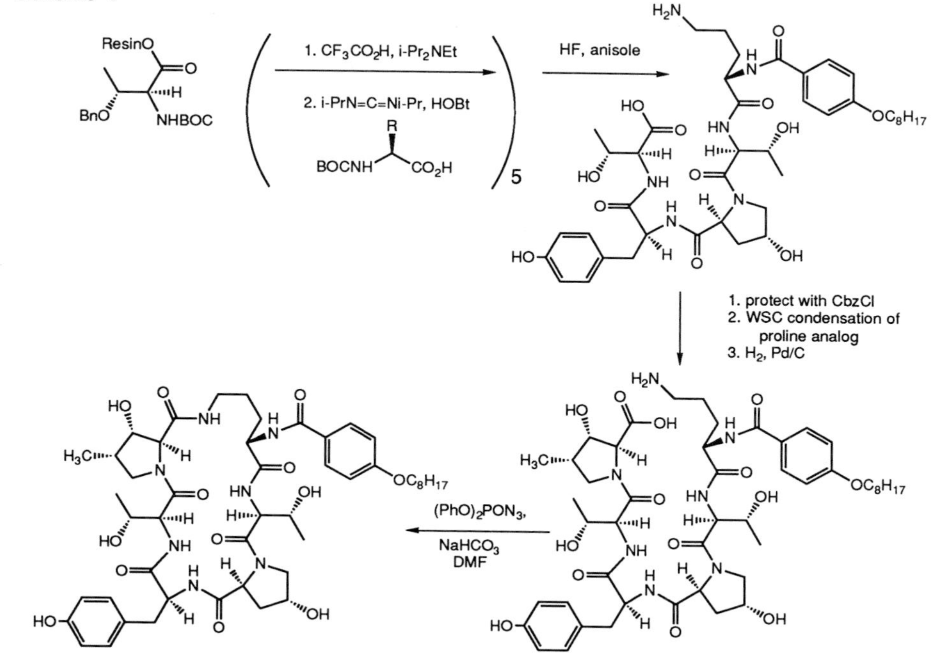

Scheme 2

occurs upon cleavage from the resin, affording a fully deprotected hexapeptide. Macrocyclization is accomplished by exposure to diphenyl phosphoryl azide and sodium bicarbonate in DMF. Analogs which contained the 3-hydroxy-4-methylproline residue were prepared by the sequence outlined in Scheme II, which makes more efficient use of this less accessible amino acid [6].

The first targets of this study were analogs wherein the dihydroxyhomotyrosine of echinocandin B was replaced by a tyrosine residue. All compounds of this type were devoid of activity, both in the in vitro antifungal assay and the glucan synthase inhibition assay, including **7**, which differs from the active echinocandin D nucleus by a CHOH group. When tyrosine was replaced by homotyrosine, both activity against the target enzyme and antifungal activity were observed, even in the simplest analog **8**, which lacks all of the proline substituents. These groups, while not essential, are not insignificant, as reintroduction of the 4-hydroxyproline (**9**) and 3-hydroxy-4-methylproline (**10**) increased potency in a stepwise fashion. Considerable latitude was possible in substitution of the latter residue, as analogs with 4-hydroxyproline and threonine in this position exhibited modest antifungal activity [6].

	7	8	9	10	cilofungin
R^1-R^3 = H		R^1= OH R^2-R^3 = H	R^1-R^2 = OH R^3 = CH$_3$		
Candida albicans MY 1055	> 128	4	4	1	0.5
C. tropicalis MY 1012	> 128	4	4	1	0.25
C. parapsilosis MY 1010	> 128	4	4	8	16
Crypt. neoformans MY 1051	> 128	>128	>128	>128	> 128
IC$_{50}$ *Candida* Glucan Synthesis	60	70	10	1	3

The four hydroxyl groups on the ornithine and homotyrosine residues of the echinocandin nucleus are considerably less important, as evidenced by the comparable activity of **10** and cilofungin **2**. The unusual hemiaminal functionality which closes the macrocycle in many of the naturally occurring echinocandins is unstable to both acid and base, and we sought methodology which would allow the selective removal of the labile hydroxyl groups in this position and in the benzylic position. Exposure of

3 to sodium cyanoborohydride in trifluoroacetic acid removed both of these more readily ionized hydroxyls, affording **11** in 56% yield.   Reduction of **3** with sodium triacetoxyborohydride in trifluoroacetic acid afforded the benzylic reduction product **12** as the major isomer in 38% yield, indicating that this hydroxyl is more labile in acid. In order to access the other mono-reduced isomer, the benzylic position was deactivated by conversion of the phenol to the methyl carbonate derivative **13**. Treatment with sodium cyanoborohydride in trifluoroacetic acid effected selective reduction at the hemiaminal, which after the removal of the carbonate protecting group, afforded **14** in 38% overall yield from **13** [7]. As anticipated from the synthetic analogs, only minor differences were observed in the antifungal activity of **3**, **11**, **12** and **14**. The analogs which lacked the hemiaminal function, **11** and **14**, did exhibit considerably greater chemical stability. While cilofungin decomposes in pH 8 buffer at room temperature with a half-life of 8.5 hr, there was essentially no loss of **11** under these conditions over 24 hr.

R = dimethylmyristoyl

13 $R^1 = OH$, $R^2 = CH_3CO$
14 $R^1 = R^2 = H$

The observation that biological activity is critically dependent upon the length of the homotyrosine resi and relatively insensitive to the removal of the chain hydroxyl groups suggested that the phenolic that the phenolic hydroxyl was important. Exposure of **3** to pentafluorophenyl trifluoromethanesulfonate in the presence of lithium hydroxide effected selective reaction at the phenol, affording triflate **15** in 49% yield. Hydrogenolysis of the sulfonate with a palladium catalyst provided the desired deshydroxy compound **16** in 99% yield. As anticipated, the antifungal activity of **16** was reduced at least sixteen-fold relative to the parent compound **3**.

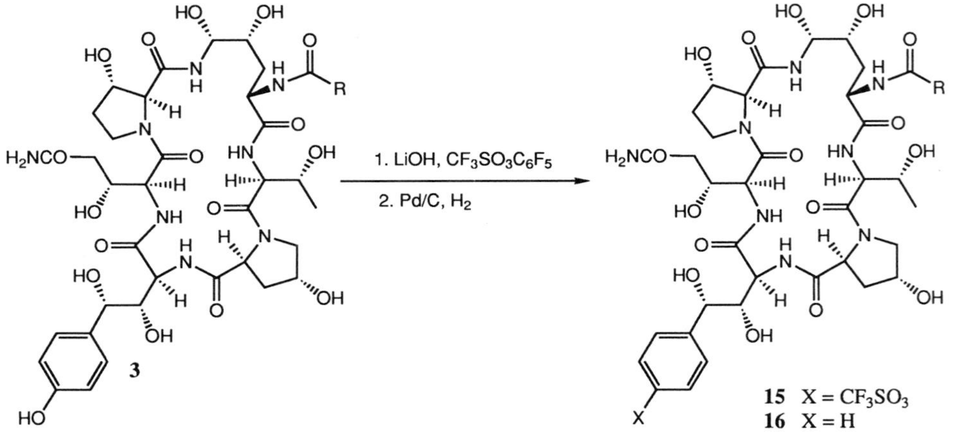

1. LiOH, $CF_3SO_3C_6F_5$

2. Pd/C, H_2

3

15 $X = CF_3SO_3$
16 $X = H$

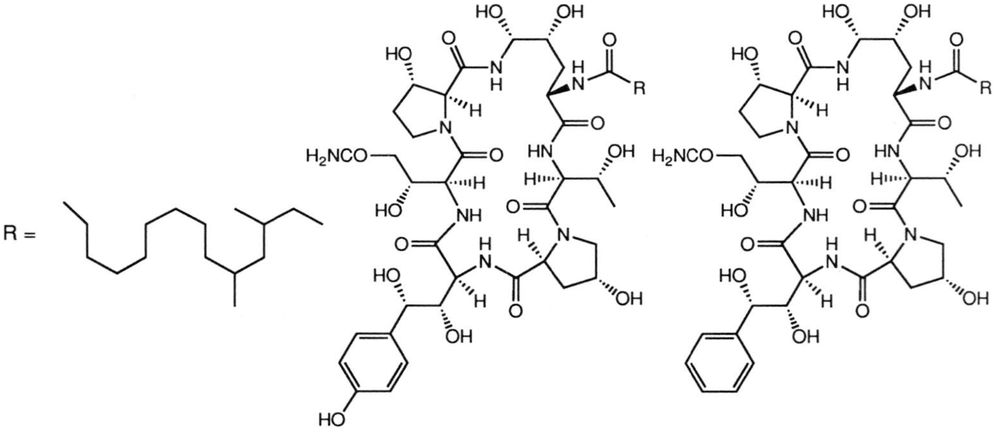

R =

	3	**16**
Candida albicans MY 1055	0.25	4
C. tropicalis MY 1012	0.06	1
C. parapsilosis MY 1010	4	> 128
Crypt. neoformans MY 1051	> 128	> 128
IC_{50} *Candida* Glucan Synthesis	0.1	> 10

These studies have established that significant variation is possible in the hexapeptide nucleus of echinocandin antifungal agents while retaining biological activity. The steric relationship of the homotyrosine phenol to the cyclic hexapeptide is an important determinant for antifungal activity.

Acknowledgements: The authors wish to acknowledge the contributions of Kenneth Bartizal, George Abruzzo, Jean Marrinan, Karl Nollstadt and Diane McFadden , who conducted the biological evaluation of the compounds described herein.

References

1. a) F. Benz, F. Knusel, H. Treichler, W. Voser, R. Nyfeler and W. Keller-Schierlein, *Helv. Chim. Acta* **57**, (1974) 2459-2477; b) C. Keller-Juslen, M. Kuhn, H.-R. Loosli, T. J. Pechter, H. P. Weber and A. v. Wartburg, *Tetrahedro*n Lett. (1976) 4147-4150.

2. C. S. Taft, and C. P. Selitrennikof, *J. Antibiot.* **43** (1990) 433-437; for a review see: J. S. Tkacz in *"Emerging Targets in Antibacterial and Antifungal Chemotherapy"*, J. Sutcliffe and N. N. Georgopapdakou Eds., Chapman and Hall, New York, 1992, pp. 495-523.

3. a) R. S. Gordee, D. J. Zeckner, L. C. Howard, W. E. Alborn and M. Debono, *Ann. N. Y. Acad. Sci.* **544** (1988) 294-301; b) M. Debono, B. J. Abbott, D. S. Fukuda, M. Barnhart, K. E. Willard, R. M. Molloy, K. H. Michel, J. R. Turner, T. F. Butler and A. H. Hunt, *J. Antibiot.* **42** (1989) 389-397 and references cited therein.

4. D. M. Schmatz, M. A. Romancheck, L. A. Pittarelli, R. E. Schwartz, R. A. Fromtling, K. H. Nollstadt, F. L. Van Middlesworth, K. E. Wilson, and M. J. Turner, *Proc. Natl. Acad. Sci. USA* **87** (1990) 5950-5954.

5. a) Y. Ohfune, N. Kurokawa, *J. Am. Chem. Soc.* **108** (1986) 6043-6045, b) D. A. Evans and A. E. Weber, *J. Am. Chem. Soc.* **109** (1987) 7151-7157.

6. R. A. Zambias, M. L. Hammond, J. V. Heck, K. Bartizal, C. Trainor, G. Abruzzo, D. M. Schmatz and K. M. Nollstadt, *J. Med. Chem.* **35** (1992) 2843-2855.

7. J. M. Balkovec and R. M. Blank, Tetrahedron Lett. 33 (1992) 4529-4532.

The Farmitalia Penem Episode

G. Franceschi, E. Perrone, M. Alpegiani, C. Battistini,
A. Bedeschi, G. Visentin, F. Zarini
Farmitalia Carlo Erba Srl, Erbamont Group
Via Carlo Imbonati, 24 - 20159 Milano, Italy

Summary: The development of ritipenem (FCE 22101, injectable antibiotic) and ritipenem acoxil (FCE 22891, orally absorbed) is a remarkable episode in the long history of penems. The evolution of synthetic methodologies leading to the selected products are presented with emphasis on the most recent efforts targeted to the ultimate production process. Price limitations imposed by the current market policy to oral agents make the direct approach to ritipenem acoxil compulsory; reduction to practice of this "reverse strategy" (prodrug first, drug second) is discussed. The ultimate frontiers of research in the original class of cephalosporin-like penems are represented by quinolonyl-penems endowed with a dual mode of action, and by a selected group of "multicharge penems" stable to renal dehydropeptidases and characterized by superior therapeutic efficacy in animal models.

The idea of gathering the reactivity determinants of penicillins and cephalosporins together into a single nucleus led the Woodward/Ciba Group [1] to effect the pioneering synthesis of penems, with initially disappointing results in terms of chemical stability and biological activity. Removal of the 7β-acylamino sidechain, which might have seemed quixotic prior to revelation by nature of "non-classical β-lactams" (clavulanic acid, olivanic acids), was the turning-point in early penem research [2]. Soon after, the discovery of thienamycin suggested grafting its peculiar 6-hydroxyethyl sidechain to the penem skeleton [3]. A favourable outcome ensued which gave fresh impetus to investigation into this class of compounds.

The conceptual contribution by the Farmitalia group [4] mostly resided in extending the analogy between penems (formally: nor-cephems) and cephalosporins by insertion of the C-3 substituents typical of the latter into the electronically equivalent C-2 position of the former. Rewardingly, this hypothesis addressed the synthesis of some excellent antibacterial agents. Among them we singled out for clinical development FCE 22101 (ritipenem) and its orally absorbed acetoxymethyl ester FCE 22891 (ritipenem acoxil) [5], which feature at C-2 the carbamoyloxymethyl group present in cefoxitin (Fig. 1). Other leading penems undergoing clinical trials are shown.

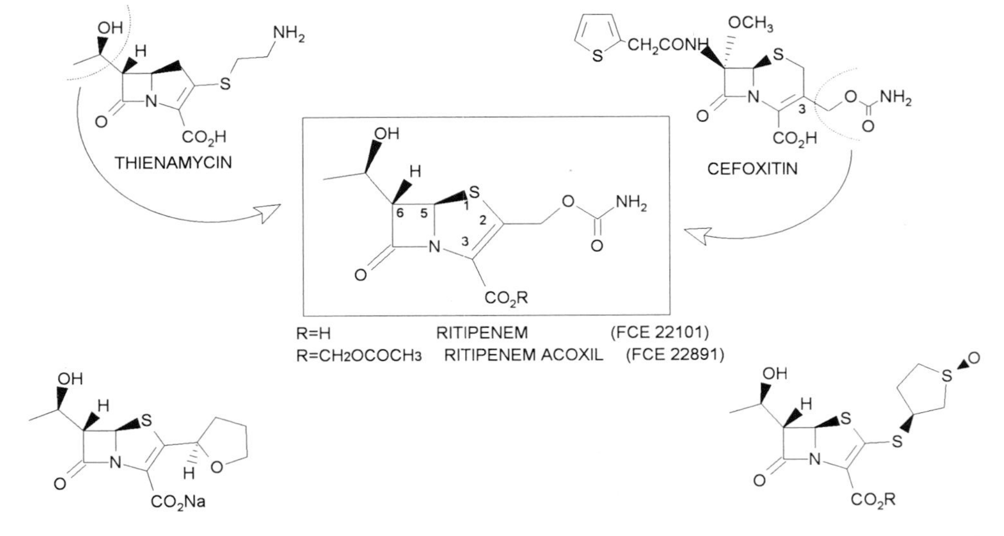
THIENAMYCIN
CEFOXITIN
R=H RITIPENEM (FCE 22101)
R=CH2OCOCH3 RITIPENEM ACOXIL (FCE 22891)
SUNTORY-YAMANOUCHI SY 5555
PFIZER Sulopenem (R=Na, CH2OCOCMe3)

Fig 1

The initial syntheses of FCE 22101 and FCE 22891, though entailing a multistep and low-yield sequence (ca. 1% from 6-APA), provided enough material to assess their remarkable biological properties. This historical route, depicted in Fig. 2, is a brief compendium of interesting reactions carried out on sensitive substrates, incorporating remarkable contributions from D. H. R. Barton (trapping with alkynes of the penicillin-derived sulfenic acid) and R. B. Woodward (thioester-phosphorane Wittig-type annulation). The 2-hydroxymethylpenem thus constructed is a pivotal intermediate for a variety of 2-CH_2X penem targets; in this instance, treatment with trichloroacetyl isocyanate and unmasking of the protected functions gave FCE 22101, easily transformed to FCE 22891 by reaction with bromomethyl acetate.

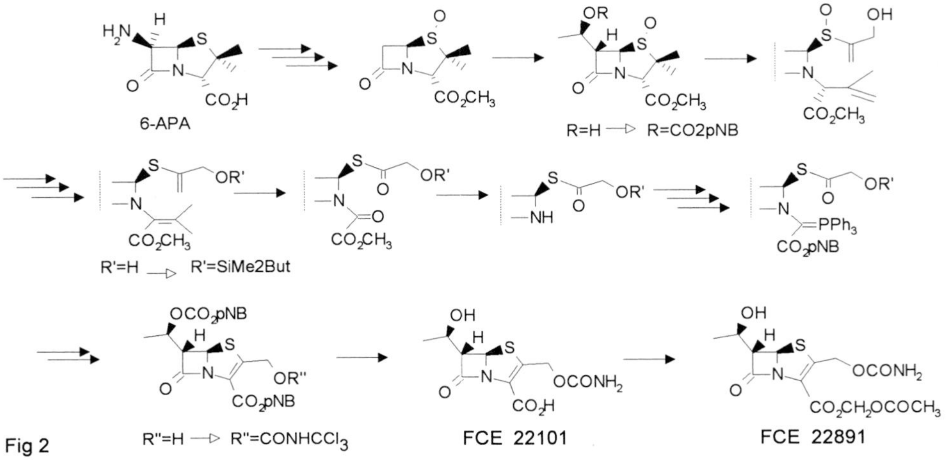
6-APA
R=H → R=CO2pNB
R'=H → R'=SiMe2But
R"=H → R"=CONHCCl3
FCE 22101
FCE 22891

Fig 2

Although aesthetically fascinating, this multistep synthesis proved impractical for large scale preparations. First, shortcuts were sought in the conversion of penicillins into azetidinyl thioesters, which avoids the tedious and unproductive removal/reconstruction of the N-appendage of secopenicillanates. Second, a real breakthrough in penem synthesis ensued from the discovery that oxalimido-thioesters, midway intermediates in the previous syntheses, could be directly cyclized upon heating in the presence of triethyl phosphite [6]. The mechanism of this reaction has been intensively investigated [7], with the involvement of carbene species and their dichotomous behaviour demonstrated [7,8] (Fig. 3). Following attack of P(OEt)$_3$ to the electrophilic oxygen of the oxalimido moiety and discharge of PO(OEt)$_3$, the fate of the generated carbene depends on the structural features of the substrate. In the case of azetidinyl-4-trithiocarbonates (Sankyo-Schering route), the thioxo group is trapped by the carbene and the resulting fused thiirane is desulfurized to the penem ring system. In the absence of the reactive thioxo moiety, intermolecular reaction with a second equivalent of P(OEt)$_3$ occurs. The resulting phosphite ylides, unusual surrogates of Woodward's phosphoranes, are likewise amenable to thermal condensation with the proximal thioester carbonyl. Under the reaction conditions, the two steps superimpose in what appears as a straightforward "dicarbonyl coupling".

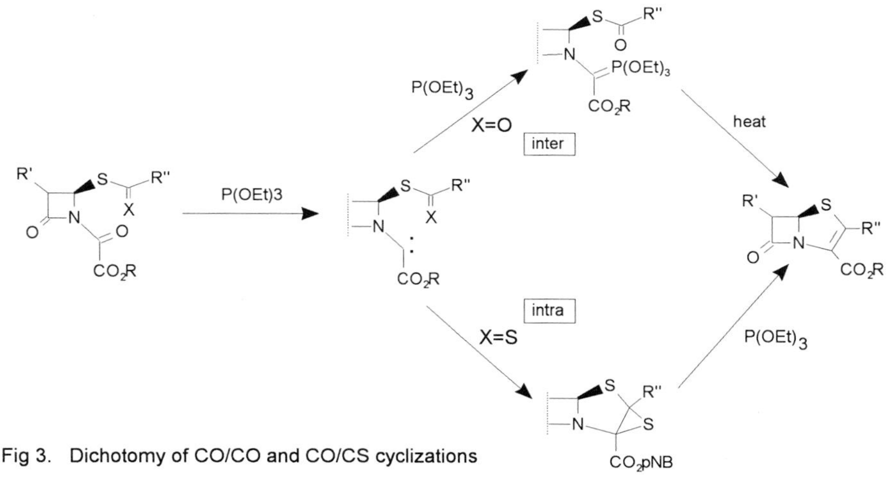

Fig 3. Dichotomy of CO/CO and CO/CS cyclizations

The third significant contribution to a more expeditious pathway from 6-APA to our target penems was a method for the direct conversion of penicillanates into azetidinyl-4-thioesters bearing a butenoate appendage at nitrogen, whose ozonolysis leads to the substrates of the dicarbonyl coupling reaction (see Fig. 4). A combination of a heavy metal salt and a strong non-nucleophilic base promotes cleavage of penams at the S-C$_2$ bond through β-elimination of a putative sulfonium species; the resulting mercaptides are then acylated *in situ* by simple addition of the appropriate acyl chloride [9,10]. These major results and other refinements (stereoselective introduction of the hydroxyethyl chain [11], temporary protection of the hydroxy group as the trimethylsilyl ether) led to a synthetic plan for FCE 22891 (Fig. 4), which clearly denotes the consistent progress made over the historical route. The bio-labile acetoxymethyl ester function can be safely utilized as a protecting group over the whole sequence, while its removal to provide the parent antibiotic FCE 22101, when desired, was efficiently accomplished by lipase-catalyzed hydrolysis [12].

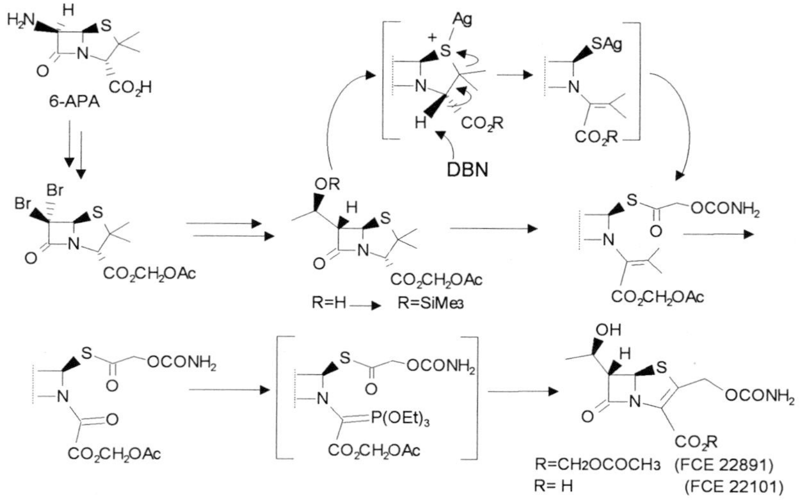

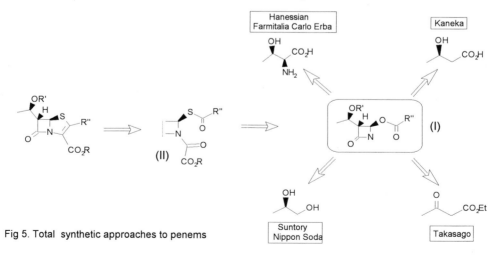

Fig 4.   Short synthesis of FCE 22891 and FCE 22101 from 6-APA

In addition to improved pathways entailing 6-APA as a chiral template, efforts were dedicated to total synthetic approaches that could benefit from exploitation of the dicarbonyl coupling reaction. In this context 3-hydroxyethyl-4-acyloxyazetidinones (I; Fig. 5) are, at one time, both targets and prominent intermediates. A fruitful collaboration with Prof. Steve Hanessian [13] allowed us to devise an expedient conversion, suitable for bulk preparations, from L-threonine to azetidinones I (R' = H or TBDMS, R'' = Ph). More recently, azetidinone I (R' = TBDMS, R'' = CH$_3$) was made commercially available for the industrial production of penems and carbapenems thanks to the contribution of three Japanese companies (Kaneka, Suntory/Nippon Soda and Takasago); as a consequence, exploitation of this synthon in the ultimate synthesis of FCE 22891 became imperative.

Fig 5. Total synthetic approaches to penems

By a new strategy (Fig. 6), ritipenem acoxil is obtained in only three steps from the azeti-dinone and two sophisticated building blocks derived from glycolic acid. The same strategy was applied to the synthesis of FCE 25199 (Fig. 7), a new orally absorbed penem featuring the bio-labile dioxolenone moiety which avoids the *in vivo* liberation of aldehydes [14].

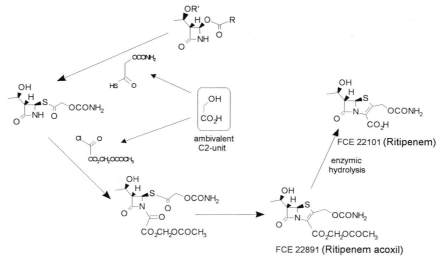

Fig 6.   Ultimate total synthesis of FCE 22891 and FCE 22101

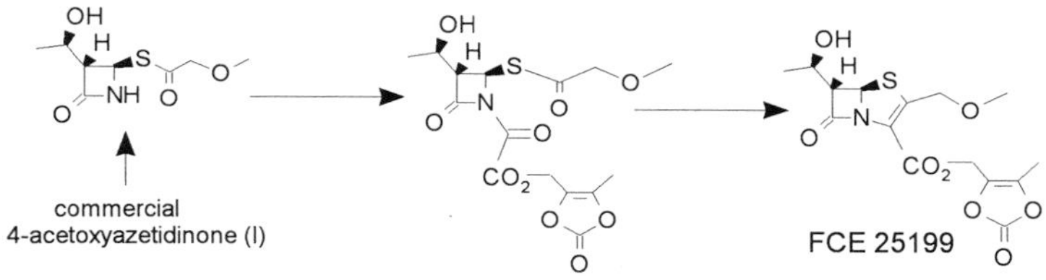

Fig 7.   Facile synthesis of a new orally absorbed penem, FCE 25199

Within the limited class of penems conceptually derived from cephalosporins, efforts to improve over ritipenem have led to two distinguished families of derivatives. The seminal observation that opening of the β-lactam ring of 2-CH_2X penems is accompanied by release of the X leaving group and of a common exomethylenethiazoline fragment [15] prompted the conception of "dual-action penems" (Fig. 8). Penems incorporating a latent antibacterial agent as the X group may discharge such agent as in the active form inside the bacterial cell. The quinolonyl-penems FCE 26600 and FCE 27070 (X= ester-linked and carbamate-linked ciprofloxacin, respectively; Fig. 9) are prominent examples of this new strategy, which impressively expands the antibacterial spectrum of penems to previously unattained levels [16].

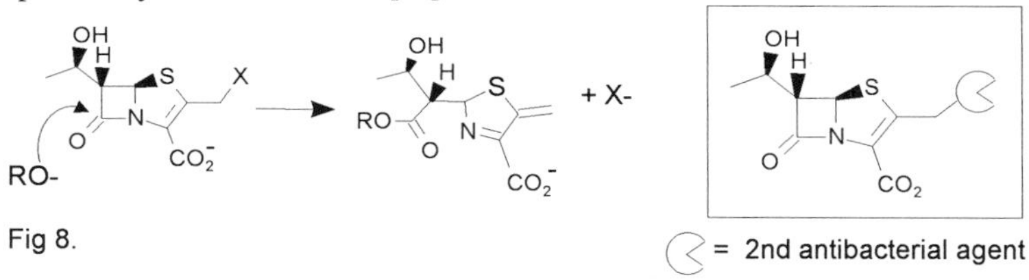

Fig 8.

= 2nd antibacterial agent

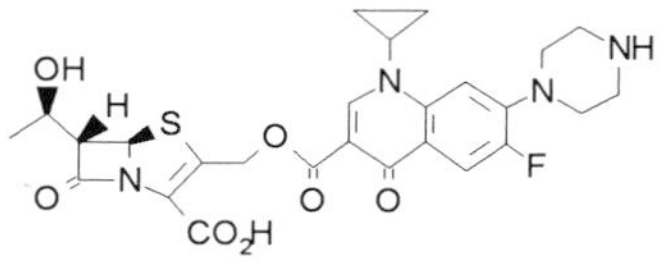

FCE 26600 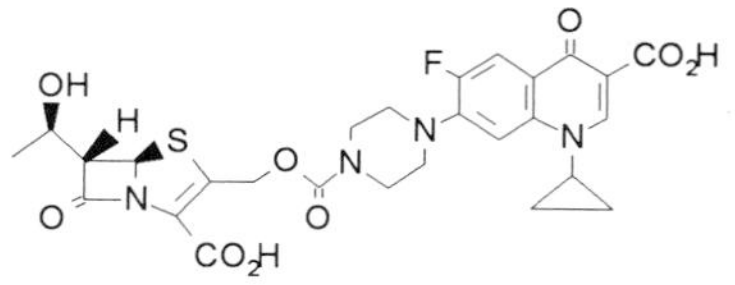   FCE 27070

Fig 9

A different stream of products resulted from the search for penems endowed with complete stability towards renal dehydropeptidases (DHP). In the 2-CH$_2$X penem class, zwitterions (X = quaternary ammonium) fulfilled this requirement [17] but lacked acceptable pharmacokinetic properties and caused convulsions in mice when tested at relatively high doses. Insertion of a phenylene spacer (2-C$_6$H$_4$CH$_2$X penems, X = quaternary ammonium) combined DHP stability with a superior pharmacokinetic profile [18], while side-effects were removed by further addition of an anionic group [19]. These "multicharge penems", exemplified by FCE 25165 and FCE 27104 (Fig. 10), proved one order of magnitude more effective than ordinary penems in treating experimental infections in mice.

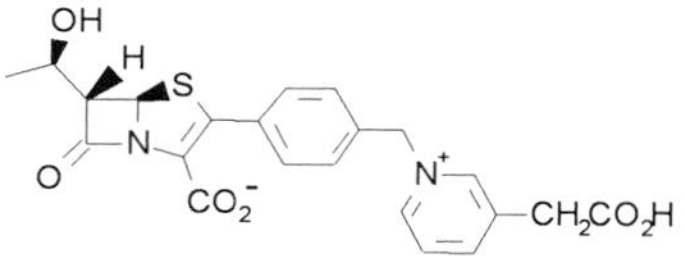

FCE 25165 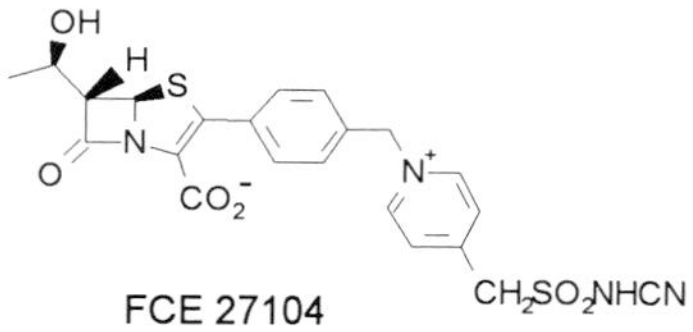   FCE 27104

Fig 10

Together with ritipenem, quaternary ammonium and quinolonyl-penems epitomize the original class of the cephalosporin-like molecules devised at Farmitalia. The thienamycin-like penems, initially investigated by Sankyo and Schering, have their prominent representative in Pfizer's sulopenem. Other interesting molecules came from the fundamental work by Ciba-Geigy and the contributions by Hoechst, Suntory and Beecham. Investigation in penems is continuing but penems, in comparison with carbapenems, offer less room for modification. Current studies on peculiar structures such as the dual-action and multicharge derivatives, among others, might bring an exciting season of research to its terminal achievements.

Acknowledgments. We thank Drs. W. CABRI, M. D'ANELLO, G.F. DALLATOMASINA for contributions to the chemical development of ritipenem acoxil and D. JABES, C. DELLA BRUNA for biological evaluation.

References

1. R. B. Woodward: *In Recent Advances in the Chemistry of β-Lactam Antibiotics.* J. Elks, Ed., The Royal Society of Chemistry, London, Special publication **28** (1977) 167-180.

2. R. B. Woodward, *Acta Pharm. Suec.* **14** Suppl. (1977) 23-25.

3. H. R. Pfaendler, J. Gosteli, R. B. Woodward, *J. Am. Chem. Soc.* **102** (1980) 2039-2043.

4. a) G. Franceschi, E. Perrone: *In Frontiers of Antibiotic Research*. H. Umezawa Ed., Academic Press (1987) 227-241. b) G. Franceschi, E. Perrone, M. Alpegiani, A. Bedeschi, C. Della Bruna F. Zarini: *In Recent Advances in the Chemistry of β-Lactam Antibiotics*. P. H. Bentley, R. Southgate Eds., The Royal Society of Chemistry, London, Special publication **70** (1989) 222-246. c) E. Perrone, G. Franceschi: *In Recent Progress in the Chemical Synthesis of Antibiotics*. G. Lukas, M. Ohno Eds., Springer-Verlag (1990) 613-703.

5. a) G. Franceschi, M. Foglio, M. Alpegiani, C. Battistini, A. Bedeschi, E. Perrone, F. Zarini, F. Arcamone, C. Della Bruna, A. Sanfilippo, *J. Antibiotics*. **36** (1983) 938-941. b) M. Foglio, C. Battistini, F. Zarini, G. Franceschi, *Heterocycles* **20** (1983) 1491-1494.

6. a) C. Battistini, C. Scarafile, M. Foglio, G. Franceschi, *Tetrahedron Lett.* **25** (1984) 2395-2398. b) A. Yoshida, T. Hoyashi, N. Takeda, S. Oida, E. Ohki, *Chem. Pharm. Bull.* **31** (1983) 768-771.

7. E. Perrone, M. Alpegiani, A. Bedeschi, F. Giudici, G. Franceschi, *Tetrahedron Lett.* **25** (1984) 2399-2402.

8. A. Afonso, F. Hon, J. Weinstein, A. K. Ganguly, *J. Am. Chem. Soc.* **104** (1982) 6138-6139.

9. M. Alpegiani, A. Bedeschi, F. Giudici, E. Perrone, G. Franceschi, *J. Am. Chem. Soc.* **107** (1985) 6398-6399.

10. M. Alpegiani, A. Bedeschi, P. Bissolino, G. Visentin, F. Zarini, E. Perrone, G. Franceschi, *Heterocycles* **31** (1990) 617-628.

11. F. Di Ninno, T. R. Beattie, B. G. Christensen, *J. Org. Chem.* **42** (1977) 2960-2965.

12. G. Franceschi, E. Perrone, M. Alpegiani, A. Bedeschi, F. Zarini: *In Penem Antibiotics*, S. Mitsuhashi, G. Franceschi, Eds., Japan Scientific Societies Press, Springer-Verlag, (1991) 3-9.

13. S. Hanessian, A. Bedeschi, C. Battistini, N. Mongelli, *J. Am. Chem. Soc.* **107** (1985) 1438-1439.

14. M. Alpegiani, A. Bedeschi, F. Zarini, C. Della Bruna, D. Jabes, E. Perrone, G. Franceschi, *J. Antibiotics* **45** (1992) 797-801.

15. a) G. Cassinelli, R. Corigli, P. Orezzi, G. Ventrella, A. Bedeschi, E. Perrone, D. Borghi, G. Franceschi, *J. Antibiotics* **41** (1988) 984-987. b) G. Visentin, E. Perrone, D. Borghi, V. Rizzo, M. Alpegiani, A. Bedeschi, R. Corigli, G. Rivola, G. Franceschi, *Heterocycles* **33** (1992) 859-891.

16. E. Perrone, D. Jabes, M. Alpegiani, B. P. Andreini, C. Della Bruna, S. Del Nero, R. Rossi, G. Visentin, F. Zarini, G. Franceschi, *J. Antibiotics* **45** (1992) 589-594.

17. E. Perrone, M. Alpegiani, A. Bedeschi, F. Giudici, F. Zarini, G. Franceschi, C. Della Bruna, D. Jabes, G. Meinardi, *J. Antibiotics* **39** (1986) 1351-1355.

18. E. Perrone, M. Alpegiani, A. Bedeschi, F. Giudici, F. Zarini, G. Franceschi, C. Della Bruna, D. Jabes, G. Meinardi, *J. Antibiotics* **40** (1987) 1636-1639.

19. R. Rossi, D. Jabes, G. Visentin, A. Bedeschi, F. Zarini: Studies on Zwitterionic and Multicharge Penems. 17th International Congress of Chemotherapy, Berlin, June 23-28 1991, abs. 2070.

Synthesis and Antimicrobial Spectrum of Some New Synthetic Monobactams

Zoltán Zubovics

Institute for Drug Research, Budapest, H-1325, P.O.B.82, Hungary

Summary: About 50 new monobactams containing quaternized nitrogen heterocycles (**3**), N-substituted 2-aminothiazol(in)es (**7**, **8**) or amino acid residues (**15**), were synthesized. The *in vitro* antibacterial activity of the most potent members (**3**) was near that of aztreonam but their *in vivo* potency was considerably weaker. Compounds of type **7**, **8** or **15** showed only moderate antibacterial activity.

Introduction

In the present paper, we report our efforts to synthesize new monocyclic β-lactams, especially monobactams with high antibacterial activity.

Monobactams exert their inhibitory activity primarily against gram-negative bacteria. Consequently, their antibacterial spectra are not as broad as the third or fourth generation cephalosporins. However, they are of great importance in the treatment of infections caused by certain bacteria, such as *Pseudomonas aeruginosa*, *Proteus vulgaris*, *Serratia marcescens*, *Haemophilus influenzae*, which are often difficult to control in clinical practice.

Chemically, a number of monobactams possessing high antibacterial activity contain structural building blocks which are similar to or identical with those occuring in the related bicyclic β-lactams, i.e. cephalosporins and penicillins.

I would like to recall briefly the most common structural elements of these compounds. The monobactam nucleus is a 3β-amino-1-sulfo-2-azetidinone ring which bears an acyl side chain attached to the 3-amino function. Acyl groups include the dipeptide chains D-Glu-D-Ala or D-Glu-L-Ala as found in the naturally occuring monobactams sulfazecin and isosulfazecin [1].

a D Sulfazecin
b L Isosulfazecin

Aztreonam : **2**

The most potent synthetic monobactams contain the well-known α-oxyimino-aminothiazolylacetyl side chains wherein the oxyimino portion bears simple alkyl or carboxyalkyl substituents and the carboxyalkyl may or may not be substituted at the α-carbon. C-3 bears an α-methoxy group in the naturally occuring monobactams mentioned above, while the synthetic derivatives have no substituent at this carbon. As for C-4, a number of compounds with various groups having either α- or β-stereochemistry have been synthesized. Thus the highly active aztreonam (**2**) [2] is a 4α-methyl derivative while carumonam [3] bears a β-carbamoyloxymethyl at C-4.

The aim of the present work was to synthesize a series of new monobactams with putative antibacterial activity. In this manuscript the chemical synthesis will be discussed and a brief summary of antibacterial activity will be given at the end.

As starting material we chose 3β-amino-4α-methyl-1-sulfo-2-azetidinone (i.e. 4α-methylmonobactamic acid) which can easily be prepared from L-threonine [4]. This monobactamic acid was coupled with various acyl side chains, most of which were substituted α-oxyimino-aminothiazolylacetyl groups. In the largest series of the present study, we prepared monobactams containing a quaternized nitrogen heterocycle. We felt this choice was justified by two facts. First, quaternized nitrogen heterocycles are well known to provide beneficial properties in penicillins and especially in cephalosporins, where such groups influence pharmacokinetics, stability to β-lactamase and/or antimicrobial potency in a favourable manner [5]. Second, Yoshimura and Nikaido [6] found that penetration of active agents via porins in the cell membranes of gram-negative bacteria is facilitated if a molecule contains one positive and two negative charges rather than one negative charge.

Only a small number of quaternized nitrogen heterocycles had been previously built into monobactams. Even in these cases, no efforts were made to prepare full series of compounds containing various substituted nitrogen heterocycles. Usually only a few examples were described whereas the larger series contained mostly non-quaternary substituents [7]. In these examples, the quaternized nitrogen heterocycles, i.e., pyridine or 1-methylpyrrolidine were incorporated into either the oxyimino substituent or C-4, or eventually into both, as indicated by the arrows. These compounds have not been reported to possess outstanding antibacterial acitivity.

Synthesis of the new monobactams

In our new compounds represented by formula **3**, the quaternized nitrogen heterocycles were attached via an acetyl bridge to the 2-amino group of the aminothiazolyl moiety. The nitrogen heterocycles were aromatic ones including pyridine, pyridazine and methylimidazole as well as saturated rings containing a tertiary nitrogen such as an N-methylated pyrrolidine, piperidine, morpholine, or 4-azaquinuclidine. The pyridine ring was substituted by one or more alkyl groups or by a hydroxyalkyl, acetyl, functionalized carboxy (ester, amide) or amino group.

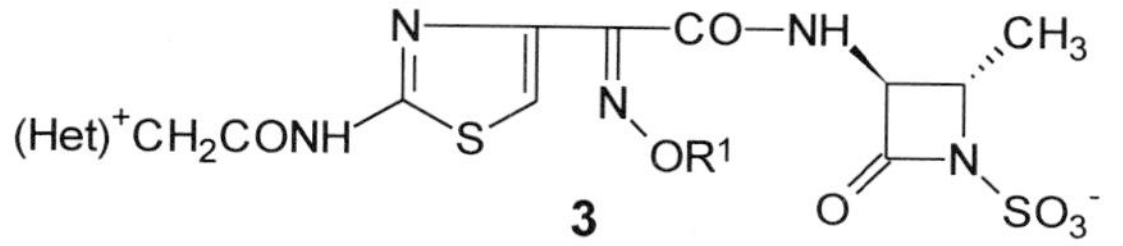
(Het)$^+$CH$_2$CONH
N
S
CO—NH
N
OR^1
CH$_3$
O
N
SO$_3^-$

3

R^1 = Me, CH_2COOH, $CHMeCOOH$, CMe_2COOH

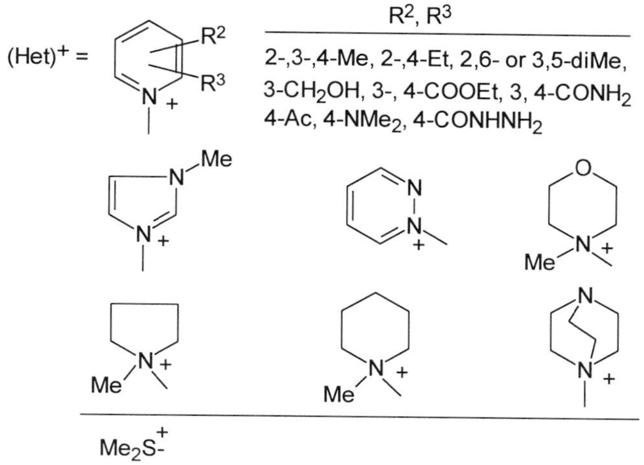
(Het)$^+$ =
R^2
R^3
N +
R^2, R^3
2-,3-,4-Me, 2-,4-Et, 2,6- or 3,5-diMe,
3-CH$_2$OH, 3-, 4-COOEt, 3, 4-CONH$_2$
4-Ac, 4-NMe$_2$, 4-CONHNH$_2$
Me
N
N +
N
N
+
O
Me
N +
N +
Me
N +
Me
N
N +
Me$_2$S$^+$-

Our strategy was to first synthesize the methoxyimino series (**3**, R^1 = methyl) using the full set of the heterocycles mentioned above. After the initial microbiological screening, the carboxyalkoxyimino analogues (**3**, R^1 = carboxyalkyl) were then prepared using those heterocycles which had given the highest antibacterial activity in the first series. In order to widen the series, a tertiary sulfur analogue containing the dimethylsulfonium group was also synthesized.

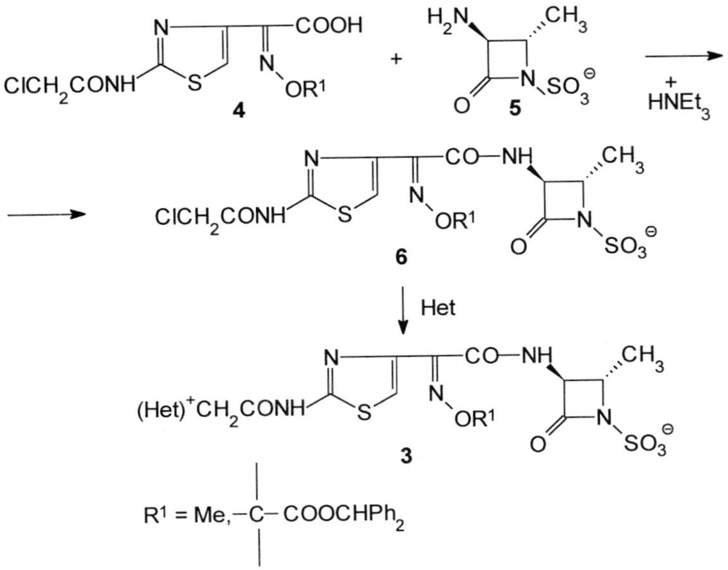

The synthesis of our new compounds started with the appropriate known oxyimino-chloroace-tamidothiazolylacetic acids **4** [8]. Chloroacetyl has been used so far only to protect the 2-amino function in similar compounds. In cases where R^1 contained a carboxy group, it was protected as its benzhydryl ester. The acids **4** were coupled with the triethylamine salt of 4α-methylmonobactamic acid (**5**) in the conventional manner [9] in DMF in the presence of DCC and HBT. The chloroacetyl deriva-tives **6** were then used to quaternize the appropriate nitrogen heterocycles to yield the target compounds of Formula **3** or, when the oxyimino group contained a protected carboxy, after deprotection in the conventional manner using trifluoroacetic acid in the presence of anisole [9].

In this manner, a series of about 40 new compounds was prepared and submitted to antibacterial testing. The *in vitro* results were promising because most of the above compounds were active against a set of gram-negative bacteria. This finding indicated that substitution at the 2-amino group of the aminothiazolyl moiety may result in potent antibacterial agents and prompted us to synthesize another series of monobactams which, unlike the above compounds, contained N-substituted aminothiazolyl groups instead of quaternized nitrogen heterocycles. Compounds **7** and **8** belonging to the methoxyimino series were prepared which included *exo*- and/or *endo*-N-substituted derivatives as shown below.

	R^1
a	H
b	Me
c	2,6-dichlorophenyl
d	2,6-dimethylphenyl

	R^1	R^2
a	H	Me
b	2,6-dichlorophenyl	Me
c	2,6-dichlorophenyl	CH_2CH_2OH

These compounds were prepared via conventional routes as follows (see next page). The appropriate thiourea **9** was condensed with ethyl 4-bromo-2-methoxyiminoacetoacetate (**10**) in the presence of dimethylaniline. Starting with monosubstituted thioureas, the cyclic thiazoles (**12**, R^2 = H) were obtained directly. Under the same conditions, N,N'-disubstituted thioureas yielded the open chain precursors **11** which were cyclized to the corresponding thiazolines (**12**, $R^2 \neq$ H) by treatment with acid. Alkaline hydrolysis and subsequent coupling (DCC and HBT in DMF) with 4α-methylmonobactamic acid triethylamine salt led to the desired compounds **7** or **8**. Alkaline hydrolysis of thiazoline **12** bearing a hydroxyethyl at N-3 did not yield the corresponding free acid but the seven membered lactone **13**. The latter was then used to acylate the monobactamic acid under mild conditions (DMF, 25 °C, 6 days).

The use of monomethylthiourea led to the 2-methylamino compound, i.e. the *exo*-isomer **7b**, while the *endo*-N-methyl derivative **8a** was prepared via the 2-tritylamino-3-methyl-precursor (**12**). The latter was prepared from N-trityl-N'-methylthiourea, and the trityl group was removed after coupling with monobactamic acid by treatment with 50 % formic acid.

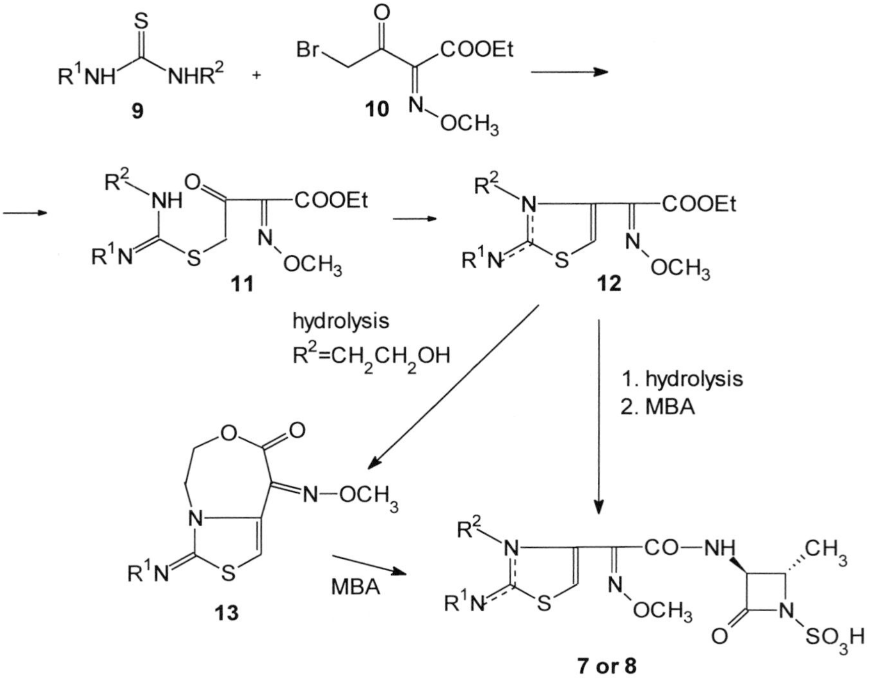

Next, the preparation of some additional new monobactams was inspired by a comparison of two monobactams known to possess high antibacterial activity, i.e. sulfazecin and aztreonam shown above. A study of their molecular models suggested that the terminal amino group in the acyl portion of sulfazecin may be located at a somewhat longer distance from the azetidinone ring than in aztreonam. This difference prompted us to synthesize a few aztreonam analogues which contained the aminothiazole moiety somewhat further from the azetidinone ring. For this purpose, the units present in sulfazecin and its isomer, i.e. D-alanine, L-alanine, and glycine were, inserted between the oxyimino-aminothiazolylacetyl group and the 3-amino group of the azetidinone.

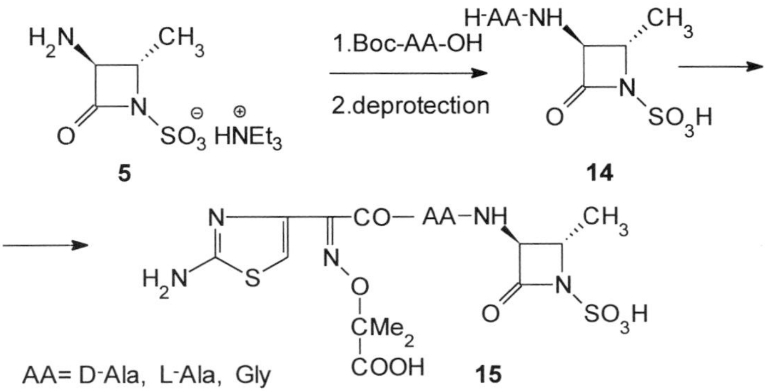

Thus, 4α-methylmonobactamic acid was coupled with the appropriate Boc-protected amino acid followed by deprotection, coupling with the corresponding carboxylic acid, and finally deprotection of the carboxy function in the oxyimino side chain to afford the desired compounds **15**.

It will be noted at this point that the geometry of the oxyimino group in all new acylated monobactams discussed in the present paper was **Z** as indicated by the chemical shift values of the thiazol(in)e C-5 proton in the ^{1}H-NMR spectra.

Biological data

The *in vitro* antibacterial activity of the new compounds was determined by the conventional serial dilution method, using aztreonam as reference compound. As expected, the growth of gram-positive bacteria was hardly affected by these compounds. On the other hand, a number of the new monobactams showed a high antibacterial potency against various gram-negative bacteria. Hence their MIC values were determined against the following selected strains : *Escherichia coli K$_{12}$, Escherichia coli 6R* (resistant to six antibiotics), *Klebsiella pneumoniae ATCC 10031, Proteus vulgaris XL, Pseudomonas pyocyanea NCTC 10490, Salmonella typhi-murium, Shigella sonnei* and *Bordetella bronchiseptica ATCC 4617.*

The *in vitro* antibacterial activity of each compound was characterized by the average of the MIC values measured against the above strains. The three subtypes of compounds in the present paper were those containing quaternized nitrogen heterocycles (**3**), those with N-substituted aminothiazol(in)es (**7**, **8**) and aztreonam analogues with inserted amino acid spacers (**15**). Only the compounds containing the quaternized nitrogen heterocycles showed excellent activity. In the methoxyimino series (**3**, R^1 = Me), the compounds containing substituted pyridines (especially 3- or 4-substituted derivatives) or methylimidazole as heterocycles were the most potent, with average MIC values against the bacteria mentioned above within the range of 0.8-4.4 μg/ml, versus 0.13 μg/ml value for aztreonam. Their carboxyalkoxyimino analogues showed somewhat higher activity : MIC = 0.7-2.0 μg/ml. On the other hand, compounds containing saturated nitrogen heterocycles were less active. As the following Table shows, the average MIC values of our most potent compounds were 6-10 times higher than that of aztreonam.

Table 1 Antibacterial activity of selected compounds **3**

Compound	R^1	Het	MIC, μg/ml n = 8	Pseudomonas MIC, μg/ml	*In vivo* ED$_{50}$, mg/kg
3a	CMe$_2$COOH	methyl-imidazole	0.78	2.05	7.0
3b	CMe$_2$COOH	3-PyCONH$_2$	0.87	8.0	5.2
3c	CMe$_2$COOH	4-PyCONH$_2$	0.84	6.0	2.3
3d	CHMeCOOH	3-PyCONH$_2$	1.29	30.0	1.6
Aztreonam			0.13	2.5	0.04

The activity of the same compounds (except for one) against 24 Pseudomonas strains was in the same order of magnitude as that of the reference compound.

With these encouraging *in vitro* data, we continued antibacterial testing *in vivo*. The activity of the above compounds against *Proteus vulgaris* XL in OF$_1$ mice after subcutaneous administration was tested. Unfortunately, *in vivo* activity was about two orders of magnitude lower than that of aztreonam.

Efforts were made to find the reasons for the low *in vivo* activity. Thus, maximal blood level and area under curve values were determined by microbiological method in CFLP mice after s.c. and i.p. administration. The obtained values were very close to those of aztreonam; hence, it was concluded that the poor *in vivo* activity was not a consequence of unsatisfactory absorption. In further experiments, aztreonam and our compounds **3a-3d** were found to be very similar with respect to β-lactamase stability, penetration via cell membrane or binding to penicillin binding proteins, respectively.

Acknowledgements: The author wishes to express his acknowledgements to all participants of this program, especially Lajos Toldy (who initiated the study) for valuable discussions, Katalin Görgényi and Zsuzsa Kurucz for their contribution to the synthetic work, to Péter Dvortsák, Gyula Jerkovich, István Pelczer and István Kövesdi for NMR spectra, Gyula Horváth for elementary analyses, István Koczka and Katalin Niszner for the design and realization of all biological measurements and last, but not least, Mária Zrinyi, Erika Száll-Molnár and István Pallagi for technical assistance.

References

1. A. Imada, K. Kitano, K. Kintaka, M. Muroi and M. Asai, *Nature* **289** (1981) 590-591
2. C. M. Cimarusti, D. P. Bonner, H. Breuer, H. W. Chang, A. W. Fritz, D. M. Floyd, T. P. Kissick, W. H. Koster, D. Kronenthal, F. Massa, R. H. Mueller, J. Pluscec, W. A. Slusarchyk, R. B. Sykes, M. Taylor and E. R. Weaver, *Tetrahedron* **39** (1983) 2577-2589.
3. S. Kishimoto, M. Sendai, S. Hashiguchi, M. Tomimoto, Y. Satoh, T. Matsuo and M. Kondo, *J. Antibiot.* **36** (1983) 1421-1424.
4. D. M. Floyd, A. W. Fritz, J. Pluscec, E. R. Weaver and C. M. Cimarusti, *J. Org. Chem.* **47** (1982) 5160-5167.
5. J. V. Uri, N. Burdash and C. M. Bendas, *Acta Microbiologica Hungarica* **35** (1988) 327-356.
6. F. Yoshimura and H. Nikaido, *Antimicrob. Ag. Chemother.* **27**(1985) 84-92.
7. see e. g. a) S. Kishimoto, T. Matsuo and M. Ochiai (Takeda), Eur. Pat. Appl. EP 93,376; *Chem. Abstr.* **100** (1984) 209515z ; b) S. Kishimoto, M. Sendai, T. Mitsumi, M. Ochiai and T. Matsuo (Takeda), Eur. Pat. Appl. EP 53,816; *Chem. Abstr.* **98** (1983) 143181u; c) U. D. Treuner (Squibb), Eur. Pat. Appl. EP 177,940; *Chem. Abstr.* **105** (1986) 190776q.
8. T. Matsuo, H. Masuya, N. Noguchi and M. Ochiai (Takeda), Eur. Pat. Appl. EP 53,815; *Chem. Abstr.* **98** (1983) 53674c.
9. R. B. Sykes, W. L. Parker, C. M. Cimarusti, W. H. Koster, A. Slusarczyk, A. W. Fritz and D. M. Floyd (Squibb), Eur. Pat. Appl. EP 48,953; *Chem. Abstr.* **97** (1982) 92116w.

Design and Synthesis of D,D Peptidase Inhibitors

Léon Ghosez*, Georges Dive #, Stéphane Dumas, Christophe Génicot,
Frank Kumli, Christopher Love and Jacqueline Marchand-Brynaert*

* Laboratoire de Chimie Organique de Synthèse
Université Catholique de Louvain
place Louis Pasteur, 1, B - 1348 Louvain-la-Neuve, Belgium
Centre d'Ingénierie des Protéines,
Université de Liège,
Institut de Chimie, B6, B - 4000 Sart-Tilman (Liège), Belgium

Summary. Cyclic Michael acceptors topologically related to penicillins or monobactams have been designed as potential alkylating inhibitors of bacterial D,D peptidases. Cyclobutene sulfonates bearing an acylamino side chain have been prepared but were unstable in aqueous solution. New methods for the preparation of highly functionalized and enantiomerically pure four-membered rings are described.

Introduction

β-Lactam antibiotics form a remarkable class of therapeutic agents which act by disrupting the biosynthesis of the bacterial cell wall. They inhibit the membrane-bound transpeptidases which are involved in the peptidoglycan cross-linking. The affinity of β–lactams for these enzymes results from their structural and conformational similarity to the peptide substrate.[1] Scheme 1 describes the mechanism proposed by Ghuysen and Frère for the interaction of β–lactam inhibitors with these serine peptidases.[2] A good inhibitor is characterized by $k_2/K \gg k_3$.

The extensive use of β-lactam antibiotics has induced bacterial resistance resulting from the production of β-lactamases.[3] These defense enzymes react with the β-lactam function of the inhibitor to form a penicilloyl enzyme which is rapidly hydrolysed to regenerate the free enzyme. New structures which could lead to more stable acyl-enzyme intermediates have been prepared and tested as potential β-lactamase resistant antibiotics.[4]

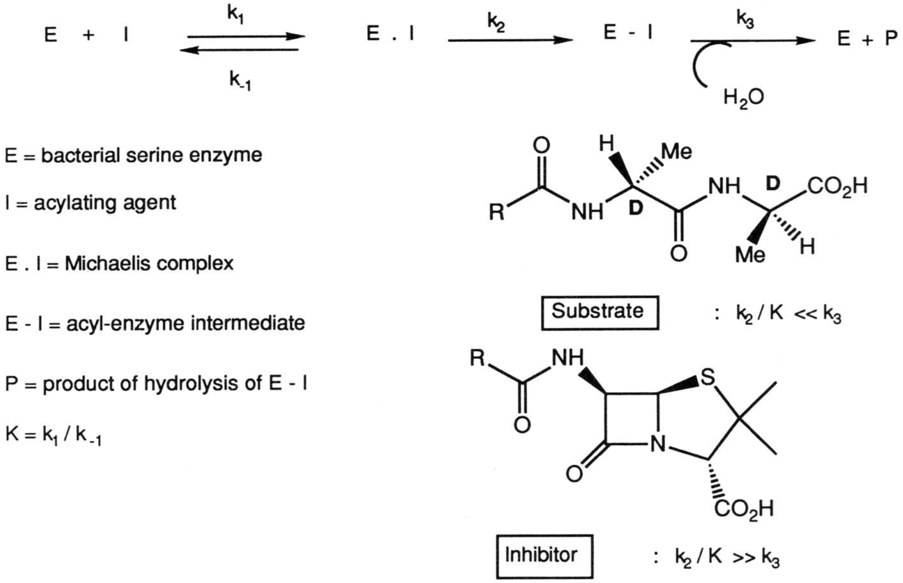

Scheme 1

Recently we proposed the use of alkylating agents susceptible to react selectively and irreversibly ($k_3 = 0$) with the target enzymes.[5] The design of alkylating inhibitors should first take into consideration the geometrical requirements for the nucleophilic attack of the hydroxyl group of the active serine residue on the electrophilic center of the inhibitor. The trajectory of attack of a nucleophile to a carbon-carbon double bond would be expected to closely approximate that on the carbonyl group of a β–lactam. This rough approximation led us to examine Michael acceptors equipped with appropriate substituents for recognition by the enzymes as potential alkylating inhibitors.

We first selected cyclobutene sulfonates as alkylating analogs of monobactams (Scheme 2). A transition state geometry for the alkylation was assumed which correctly placed the side chains necessary for the recognition step. All degrees of freedom of monobactam (SQ 26324) and the related cyclobutene sulfonate were fully optimized by MNDO calculations. The energetically accessible conformations of both structures were superimposed and best fitted using the BMFIT program. Scheme 2 shows that a good fit was only observed when X = H or F. Since we did not expect a β–fluorocyclobutene sulfonate to exhibit sufficient chemical stability in aqueous solution, we selected the parent cyclobutene sulfonate as our first target.

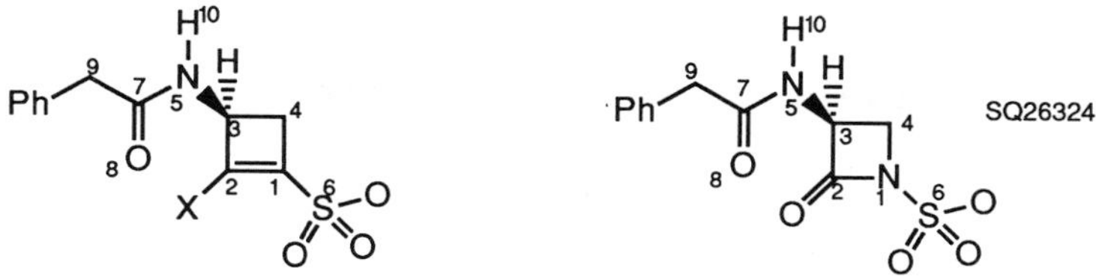

Atoms		X = H	X = F	X = OH
(* Atoms superposed)	1 *	0.09	0.09	0.09
	2 *	0.07	0.07	0.06
	3 *	0.02	0.01	0.01
	4 *	0.05	0.04	0.04
	5 *	0.04	0.04	0.04
	6	0.66	0.65	0.64
	7	0.08	0.11	1.14
	8	0.10	0.24	2.37
	9	0.16	0.11	0.90
	10	0.10	0.14	0.88

Distances between corresponding atoms (A)

Scheme 2

Synthesis of 3-Acylamino-1-Cyclobuten-1-Sulfonate

Our synthetic plan is outlined in Scheme 3. It is based on the recent observation that the central carbon-carbon bond of 1-bicyclobutyl phenyl sulfone can be cleaved with phenylselenyl azide.[6]

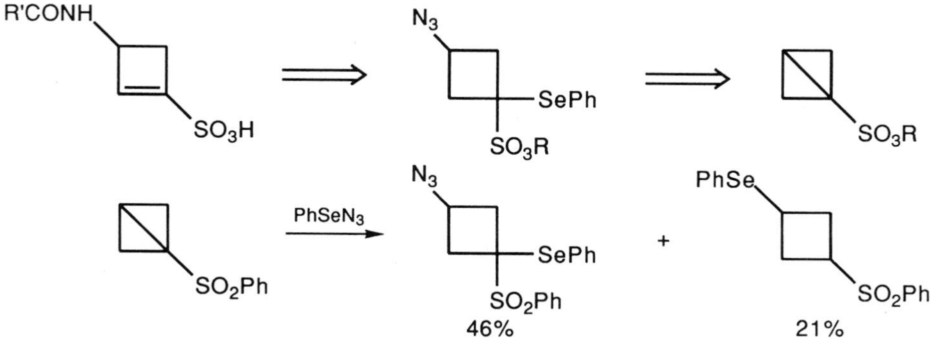

Scheme 3

Epichlorhydrin was readily transformed into sulfonate **1** which, in a one-pot sequence, gave the bicyclobutyl sulfonate **2** (Scheme 4). Both intermediates **3** and **4** have been isolated and characterized in separate experiments.

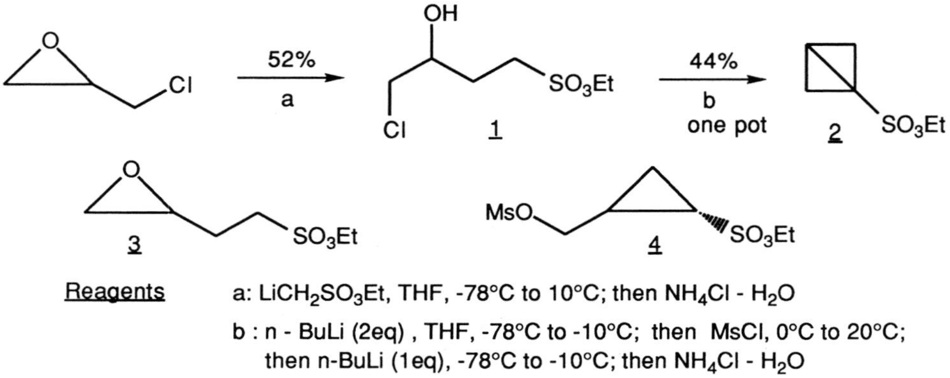
52%
a
OH
Cl
1
SO3Et
44%
b
one pot
2
SO3Et
3
SO3Et
MsO
4
SO3Et
Reagents
a: LiCH2SO3Et, THF, -78°C to 10°C; then NH4Cl - H2O
b : n - BuLi (2eq) , THF, -78°C to -10°C; then MsCl, 0°C to 20°C;
then n-BuLi (1eq), -78°C to -10°C; then NH4Cl - H2O

Scheme 4

The central carbon-carbon bond of the bicyclobutyl sulfonate **2** was readily cleaved by electrophilic reagents (Scheme 5). The reaction with *in situ* generated phenylselenyl azide selectively yielded cyclobutyl sulfonate **5** which, in a few steps, was readily converted into the target molecule **6** (Scheme 6). This could be purified and characterized in aprotic solvents but was too unstable in water or alcohols to allow for biological evaluation.

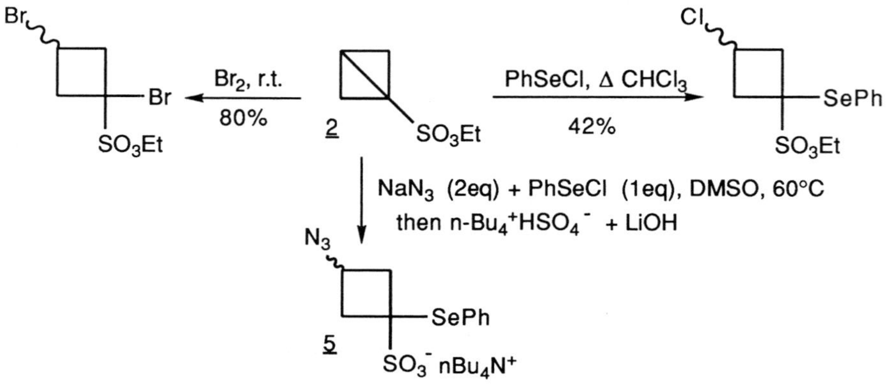
Br
Br
SO3Et
Br2, r.t.
80%
2
SO3Et
PhSeCl, Δ CHCl3
42%
Cl
SePh
SO3Et
NaN3 (2eq) + PhSeCl (1eq), DMSO, 60°C
then n-Bu4+HSO4- + LiOH
N3
SePh
5
SO3- nBu4N+

Scheme 5

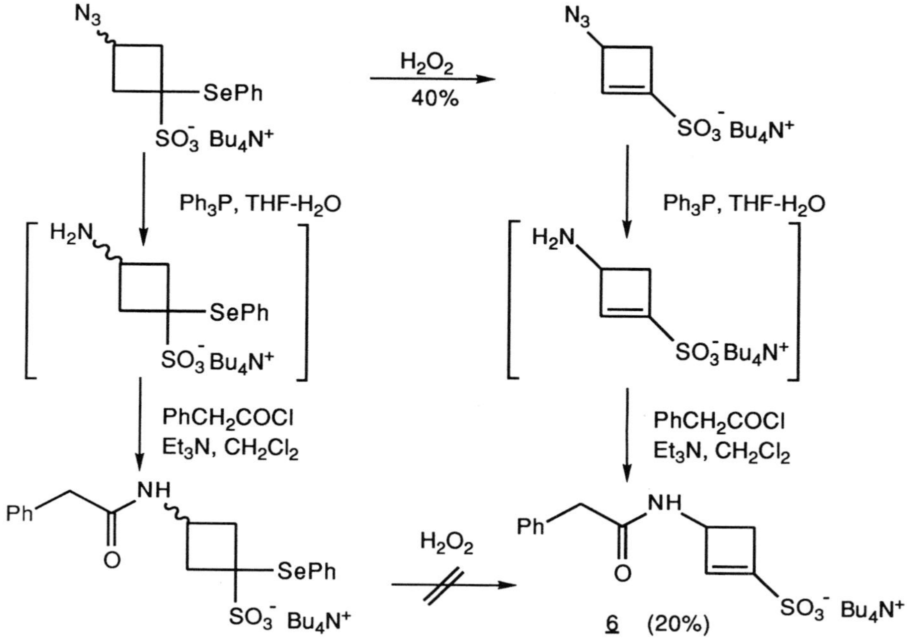

Scheme 6

Synthesis of Cyclobutenones and γ-Butenolides

Cyclobutenones and butenolides were also considered as potential alkylating analogs of the β-lactam antibiotics. This led us to examine new synthetic methods towards these classes of compounds.

We have reported earlier [7] on an efficient synthesis of cyclobutenones by [2 + 2] cycloaddition of keteniminium salts to acetylenes (Scheme 7). It was found that, in contrast with other enones,

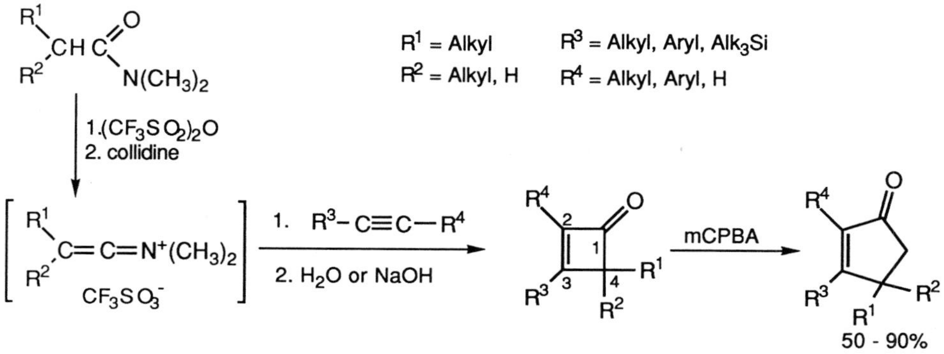

R^1 = Alkyl R^3 = Alkyl, Aryl, Alk$_3$Si
R^2 = Alkyl, H R^4 = Alkyl, Aryl, H

50 - 90%

Scheme 7

cyclobutenones under a Baeyer-Villiger oxidation gave cleavage of the C1-C4 bond. This reversal of migratory aptitude of the two C-C = O σ bonds probably results from an allylic stabilization of a developing cationic center at C4 in a late transition state.

This highly convergent [2 + 2 + 1] strategy toward butenolides was expected to be readily applied to the preparation of enantiomerically pure butenolides. The first examples of asymmetric synthesis of cyclobutenones are shown in Scheme 8. A pyrrolidine with C2 symmetry was used as chiral auxiliary [8] in order to avoid the formation of diastereoisomeric keteniminium salts which would react with opposite facial selectivity. The low yields of cycloadducts probably resulted from a cleavage of the four-membered ring during the hydrolysis step.

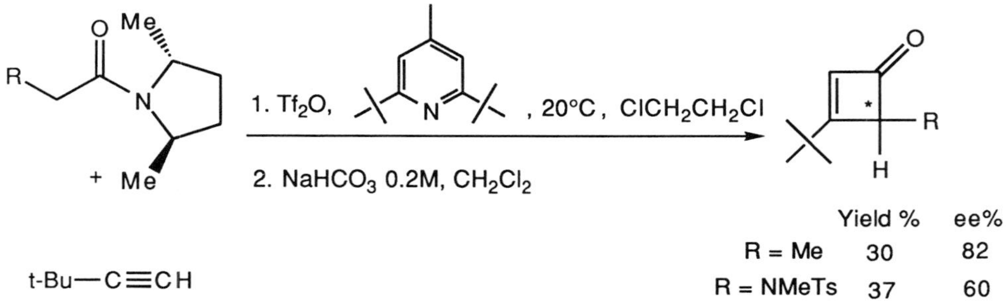
1. Tf2O, , 20°C, ClCH2CH2Cl
2. NaHCO3 0.2M, CH2Cl2
t-Bu—C≡CH
Yield % ee%
R = Me 30 82
R = NMeTs 37 60

Scheme 8

We therefore examined the synthesis of functionalized cyclobutanones which could serve as precursors of cyclobutenones (Scheme 9).

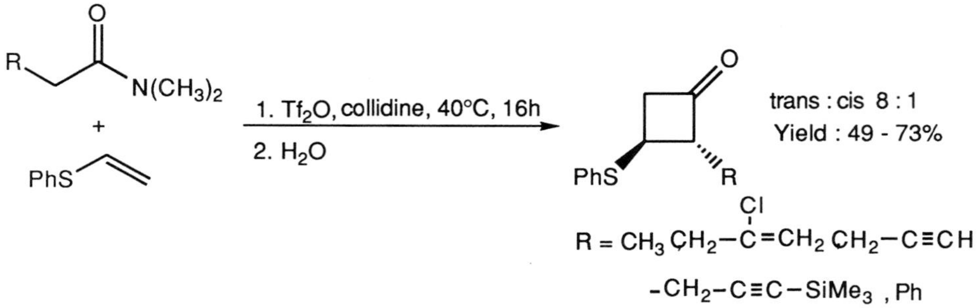
N(CH3)2
PhS
1. Tf2O, collidine, 40°C, 16h
2. H2O
trans : cis 8 : 1
Yield : 49 - 73%
R = CH3 CH2–C(Cl)=CH2 CH2–C≡CH
-CH2–C≡C–SiMe3 , Ph

Scheme 9

Keteniminium salts bearing various unsaturated functional groups readily cycloadded to phenyl vinyl sulfide to give the corresponding cyclobutanones. Selective manipulations of the various functionalities were easily performed (Scheme 10). Elimination of phenylsulfenic or phenylsulfinic acid from **8** or **7** would be expected to provide the corresponding reactive cyclobutenones. This is presently being examined.

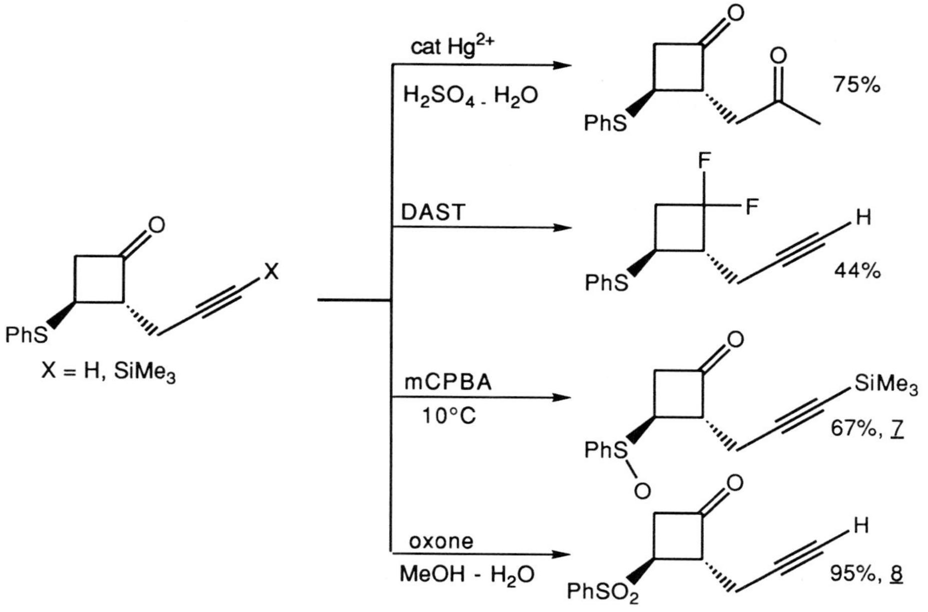

Scheme 10

Asymmetric synthesis of cyclobutanones and γ-lactones

Chiral keteniminium salts are ideal reagents for the asymmetric synthesis of cyclobutanones : (a) they are very reactive even to olefins bearing no activating groups, (b) they yield cyclobutaniminium salts which are easily purified as perchlorate salts, (c) they readily hydrolyze to cyclobutanones and a chiral amine which can be recovered.

Model experiments with keteniminium salts derived from chiral isobutyramides demonstrated the high potential of the method.[9] More recently we have examined asymmetric [2 + 2] cycloadditions of a keteniminium salt derived from N-tosylsarcosinamide (Scheme 11). [10]

The sarcosinamide derived from prolinol methyl ether (Scheme 11, R^1 = CH_2OMe, R^2 = H) could not be used here because it was expected to generate two diastereoisomeric keteniminium salts which should react with opposite facial selectivities. The formation of two diastereoisomeric keteniminium salts could be suppressed by using keteniminium salts derived from a pyrrolidine with C_2 symmetry (Scheme 11, R^1 = R^2 = CH_3). It was indeed found that keteniminium salt **9** cycloadded in good yields to various 1,2-disubstituted olefins (Scheme 12).

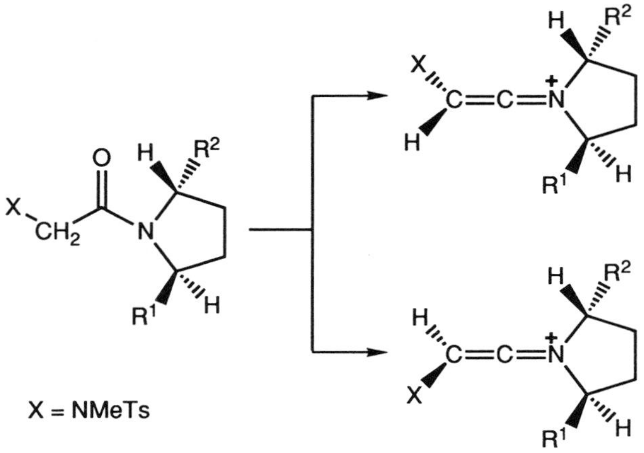

Scheme 11

Baeyer-Villiger oxidation of the hydrolyzed adducts proceeded regiospecifically to give the γ-lactones. Here the NTsMe group could be readily exchanged by treatment with water or an alcohol in the presence of an acid to give the corresponding lactol or its ether.

The cycloadditions proceeded with high facial selectivities in the case of 1, 2 disubstituted olefins. The γ-lactones were obtained in high enantiomeric purities. A single recrystallization gave the enantiomerically pure lactones.

Cis-butene only gave a trans adduct as a result of a fast base-catalysed epimerization of the primarily formed cis-adduct. Terminal olefins such as 1-hexane reacted with poor facial selectivities.

We are presently using this strategy to build molecules which would fit the criterion mentioned above to interact with the target enzymes.

Acknowledgements

This work was generously supported by the Ministère de la Région Wallonne, Direction de l'Administration de l'Energie et Technologies nouvelles, SmithKline Beecham, the Institut pour l'Encouragement de la Recherche Scientifique dans l'Industrie et l'Agriculture, the Fonds National de la Recherche Scientifique and the Ministère de l'Education et de la Recherche Scientifique de la Communauté Française de Belgique (Action Concertée 86/91-84).

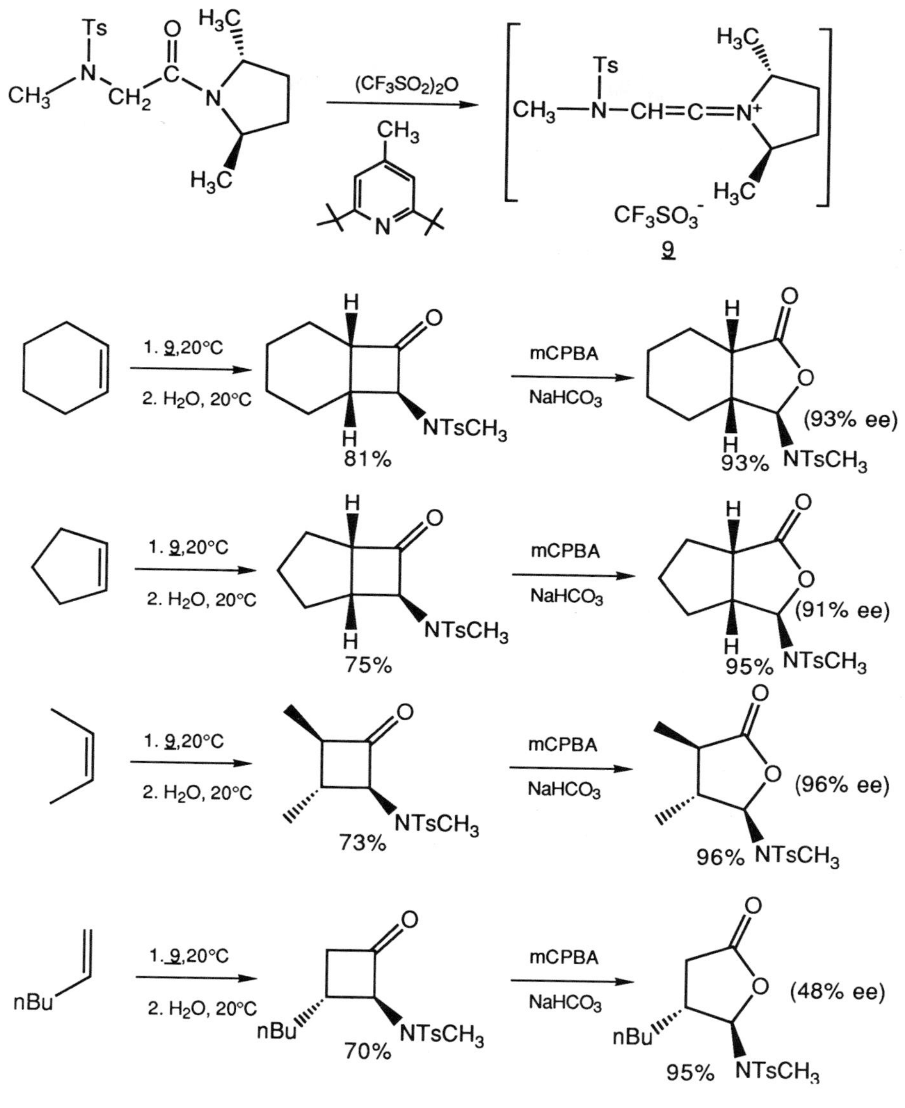

Scheme 12

References

1. D. J. Waxman, J. L. Stroninger in *"Chemistry and Biology of β-lactam Antibiotics"* (R. B. Morin, M. Gorman, eds), vol. 3, Academic Press, New York, 1982, pp 209-285.

2. J. M. Frère, B. Joris *CRC Critical Rev. Microb.*, **11** (1985) 299.

3. A. L. Fink , *Bioorg. Chem. Enzym. Catal.*, **1992**, 41.

4. J. Marchand-Brynaert, L. Ghosez, in *"Recent Progress in the Chemical Synthesis of Antibiotics"* (G. Lukacs, M. Ohno, eds), Springer Verlag, Berlin Heidelberg, 1990, pp 727-794.

5. a) J. Marchand-Brynaert, Z. Bounkhala-Khrouz, B. J. van Keulen , H. Vanlierde, L. Ghosez, *Isr. J. Chem.*, **29** (1989) 247;
   b) J. Marchand-Brynaert, Z. Bounkhala-Krouz, H. Vanlierde, L. Ghosez, *Heterocycles*, **30** (1990) 971;
   c) J. C. Carretero, J. Davies, J. Marchand-Brynaert , *Bull. Soc. Chim. Fr.*, **127** (1990) 835;
   d) J. Marchand-Brynaert, D. Ferroud, B. Serckx-Poncin, L. Ghosez, *Bull. Soc. Chim. Belg.*, **99** (1990) 1075;
   e) J. Marchand-Brynaert , J. Davies, L. Ghosez , *Heterocycles*, **33** (1992) 313.

6. Y. Gaoni, *Tetrahedron* , **45** (1989) 2819.

7. C. Schmit , S. Sahraoui-Taleb, E. Differding, C.-G. Dehasse-De Lombaert, L. Ghosez, *Tetrahedron Letters*, **25** (1984) 5043.

8. L.-y. Chen, L. Ghosez, *Tetrahedron Letters* **31** (1990) 4467.

9. a) C. Houge, A.-M. Frisque-Hesbain, A. Mockel, L. Ghosez, *J. Am. Chem. Soc.*, **104** (1982) 2920;
   b) H. Saimoto, C. Houge, A.-M. Frisque-Hesbain, A. Mockel, L. Ghosez., *Tetrahedron Letters* **24** (1983) 2251;
   c) L.-y.Chen , L. Ghosez, *Tetrahedron Asymmetry*, 1991, **2** (1991) 1181.

10. a) C. Génicot , B. Gobeaux, L. Ghosez, *Tetrahedron Letters* **32** (1991) 3827;
    b) C. Génicot , L. Ghosez, *Tetrahedron Letters* **32** (1992) 7357.

Recent Developments in the Chemistry of Quinolones

Daniel Bouzard

Bristol Myers Squibb Research Institute
Marne-la-Vallée - 77422 France

Summary : The resurgence of interest in quinolones and related compounds has resulted in an enormous amount of research on new structural modifications to improve the overall spectrum of antibacterial activity, bioavailability and safety. This review summarizes the most relevant chemistry published during the last three years.Only minor improvments have been made on the synthesis of the skeleton. New substitutions have been focused on the C-5 position and on 7-alkenyl or cycloalkenyl derivatives. Attention has also been concentrated on enantioselective synthesis of N-1 or C-7 substituted chiral derivatives.
Temafloxacin, tosufloxacin and sparfloxacin have emerged as clinically useful new compounds and have been marketed or are in advanced clinical development.

The quinolones are a class of purely synthetic antibacterials and no natural analogues have been found so far. The latest generation of compounds available for therapeutic uses or in advanced clinical development have a broad spectrum, good bioavailability and good safety profile.

The quinolones are simple molecules and many of the possible chemical modifications have been made around the chemical skeleton. Detailed structure-activity relationships have been reviewed recently [1]. As shown in figure 1, the effects of a substituent are multiple and greatly affect *in vitro* and *in vivo* activity, bioavailability and toxicity.

The chemistry of quinolones presented below is limited to the most recent developments published during the last 3 years, and as far as possible, the examples have been selected as the most relevant among the huge amount of work published.

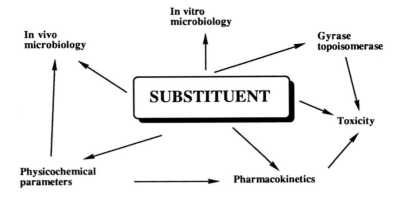

Figure 1

1- Synthesis of the Nucleus

Only minor modifications have been recently made on the synthesis of the skeleton of the quinolones. The most versatile method was published by Bayer' chemists in the late 80's [2], (figure 2). An enamine was directly condensed with an acid chloride; subsequent cyclisation afforded the quinolone nucleus. This elegant method, originally described for quinolones (Z=CH), worked also in the naphthyridone series.

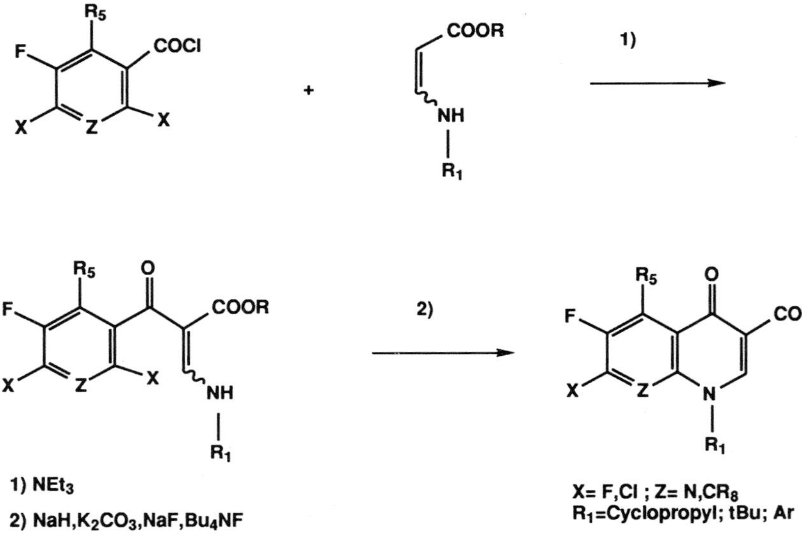

1) NEt$_3$

2) NaH, K$_2$CO$_3$, NaF, Bu$_4$NF

X= F, Cl ; Z= N, CR$_8$
R$_1$ = Cyclopropyl; tBu; Ar

Figure 2

2-Aza Derivatives

With the aim of improving antibacterial activity, the 2-aza analogues of naphthyridine derivatives were prepared [3], (figure 3). Condensation of the acid chloride with ethyl diazoacetate afforded α-diazo-β-ketoesters which reacted with tributyl- or tricyclohexylphosphine to give the intermediary labile triphenylphosphazine. Reductive cyclization proceeded successfully to the pyridazone. The best yields have been obtained when Y = F since a fluorine atom is more reactive than a chlorine atom. Treatment of the pyridazone with ethyl iodide in N ,N-dimethylformamide gave alkylation at N-1. Finally, acidic hydrolysis furnished the carboxylic acid.

1) N_2CHCO_2Et / 55°C

2) R_3P / $IsoPr_2O$

3) tert-BuOK / dioxane

4) RI or ROTs / K_2CO_3 in DMF

X = Cl, $p\,CH_3C_6H_4S$

Y = Cl, $p\,CH_3C_6H_4S$, F

R = Et, FCH_2CH_2

2- New Substitutions

7-Alkenyl or Cycloalkenyl Quinolone Derivatives

Most of the heterocycles that have been evaluated at C-7 are linked to the aromatic nucleus *via* a nitrogen-carbon bond and the incorporation of carbon-bonded aliphatic heterocycles or carbocycles at C-7 of the quinolone or naphthyridine nucleus has received less attention. Recently Parke-Davis' chemists have published [4-6] a methodology that allowed the direct attachment of a large variety of unsaturated cyclicand acyclic residues to a fully functionalized naphthyridine substrate (figure 4).

The reaction of suitable 7-activated quinolones with 1,1-dimethylethyl-(3-(tri-*n*-butylstannyl)-2-cyclopenten-1-yl) carbamate provided the coupled product. The coupling reactions were generally carried out using 1.0-1.5 equivalents of the tin reagent and 2 mol % of either bis(triphenylphosphine)dichloropalladium(II) or tetrakis(triphenylphosphine)palladium(0) as the catalyst. In general, these compounds

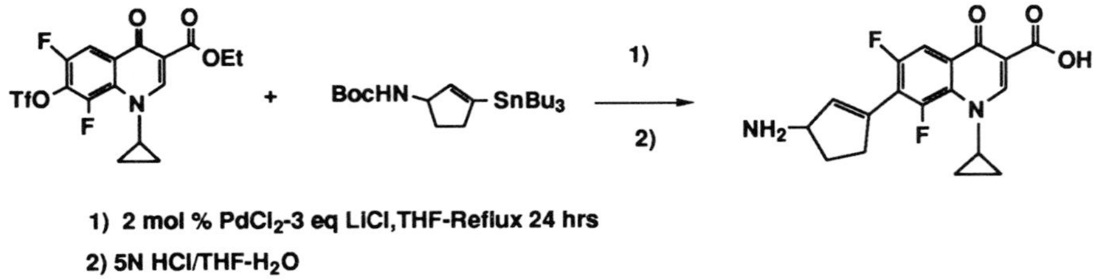

1) 2 mol % PdCl$_2$-3 eq LICl,THF-Reflux 24 hrs
2) 5N HCl/THF-H$_2$O

Figure 4

displayed activity comparable to their nitrogen-linked counterparts, suggesting that the attachment of the side chain to the naphthyridine nucleus does not have to be through a nitrogen atom to maintain potent microbiological activity.

5 -Substituted Naphthyridones

The most common aproach to the synthesis of 5-substituted naphthyridones started with the corresponding 4-substituted nicotinic acids [7,8]. Deprotonation of the 2,6-dichloro-5-fluoronicotinic acid with 2 equivalents of lithium diisopropylamide or methyllithium followed by addition of the suitable reagent afforded the desired nicotinic acids in good yield (figure 5).

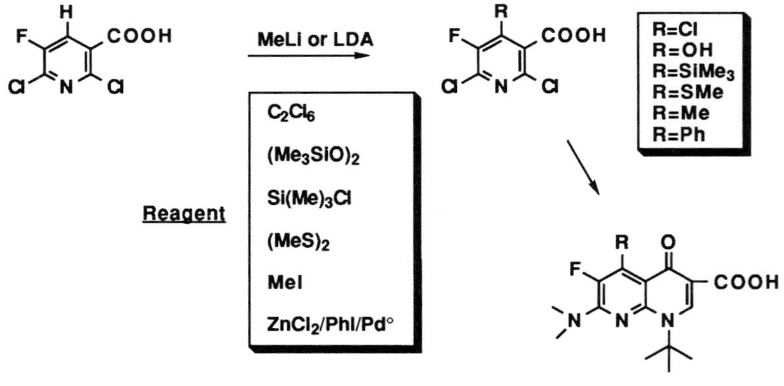

Figure 5

Following conventional chemistry the 4-methylnicotinic acid derivatives gave a series of 5-methyl naphthyridone derivatives which possessed excellent broad-spectrum activity against Gram-positive and Gram-negative pathogens as well as good pharmacokinetic properties. 2,6-dichloro-4-formyl-5-fluoronicotinic acid was readily prepared by alkylation of the lithio intermediate with dimethylformamide, but reduction with sodium borohydride gave only the undesired lactone [9]. All attempts to open this lactone were unsuccesful (figure 6).

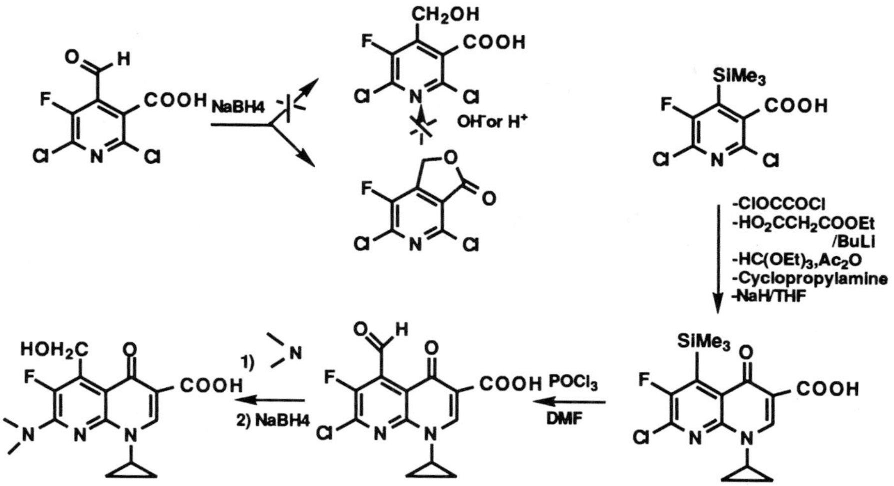

Figure 6

Another approach based on the introduction of the formyl group at a later stage *via* the 5-trimethylsilynaphthyridone was then selected. *Ipso* substitution of the trimethylsilyl group with the Vilsmeier reagent gave the 5-formyl derivative, which was easily reduced to the corresponding 5-hydroxymethyl derivative.

5-amino-substituted quinolones were reported (9) as potent *in vitro* and *in vivo* antibacterial derivatives. Sparfloxacin was the best representative of the series. Synthesis of the naphthyridone analogues was very recently described following the original methodology (figure 7).

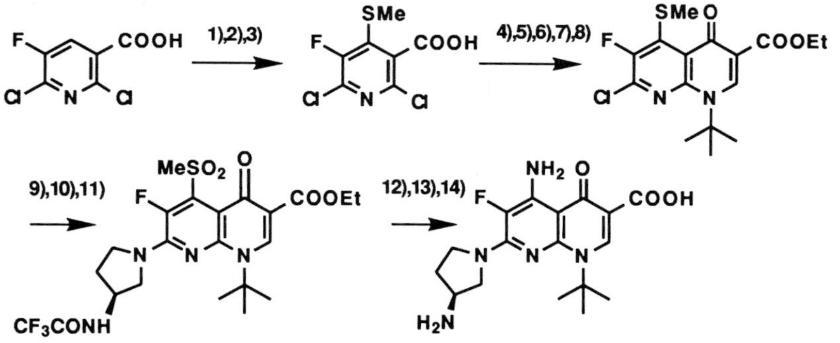

1): MeLi or LDA; 2) : MeSSMe; 3) : H⁺; 4) : (COCl)₂, CH₂Cl₂, DMF; 5): nBuLi, EtOCOCH₂COOH, THF, -75°C (44%); 6) : HC(OEt)₃, Ac₂O
7) : tBuNH₂, EtOH (57%) ; 8) : NaH, Dioxane (78%);9) : 3-(S)-AminoPyrrolidine,DBU, CH3CN, 60°; 10) : (CF₃CO)₂, CH₂Cl₂;11) CF₃COOH
H₂O₂, r.t; 12) : Ph-CH₂-NH₂, trichloroethylene, reflux; 13) : H₂ - Pd/C, EtOH; 14) : NaOH, reflux then HCl,EtOH

Figure 7

The synthesis started from 2,6-dichloro-5-fluoronicotinic acid which ,after deprotonation and reaction with dimethyldisulfide, gave the 4-methylthio derivative, which was converted to the 5-methylthio-quinolone following conventional methodology. At that stage, aminopyrrolidine was condensed with the 5-methylthio intermediate and the amino group was protected via trifluoroacetylation. The methylthio group was oxidized to a methylsulfonyl restidue to serve as a leaving group. Condensation of benzylamine, hydrogenolysis, and finally deprotection gave the desired 5-aminonaphthyridone.

5-Methyl Quinolones

By analogy with the naphthyridone series, 5-methylquinolones were prepared from the corresponding toluic acids [10] (figure 8).

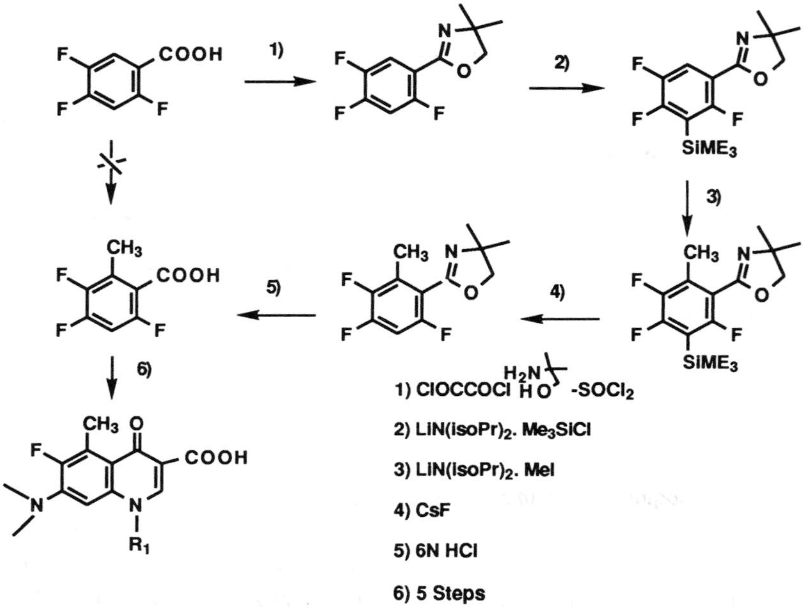

1) ClOCCOCl $\overset{H_2N}{\underset{HO}{\bigvee}}$ -SOCl₂

2) LiN(isoPr)₂. Me₃SiCl

3) LiN(isoPr)₂. MeI

4) CsF

5) 6N HCl

6) 5 Steps

Figure 8

In order to produce the desired isomer the anion, generated from the trifluorooxazoline by lithium diisopropylamide was reacted with trimethylsilyl iodide to afford trimethylsilyloxazoline. This was treated with another equivalent of base and quenched with methyl iodide. Removal of the trimethylsilyl group with cesium fluoride in wet dimethylformamide gave the methyloxazoline which was hydrolyzed with 6N hydrochloric acid to 2-methyl 3,4,6-trifluoro benzoic acid. Elaboration of the desired quinolone was achieved using standard methodologies.

2-Substituted Quinolones

A general method for preparing a wide variety of C-2 substituted quinolones was published in 1989 by J.S. Kiely et al.. [11] This method has the advantage of providing multiple analogs from a common intermediate via its reaction with a Grignard reagent (figure 9). Carbon nucleophiles added to the C-2 position in a Michael fashion afforded the C-2 substituted adducts with good selectivity. Addition of phenylselenyl chloride folloed by oxidation and *in situ* elimination readily produced the desired 2,3-unsaturated ester. All tof he 2-substituted quinolones had no antimicrobial activity, in contrast to the known activity of the C-2 to N-1 bridged analogs, but in agreement with the reported inactivity of the previously known C-2 methylated analogs.

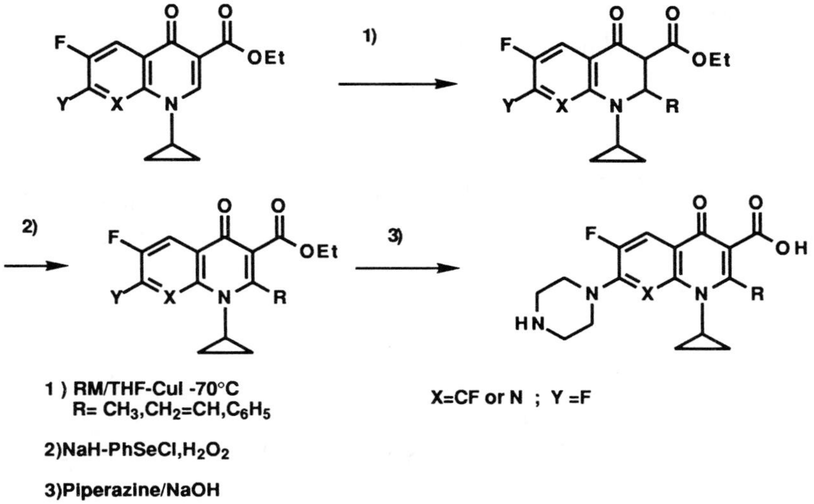

1) RM/THF-CuI -70°C
   R= CH₃,CH₂=CH,C₆H₅

2)NaH-PhSeCl,H₂O₂

3)Piperazine/NaOH

X=CF or N ; Y =F

Figure 9

Tetrahydroisothiazoloquinolones and Naphthyridones

D.T.W.Chu *et al.* [12], discovered that tetrahydroisothiazoloquinolones possessed antibacterial activity 4 to 10 times more greater than quinolones. Recently, they published a practical synthesis of these derivatives (figure 10). The strategy required the condensation of a β-ketoester with an iminochloro-thioformate which was prepared in a "one-pot" procedure from cyclopropyl isothiocyanate and thiophenol followed by reaction with phosphorous pentachloride. Oxidation with *m* -chloroperbenzoic acid yielded the 2-phenylsulfinyl derivative which was displaced with sodium hydrosulfide to a 2-mercapto intermediate. Without purification, this was reacted with hydroxylamine-*O* -sulfonic acid to give a hydrosulfamine derivative which was cyclized *in situ* to yield the desired heterocycle.

Regioselective Displacement Reactions of-5,6,7,8,Tetrafluoro Quinolones

In order to study the effects of several substituents in the 5-position of the quinolone derivatives, a further fluorine atom at C-5 of the 6,8 difluoroquinolone ring system was introduced. Conversion of the C-5 and/or C-7 fluorine atom(s) into other functional groups was performed in the hope of developing new agents with improved antibacterial activity [13].

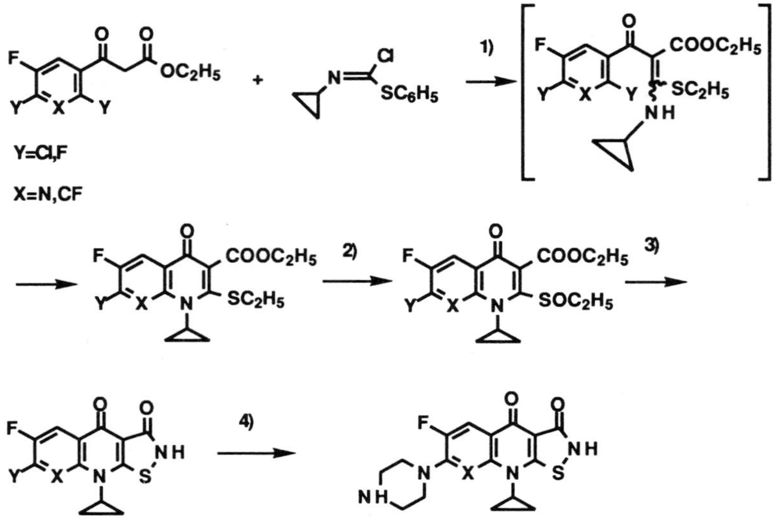

1)NaH,DMF 2)mCPBA,CH₂Cl₂ 3) NaSH,THF-H₂NOSO₃H,NaHCO₃ 4)Piperazine ,Pyridine

Figure 10

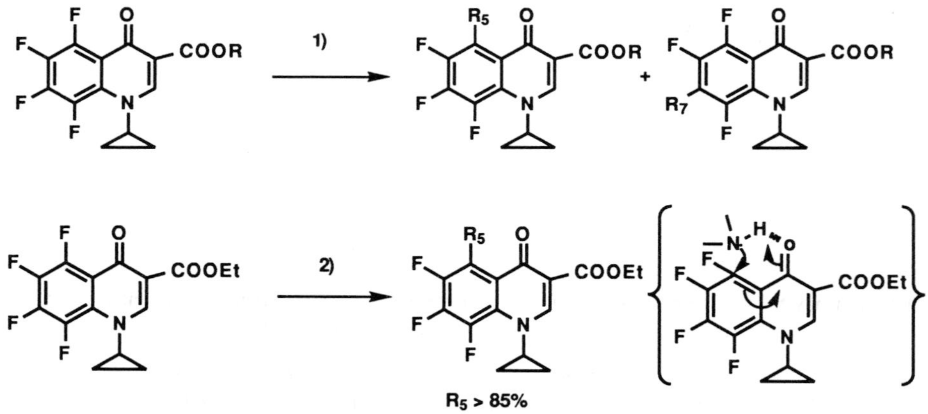

R₅ > 85%

1) ⟩NH or EtO⁻ in solvent

2) ⟩NH   in toluene

Figure 11

Displacement reactions of the tetrafluoro substituted quinolone are predicted to proceed preferentially at C-5 and/or C-7. Displacement reactions of the ester and its carboxylic acid with amines in a solvent produced a mixture of C-5 and C-7 substituted derivatives in both cases. In order to find reaction conditions suitable for the regioselectivity, the influence of the solvent, especially its polarity, was examined. A non-polar solvent provided predominantly the C-5 substituted product. The hydrogen bonding between the amine hydrogen and the C-4- oxo group probably plays an important role ; the lone pair of the amine nitrogen would preferentially attack at C-5. A non-polar solvent makes this reaction more favorable than a polar one (figure 11). In a polar solvent ,the C-7 orientation became quite favorable and the use of the boron-chelated compound with a polar solvent caused a remarkably high regioselectivity at C-7 (figure 12).

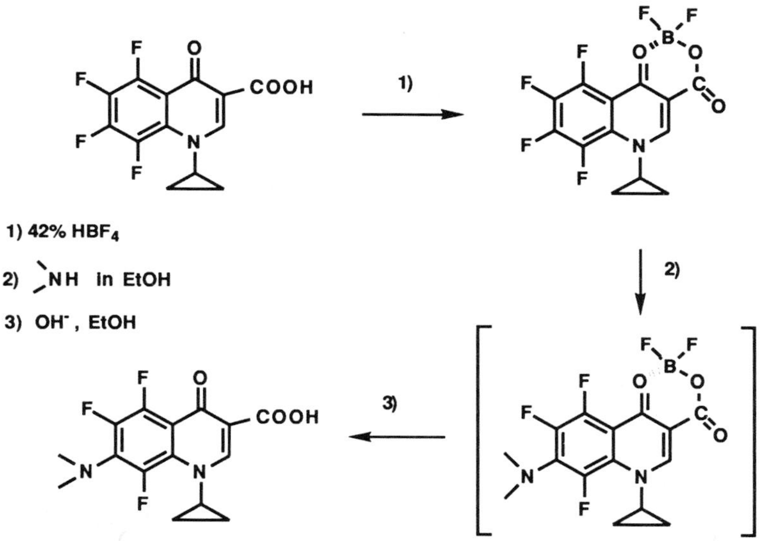

1) 42% HBF₄

2) NH In EtOH

3) OH⁻, EtOH

Figure 12

3- Chiral Derivatives

(S)-Ofloxacin

Ofloxacin has been developed as a highly active new quinolone antibacterial agent against Gram-positive and Gram-negative pathogens. Chemically, it is characterized by a tricyclic structure with a methyl group at the C-3 position of the oxazine ring, thus providing an asymmetric center at this position. The two optical isomers have been reported and the *(S)* -enantiomer was 8 to 128 times more potent than its counterpart *(R)* . The first method to access both enantiomers was based on resolution of synthetic intermediates. However, this method was wasteful since one of the two optical isomers was useless. A new method based on asymmetric reduction of an acyclic amine with chiral reagents was investigated [1] (figure 13). A cyclic enamine was prepared in a manner similar to that reported previously. Chiral sodium triacyloxyborohydrides that had been reported as excellent reducing agents of cyclic amines were effective for the asymmetric reduction of the enamine with a high degree of enantioselectivity (95 % e.e.). Crystallization of the *(R*) (-)-camphor-10-sulfonic acid, followed by treatment of the salt with aqueous sodium hydroxide, yielded the optically pure enantiomer which was easily converted to *(S*)-ofloxacin in 5 steps.

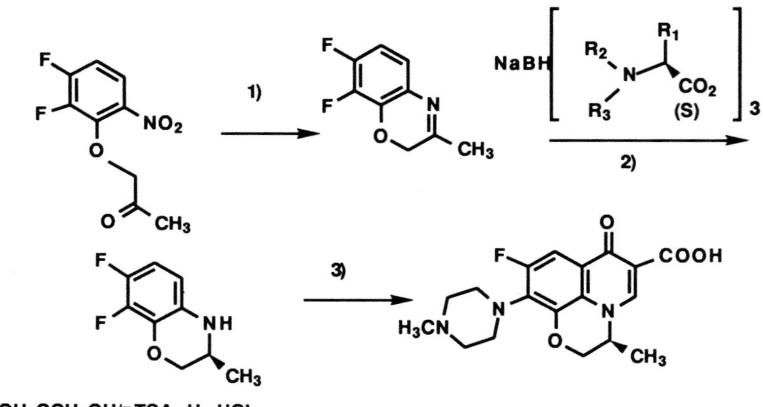

1) HOH₂CCH₂OH/pTSA- H₂-HCl

1) $HOH_2CCH_2OH/pTSA- H_2-HCl$

2) $CH_2Cl_2/-40°C-R^1R^2=-(CH_2)_3-;R^3=COO_2CH_2CHMe_2$

3) 5 Steps (14)

Figure 13

DU - 6859

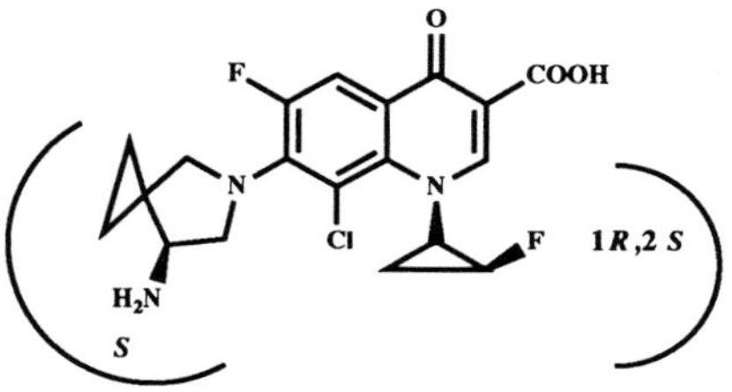

Quite recently, DU - 6859 was found as a new generation of quinolonecarboxylic acids exhibiting excellent antibacterial activtiy and few side effects [15]. This molecule is characterized by a *cis* -oriented (1R,2S)-2-fluorocyclopropylamino moiety at N-1 and (3S)-3-amino-4-spirocycloalkylpyrrolidine at C-7. Further development of DU - 6859 needed an efficient and short synthesis of both new chiral groups.

Synthesis of the fluorocyclopropylamine moiety [16] involved formation of a diphenyl-2-oxazolidin- one which reacted with dimethoxyethane to give a mixture of diastereoisomers. Upon thermal elimination of methanol under reduced pressure, the N-vinyloxazolidone was obtained (figure 14). Reaction of a zinc- monofluorocarbenoid with this N-vinyloxazolidone was next studied. It was found that the addition of 1,2-diethoxyethane as a bidentate ligand and molecular sieves gave the best yield (88%) and *cis*- diastereoselectivity (1+2) 90%. The remarkable diastereofacial selectivity may be explained by the attack of the zinc monofluorocarbenoid from the less hindered side of the vinyl oxazolidinone (NOE was observed between Ha and Hb). 1 was easily isolated by chromatography in 50% yield. Reductive removal of the oxazolidone moiety was readily accomplished by hydrogenolysis in the presence of palladium.

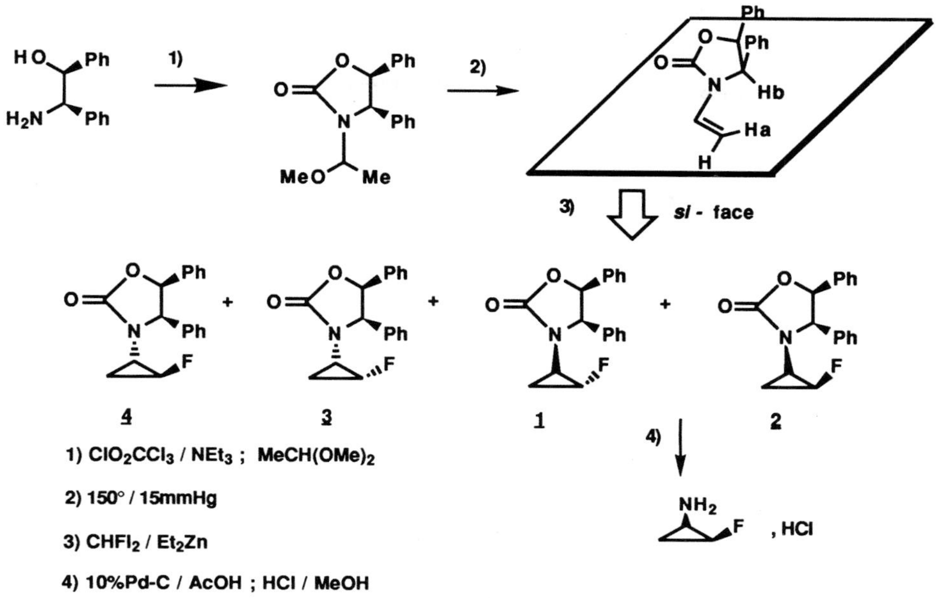

1) ClO_2CCl_3 / NEt_3 ; $MeCH(OMe)_2$

2) 150° / 15mmHg

3) $CHFI_2$ / Et_2Zn

4) 10%Pd-C / AcOH ; HCl / MeOH

Figure 14

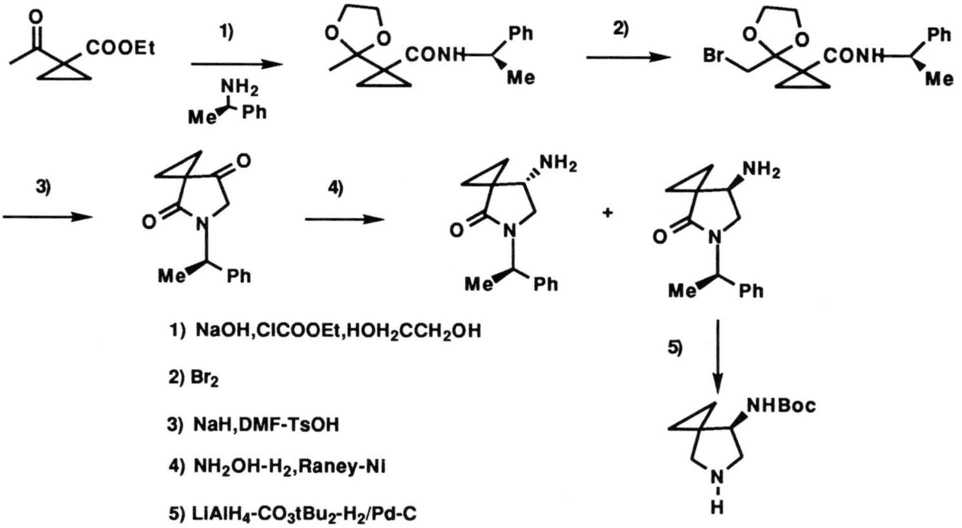

1) $NaOH$,$ClCOOEt$,HOH_2CCH_2OH

2) Br_2

3) NaH,DMF-$TsOH$

4) NH_2OH-H_2,$Raney$-NI

5) $LiAlH_4$-CO_3tBu_2-H_2/Pd-C

Figure 15

The spirocycloalkylpyrrolidine moiety was prepared as depicted in figure 15. Finally, DU-6859 was elaborated from both precursors following classical methodology used in quinolone chemistry (figure 16).

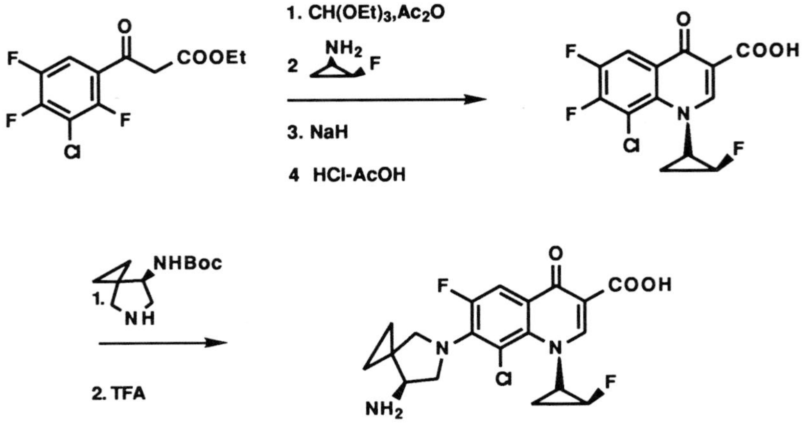

Figure 16

7-Diazabicycloheptane Substituted Naphthyridines

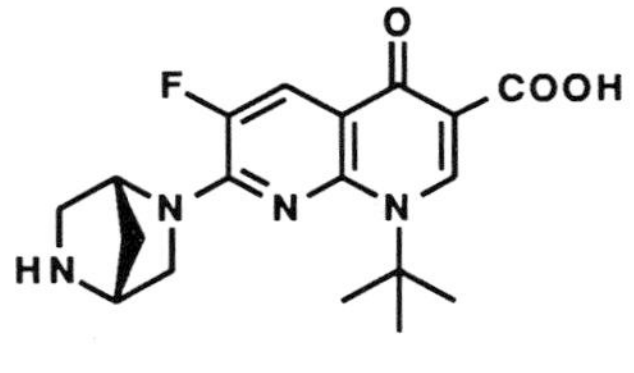

BMY 40062

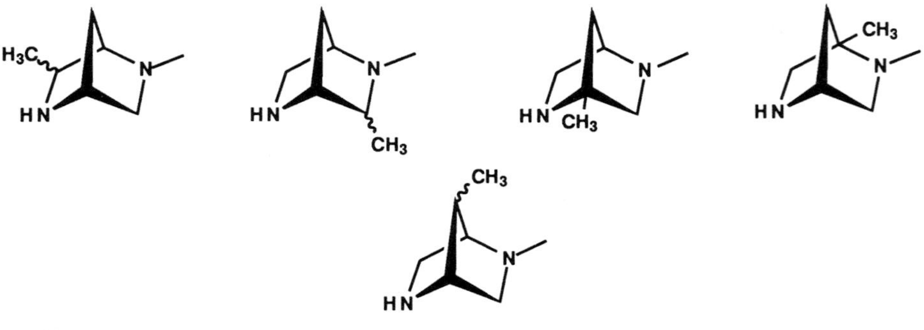

Figure 17

BMY 40062 is a recently discovered naphthyridine with a t-butyl group at N-1 and a (1*R*, 4*S*,)-2,5-diazabicycloheptane at C-7 was recently discovered [17]. This broad-spectrum compound showed improved bioavailability in animals and man. However, it was found to induce pseudoallergic type reactions after oral administration in man. In the anesthetized dog, i.v administration of BMY 40062 was shown to release histamine, causing severe hypotension. In order to decrease this undesirable effect, the bridged piperazine moiety was C-methylated in several positions (figure 17).

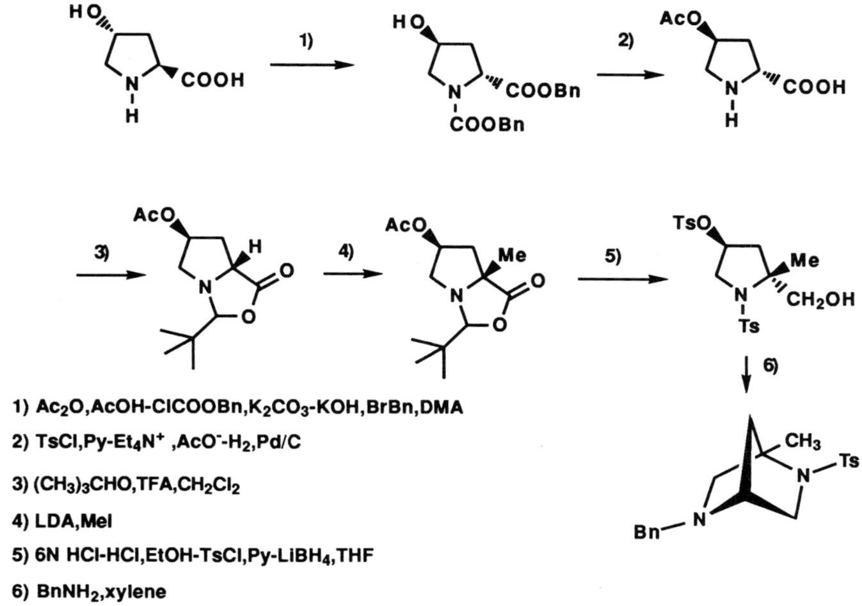

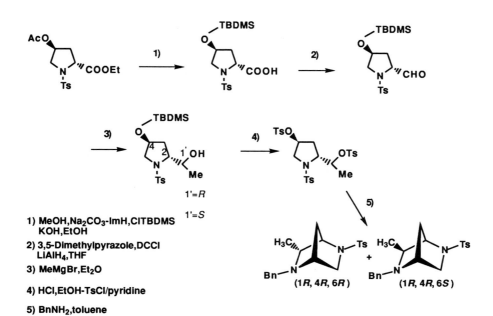

HO
COOH
N
H
1)
HO
COOBn
N
COOBn
2)
AcO
COOH
N
H
3)
AcO
H
N
O
O
4)
AcO
Me
N
O
O
5)
TsO
Me
CH2OH
N
Ts
6)
CH3
N
Ts
Bn
N
1) Ac2O,AcOH-ClCOOBn,K2CO3-KOH,BrBn,DMA
2) TsCl,Py-Et4N+ ,AcO--H2,Pd/C
3) (CH3)3CHO,TFA,CH2Cl2
4) LDA,MeI
5) 6N HCl-HCl,EtOH-TsCl,Py-LiBH4,THF
6) BnNH2,xylene

Figure 18

AcO
COOEt
N
Ts
1)
TBDMS
O
COOH
N
Ts
2)
TBDMS
O
CHO
N
Ts
3)
TBDMS
O
4
2
1'
OH
N
Ts
Me
1'=R
1'=S
4)
TsO
OTs
N
Ts
Me
5)
H3C
N
Ts
Bn
N
(1R, 4R, 6R)
+
H3C
N
Ts
Bn
N
(1R, 4R, 6S)
1) MeOH,Na2CO3-ImH,ClTBDMS
KOH,EtOH
2) 3,5-Dimethylpyrazole,DCCI
LiAlH4,THF
3) MeMgBr,Et2O
4) HCl,EtOH-TsCl/pyridine
5) BnNH2,toluene

Figure 19

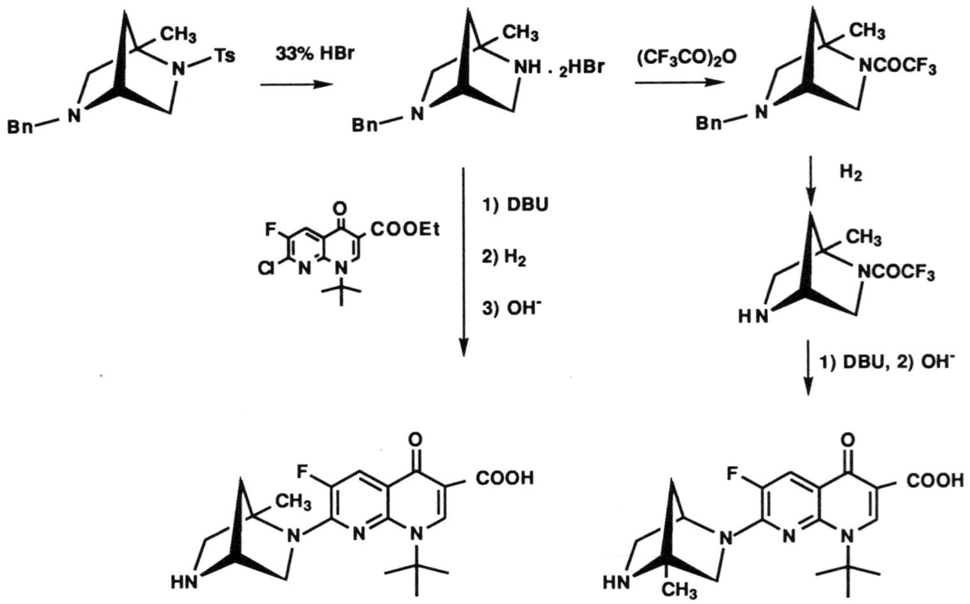

Figure 20

Synthesis [18-19] of C-1 and C-3 methylated diazabicycloheptane is shown in figures 18 and 19 respectively. In both cases *trans*-4-hydroxy -L-proline is the commercially available.starting material Selective protection-deprotection of methyldiazabicycloheptane derivatives allowed unamiguous preparation of the different possible isomers (figure 20).

3 -Dual Penems and Carbapenems

When cephalosporins react with bacterial enzymes, opening the β-lactam ring leads to liberation of the 3'-substituent, if that substituent can function as a leaving group. When the leaving group possesses antibacterial activity of its own, the cephalosporin should exhibit a dual mode of action. Cephalosporins in which antibacterial quinolones are ester-linked to the cephalosporin 3'-position have been described recently [20]. These compounds showed excellent broad-spectrum antibacterial activity which reflects both cephalosporin and quinolone -like contributions. This suggests that these bifunctional compounds can act as targeted prodrugs for the delivery of quinolones at or near the site of action. Penems and carbapenems are two other classes of β-lactam antibacterials which can accomodate a vinylogous-linked quinolone that would be released upon ring-opening of the β-lactam. For the synthesis of the carbamate-linked dual-action carbapenem, the azetidinone part was chloro-formylated and coupled to silylated ciprofloxacin. Subsequent treatment with tetrakis(triphenylphosphine)palladium(0) yielded the carbamate-linked dual-action carbapenem (figure 21).

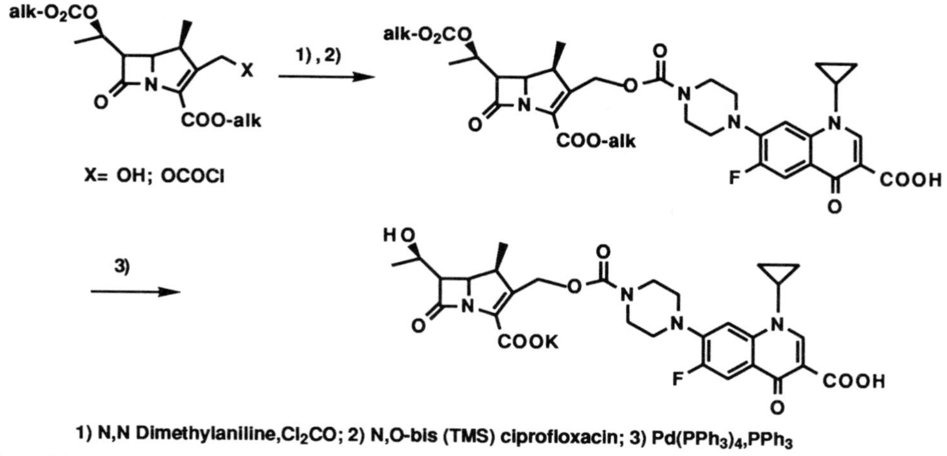

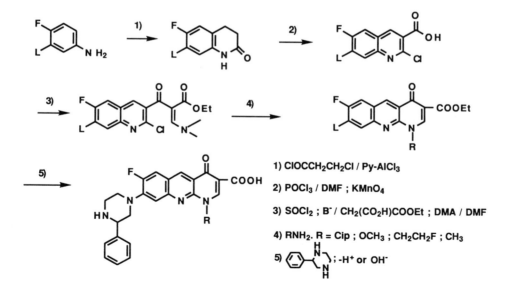

1) N,N Dimethylaniline,Cl₂CO; 2) N,O-bis (TMS) ciprofloxacin; 3) Pd(PPh₃)₄,PPh₃

Figure 21

Benzonaphthyridone

The synthesis of benzonaphthyridones a chemical family which was recently described, has identified a new class of antibacterials with potent *in vitro* and *in vivo* activity against Gram-positive bacteria [21]. The synthesis of benzonaphthyridones involved a 2-chloroquinolone as a key intermediate from which various derivatives were prepared (figure 22).

1) ClOCCH₂CH₂Cl / Py-AlCl₃

2) POCl₃ / DMF ; KMnO₄

3) SOCl₂ ; B⁻ / CH₂(CO₂H)COOEt ; DMA / DMF

4) RNH₂. R = Cip ; OCH₃ ; CH₂CH₂F ; CH₃

5) ; -H⁺ or OH⁻

Figure 22

6-Conclusion

During the last decade, world-wide attention has been given to the synthesis of quinolone derivatives as potential new antibacterial agents. These efforts resulted in the identification of clinically important agents which invariably contain a condensed N-substituted 4-pyridone-3-carboxylic acid moiety. With the exception of compounds of type A ,the above mentioned structural element is present in all potent DNA gyrase inhibitors.

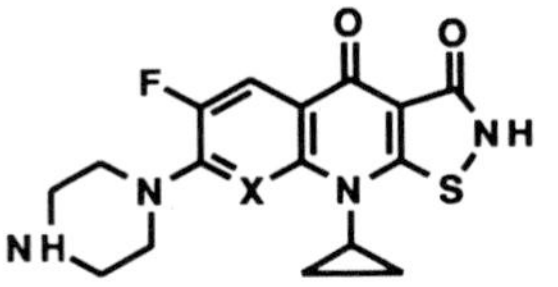

The antibacterial potency of the 4-quinolones is known to be influenced by many factors, several of which have been examined in recent structure-activity studies . All positions on the quinolone ring have been investigated and optimized. Very potent broad spectrum derivatives have been identified which are certainly the most powerful antibacterial agents never synthesized. Nevertheless these compounds have quite often exhibited other properties which caused toxicity and preclude their clinical uses. The most important and relevant toxicity concerns are

    -cytotoxicity linked to the mechanism of action of quinolones,
    (selectivity for DNA gyrase / topoisomerase II),
    -skin and photosensitivity,
    -bone and joint toxicity,
    -crystalluria.

Among the last generation of quinolones, temafloxacin, tosufloxacin and sparfloxacin (figure 23) have emerged as new antibacterial agents useful in human therapy.

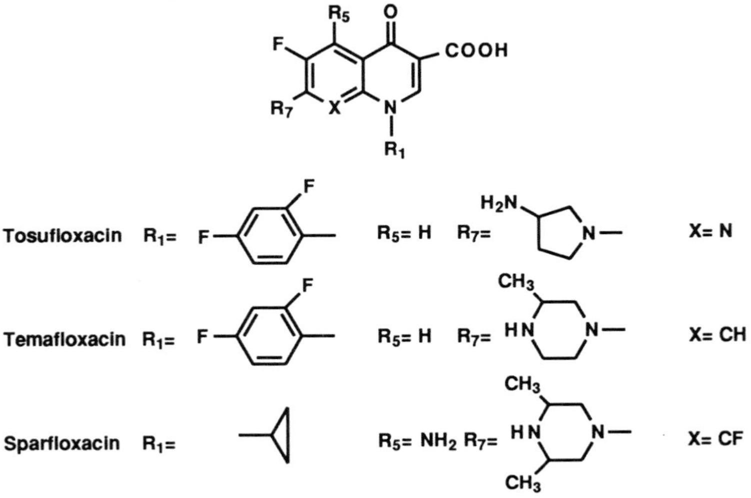

Figure 23

References

1. L. A. Mitscher, P. V. Devasthale, R. M. Zavod: The Quinolones G. C.Crumpin Ed., Springer Verlag.London 1990, 115-146.

2. K. Grohe, H. Heitzer *Liebigs Ann. Chem* **1987**, 29.

3. T. Miyamoto and J. Matsumoto *Chem. Pharm. Bull;* **38** (1990) 3359.

4. E. Laborde, J. S. Kiely, L. E. Lesheski and M. C. Schroeder *J. Heterocyclic Chem.* **28** (1991) 191.

5. E. Laborde. L. E. Lesheski and J. S. Kiely *Tetrahedron Lett.* **31** (1990) 1837.

6. J.S . Kiely, E. Laborde, L. E. Lesheski and R. A. Bucsh *J. Heterocyclic Chem* . **28** (1991) 1581.

7. D. Bouzard, P. Di Cesare, M. Essiz, J. P. Jacquet, P. Remuzon, R. E. Kessler and J. Fung-Tomc *J. Med. Chem.* **35** (1992) 518.

8. P. Di Cesare, D. Bouzard, JP. Jacquet, A. Aulombard, P. Hoffmann *J. Org. Chem;* (submitted)

9. D. Bouzard, P. Di Cesare, P. Hoffmann *Drugs under exp. and cli. Research* (in press); P.Remuzon, D. Bouzard, P. DiCesare, C. Dussy, J. P. Jacquet and A. J aegly *J. Heterocyclic Chem.* **29** (1992) 98.

10. S. E. Hagen and J. M. Domagala *J. Heterocyclic. Chem.* **27** (1990) 1609.

11. J. S. Kiely, S. Huang and L. E. Lesheski *J. Heterocyclic Chem.* **26** (1989) 1675.

12. D. T. W. Chu and A. K. Claiborne *J. Heterocyclic Chem.* **27** (1990) 1191.

13. K. Shibamori, H. Egawa, T. Miyamoto, Y. Nishsmura, A. Itokawa, J. Nakano and J. Matsumoto *Chem. Pharm. Bull.* **38** (1990) 2390.

14. S. Atarashi, H. Tsurumi, T. Fujiwara and I. Hayakawa *J. Heterocyclic Chem.* **28** (1991) 329.

15. I. Hayakawa, S. Atarashi, Y. Kimura, K. Kawakami, T. Saito, T. Yafune, K. Sato, K. Une, M. Sato 31st ICAAC Chicago ,1991, Abstract 1504.

16. O. Tamara, M. Hashimoto, Y. Kobayashi, T. Katoh, K. Nakatani, M. Kamada, I. Hayakawa, T. Akiba and S. Terashima *Tetrahedron Lett.* **33** (1992) 3483.

17. D. Bouzard, P. Di Cesare, M. Essiz, J. P. Jacquet, J. R. Kiechel, P. Remuzon, A. Weber. *J. Med. Chem.* **33** (1990) 1344.

18. P. Remuzon, M. Massoudi, D. Bouzard and J. P. Jacquet *Heterocycles* **34** (1992) 679.

19. P. Remuzon, D. Bouzard, C. Dussy, J. P. Jacquet and M. Massoudi *Heterocycles* **34** (1992) 241.

20. A. J. Corraz, S. L. Dax, N. K. Dunlap, N. H. Georgopapadakou, D. D. Keith, D. L. Pruess, P. L. Rossman, R. Then, J. Unowsky and C. Wei *J. Med. Chem* **35** (1992) 1828.

21. M. Barreau, M. Antoine, J. L. Benichon, J. F. Desconclois, P. Girard, M. Robin, S. Wentzler, G. Picaut ICAAC Chicago 1992, abstract 369.

Cyclopentanoid Antibiotics: New Syntheses Based on Organophosphorous and Organosulfur Reagents

Marian Mikołajczyk

Centre of Molecular and Macromolecular Studies, Polish Academy of Sciences, 90-363 Łódź, Sienkiewicza 112, Poland

Summary. The focus of this report deals with the synthesis of cyclopentanoid antibiotics. More recent results on the synthesis of methylenomycin B (**3**) are discussed. The utility of the intramolecular carbenoid cyclization reaction for the construction of α-phosphoryl-cyclopentenones and α-phosphoryl-cyclopentanones as precursors of methylenomycin B (**3**) and sarkomycin (**4**) is demonstrated. Chemo-enzymatic synthesis of optically active sarkomycin is also presented. A general approach to the synthesis of cycloalkenones from bis-β-ketophosphonates is described and its synthetic utility is exemplified by the total synthesis of optically active isoterrein (**5**).

Introduction

In recent years a major focus of the author's laboratory has been the efficient preparation of 1,4-dicarbonyl compounds and functionalized cyclopentenones and cyclopentanones using organic phosphorus and sulfur compounds as reagents. [1] The 2-cyclopentenone or cyclopentanone moiety is part of several important naturally occurring products. Among them, a group of antibiotics known as the cyclopentanoid class attracted considerable attention because of a wide spectrum of biological activity and structural diversity. For example, methylenomycin A (**1**), desepoxy-4,5-didehydromethylenomycin A (**2**) and methylenomycin B (**3**), which have recently been isolated from the culture both of *Streptomyces* species, [2-5] show a high activity against broth gram-positive and gram-negative bacteria. Moreover, methylenomycin A (**1**) was active against Lewis lung carcinoma in mice. [6]

A closely related cyclopentanoid antibiotic is sarkomycin (**4**) which was first isolated by Umezawa et al. [7] in 1953 and its structure established in 1955. [8] In addition to antibacterial and antiphage properties, sarkomycin also shows antitumor activity. [9] In spite of the chemical instability of sarkomycin and some problems in its storage, the pharmacological studies of this antibiotic led to marketing in the USA of a preparation containing **4** as an antitumor drug.

From the point of view of structural similarity, terrein (**5**) may also be put into the class of cyclopentanoids. Terrein (**5**) was isolated as a metabolite of Aspergillus terreus by Raistrick and Smith [10] in 1935 and its structure was established twenty years later by Grove [11] and Barton. [12] In this case the cyclopentanone ring contains two asymmetric carbon atoms at C(4) and C(5). At the beginning of our work the only syntheses of racemic **5** have been reported. [13-15] Recently, Altenbach and Holzapfel [16] reported the synthesis of (+)-terrein.

Although the structures of all of the above mentioned cyclopentenones and cyclopentanones are deceptively simple, their synthesis is not trivial. This is due to the chemical instability of these compounds as well as some of the synthetic intermediates. Therefore, cyclopentanoid antibiotics became attractive synthetic targets in numerous laboratories. As part of our program on the application of phosphorus and sulfur compounds in organic synthesis we have also been engaged for a few years in the preparation of these compounds. The aim of the present account is to summarize the recent results obtained in the author's laboratory on the synthesis of methylenomycin B (**3**), sarkomycin (**4**) and terrein (**5**).

Syntheses of Methylenomycin B

In our initial studies on the synthesis of 1,4-diketones, a general strategy was developed for the construction of the 1,4-dicarbonyl skeleton and 2-cyclopentenone unit which involves the Horner-Wittig reaction of the properly substituted α–phosphoryl sulfides (for example **6**) with mono- or 1,3-di-carbonyl compounds. [17] According to this approach, 2,3-dimethyl-cyclopenten-2-one (**7**) - a precursor of methylenomycin B - was successfully prepared. The exocyclic methylene function was then introduced to the cyclopentenenone ring in **7** according to the procedure of Jernow et al., affording methylenomycin B (**3**) in 16% overall yield. [18]

Since our first synthesis of **3** was not efficient, we devised a new approach to this target starting from the readily available diethyl 2-oxo-propanephosphonate (**8**). [19] Its Horner-Wittig reaction with methylthioethanal gave the corresponding vinyl ketone to which a propionyl anion equivalent was added in a 1,4-fashion. Base-catalyzed cyclization of the 1,4-diketone led to the formation of the cyclopentenone **9**. Oxidative elimination of the methylthio group in **9** afforded methylenomycin B (**3**) in 26% overall yield.

Taking into account that the Horner-Wittig reaction is a well-established method for olefinic bond formation, it was expected that this reaction would also be suitable for the introduction of the exocyclic α-methylene group into the cyclopentenone ring. In accord with such a strategy, the α-phosphoryl cyclopentenone **11** should be a precursor of **3**. Its synthesis has been accomplished starting from diethyl 2-

oxobutanephosphonate (10) in three steps including a) addition of n-propanal to the keto-protected phosphonate, b) oxidation of the α-hydroxy - adduct and c) base-catalyzed cyclization of the phosphorylated 1,4-diketone 12 to 11. In the final step, the Horner-Wittig reaction of 11 performed under very mild conditions gave methylenomycin B (3) in 36% overall yield. [20] Scheme 1 depicts the above mentioned synthetic approaches to methylenomycin B (3).

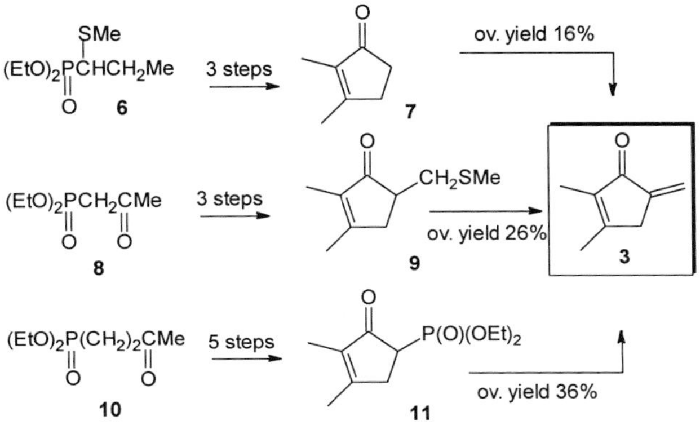

Scheme 1

Recently, a more effective and shorter synthesis of 3 from the β-ketophosphonate 8a (see Scheme 2) has been devised. [20] The key step in this synthesis involves alkylation of the α-phosphonate anion derived from 8a with 1-bromo-2-methoxy-prop-2-ene as a synthetic equivalent of bromoacetone. The monoalkylated product upon acidic hydrolysis gave the desired phosphorylated 1,4-diketone 12 upon acidic hydrolysis. Two already elaborated reactions, i.e. the base-catalyzed cyclization of 12 to 11 and the Horner-Wittig reaction of the latter, completed the synthesis of methylenomycin B (3) in 36% overall yield.

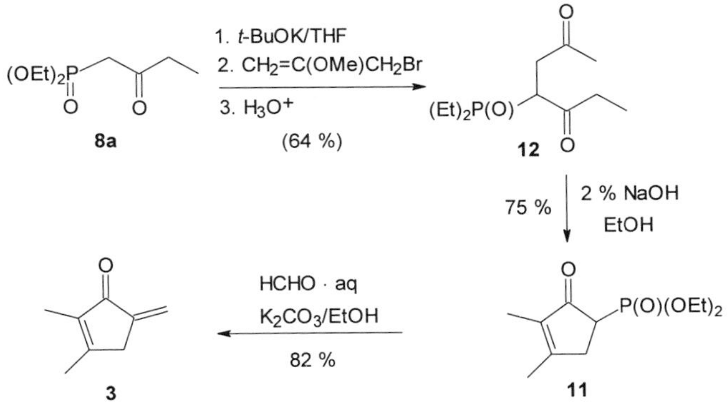

Scheme 2

A characteristic feature of the synthetic approaches to methylenomycin B (3) presented above is that the cyclopentenones 7,9 and 11 - precursors of 3 - are prepared by the intramolecular base-catalyzed cyclization of the corresponding 1,4-diketones. In this regard, the synthesis of methylenomycin B (3) shown in Scheme 3 is completely different because the construction of the cyclopentenone ring is based on the intramolecular carbenoid cyclization. [21]

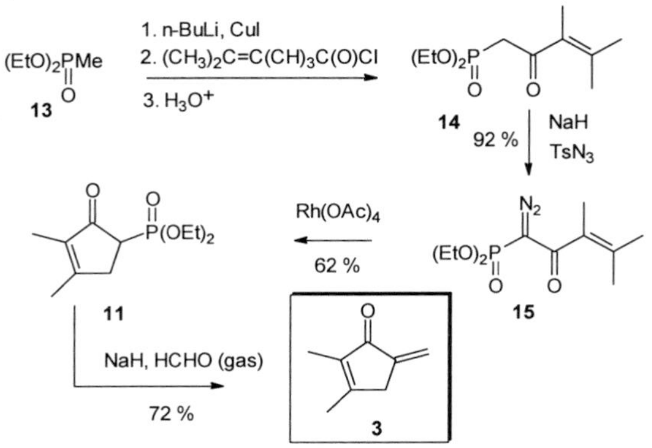

Scheme 3

In this synthesis of **3**, diethyl methanephosphonate **13** was used as a substrate. In the first step, the lithio-copper salt of **13** was acylated with 2,3-dimethyl-2-butenoyl chloride to give the corresponding β-ketophosphonate **14**. In the next step of the synthesis, **14** was transformed into α-diazo derivative **15** under typical conditions of the diazo-transfer reaction. The latter compound underwent intramolecular carbenoid cyclization in the presence of catalytic amounts of rhodium (II) acetate to give the cyclopentenone **11**. To complete the synthesis of **3**, **11** was treated with sodium hydride and then with gaseous formaldehyde. The total yield of methylenomycin B (**3**) from **13** was 24%.

Synthesis of Racemic and Enantiomerically Pure Sarkomycin

In continuation of our work on the synthesis of cyclopentanoid antibiotics, we selected sarkomycin (**4**) as a second target. Since sarkomycin like methylenomycin B (**3**) contains the reactive exocyclic α-methylene moiety, our strategy was to synthesize 2-diethoxyphosphoryl-3-carboxy-cyclopentanone **19** as a key intermediate which should be converted into sarkomycin (**4**) via the Horner-Wittig reaction. Moreover, as in the case of the synthesis of methylenomycin B shown in Scheme 3, we decided to utilize the intramolecular carbenoid cyclization for the construction of the suitably substituted cyclopentanone ring.

A total synthesis of (±)-sarkomycin (**4**) from diethyl 2-oxopropanephosphonate (**8**) is shown in Scheme 4 and is discussed briefly below. [22]

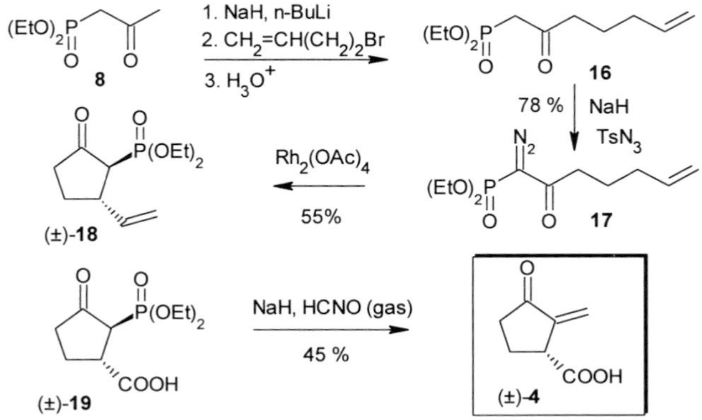

Scheme 4

The dianion generated from **8** was reacted with homoallyl bromide to give the β-ketophosphonate **16** which, on treatment with sodium hydride and tosyl azide, afforded the corresponding α-diazophosphonate **17**. The next important step in the synthesis of **4** was the intramolecular cyclization of **17** catalyzed by rhodium (II) acetate leading to 2-diethoxyphosphoryl-3-vinyl-cyclopentanone **18**. Ozonolysis of the vinyl moiety in **18** and subsequent oxidation of the corresponding aldehyde with Jones reagent resulted in the formation of the cyclopentanone **19**, a precursor of sarkomycin (**4**). To complete the preparation of **4**, the Horner-Wittig reaction of **19** with formaldehyde was carried out. After the usual work-up and column chromatography (±)-sarkomycin (**4**) was obtained in 12% overall yield.

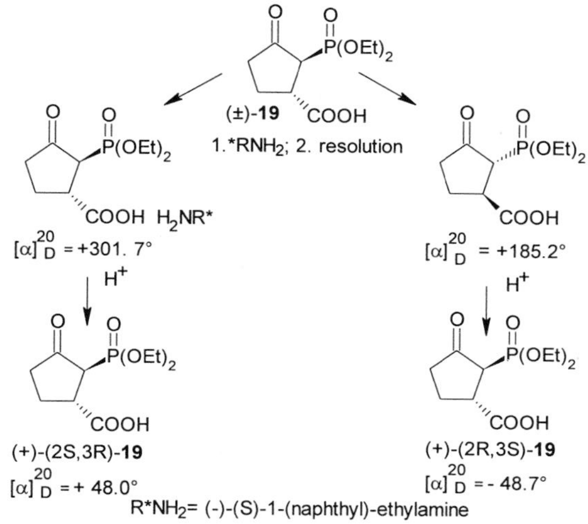

Scheme 5

Although sarcomycin (**4**) has been the target of numerous synthetic efforts, there are only a few reports describing the synthesis of a chiral compond. [23] Taking into account that the sarcomycin precursor **19** is fairly stable (in contrast to sarcomycin itself) and contains the carboxylic group as a resolving handle, it was possible to resolve the racemic cyclopentanone **19** via diastereoisomeric salts. To this end the diastereomeric salts with (-)-(*S*)-1-(1-naphthyl)-ethylamine were obained. Their resolution was effected by flash chromatography using ethyl acetate-hexane as eluent. The enantiomerically pure acids **19** were isolated from these salts by chromatorgrphy using ion exchange. Some experimental data concerning resolution are given in Scheme 5.

It is interesting to point out that the partially resolved samples of the carboxy-cyclopentanone **19** show enantiomeric non-equivalence in ^{31}P-NMR spectra measured in benzene. This new example of chiral self-discrimination may be rationalized in terms of the short-lived diastereomeric dimers shown below.

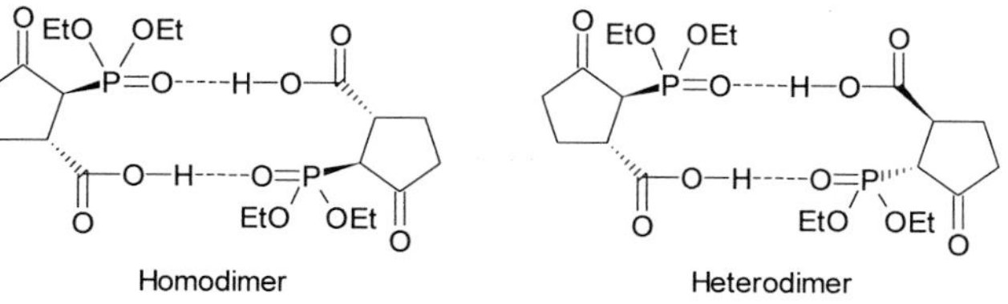

Homodimer   Heterodimer

Another, simple approach to optically active acids **19** was based on enzymatic hydrolysis of (±)-2-diethoxyphosphoryl-3-carbomethoxy-cyclopentanone (**20**). This reaction, when carried out to ca. 50% conversion, afforded optically active acid **19** and unreacted ester **20** having opposite configuration at the carbon atom C(3). The most efficient kinetic enzymatic resolution was observed with α-chymotrypsin. In this case enzymatic hydrolysis gave the acid (-)-**19** with 77% ee.. The results of enzymatic hydrolysis of (±)-**20** are collected in Table 1.

Table 1. Preparation of Optically Active Acid **19** by kinetic Enzymatic Resolution of Methyl Ester **20**

Enzyme	pH	Conversion [%]	Acid $[\alpha]_D^{20}$	ee [%]*
Pig liver esterase	7	47	+5.6	11.5
Pig liver esterase	8	47	+7.2	14.8
α-Chymotrypsin	8	50	-37.4	76.8
Pronase from Streptomyces griseus	8	55	-8.1	16.6
Pronase from Bacillus subtilus	8	50	-16.8	34.5
Lipase from Pseudomonas fluorescens	8	no reaction	-	-
Lipase from Mucor javanicus	8	45	-1.2	2.5

* Calculated on the basis of $[\alpha]_D^{20} = -32.5°$ given by Umezawa et al. [9]

With the enantiomeric acids **19** in hand, the last and crucial step of the synthesis of optically active sarkomycin (**4**) was examined. It was found that the Horner-Wittig reaction of (+)-**19** and (-)-**19** with formaldehyde gave (-) and (+)-sarkomycin (**4**), respectively. Comparison of the ee values of the starting acids **19** with those of sarkomycin **4** indicates that the Horner-Wittig reaction proceeds practically without racemization at the asymmetric carbon atom C(3).

$[a]_D^{20} = +48.0°$ (ee 98.5 %) $[a]_D^{20} = -30.9°$ (ee 95.1 %)

$[a]_D^{20} = -37.4°$ (ee 76.8 %) $[a]_D^{20} = +24.1°$ (ee 74.1 %)

*Calculated on the basis of $[\alpha]_D^{20} = -32.5°$ given by Umezawa et al. [9]

Synthesis of Optically active Isoterrein

Before discussing our synthesis of isoterrein (**5**), it seems desirable to briefly present a general approach to cycloalkenones utilizing bis-β-ketophosphonates **21** as substrates. Bis-β-ketophosphonates **21** are a practically unknown class of compounds and, as we found, may easily be prepared from dicarboxylic acid esters and α-phosphonate carbanions. As expected, they undergo intramolecular Horner-Wittig reaction to give cycloalkenones **22** in good yields.

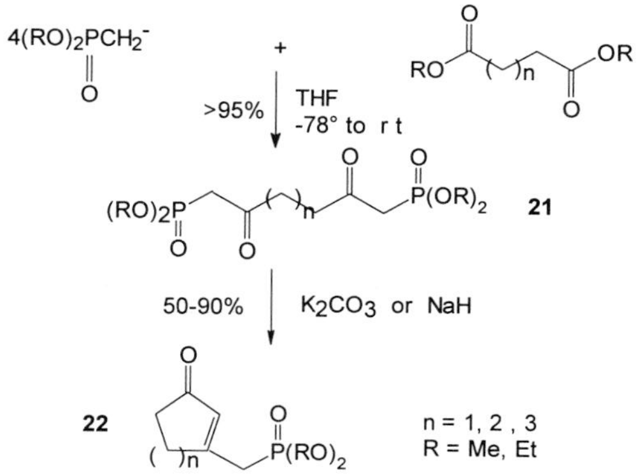

Due to the presence of the phosphonate moiety, cycloalkenones **22** may be used for further transformations. For example, proton abstraction from **22** leads to the corresponding anion exhibiting typical ambivalent reactivity as a consequence of the negative charge distribution between three atoms: α-phosphonate carbon, α-carbonyl carbon and oxygen (see Scheme 7). It is worth noting that the reaction of this anion with carbonyl compounds gives the Horner-Wittig reaction products.

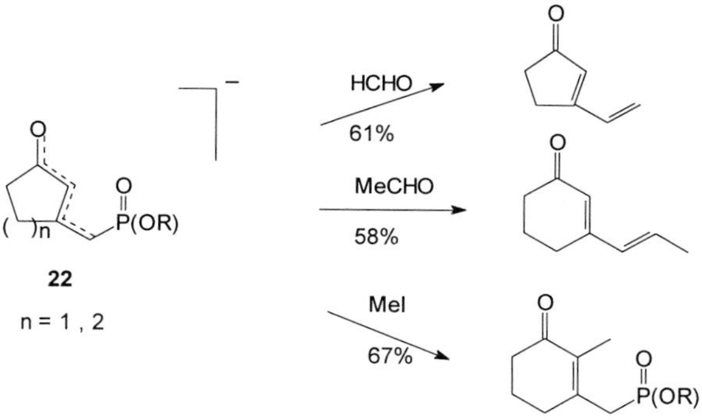

Scheme 7

Synthetic utility of cyclopentenones of the type **22** (n = 1) was illustrated by Altenbach and Holzapfel [16] by the synthesis of optically active (+)-terrein (**5**) from (-)-(R,R)-dimethyl tartrate (**23**), which has the same chirality of the diol moiety.

In our independent studies [24] the protected derivative **24** of (-)-dimethyl tartrate **23** was treated with the dimethyl methanephosphonate anion to give the corresponding bis-β-ketophosphonate **25**.

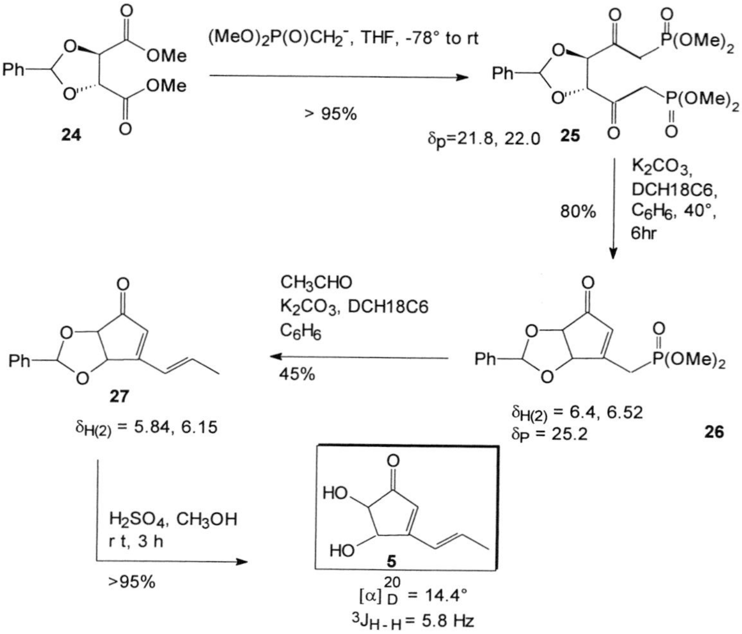

Scheme 8

Base-catalyzed cyclization of **25** afforded the cyclopentenone **26**. The anion of the latter upon treatment with acetaldehyde gave the expected Horner-Wittig reaction product **27**. In the final step of the synthesis, the acetal protecting group was removed under acidic conditions and terrein **5** was obtained in ca. 35% overall yield.

Unexpectedly, the synthetic terrein obtained as shown in Scheme 8 exhibited different properties from those of the natural (+)-terrein. First of all, the optical rotation of our product was -14.4°, whereas that of the isolated terrein is +185°. [10] Moreover, the coupling constant between protons at C(4) and C(5) of 5.8 Hz is similar to that reported for racemic isoterrein **5** (6.0 Hz)15 and different from that of natural terrein (2.6 Hz). [12] Therefore, based on the coupling constant values, it is reasonable to assume that our synthesis resulted in the formation of optically active isoterrein **5** having the *cis* configuration of the diol moiety. Most probably, the *trans* to *cis* isomerization at the asymmetric carbon atoms takes place during the intramolecular Horner cyclization of **25** to **26**. A similar isomerization has been observed by Bestmann and Roth [25] during cyclization using phosphonium ylides.

The optical purity and absolute configuration of (-)-isoterrein obtained as shown in Scheme 8 is currently investigated in our laboratory.

Acknowledgement. The author is indebted to his coworkers, whose names appeared in the references, for their valuable contribution to realization of this research program. Since 1991 this program was financially supported by the State Committee for Scientific Research (in part within the Grant No 206189101).

References

1. For a recent summary account see: M. Mikolajczyk, *Reviews on Heteroatom Chemistry* **2** (1989) 19-39.

2. T. Haneishi, N. Kitahara, Y. Takiguchi, M. Arai, S. Sugawara, *J. Antibiot.* **27** (1974) 386-392.

3. T. Haneishi, A. Terahara, M. Arai, T. Hata, C. Tamura, *J. Antibiot.* **27** (1974) 393-399.

4. T. Haneishi, A. Terahara, K. Hamono, M. Arai, *J. Antibiot.* **27** (1974) 400-407.

5. U. Horneman, D. A. Hopwood, *Tetrahedron Lett.* (1978) 2977-2800.

6. A. Terahara, T. Haneishi, M. Arai, *Heterocycles* (1979) 353-358.

7. H. Umezawa, T. Takeuchi, K. Nitta, T. Yamamoto, S. Yamaoka, *J. Antibiot., Ser. A* **6** (1953) 101-106.

8. I. R. Hooper, L. C. Cheney, M. J. Gron, O. B. Fardig, D. A. Johnson, D. L. Johnson, F. M. Patermiti, H. Schmitz, W. B. Wheatley, *Antibiot. Chemother.*, (Washington D. C.) **5** (1955) 585-95.

9. H. Umezawa, T. Yamamoto, T. Takeuchi, T. Osato, Y. Okami, S. Yamaoka, T. Okuda, K. Nitta, K. Yagishita, R. Utahara, S. Umezawa, *Antibiot. Chemother.* (Washington D. C.) **4** (1954) 514-20.

10. H. Raistrick, G. Smith, *Biochem. J.* **29** (1935) 606-611.

11. J. F. Grove, *J. Chem. Soc.* (1954) 4693-4694.

12. D. H. R. Barton, E. Miller, *J. Chem. Soc.* (1955) 1028-1029.

13. J. Auerbach, S. M. Weinreb, *J. Chem. Soc. Chem. Commun.* (1974) 298-299.

14. D. H. R. Barton, L. A. Hulshof, *J. Chem. Soc. Perkin 1* (1977) 1103-1106.

15. A. J. M. Klunder, W. Bos, B. Zwanenburg, *Tetrahedron Lett.* **22** (1981) 4557-4560.

16. H. -J. Altenbach, W. Holzapfel, *Angew. Chem.* **102** (1990) 64-65.

17. M. Mikołajczyk in H. Nozaki (ed.) *Current Trends in Organic Synthesis.* Pergamon Press, Oxford and New York, 1983, pp. 347-358.

18. M. Mikołajczyk, S. Grzejszczak, P. Łyżwa, *Tetrahedron Lett.* **23** (1982) 2237-2240.

19. M. Mikołajczyk, P. Bałczewski, *Synthesis* (1987) 659-661.

20. M. Mikołajczyk, A. Zatorski, *J. Org. Chem.* **56** (1991) 1217-1223.

21. M. Mikołajczyk, R. Żurawiński, *Synlett* (1991) 575-576.

22. M. Mikołajczyk, R. Żurawiński, P. Kiełbasiński, *Tetrahedron Lett.* **30** (1989) 1143-1146.

23. G. Helmchen, K. Ihrig, H. Schindler, *Tetrahedron Lett.* **28** (1987) 183-186 and references therein.

24. M. Mikołajczyk, M. Mikina, unpublished results.

25. H. J. Bestmann, D. Roth, *Angew. Chem.* **29** (1990) 99-100.

Synthesis of Antimycotically Active Dipeptides

Hanno Wild

Bayer AG, Chemistry Science Laboratories Pharma

P.O. Box 101709, 5600 Wuppertal 1, Germany

Summary: The enantioselective and diastereoselective synthesis of the antimycotically active dipeptides chlorotetaine, bromotetaine, bacilysin and of the unusual amino acid anticapsin are reported. All natural products posses the (*S*)-configuration at C-1' of the substituted cyclohexenyl residues of the *C*-terminal amino acids, which contradicts the assignments in the literature.

Introduction

Chlorotetaine and bromotetaine are natural products recently isolated from *Bacillus subtilis* strain BGSC 1E2 [1]. They are antimycotically active against a broad range of fungi, most probably by irreversibly inhibiting glucosamine-6-phosphate synthetase and in this way interfering with cell wall biosynthesis.

The compounds are dipeptides of alanine and an unusual and new halogen containing amino acid. The chiral C-atom in the cyclohexenyl residue (C-1' position) has been assigned the (*R*)-configuration by analysis of the circular dichroism (CD) spectrum of chlorotetaine.

CHLOROTETAINE (X = Cl)
BROMOTETAINE (X = Br)

As larger quantities of the compounds were required for complete evaluation of their biological properties, a general synthesis of chlorotetaine and bromotetaine was developed, which also should allow the introduction of other substituents of interest.

In planning the synthesis, it had to be borne in mind that the dipeptides are stable only in solution at a pH between 3-5. Above pH 7 the biological activity decreases rapidly, especially on heating [1]. The main reaction in alkaline media is the intramolecular 1,4-addition of the amide to the enone system, with formation of a 6-oxo-perhydroindole. Since the risk of this side reaction is present during all intermediate steps of the synthesis, the enone should not be formed until the latest possible moment. Further synthetic problems are the possibility of the haloenone to aroma-tize under more drastic conditions and the control of the relative stereochemistry of the cyclohexenyl residue.

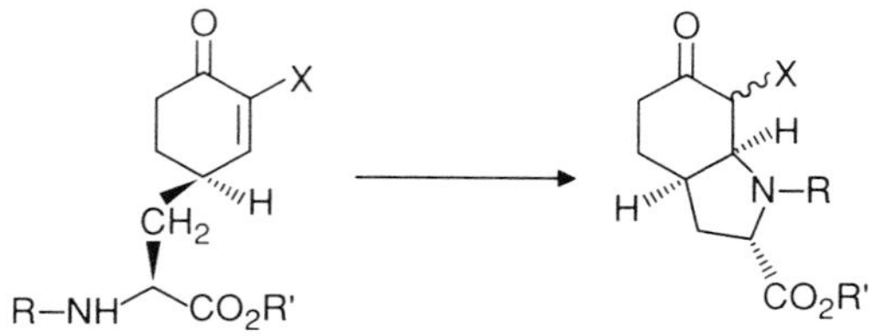

Synthesis of Chlorotetaine and Bromotetaine[2]

Retrosynthetically, the chloroenone of chlorotetaine can be derived from a silyl enol ether, which itself would be the product of a known diastereoselective deprotonation-silylation sequence of a 4-substituted cyclohexanone. The α-amino acid is built up by an enantioselective 1,6-addition of a chiral amino acid synthon to 4-methylene-2-cyclohexenone (Scheme 1).

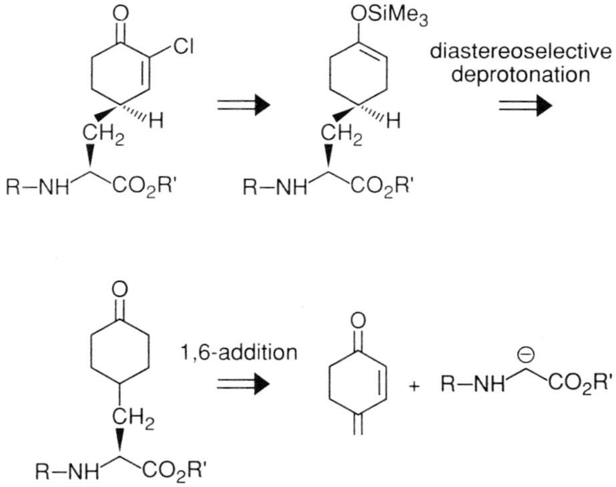

Scheme 1

4-Methylene-2-cyclohexenone is available by a two-step sequence starting from p-anisalcohol: Birch-reduction followed by acidic hydrolysis enol ether and concurrent water elimination of water (Scheme 2) [3]. As the α-amino acid, synthon a bislactim ether was chosen [4]. 1,6-Addition of the cuprate of the bislactim ether was highly stereoselective to give the 3-enone after kinetic protonation in moderate yield as a single diastereomer together with some of the undesired 1,4-adduct. After hydrogenation of the double bond, the desired configuration at C-1' was produced as described by Koga [5] and Simpkins [6] by diastereoselective deprotonation with lithium (R,R)-bis-(phenylethyl)amide. By analogy with the reaction of simple 4-alkylcyclohexanones, the principal diastereomer should possess the indicated configuration. The diastereoselectivity of the reaction was somewhat variable (de = 60-80%); however, other chiral bases used and lower temperatures did not improve the selectivity.

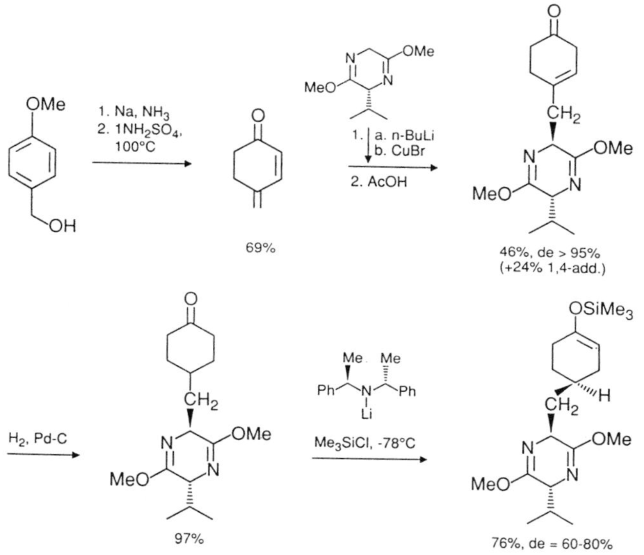

Scheme 2

To generate the necessary 2-chloroenone system, the silyl enol ether was first converted into a diazo compound (scheme 3). Interestingly, the lithium enolate was obtained by treatment of the silyl enol ether with excess methyl lithium at 0 °C without concurrent deprotonation of the bis-lactim ether.

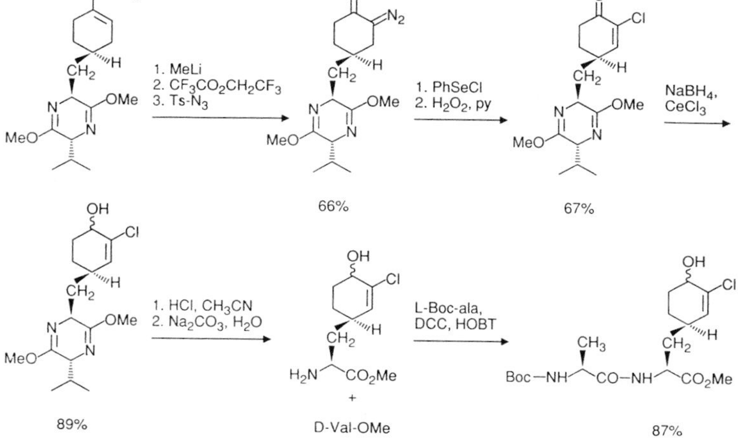

Scheme 3

C-Acylation of the enolate with trifluoroethyl trifluoroacetate [7] followed by reaction with p-tosyl azide gave the α-diazo cyclohexanone. This compound reacted smoothly with one equivalent of phenyl selenyl chloride to an intermediate which, after oxidative elimination, gave rise to the desired 2-chloroenone [8]. Before hydrolysis of the bislactim ether, the enone had to be reduced to the allylic alcohol because of the high tendency of the free amine to undergo an intramolecular 1,4-addition as mentioned before. After reduction, the hydrolysis was uneventful. The crude product was coupled with L-Boc-alanine and the desired dipeptide purified by chromatography on silica gel.

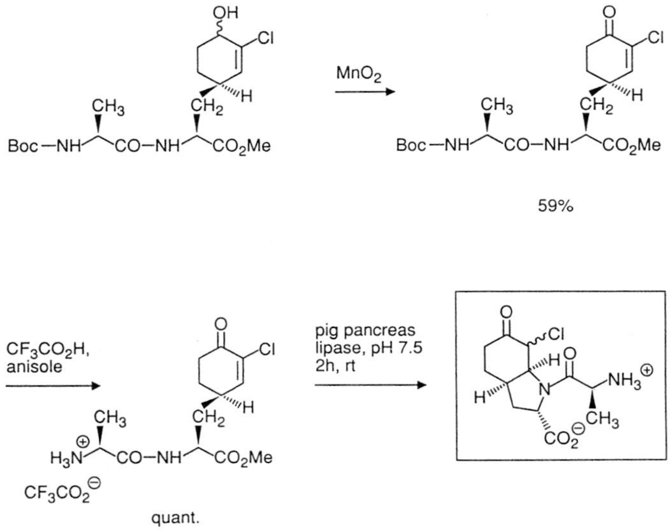

Scheme 4

This dipeptide was reoxidized to the cyclohexenone in a good yield by manganese dioxide (scheme 4). At this stage, small quantities of the undesired diastereomer were easily removed by chromatography. After cleavage of the Boc protecting group, the methyl ester was saponified with porcine pancreas lipase at pH 7.5 and ambient temperature. To our great disappointment, despite these mild conditions not a trace of the desired free dipeptide was obtained. Rather, a diastereomeric mixture of the octahydroindole arising from intramolecular 1,4-addition was isolated.

Parallel synthesis of the epimeric dipeptide having the (*S*)-configuration in the cyclohexenyl residue was undertaken (Scheme 5). Deprotonation of the cyclohexanone in the presence of trimethyl-silyl chloride by (*S,S*)-bis(phenylethyl)amide yielded the desired epimeric silyl enol ether with the same selectivity. The protected dipeptide was elaborated in a reaction sequence completely analogous to that of its (*R*)-epimer. Cleavage of the Boc protecting group, followed by lipase-catalyzed saponification and reverse-phase chromatography to separate decomposition products, this time yielded a dipeptide which was in every respect (NMR, HPLC and most significantly CD) identical to a sample of natural chlorotetaine kindly provided to us by *Prof. Jung*. On the premises that the diastereoselective deprotonation did occur with the expected selectivity, the configuration of chlorotetaine at the C-1' position has to be corrected to (*S*).

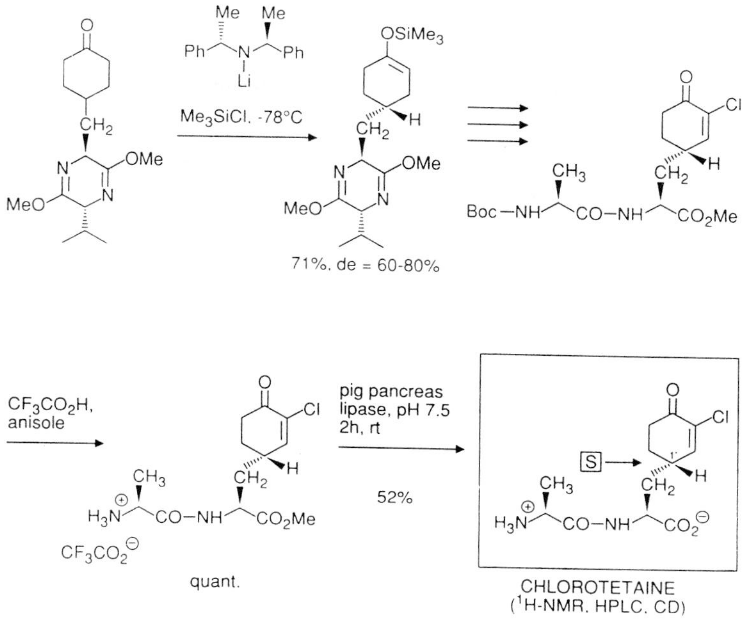

Scheme 5

By substituting phenyl selenyl chloride with phenyl selenyl bromide, with the (S)-α-diazo-cy-clohexanone, is converted by the same reaction sequence to bromotetaine (scheme 6). In this case, the identity of the natural and the synthetic product was shown by comparison of their NMR spectra. Most significantly, their ^{13}C-NMR spectra were superimposable with differences of chemical shifts less than 0.1 ppm.

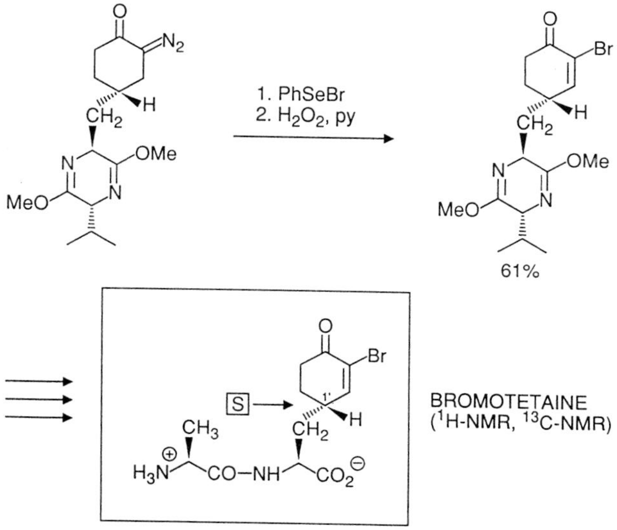

Scheme 6

The higher stability of chlorotetaine compared to its epimer is easy to explain (scheme 7): whereas cyclization of the (*R*)-epimer leads to a perhydroindole with an *exo*-carboxyl substituent, the same reaction starting from chlorotetaine is made more difficult by the formation of a product with an *endo*-carboxyl residue. Under the conditions of the enzymatic saponification reaction (pH 7.5, 23°C, no enzyme), chlorotetaine has a half-life of 17.5 h whereas the methyl ester is much more reactive with a half-life of only 30 min. Fortunately, the enzymatic saponification reaction is slightly faster. However, the half-life of the epimeric methyl ester under these conditions is less than 5 min, which does explain that no trace of the free dipeptide could be isolated.

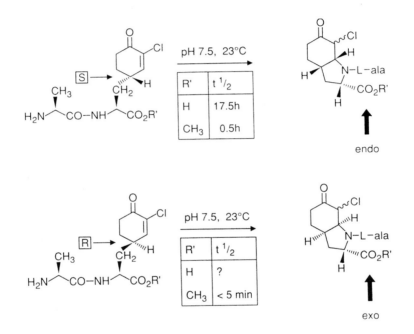

Scheme 7

That the selective deprotonation of the cyclohexanone had actually followed the expected course was demonstrated beyond question by a X-ray structure analysis of the *N*-trityl derivative of the *C*-terminal amino acid of chlorotetaine (Scheme 8). The asymmetric C-atom in the cyclohexene ring of this compound, and in chlorotetaine, was shown by this analysis to posses an (*S*)-configuration. The amino acid residue in the crystal is in the axial position. However, this is a special effect occuring in this compound, because it was shown by ^{1}H-NMR spectroscopy that the C-1' substituent of chlorotetaine in solution (D$_2$O) is equatorial.

Applying the reversed octant rule, this conformation together with the (*S*)-configuration is completely in accord with the observed negative Cotton effect in the CD spectrum. The originally determined (*R*)-configuration resulted from a misinterpretation of the CD spectrum, conditioned by comparison of chlorotetaine with the related compounds bacilysin and anticapsin (see next).

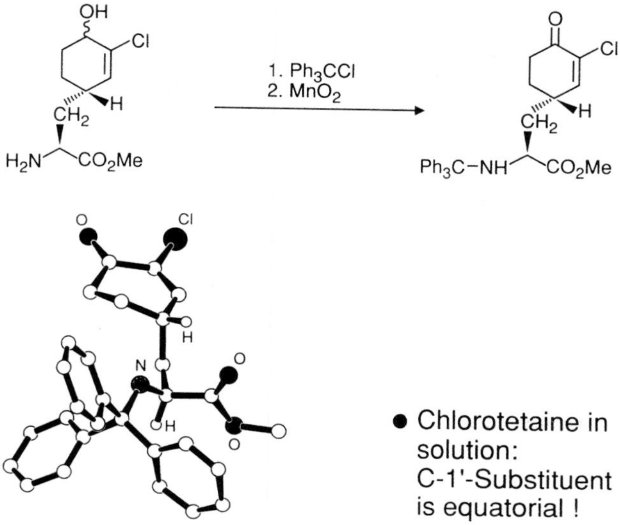

- Chlorotetaine in
  solution:
  C-1'-Substituent
  is equatorial !

Scheme 8

Synthesis of Anticapsin and Bacilysin

Two compounds which are closely related to chlorotetaine/bromotetaine and which have been known for more then twenty years are the dipeptide bacilysin (tetaine) [9] and its *C*-terminal amino acid anticapsin [10]. Both compounds are antimycotically and antibacterially active and it has been demonstrated that they are inhibitors of the glucosamine-6-phosphate synthetase [11]. The stereochemistry of the epoxy-cyclohexanone was determined by CD (positive Cotton effect) and the *trans*-relationship of the epoxide and the C-1' substituent was derived from the coupling constants of the adjacent protons (H-1' and H-2') in the ^{1}H-NMR spectrum. By comparison with published data, coupling constants of less than 1 Hz should be typical for a *trans*-relationship, whereas *cis*-protons have coupling constants of 2-4 Hz.

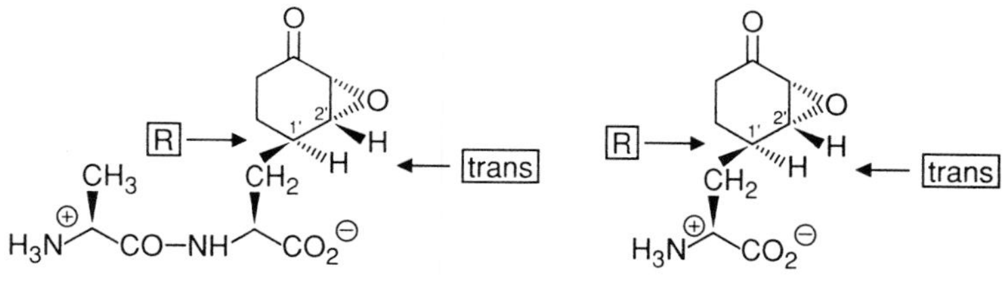

BACILYSIN (TETAINE)   ANTICAPSIN

With the structures of chlorotetaine and bromotetaine proven, the question arose whether bacilysin and anticapsin actually have the (R)-configuration at C-1' have a different relative stereochemistry than the haloenone dipeptides.

Although there are three syntheses of anticapsin already published [12-14], none of them controls the stereochemistry at C-1' and all include separation of 1:1 diastereomeric mixtures. However, one synthesis seems to control the *trans*-relationship between the epoxide and C-1' substituent [14].

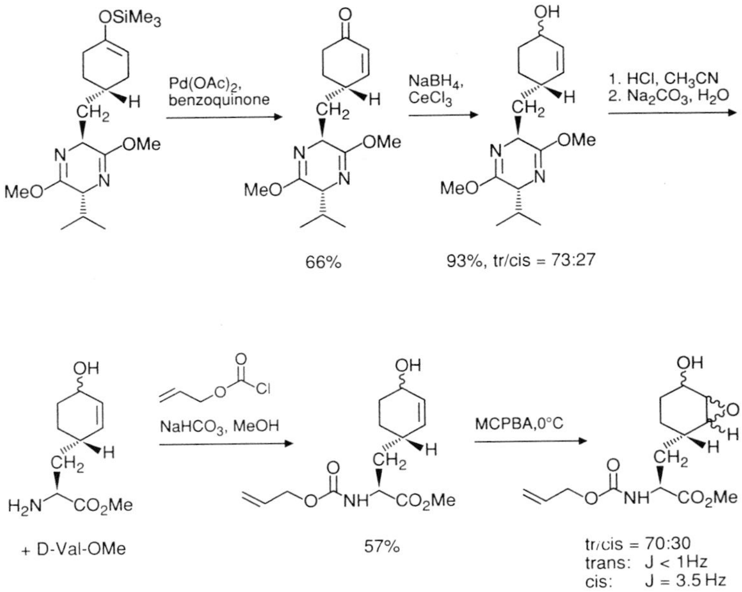

OSiMe₃

Pd(OAc)₂, benzoquinone

NaBH₄, CeCl₃

OH

1. HCl, CH₃CN
2. Na₂CO₃, H₂O

66%

93%, tr/cis = 73:27

OH

+ D-Val-OMe

NaHCO₃, MeOH

57%

MCPBA, 0°C

OH

tr/cis = 70:30
trans: J < 1 Hz
cis: J = 3.5 Hz

Scheme 9

In order to prove the stereochemistry of anticapsin and bacilysin, a synthesis of these two compounds was undertaken starting with the silyl enol ethers of known configuration from the synthesis of chlorotetaine and bromotetaine (scheme 9). The silyl enol ether which should yield the final product with the (S)-configuration at C-1' was first converted to the cyclohexenone [15]. The ketone was then reduced with sodium borohydride in the presence of cerium trichloride to yield the allylic alcohol as a *trans/cis*-mixture of 73:27. This is the expected ratio in the reduction of 4-substituted cyclohexenones [16]. Acidic hydrolysis, protection of the amine as the allyl carbamate and selective epoxidation with MCPBA gave a 70:30-mixture of two diastereomers. As the epoxidation with MCPBA should occur selectively *cis* to the allylic alcohol, the major isomer should have a *trans*-relationship between the epoxide and C-1' substituent. In fact, the ¹H-NMR spectrum of the mixture did not show any coupling constant for the major (*trans*) isomer and a small coupling constant for the two protons of the minor (*cis*) isomer, well in accord with the literature precedent.

However, after oxidation of the alcohol with chromium trioxide a separable mixture of diastereomers was obtained whose ¹H-NMR spectra were quite confusing (scheme 10). Now the *cis*-protons of the minor product did not couple with each other, whereas the *trans*-protons of the major isomer did. Palladium-catalyzed deprotection of the amine of the major isomer under slightly basic conditions did not allow the isolation of the primary amine, but led directly to the perhydroindole. This reaction can only occur easily when the epoxide and the C-1' substituent are *trans*. However, deprotection of the minor isomer followed by mild saponification gave an amino acid which was in every respect (^{1}H-NMR, ^{13}C-NMR, α_D, CD) identical to a sample of natural anticapsin kindly provided by *Eli Lilly and Co.*.

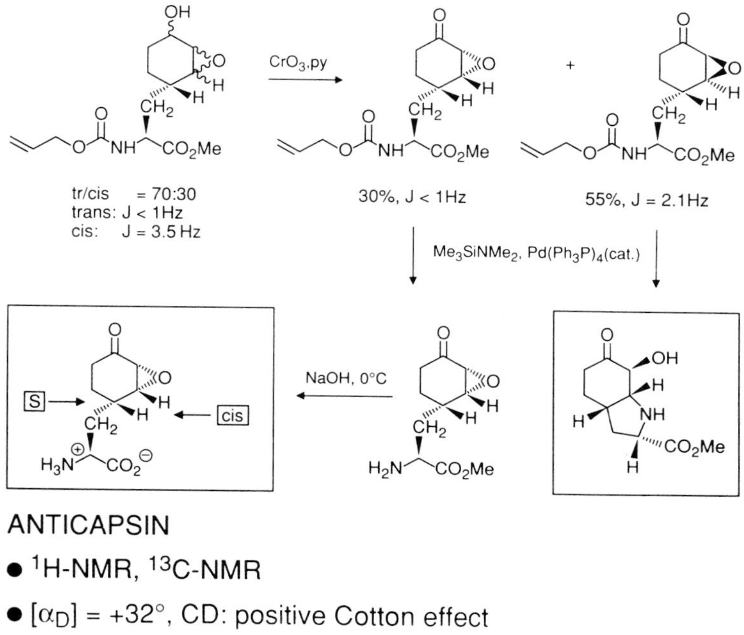

tr/cis   = 70:30
trans: J < 1 Hz
cis: J = 3.5 Hz

30%, J < 1 Hz

55%, J = 2.1 Hz

Me₃SiNMe₂, Pd(Ph₃P)₄(cat.)

NaOH, 0°C

ANTICAPSIN

● ¹H-NMR, ¹³C-NMR

● $[\alpha_D] = +32°$, CD: positive Cotton effect

Scheme 10

This proved that anticapsin, like chlorotetaine and bromotetaine, has the (*S*)-configuration at the C-1' position, in contradiction to the assignment reported in the literature [10]. The relationship between the epoxide and the adjacent substituent is *cis* and not *trans*, which means that the originally determined absolute configuration of the epoxide, which is responsible for the positive Cotton effect in the CD spectrum, was correct.

During synthesis of the 1'-(*R*)-isomer of anticapsin starting with the epimeric silyl enol ether, the same observations were made as in the synthesis of anticapsin itself (scheme 11). The major isomer after oxidation of the alcohol to the ketone has a coupling constant between the *trans*-protons and, after deprotection, only the perhydroindole was isolated. After the two-step deprotection sequence, the minor isomer yielded an amino acid whose ¹H-NMR spectrum was almost identical to that of anticapsin, but which could easily be distinguished from anticapsin by its ¹³C-NMR spectrum, its negative optical rotation and a negative Cotton effect in the CD spectrum.

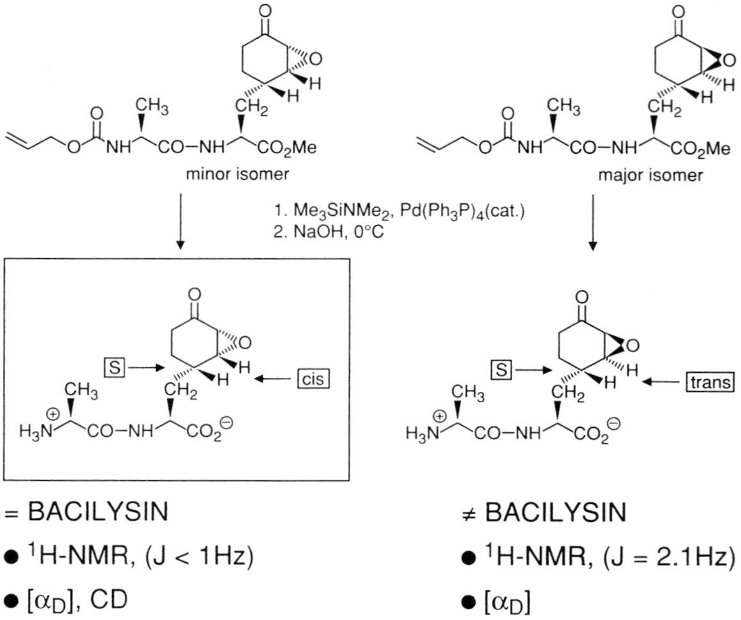

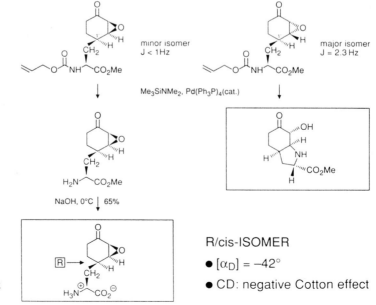

Scheme 11

Using the same methodology, the dipeptide bacilysin and three of its diastereomers were synthesized only by substituting the Alloc protecting group by a protected L-alanine (scheme 12). Again after oxidation and deprotection, the minor isomer led to the natural product. In this case,

Scheme 12

both *cis*- and both *trans*-products were obtained, but only the (*S*)-*cis* diastereomer possess spectral and optical data which were in accord with the data published for bacilysin. Also in the final products, both *trans*-compounds showed a small coupling constant for the *trans*-protons in the ^{1}H-NMR spectrum.

Up to now we have not been able to get an X-ray structure of any compounds in the bacilysin/- anticapsin series. Therefore, some model studies were undertaken to secure the relative stereochemistry of the epoxide in the cyclohexanone ring.

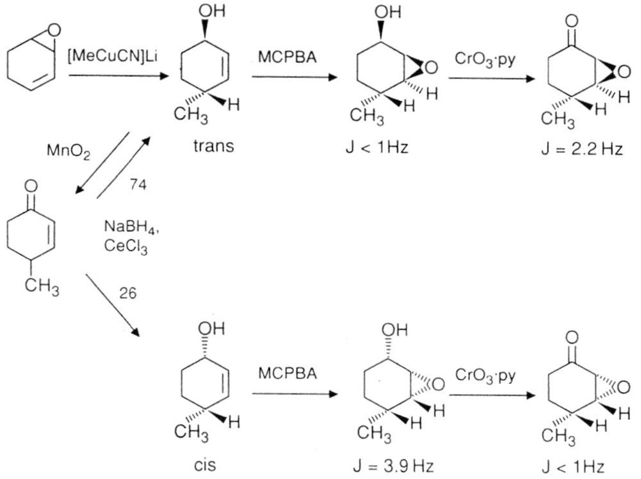

Scheme 13

To this end, cyclohexadiene monoepoxide was opened with a methylheterocuprate, a reaction known to yield *trans*-1-hydroxy-4-methyl-2-cyclohexene [17] (scheme 13). Epoxidation gave a single allylic alcohol, whose *trans*-protons did not couple in the ^{1}H-NMR spectrum. Oxidation of the alcohol yielded the ketone and, as seen before, this compound did have a small coupling constant between the *trans*-protons. In order to get into the *cis*-series, the allylic alcohol was oxidized to the ketone and subsequently rereduced to a separable 74:26-mixture of *trans*- and *cis*-alcohols. In this *cis*-series, the epoxyalcohol and the epoxyketone behave in an opposite manner. The alcohol does have and the ketone does not have an observable coupling between the *cis*-protons. This exactly parallels the behaviour of the anticapsin/bacilysin-precursors.

A final comparison was made by opening the model epoxyketones using chlorodimethylsulfonium generated *in situ* by mixing trimethylsilyl chloride and DMSO[18] (scheme 14). The *trans*-compound gave a chlorohydrin with all substituents equatorial, whereas the *cis*-derivative yielded the expected product with two axial substituents, as shown by the coupling constants in the ^{1}H-NMR spectrum. The same reactions performed with the anticapsin precursor and its diastereomer yielded chlorohydrins whose spectral data were perfectly in accord with the *cis*-stereochemistry for the anticapsin precursor and the *trans*-relationship for its diastereomer.

The only unsettled question is why one of the syntheses of anticapsin already mentioned did arrive at the correct final product although the synthetic sequence should have secured a *trans*-relationship between epoxide and C-1' substituent [14]. Research to answer this question is presently under way.

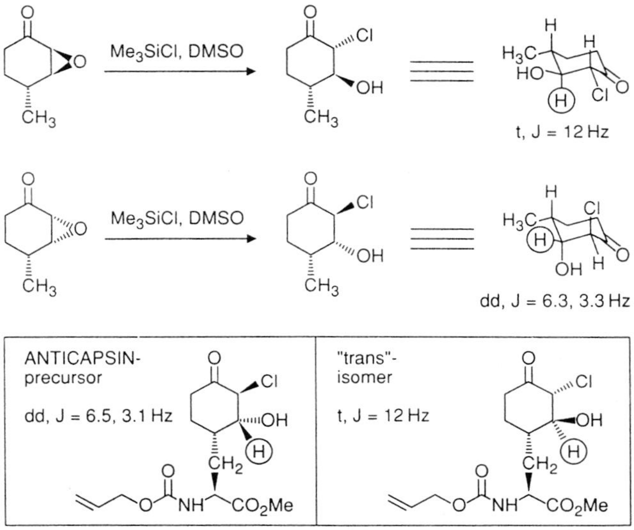

Me$_3$SiCl, DMSO

t, J = 12 Hz

Me$_3$SiCl, DMSO

dd, J = 6.3, 3.3 Hz

ANTICAPSIN-precursor	"trans"-isomer
dd, J = 6.5, 3.1 Hz	t, J = 12 Hz

Scheme 14

A last point remains to be mentioned (scheme 15): In all ^{1}H-NMR spectra of the free betaines of bacilysin, anticapsin and their diastereomers which are recorded in D$_2$O, a second component is visible. This component can be seen in the original spectrum of anticapsin published in 1970 [19]. Most probably this component, which is always present to an extent of about 20% even in highly purified material, is the hydrate of the epoxyketone. Again the change from an sp^2- to an sp^3-center in the cyclohexane ring leads to a switch of the coupling constants of the *cis*- and the *trans*-protons respectively, as already shown for the epoxy alcohols.

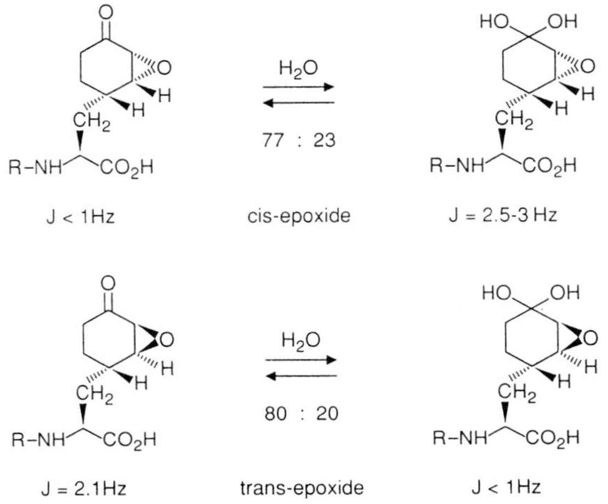

H$_2$O

77 : 23

R–NH CO_2H

J < 1 Hz   cis-epoxide   J = 2.5-3 Hz

H$_2$O

80 : 20

J = 2.1 Hz   trans-epoxide   J < 1 Hz

Scheme 15

In summary, our work for the first time proves the stereochemistry of the antimycotically active natural products chlorotetaine, bromotetaine, bacilysin and anticapsin. The configuration at the chiral center at the C-1' position of all compounds has to be revised to (*S*). The epoxide in the cyclohexanone ring of bacilysin and anticapsin is *cis* to the C-1' substituent.

Acknowledgments: I thank *Dr. L. Born* (X-ray), *Dr. D. Gondol* (NMR), *Dr. R. Grosser* (CD), *Dr. P. Schmitt* (NMR) and *Dr. C. Wünsche* (MS) for the physical measurements, *Prof. G. Jung* for providing a sample of natural chlorotetaine and spectral data of bromotetaine, *li. Lilly and Co.* for providing a sample of natural anticapsin and *A. Urban* for conducting a major part of the experimental work. I thank my colleagues at the *Chemical Science Laboratories Pharma* for many helpful discussions.

References

1. Chlorotetaine: C. Rapp, G. Jung, W. Katzer, W. Loeffler, *Angew. Chem.* **100** (1988) 1801; *Angew. Chem. Int. Ed .Engl.* **27** (1988) 1733; bromotetaine: G. Jung, personal communication.

2. H. Wild, L. Born, *Angew.Chem.* **103** (1991) 1729; *Angew. Chem. Int. Ed. Engl.* **30** (1991) 1685.

3. A.J. Birch, *J. Proc. R. Soc .N.S.W.* **83** (1949) 245.

4. U. Schöllkopf, D. Pettig, E. Schulze, M. Klinge, E. Egert, B. Benecke, M. Noltemeyer, *Angew. Chem.* **100** (1988) 1238; *Angew. Chem. Int. Ed. Engl.* **27** (1988) 1194.

5. R. Shirai, M. Tanaka, K. Koga, *J. Am. Chem. Soc.* **108** (1986) 543.

6. C. M. Cain, R. P. C. Cousins, G. Coumbarides, M. S. Simpkins, *Tetrahedron* **46** (1990) 523.

7. R. L. Danheiser, R. F. Miller, R. G. Brisbois, S. Z. Park, *J. Org. Chem.* **55** (1990) 1959.

8. D. J. Buckley, M. A. McKervey, *J. Chem. Soc.P erkin Trans. 1* **1985**, 2193.

9. J. E. Walker, E.-P. Abraham, *Biochem. J.* **118** (1970) 563.

10. N. Neuss, B. B. Molloy, R. Shah, N. DeLaHiguera, *Biochem. J.* **118** (1970) 571.

11. H. Chmara, H. Zähner, E. Borowski, S. Milewski, *J. Antibiotics* **37** (1984) 652.

12. R. W. Rickards, J. L. Rodwell, K. J. Schmalzl, *J. Chem. Soc. Chem. Commun.* **1977**, 849.

13. B. C. Laguzza, B. Ganem, *Tetrahedron Lett.* **22** (1981) 1483.

14. M. Souchet, M. Baillarge, F. LeGoffic, *Tetrahedron Lett.* **29** (1988) 191.

15. Y. Ito, T. Hirao, T. Saegusa, *J. Org. Chem.* **43** (1978) 1011.

16. J.-L. Luche, L. Rodriguez-Hahn, P. Crabbe, *J. Chem. Soc.; Chem. Commun.* **1978**, 601; W. Sucrow, G. Rädeker, *Chem. Ber.* **121** (1988) 219.

17. J. P. Marino, N. Hatanaka, *J. Org. Chem.* **44** (1979) 4467.

18. F. Ghelfi, R. Grandi, U. M. Pagnoni, *J. Chem. Res. (S)* **1988**, 200.

19. R. Shah, N. Neuss, M. Gormann, L. D. Boeck, *J. Antibiotics* **23** (1970) 613.

Cyclosporins: Novel Insights into Structure Activity Relationships and Mechanism of Action

Hans Fliri

SANDOZ PHARMA AG, Preclinical Research Laboratories

Department of Immunology, CH-4002 Basle

Summary: The immunosuppressive activity of cyclosporin requires in a first step binding to cyclophilin, followed by binding of the binary complex to the phosphatase calcineurin and inhibiting its enzyme activity. Non-immunosuppressive cyclosporin derivatives of comparable high affinity to cyclophilin fail to mediate the calcineurin interaction. An analogous structure-activity relationship has previously been established for the immunosuppressive macrolide FK-506.

Introduction.

When cyclosporin (Sandimmune[R], **1**) was introduced in 1983 as a key therapeutic agent to suppress rejection of transplanted organs, little was known about the molecular mechanism involved in its mode of action. Since then, not only has our understanding of the physiology and function of T lymphocytes been expanded, the availability of cyclosporin and, later, of the macrolide FK-506 and its long known structural relative rapamycin has been instrumental in dissecting individual steps of lymphocyte signalling and helping understand these mechanisms at the molecular level.

For clarity reasons, a short discussion on the macrolides will be included. Research on their biochemical mechanism of action has had fundamental impact and was indispensable in bringing our level of understanding of cyclosporins to its present state.

Before discussing cyclosporin and its mode of action in more detail, let me briefly review how T cells recognize and respond to antigen (fig.1). T cells, which together with B cells and the antibodies secreted by them represent the specific arm of the immune system, recognize antigen when it is presented to them by MHC molecules on the surface of an antigen presenting cell. The T cell surface molecule mediating this structure-specific recognition is called the T cell receptor. Recognition of antigen triggers a signal from the T cell receptor which is transmitted into the cytoplasm and passed on into the nucleus. Besides Ca^{2+}-dependence, only a few details are known of how this signal is transmitted. Following this signal, the transcription of several lymphokine genes is initiated; for the following discussion, that of interleukine-2 (IL-2) is the most relevant one. After translation of mRNA into protein, the lymphokines are secreted from the cell and exert their actions on other cells expressing the appropriate receptors. IL-2 also binds in an autocrine fashion to its receptor on the cell membrane, creating a new signal. This signal, originating from the IL-2 receptor, is not Ca^{2+}-dependent and ultimately leads to cell division, proliferation and differentiation, creating a number of offspring cell populations with specialized functions. The over all consequence of these events represents an antigen-specific immune response.

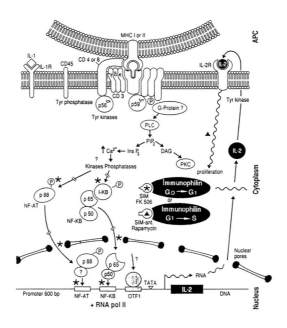

Figure 1

It was shown early on that cyclosporin could inhibit T cell proliferation induced by antigen and mitogenic lectins but had little effect on proliferative responses induced by IL-2. Results from several laboratories convincingly demonstrate that the inhibitory effects of cyclosporin primarily consist in inhibition of lymphokine gene transcription [1].

Biochemical mechanism of action of cyclosporin: Part 1.

In 1984, isolation of an 18 kD cytosolic protein from bovine thymus specifically binding cyclosporin was reported by Handschumacher [2]. The same group showed that cyclophilin occurs abundantly in two isoforms (up to 0.4% of total cytosolic protein) in all tissues [3]. Depending on the conditions to measure binding, a KD of 0.2 to $20x10^{-8}$ has been reported for the interaction of cyclosporin and cyclophilin [2-4]. Soon after its discovery, cyclophilin was cloned [5] and expressed [6] in the Sandoz laboratories.

Independent investigations on enzymes catalyzing refolding of denatured proteins led to a rediscovery of cyclophilin as an enzyme catalyzing *cis-trans* isomerization of peptidyl-prolyl bonds [7,8] (fig. 2). This enzyme activity (later termed "rotamase" activity [9]) is susceptible to inhibition by cyclosporin.

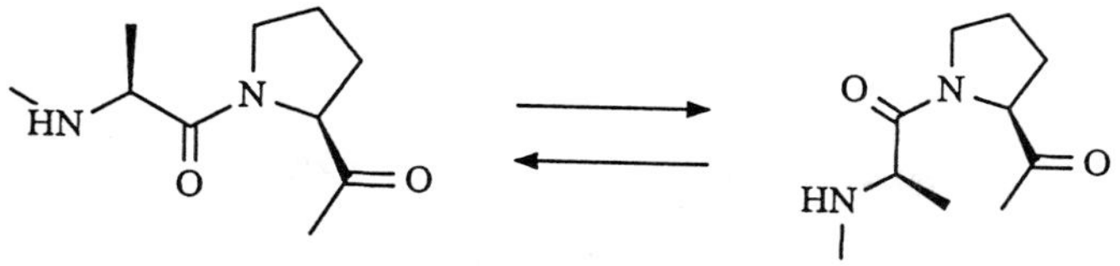

trans   cis

Figure 2

During this period, a series of publications in the biochemical literature suggested a functional relationship between folding and unfolding of proteins and their intracellular topology. Proteins, in order to cross the membranes of intracellular organelles, do so only after unfolding and must be refolded after the transit through the membrane [10]. Suddenly, there was a protein exhibiting catalytic folding activity! Further evidence, albeit circumstantial, for a role of cyclophilin(s) in protein folding and/or topology was provided by the discovery of the ninaA gene in Drosophila, which apparently has a crucial function for the proper folding and transport of rhodopsin in Drosophila visual cells and which has high homology to cyclophilin [11-13].

At this point, it seemed logical to assume that immunosuppression by cyclosporin would somehow be caused by inhibition of cyclophilin's rotamase activity. However, a number of facts were not compatible with this notion:

a) Some cyclosporin analogues, e.g. 2, were discovered which bound equally well as cyclosporin A to cyclophilin and were powerful rotamase inhibitors; nevertheless, they were devoid of immunosuppressive properties.

b) Other cyclosporin derivatives, such as 3, which only weakly bound to cyclophilin, yet demonstrated moderate to good immunosuppressive activity (fig. 3).

c) Stoichiometric considerations suggested that at IC_{50} immunosuppressive concentrations, only 1 to 10% of total cyclophilin is bound to cyclosporin, suggesting that it is the formation of the complex rather than inhibition of some function which is responsible for immunosuppression.

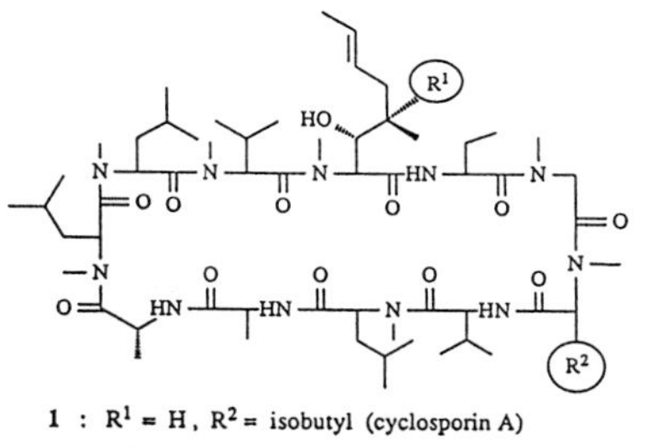

1 : R^1 = H, R^2 = isobutyl (cyclosporin A)

2 : R^1 = H, R^2 = isopropyl (MeVal^{4}Cs)

3 : R^1 = Me, R^2 = isobutyl (dimethyl-Bmt^{1}Cs)

	Immunosuppression	Cyclophilin Binding
1	+++	+++
2	none	+++
3	+	+/-

Figure 3

The first experiments convincingly demonstrating that growth inhibitory effects of cyclosporin are rather due to a "gain of function" by complex formation with cyclophilin rather than inhibition of a function were provided by Tropschug [14]: Mutants of either *Neurospora* crassa or *Saccharomyces* cerevisiae resistant to cyclosporin were either lacking cyclophilin altogether or the protein had lost its affinity to cyclosporin.

Macrolides: FK-506 and rapamycin.

The discovery of FK-506 (**4**, cf. fig. 4) in 1987 quickly raised widespread attention: In several experimental models of immunosuppression, FK-506 exhibited a strikingly similar spectrum of activity as cyclosporin but was found to be much more potent [15]. The similarity to cyclosporin in immunosuppressive activity prompted a search for cytosolic FK-506 binding proteins (FKBP's). In 1989, the first FKBP was isolated, purified and shown to also possess peptidyl-prolyl rotamase activity which is inhibited by FK-506 [16-18] but not by cyclosporin. Because of its structural similarity to FK-506, the long known antibiotic rapamycin (**5**), recognized as an immunosuppressant during the initial investigations, was rediscovered [19,20]. A series of remarkable findings followed: 1) Rapamycin inhibits T cell proliferation at sub-nanomolar concentrations, equalling FK-506 in potency [21], 2) it inhibits a later step in the activation pathway [22], 3) FK-506 and rapamycin are reciprocal antagonists [23], 4) both rapamycin and FK-506 bind with very similar activity to FKBP and inhibit its rotamase activity [24], 5) a synthetic ligand (506BD, **6**) binding to FKBP (506BD) could block the effects of both rapamycin and FK-506 but not of cyclosporin [25].

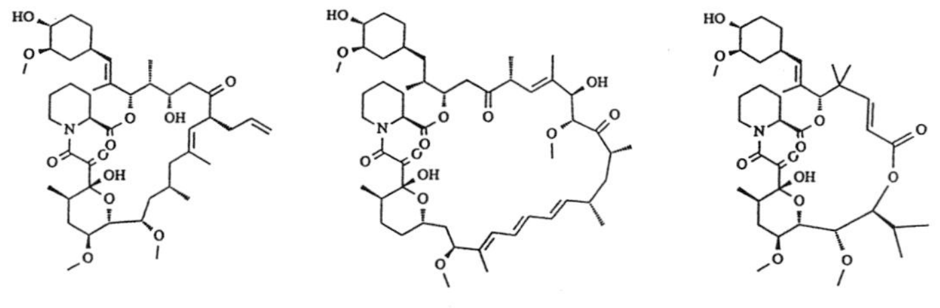

Figure 4   4   5   6

These results suggested again a "gain of function" of FKBP by binding of FK-506 rather than inhibition. The search for a possible target of the FK-506/FKBP complex culminated in the discovery of the known phosphatase calcineurin as a common target for the complex of FK-506/FKBP and of cyclosporin/cyclophilin, but not of rapamycin/FKBP by S.L.Schreiber and collaborators [26]. Remarkably, both complexes, composed of structurally dissimilar ligands and different binding proteins, compete for the same binding site on calcineurin and inhibit its phosphatase activity!

Subsequently, a unifying picture for the biochemical mode of action of the macrolides was proposed [27]: FK-506 and rapamycin bind to FKBP via their common part of the molecule (the "binding domain"). Binding of ligand alters the conformation of both ligand and protein in such a way that the FK-506/FKBP complex but not the rapamycin/FKBP complex acquires affinity to calcineurin, ultimately resulting in inhibition of calcineurin's phosphatase activity. The part of the macrolide molecules mediating the calcineurin interaction has been termed "effector domain" (fig. 5). A target for the rapamycin/FKBP complex has not been conclusively identified [28].

FK 5068
IMMUNOPHILIN BINDING DOMAIN
EFFECTOR DOMAIN
RAPAMYCIN

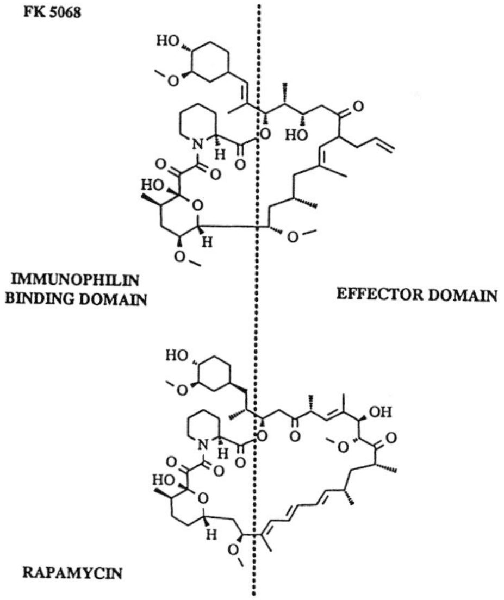

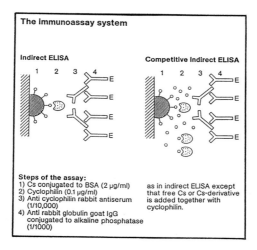

Figure 5

Biochemical mechanism of action of cyclosporin: Part 2.

An extensive comparative evaluation of cyclosporin-cyclophilin binding studies, combined with the development of a highly sensitive functional assay for IL-2 gene activation, revealed recently an identical mode of action for cyclosporin and FK-506. For cyclophilin binding studies, a competitive ELISA system was used involving plate bound cyclosporin ligand and soluble cyclophilin; for detection, a rabbit anti-cyclophilin antibody and goat anti-rabbit antiserum was used (fig. 6).

The immunoassay system
Indirect ELISA
Competitive Indirect ELISA
1 2 3 4
E
Steps of the assay:
1) Cs conjugated to BSA (2 µg/ml)
2) Cyclophilin (0.1 µg/ml)
3) Anti cyclophilin rabbit antiserum (1/10,000)
4) Anti rabbit globulin goat IgG conjugated to alkaline phosphatase (1/1000)
as in indirect ELISA except that free Cs or Cs-derivative is added together with cyclophilin.

Figure 6

As a functional assay for IL-2 gene transcription, a reporter gene assay was established [29] (fig.7): A DNA construct consisting of the coding region for the enzyme beta-galactosidase under the transcriptional control of the human IL-2 promoter was integrated into an inducible T cell line. Upon activation, this transfected cell line will produce not only IL-2 but also the enzyme beta-galactosidase which can be rapidly quantified using a fluorogenic substrate in a highly reproducible manner. This assay detects any substance which inhibits the activation of the IL-2 gene with much higher sensitivity than proliferation-based tests [30].

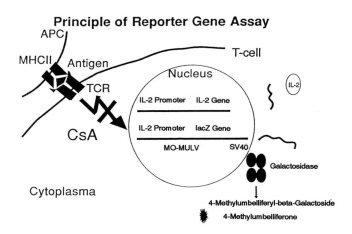

Figure 7

A comparative evaluation of a large series of cyclophilin-binding cyclosporin derivatives in this test revealed analogues of equal affinity to cyclophilin yet failing to inhibit IL-2 transcription (fig. 8). Would such compounds be cyclosporin-antagonists? Addition of increasing molar excesses of SDZ 220-384 (2) to an IC_{95} concentration of cyclosporin A in a reporter gene assay leads to a progressive reversal of the inhibition of the reporter gene activity (fig. 9). Using a dilution series of cyclosporin A in this experiment shows that SDZ 220-384 (2) acts as a classical competitive antagonist (Kd ca. 150 nM).

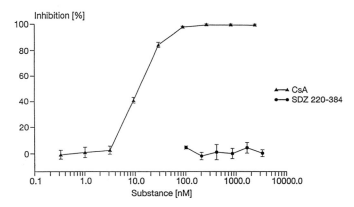

Figure 8

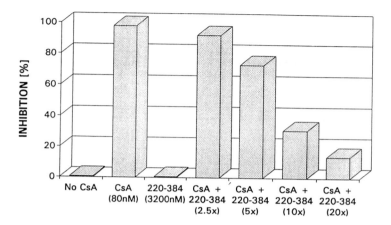

Figure 9

These results provide the formal proof that binding of cyclosporins to cyclophilin is necessary but not sufficient for immunosuppression and that like FK-506, cyclosporin must have two different functionally relevant domains: one domain required for cyclophilin binding ("binding-domain") and another one mediating binding of the complex to calcineurin, i.e. an "effector-domain". The amino acids encompassing the binding domain were identified first by an immunochemical approach [31] and later supported by the collaborative NMR investigation of K.Wuethrich and the Sandoz group [32] and, independently, by S.W.Fesik [33]. Recently, the X-ray crystal structure of the cyclosporin-cyclophilin complex was solved by the Sandoz group [34], confirming earlier NMR results that the binding domain involves amino acids 1,2,9,10, and 11 whereas the structure-activity analysis of cyclophilin binding and inhibition of IL-2 transcription indicates a role in amino acids 4-8 for effector function.

The seemingly paradoxical fact that there exist cyclosporin derivatives such as dimethyl-Bmt1-Cs (3) which have low affinity to cyclophilin but yet are immunosuppressants of moderate potency can best be explained by assuming a conformational transition of the cyclosporin-cyclophilin complex upon binding to calcineurin. Formation of this "supra-molecule" induces a cyclophilin conformation where 3 can be more easily accomodated than in the binary complex [35].

Synthesis of the cyclosporin derivatives.

The cyclosporin derivatives described here were synthesized by two principal methodologies, one consisting in a solid-phase synthesis of a linear undecapeptide followed by cyclization in solution as described by Wenger [36], the second by semi-synthetic modification of natural products. To this end, methods were developed to selectively open the cyclosporin macrocycle at defined positions.

Peptide bonds to beta-hydroxy amino acids such as serine and threonine can be activated to cleavage by oxazolidone formation with N,N'-carbonyl diimidazole (fig. 10). Cleavage of the peptide bond is effected with sodium methoxide in methanol. Applying this principle to natural cyclosporin derivatives of threonine or serine (e.g. cyclosporin C or (D)-serine-8-cyclosporin), linear cyclosporin derivatives can be obtained. Removal of one or more amino acids by Edman degradation creates shortened peptide fragments which can be used to assemble novel cyclosporins.

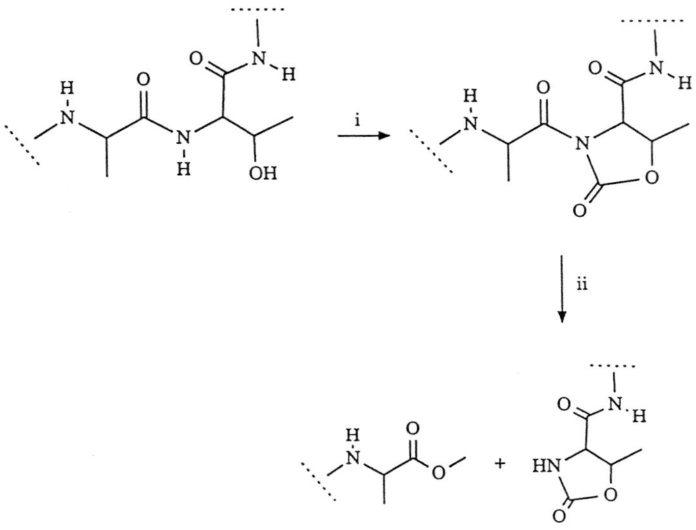

Figure 10

Alternatively, treatment of cyclosporin with phosphorous pentasulfide (or equivalent reagents) leads to mixtures of products where one or more amide bonds have been converted to thioamides [37]. These thioamides can be selectively cleaved to either create linear undecapeptide fragments (from monothio amides) or to cleave the macrocycle at two different positions, yielding two oligopeptide fragments which again can be reconstituted to novel cyclosporin derivatives (fig. 11).

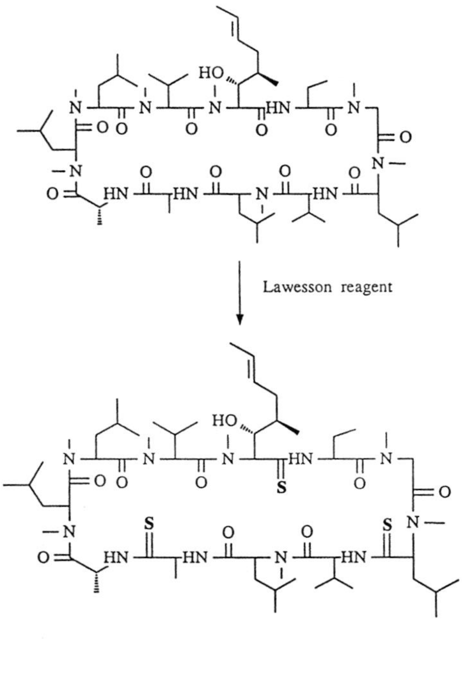

Figure 11

Conclusion.

Cyclosporin and FK-506 exert their respective immunosuppressive effects by binding to cytosolic receptors (immunophilins). The binary complexes in turn bind to and inhibit the activity of the serine/threonine phosphatase calcineurin. That calcineurin is indeed the target relevant for immunosuppression is evidenced by recent observations [38,39]: Over-expression of calcineurin in T cells renders the cells hyper-responsive to antigen and more resistant to cyclosporin and Fk-506. The next question, what is the substrate/target of calcineurin which is dephosphorylated during T cell activation, can not be answered at this time but will certainly not remain hidden for long.

Acknowledgement: Besides the referenced work, all the research described in this article was carried out at the Preclinical Research Laboratories, SANDOZ PHARMA AG., Basle, by the following scientists: G. Baumann, R. Breckenridge, P. Bollinger, M.K. Eberle, P.C. Hiestand, J. Kallen, S. Ko, R.M. Movva, V. Quesniaux, M.H. Schreier, Z.L. Su, M.D. Walkinshaw, H.P. Weber, R.M. Wenger, H. Widmer, G. Zenke, M. Zurini, aided by the skillful and dedicated help of many technical associates.

References

1. For a recent review see N. H. Sigal, F. J. Dumont: Pharmacologic probes of lymphocyte signal transduction. *Annu. Rev. Immunol.* **10** (1992) 519-560.

2. R. E. Handschumacher, M. W. Harding, J. Rice, R. J. Drugge, D. W. Speicher: Cyclophilin. A specific cytosolic binding protein for cyclosporin A. *Science* **226** (1984) 544-547.

3. A. J. Koletsky, M. W. Harding, R. E. Handschumacher: Cyclophilin. Distribution and variant properties in normal and neoplastic tissue. *J. Immunol.* **137** (1986) 1054-1059.

4. J. L. Kofron, P. Kuzmic, V. Kishore, E. Colon-Bonilla, D. H. Rich: Determination of kinetic constants for peptidyl-prolyl *cis-trans* isomerases by an improved spectrophotometric assay. *Biochemistry* **30** (1991) 6127-6134.

5. B. Haendler, W. R. Hofer, E. Hofer: Complementary DNA for human T-cell cyclophilin. *EMBO J.* **6** (1987) 947-950.

6. B. Haendler, R. Keller, P. C. Hiestand, H. P. Kocher, G. Wegmann, R. N. Movva: Yeast cyclophilin. isolation and characterization of the protein, cDNA and gene. *Gene* **83** (1989) 39-46.

7. G. Fischer, L. B. Wittmann, K. Lange, T. Kiefhaber, F. X. Schmidt: Cyclophilin and peptidyl-prolyl *cis-trans* isomerase are probably identical proteins. *Nature* **337** (1989) 476-478.

8. N. Takahashi, T. Hayano, M. Suzuki: Peptidyl-prolyl *cis-trans* isomerase is the cyclosporin A binding protein cyclophilin. *Nature* **337** (1989) 473-475.

9.   M. W. Harding, A. Galat, D.E. Uehling, S. L. Schreiber: A receptor for the immunosuppressent FK-506 is a *cis-trans* peptidyl-prolyl isomerase. *Nature* **341** (1989) 758-60.

10.   K. P. Baker, G. Schatz: Mitochondrial proteins essential for viability mediate protein import into yeast mitochondria. *Nature* 349 (1991) 205-208.

11.   B. H. Shieh, M. A. Stamnes, S. Seavello, G. L. Harris, C. S. Zuker: The ninaA gene required for visual transduction in Drosophila encodes a homologue of cyclosporin A-binding protein. Nature **338** (1989) 67-70.

12.   S. Schneuwly, R. D. Shortridge, D. C. Larrivee, T. Ono, M. Ozaki, W. L. Pak: Drosophila ninaA gene encodes an eye-specific cyclophilin (cyclosporine A binding protein). *Proc. Natl. Acad. Sci. USA* **86** (1989) 5390-94.

13.   M. A. Stamnes, B.-H. Shieh, L. Chuman, G. L. Harris, C. S. Zuker: The cyclophilin homolog ninaA is a tissue-specific integral membrane protein required for the proper synthesis of a subset of Drosophila rhodopsins. *Cell* **65** (1991) 219-27.

14.   M. Tropschug, I. B. Barthelmess, W. Neupert: Sensitivity to cyclosporin A is mediated by cyclophilin in Neurospora crassa and Saccharomyces cerevisiae. *Nature* **342** (1989) 953-55.

15.   T. Kino, H. Hatanaka, M. Hashimoto, T. Goto, M. Okuhara, M. Kohsaka, H. Aoki, H. Imanaka: FK-506, a novel immunosuppressant isolated from a Streptomyces. I. Fermentation, isolation, physico-chemical and biological characteristics. *J. Antibiot.* **40** (1987) 1249-55.

16.   J. J. Siekierka, M. J. Staruch, S. H. Hung, N. H. Sigal: FK-506, a potent novel immunosuppressive agent, binds to a cytosolic protein which is distinct from the cyclosporin A-binding protein, cyclophilin. *J. Immunol.* **143** (1989) 1580-83.

17.   J. J. Siekierka, S. H. Hung, M. Poe, C. S. Lin, N. H. Sigal: A cytosolic binding protein for the immunosuppressant FK-506 has peptidyl-prolyl isomerase activity but is distinct from cyclophilin. *Nature* **341** (1989) 755-57.

18.   M. W. Harding, A. Galat, D. E. Uehling, S. L. Schreiber: A receptor for the immunosuppressent FK-506 is a *cis-trans* peptidyl-prolyl isomerase. *Nature* **341** (1989) 758-60.

19.   C. Vezina, A. Kudelski, S. N. Sehgal: Rapamycin (AY-22,989), a new antifungal antibiotic. I. Taxonomy of the producing streptomycete and isolation of the active principle. *J. Antibiot.* **28** (1975) 721-26.

20.   R. R. Martel, J. Klicius, S. Galet: Inhibition of the immune response by rapamycin, a new antifungal antibiotic. *Can. J. Physiol. Pharmacol.* **55** (1977) 48-51.

21.   N. H. Sigal, C. S. Lin, J. J. Siekierka: Inhibition of human T-cell activation by FK-506, rapamycin, and cyclosporine A. *Transplant. Proc.* **23** (1991) 1-5.

22.   F. J. Dumont, M. J. Staruch, S. L. Koprak, M. R. Melino, N. H. Sigal: Distinct mechanisms of suppression of murine T-cell activation by the related macrolides FK-506 and rapamycin. *J. Immunol.* **144** (1990) 251-58.

23.   F. J. Dumont, M. R. Melino, M. J. Staruch, S. L. Koprak, P. A. Fischer, N. H. Sigal: The immunusuppressive macrolides FK-506 and rapamycine act as reciprocal antagonists in murine T-cells. *J. Immunol.* **144** (1990) 1418-24.

24.   F. J. Dumont, P. A. Fischer, M. J. Staruch, N. H. Sigal: Kinetic analysis of the reversal of FK-506-mediated immunosuppression by rapamycin. *FASEB J.* **4** (1990) A2038 (Abstr.).

25.   B. E. Bierer, P. K. Somers, T. J. Wandless, S. J. Burakoff, S. L. Schreiber: Probing immunosuppressant action with a nonnatural immunophilin ligand. *Science* **250** (1990) 556-59.

26.   J. Liu, J. D.,Jr. Farmer, W. S. Lane, J. Friedman, I. Weissman, S. L. Schreiber: Calcineurin is a common target of cyclophilin-cyclosporin A and FKBP-FK-506 complexes. *Cell* **66** (1991) 807-15.

27.   Discussed in detail in ref. 1.

28.   For a recent discussion see S.L. Schreiber: Immunophilin-sensitive protein phosphatase action in cell signalling pathways. *Cell* **70** (1992) 1-20.

29.   P. S. Mattila, K. S. Ullmann, S. Fiering, E. A. Emmel, M. McCutcheon, G. R. Crabtree, L.A. Herzenberg: The actions of cyclosporin A and FK506 suggest a novel step in the activation of T lymphocytes. *EMBO* **9** (1990) 4425-4433.

30.   G. Baumann, G. Zenke, R. Wenger: Molecular Mechanisms of Immunosuppression. *Autoimmunity* **5** (Suppl.A) (1992) 67-72

31.   V. F. Quesniaux, M. H. Schreier, R. M. Wenger, P. C. Hiestand, M. W. Harding, R. M. Van Regenmortel: Cyclophilin binds to the region of cyclosporin involved in its immunosuppressive activity. *Eur. J. Immunol.* **17** (1987) 1359-65.

32.   C. Weber, G. Widmer, B. Von Freyberg, R. Traber, W. Braun, H. Widmer, K. Wuethrich: The NMR structure of Cyclosporin bound to cyclophilin in aqueous solution. *Biochemistry* **30(26)** (1991) 6563-6574.

33.   S. W. Fesik, R. T. Gampe, H. L. Eaton, G. Gemmecker, E. T. Olejniczak, P. Neri, D. A. Egan, R. Edalji, R. Simmer, R. Helfrich, J. Hochlowski, M. Jackson: NMR Studies of [U-^{13}C]cyclosporin A bound to cyclophilin: Bound conformation and portions of cyclosporin involved in binding. *Biochemistry* **30** (1991) 6574-6583.

34.   G. Pflügel, J. Kallen, T. Schirmee, J. N. Jansonius, M. G. M. Zurini, M. D. Walkinshaw: The x-ray structure of a decameric cyclophilin-cyclosporin crystal complex. *Submitted to Nature* (1992)

35.   M. K. Rosen, P. J. Belshaw, D. G. Alberg, S. L. Schreiber: The conformation of cyclosporin

A bound to cyclophilin is altered (once again) following binding to calcineurin: An analysis of receptor-ligand-receptor interactions. *Bioorganic & Medicinal Chem. Lett.* **2** (1992) 747-753.

36.   R. M. Wenger: Synthesis of cyclosporine. *Helv.Chim.Acta* **67** (1984) 502-

37.   a) P. Bollinger, Preclinical Research Laboratories, SANDOZ PHARMA AG., unpublished work.

b) D. Seebach, S. Y. Ko, H. Kessler, M. Koeck, M. Reggelin, P. Schmieder, M. D. Walkinshaw, J. J. Boelsterli, D. Bevec: Thiocyclosporins: Preparation, solution and crystal structure, and immunosuppressive activity. *Helv.Chim.Acta* **74** (1991) 1953-1990.

38.   N. A. Clipstone, G. R. Crabtree: Identification of calcineurin as a key signalling enzyme in T-lymphocyte activation. *Nature* **357** (1992) 695-697.

39.   S. J. O'Keefe, J. Tamura, R. L. Kincaid, M. J. Tocci, E. A. O'Neill: FK-506 and CsA-sensitive activation of the interleukin-2 promoter by calcineurin. *Nature* **357** (1992) 692-694.

Conformational Analysis in Solution of Antibiotics with Immunosuppressive Activity

Horst Kessler and Dale F. Mierke

Organisch-Chemisches Institut, Technische Universität München

Lichtenbergstr. 4, D - 8046 Garching, Germany

Summary: The conformational analysis of the immunosuppressive drugs cyclosporin A (CsA) and FK 506 in $CDCl_3$ solution by modern NMR methodology and molecular dynamics calculations is briefly described. The impression of one rigid structure of CsA, which might be assumed from the identical backbone conformations observed in $CDCl_3$, C_6D_6, and THF solutions and in the crystal, is falsified from the many conformations seen in more polar solvents. The change in conformation resulting from the addition of LiCl to the THF solution and the identical changes observed when CsA is bound to the biological receptor, together with kinetic measurements of binding, illustrate an induced fit model for this interaction. The fiction of "the conformation", at least for partially flexible molecules, is discussed.

Introduction

It is the aim of pharmaceutical chemists to understand the interaction of biologically active compounds with receptors at the molecular level.[1] In most cases the structure of the receptors are not known and molecular investigations are restricted to the substrate alone. Hence, a first step in the above mentioned direction are structural studies of biologically active compounds. In this respect tremendous strides have been made in the last decade by NMR spectroscopy, which allows for the study of the conformation of interesting compounds in solution and therefore the determination of the dependence of the 3D structures on environment. Our knowledge about structure and internal mobility is based on experimental observations from NMR spectroscopy (i.e. intramolecular distances between protons, bond angle information, orientations of groups) which are combined with molecular dynamics calculations. Often X-ray can provide additional and alternative information about the structure of the molecules in the crystal.

To demonstrate the scope and limitations of such studies we present here our studies of cyclosporin A (CsA) [2] and FK 506 [3] (Figure 1), two antibiotics which exhibit immunosuppressant activity. The former is used as a drug to prevent graft rejection in organ transplantations with tremendous success. The latter was recently found to have similar activities and is presently in clinical trials. Chemically, CsA is a cyclic undecapeptide in which seven of the eleven amide bonds are *N*-methylated. The main feature is the unsusual amino acid *N*-methyl-butenyl-methyl-threonine (MeBmt) and four *N*-methyl-leucines.[4] Altogether CsA is highly lipophilic and not soluble in water. In contrast FK 506 is a peptide-macrolide which contains only one amide unit (Figure 1). FK 506 also has explicit lipophilic properties.

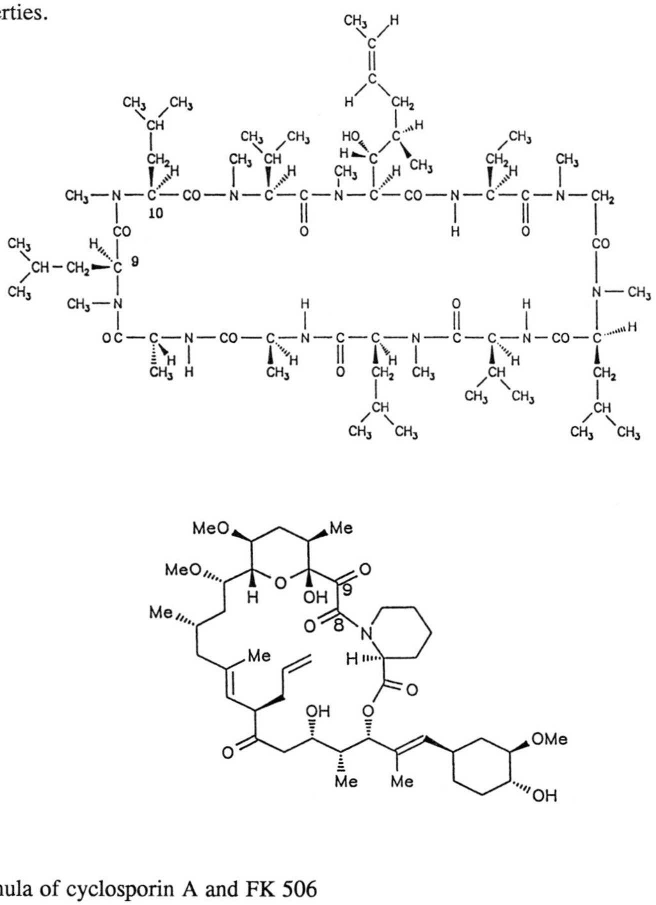

Figure 1. Formula of cyclosporin A and FK 506

In an NMR spectrum every chemically different, magnetically active nucleus (^{1}H, ^{13}C, ^{15}N) is represented by a signal (often consisting of several lines) of distinct chemical shift. The first step in the NMR spectroscopic structure determination is the assignment of these signals to the molecular constitution, i.e. the elucidation of the connectivity of all atoms. This is preferentially achieved by using J-coupling constants between the nuclei as this phenomenon is transmitted via chemical bonds. Hence, assignment means a direct proof of the compounds entire chemical structure.[5] The next step is the determination of through space distances by measuring cross relaxation rates (expressed by initial build-up rates of nuclear Overhauser effects, NOE, between protons).[6] The set of intramolecular distances is then converted into a three dimensional structure by distance geometry calculations using only the molecular constitution and experimental data.[7] Often several families of structures compatible with the experimental data result from these calculations. Their energies and mobilities are elucidated in molecular dynamics calculations,[8] which yield a time dependent structure at a certain temperature and over a given time span, typically 100 ps. It is important to note that the solvent has to be included explicitly in these calculations to obtain reasonable structures. Using these procedures we have to understand that most structures contain more or less flexible parts and the representation of "the structure" in a nicely colored picture is to a large extent fiction. However, the presentation of a time-dependent structure is not straightforward. The commonly used superposition of many structures greatly depends on the atoms superimposed, choosing different atoms produces greatly different results. Here we present the averaged structure (obtained as a mean over all structures along a trajectory, followed by energy minimization), with the proviso that the dynamics, often indicated by standard deviations of dihedral angles, are kept in mind.

Cylosporin

The structure of cyclosporin A in $CDCl_3$ solution obtained in the above described manner [9] is represented in Figure 2.

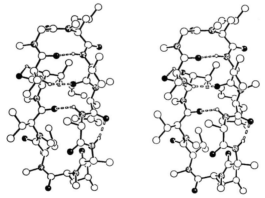

Figure 2.   Stereoplot of the mean structure (energy minimized) of cyclosporin A in $CDCl_3$ over a trajectory of 100 ps free of MD in solution.

This structure consists of a slightly twisted ß-sheet involving three hydrogen bridges from Val5NH, Abu2NH and Ala7NH and a ßII' turn involving Sar3-MeLeu4. The other part of the molecule forms a loop, which includes a *cis* peptide bond between MeLeu9 and MeLeu10. All other peptide bonds are trans oriented.

A very similar structure has already obtained previously using less sophisticated methods.[10] The "frozen structure" in the crystal obtained by X-ray is virtually identical to this solution structure, which is also found in C_6D_6 or THF solution.[11] This generates the impression that CsA is a rigid molecule and we might assume that the structure of the molecule when receptor bound is identical. However, both speculations are wrong. A solution of CsA in methanol exhibits more than four different conformations undergoing slow exchange on the NMR chemical shift time scale.[12] In DMSO solution we have identified more than seven conformations.[5]

Recently the structure of CsA at its biological receptor, the *cis-trans* peptide proline isomerase, cyclophilin has been determined by two groups. It has been demonstrated that a drastic conformational [13] change occurs on binding involving a *cis-trans* isomerization of the MeLeu⁹-MeLeu¹⁰ peptide bond. A similar change can be observed when CsA is complexed with lithium ions.[14] In fact the conformation of CsA with lithium ions (Figure 3) turned out to be identical to the receptor bound conformation around the site where binding to cyclophilin occurs.[14]

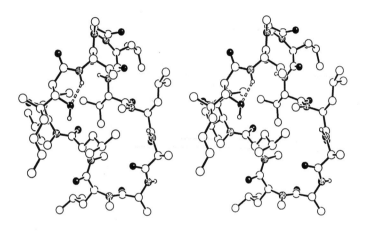

Figure 3. Conformation of the lithium complex of CsA

It is important to note that detailed kinetic studies of the binding of CsA to cyclophilin exhibits an activation barrier which corresponds to the *cis-trans* isomerization barrier, when solvents such as CDCl₃ or THF are used.[14] However, the addition of LiCl to a THF solution results in the disappearance of this distinct barrier in complete agreement with the observation of the all-*trans*-conformation of the CsA-LiCl-complex.[14] Kinetically, a second rate process follows, most probably involving the fine tuning of side chain which must interlock the complex formation (a similar process is observed in protein folding in the long time regime).[15] In DMSO solution, the inhibition of the *cis-trans*-proline isomerase (cyclophilin) is kinetically very slow in agreement with the multiple conformations observed in this solvent.[14]

In conclusion, it is shown that CsA, although prefering one conformation in CDCl₃ and in the crystal, is a flexible molecule, which can change its structure under different environmental conditions. There is also a conformational change on binding to the receptor cyclophilin. Together with kinetic studies this represents a well defined example of induced fit in biological recognition. It further demonstrates that in conformation-activity relationships, one has to be careful to check the softness of the conformation (easiest by different environments). For drug research, the design of structurally more restrained analogues (cyclization to smaller and more rigid molecules) is the way to go.[16]

Knowing the structure of the CsA cyclophilin complex does not end the story. The biological response of this complex results from further interaction with calcineurin and other proteins at the nucleus of the T lymphocyte.[17] This important process is under investigation in several groups.

Dynamics of Cyclosporin

NMR spectroscopy provides additional information of internal molecular dynamics. These motions, which are actually flipping processes among different conformational states, may occur on different time scales. Processes slower than 0.1 sec yield distinct spectral sets of each conformer in the NMR spectrum.[18] Faster processes may be detected by $T_{1\rho}$ measurements and lowering the temperature.[19] Often the set of experimental data cannot be attributed to a single conformation and an equilibrium between several conformations (in the simplest case: two structures) equilibrating fast on the NMR chemical shift time scale has to be assumed.[20] This is often found for side chains of amino acids and expressed in the homo- and heteronuclear J-coupling constants.[21] A detailed analysis requires the diasterotopic assignment of prochiral groups (e.g. the two geminal ß-protons of the four N-methylleucines in CsA or the geminal methyl groups of valine) which can be obtained from several J-coupling constants and sometimes from NOE values.[5,10] The analysis of side chain conformations in cyclosporin shows for example that MeLeu[4] and MeLeu[9] prefer one conformation about the α,β bond ($\chi 1$ angle according to the peptide nomenclature) whereas MeLeu[6] populates two rotamers in a ratio of 30:70 and MeLeu[10] all three staggered rotamers (49:35:16). This holds true for $CDCl_3$ [9,21] solution but in the lithium complex the situation is quite different: for MeLeu[6] and MeLeu[10] the conformation with $\chi 1 = -60°$ dominates, whereas MeLeu[4] and MeLeu[9] populate all three rotamers to a considerable extent.[13]

Mobility can also be analyzed from MD calculations. There is no preset procedure for determining the presence of different conformations exchanging rapidly on the NMR timescale. Our approach has been to start with the assumption that *one predominate* conformation exists (of course, the conformation is considered to be flexible, plus or minus 10° for each of the dihedral angles). The experimental restraints are then applied and the single conformation is optimized to arrive at no violations of the restraints. It is important to try to obtain absolute agreement, for even the smallest deviations may be an indication of multiple conformations. If there are deviations, even after reexamination of the experimental data and in some cases additional experiments, the next step is to try to fit the experimental data to multiple conformations.[20] The structure refinement of CsA in chloroform solution with the application of 117 NOEs and 63 coupling constants (see below), quite clearly indicated the presence of one predominant backbone conformation in this solvent.

In the structure refinement, it is very important to incorporate as much experimental data as possible and in addition, to try to mimic the environment in which the experimental measurements were carried out. One of the most natural, recent improvements is to use the same solvent for NMR experiments and MD calculations. We have carried out simulations in a number of different solvents, including carbon tetrachloride,[22] dimethyl sulfoxide,[23,24] chloroform,[25,26] and methanol.[27] These simulations are costly; the solvent will increase the number of atoms by an order of magnitude. The consequences of this are that the number as well as the length of the simulations is reduced (it is common to carry out many calculations with many different starting conformations). In the solvent simulations, the number of starting structures is limited. However, the results from the solvent simulations are so much better that the limitations are justified. As computer technology improves, these consequences will become less important. The *in vacuo* effects that often plague MD simulations are no longer observed.[24] In addition, the interactions between the peptide and the solvent can be examined. These interactions (i.e. the formation of intermolecular hydrogen bonds) play an important role in the stabilization of the conformation of the peptide. The solvent is also important for the dynamics of the molecule. With the assumption that the solvent simulation is a good mimetic of the solution, the simulation should provide insight into the motions and dynamics of the peptide. The results from the simulations can then be compared to the NMR measured parameters (T_1, T_2 and NOEs). To examine this idea, we have recently carried out a long (1 nanosecond) simulation of CsA in chloroform. From the trajectory the correlation functions of the proton-proton vectors (for which NOEs are observed) have been calculated.[28] From these simulations, the effects of different motions on the observed NOEs can be accounted for, allowing for the development of more realistic distance restraints.[29] Very similar calculations have recently been published for antamanide, a cyclic deca-peptide.[30]

As already mentioned above, another experimental parameter is the coupling constant, which supplies conformational preferences about dihedral angles. However, the method has been scarcely used directly in structure refinements because of the many possible answers (up to four) from the Karplus equation. This is illustrated for a $3J_{HN-H\alpha}$ coupling constant of 8.0 Hz in Figure 4.

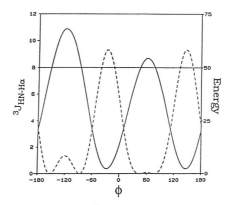

Figure 4.   A plot of the Karplus curve of the $^3J_{HN-H\alpha}$ coupling constant using A, B and C coefficients of 9.4, -1.5 and 0.4, respectively (solid line). The four dihedral angles resulting from a coupling constant of 8.0 Hz are illustrated (horizontal line). The penalty function used in the simulations, for a coupling constant of 8.0 Hz is shown (dashed line).

The coupling constant of 8.0 Hz allows values of 44°, 76°, -87° and -152° for the angle ϕ. As can be seen in the figure, coupling constants of greater than 8.7 Hz produce only two answers (centered about -120°), it is therefore common practice to only utilize those coupling constants. On the other hand, if more than one coupling constant describing a certain dihedral angle is measured, then some or all of the ambiguities disappear.[31,32] The coupling constants that can be easily measured for the dihedral angle ϕ of a peptide backbone are shown in Figure 5.

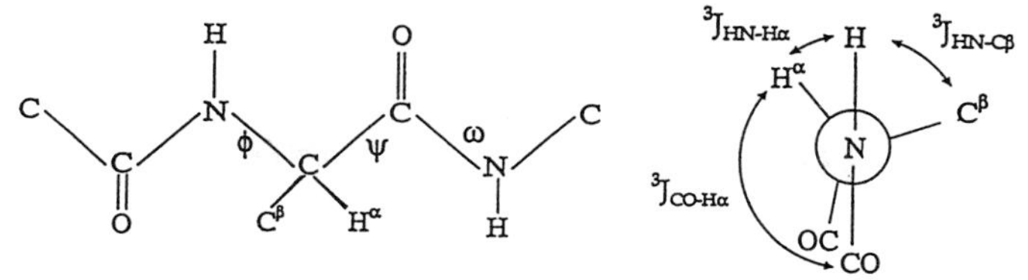

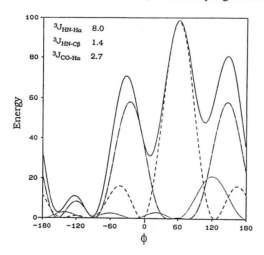

Figure 5. Illustration of the backbone dihedral angles (left) and the coupling constants resulting in information about the dihedral angle ϕ.

These coupling constants can be used directly in a penalty function included in the MD force field.[33] The resulting penalty function for the heteronuclear coupling is included in Figure 4 (dashed line). If the torsional angle is in agreement with the experimental coupling constant, no additional energy is added (i.e. for the 8 Hz coupling there are minima at 44°, 76°, -87° and -152°). The advantage of this procedure is that additional coupling constants, which are experimentally determined, can be added in a similar manner.

For CsA we have measured these coupling constant for many of the residues.[34] An example of the resulting penalty function is shown in Figure 6. It is shown that with the additional two heteronuclear coupling constants ($^3J_{HN-C\beta}$ and $^3J_{CO-H\alpha}$) the previous two minima with positive ϕ values resulting from the homonuclear coupling constants alone are removed. In addition, the remaining two minima are not of the same energy, with the minimum at -87° slightly favored. The ϕ value from the distance restrained MD simulations (without coupling restraints) was -104°.[9]

Figure 6. Potential energy profile as a function of the dihedral angle ϕ. The energy is plotted for $^3J_{NH-H\alpha}$ (dot-dash), $^3J_{HN-C\beta}$ (dot), $^3J_{CO-H\alpha}$ (dash) and for the sum of the three couplings (solid).

FK 506

The 23-membered macrocyclic lactone FK 506 contains one amide bond at the acyl-pipecolic acid moiety which can exist in a *cis* and/or a *trans* configuration. The *cis* isomer is found in the solid state.[3] In CDCl$_3$ solution a ratio of 2:1 (*cis/trans*) is found, whereas in the structure of FK 506 when it is bound to its receptor the FK binding protein (FKBP) the amide bond is *trans*.[35,36]

The conformation of both isomers of FK 506 in CDCl$_3$ solution were determined by NMR spectroscopy and MD calculations.[37] For a reliable analysis, the diastereotopic assignments of the protons at the three methylene groups in the macrocyclic ring was crucial. This assignment could not be obtained from homonuclear proton coupling constants and NOE values alone. Hence, a heteronuclear 3D NMR technique, the non-decoupled folded HMQC-TOCSY,[32] was used to obtain more than 120 different heteronuclear long range coupling constants from one single spectrum.[37]

We will not discuss here these techniques or the results of the conformational analysis of FK 506. Suffice it to say that the determination of structures of medium sized molecules often requires the whole arsenal of modern multidimensional NMR techniques and the careful use of selected calculational procedures.

Conclusion

The conformation in CDCl$_3$ solution of the immunosuppressants cyclosporin A and FK 506 were determined by NMR spectroscopy and MD calculations. Both compounds adopt only a few (CsA: one, FK 506: two) backbone conformations in this solvent, but in both cases the structure of these drugs is changed when bound to their biological receptors. This corresponds to an induced fit. For drug design, this must always be kept in mind and more conformationally restricted analogues are required to derive the bioactive conformation directly from the conformation in solution.

Acknowledgement: Financial support by the Deutsche Forschungsgemeinschaft and the Fonds der Chemischen Industrie is gratefully acknowledged. D. M. thanks the Fulbright Commission for a research fellowship.

References

1. See e.g. Computer-Aided Drug Design (T. J. Perun, C. L. Propst, Eds.), Marcel Dekker, New York, 1989.
2. R. Traber, M. Kuhn, A. Rüegger, H. Lichte, H.-R. Loosli, A. von Wartburg: Die Struktur von Cyclosporin C. *Helv. Chim. Acta* **60** (1977) 1247-1255.
3. H. Tanaka, A. Kuroda, H. Marusawa, H. Hatanaka, T. Kino, T. Goto, M. Hashimoto, T. Taga: Structure of FK 506: A Novel Immunosuppressant Isolated from Streptomyces. *J. Am. Chem. Soc.* **109** (1987) 5031-5033.
4. R. M. Wenger: Synthesis of Cyclosporin and Analogues: Structural Requirements for Immunosuppressive Activity. *Angew. Chem. Int. Ed. Engl.* **24** (1985) 77-85.
5. H. Kessler, H.-R. Loosli, H. Oschkinat: Peptide Conformation. Part 30. Assignment of the ^{1}H-, ^{13}C-, and ^{15}N-NMR Spectra of Cyclosporin A in CDCl$_3$ and C$_6$D$_6$ by a Combination of Homo- and Heteronuclear Two-Dimensional Techniques. *Helv. Chim. Acta* **68** (1985) 661-681.
6. D. Neuhaus, M. Williamson "The Nuclear Overhauser Effect". VCH Weinheim, 1989.
7. G. M. Crippen, T. F. Havel "Distance Geometry and Molecular Conformation". John Wiley, New York, 1988.

8.   W. F. van Gunsteren, H. J. C. Berendsen: Computer Simulation of Molecular Dynamics: Methodology, Applications, and Perspectives in Chemistry. *Angew. Chem. Int. Ed. Engl.* **29** (1990) 992-1923.

9.   H. Kessler, M. Köck, T. Wein, M. Gehrke: Reinvestigation of the Conformation of Cyclosporin A in Chloroform. *Helv. Chim. Acta* **73** (1990) 1818-1832.

10. H.-R. Loosli, H. Kessler, H. Oschkinat, H.-P. Weber, T.J. Petcher, A. Widmer: Peptide Conformations. Part 31. The Conformation of Cyclosporin A in the Crystal and in Solution. *Helv. Chim. Acta* **68** (1985) 682-704.

11. H. Kessler, M. Gehrke, J. Lautz, M. Köck, D. Seebach, A. Thaler: Complexation and Medium Effects on the Conformation of Cyclosporin A Studied by NMR Spectroscopy and Molecular Dynamics Calculations. *Biochem. Pharmacol.* **40** (1990) 169-173.

12. S. Y. Ko, C. Dalvit: Conformation of Cyclosporin A in Polar Solvents. *Int. J. Pept. Prot. Res.* **40** (1992) 380-382.

13. (a) C. Weber, G. Wider, B. von Freyberg, R. Traber, W. Braun, H. Widmer, K. Wüthrich: The NMR Structure of Cyclosporin A Bound to Cyclophilin in Aqueous Solution. Biochemistry **30** (1991), 6563-6574. (b) S. W. Fesik, R. T. Gampe, Jr., H. L. Eaton, G. Gemmecker, E. T. Olejniczak, P. Neri, T. F. Holzman, D. A. Egan, R. Edalji, R. Simmer, R. Helfrich, J. Hochlowski, M. Jackson: MNR Studies of [U-^{13}C]Cyclosporin A Bound to Cyclophilin: Bound Conformation and Portions of Cyclophilin Involved in Binding. Biochemistry **30** (1991) 6574-6583.

14. (a) M. Köck, H. Kessler, D. Seebach, A. Thaler: Novel Backbone Conformation of Cyclosporin A: The Complex with Lithium Chloride. *J. Am. Chem. Soc.* **114** (1992) 2676-2686. (b) J. L. Kofron, P. Kuzmic, V. Kishore, G. Gemmecker, S. W. Fesik, D. H. Rich: Lithium Chloride Pertubation of *Cis-Trans* Peptide Bond Equilibria: Effect on Conformational Equilibria in Cyclosporin A and on Time-Dependent Inhibition of Cyclophilin. *J. Am. Chem. Soc.* **114** (1992) 2670-2675.

15. P. A. Evans, R. A. Kautz, R. O. Fox, C. M. Dobson: A Magnetization-Transfer Nuclear Magnetic Resonance Study of the Folding of Staphylococcal Nuclease. *Biochemistry* **28** (1989) 362-370.

16. H. Kessler: Peptide Conformations. Part XIX. Conformational and Biological Activity of Cyclic Peptides. *Angew. Chem. Int. Ed. Engl.* **21** (1982) 512-523.

17. M. K. Rosen, S. L. Schreiber: Natural Products as Probes of Celluar Function: Studies of Immunophilins. *Angew. Chem. Int. Ed. Engl.* **31** (1992) 384-400.

18. H. Kessler: Detection of Hindered Rotation and Inversion by NMR Spectroscopy. *Angew. Chem. Int. Ed. Engl.* **9** (1970) 219-235.

19. K. D. Kopple, Y. S. Wang, A. G. Cheng, K. K. Bhandary: Conformation of Cyclic Octapeptides. 5. Crystal Structure of Cyclo(Cys-Gly-Pro-Phe)2 and Rotating Frame Relaxation ($T_{1\rho}$) NMR Studies of Internal Mobility in Cyclic Octapeptides. *J. Am. Chem. Soc.* **110** (1988) 4168-4176.

20. H. Kessler, C. Griesinger, J. Lautz, A. Müller, W. F. van Gunsteren, H.J.C. Berendsen: Conformational Dynamics Detected by Nuclear Magnetic Resonance NOE-Values and J Coupling Constants. *J. Am. Chem. Soc.* **110** (1988) 3393-3396.

21. H. Kessler, C. Griesinger, K. Wagner: Peptide Conformations 42. Conformation of Side Chains in peptides Using Heteronuclear Coupling Constants Obtained by Two-Dimensional NMR Spectroscopy. *J. Am. Chem. Soc.* **109** (1987) 6927-6933.

22. J. Lautz, H. Kessler, W. F. van Gunsteren, H.-P. Weber, R. M. Wenger: On the Dependence of Molecular Conformation on the Type of Solvent Environment: A Molecular Dynamics Study of Cyclosporin A. *Biopolymers* **29** (1990) 1669-1687.

23. D. F. Mierke, H. Kessler: Molecular Dynamics with Dimethyl Sulfoxide as a Solvent. Conformation of a Cyclic Hexapeptide. *J. Am. Chem. Soc.* **113** (1991) 9466-9470.

24. M. Kurz, D. F. Mierke, H. Kessler: Calculation of Molecular Dynamics for Peptides in Dimethyl Sulfoxide: Elimination of Vacuum Effects. *Angew. Chem. Int. Ed. Engl.* **31** (1992) 210-212.

25. W. L. Jorgensen, J. M. Briggs, M. L. Contreras: Relative Partition Coefficients for Organic Solutes from Fluid Simulations. *J. Phys. Chem.* **94** (1990) 1683-1686.

26. R. Konat, D. F. Mierke, H. Kessler: Synthesis and Conformational Analysis of Hymenistatin. *J. Am. Chem. Soc.* (1992) submitted.

27. W. L. Jorgensen: Transferable Intermolecular Potential Functions. Application to Liquid Methanol including Internal Rotation. *J. Am. Chem. Soc.* **103** (1981) 335-350.

28. D. F. Mierke, H. Kessler: Improved Molecular Dynamics Simulations for the Determination of Peptide Structures. *Biopolymers* (1992) in press.

29. T. M. G. Koning, R. Boelens, G. A. van der Marel, J. H. van Boom, R. Kaptein: Structure Determination of a DNA Octamer in Solution by NMR Spectroscopy. Effects of Fast Local Motions. *Biochemistry* **30** (1991) 3787-3797.

30. R. Brüschweiler, B. Roux, M. Blackledge, C. Griesinger, M. Karplus, R. R. Ernst: Influence of Rapid Intramolecular Motion on NMR Cross-Relaxation Rates. A Molecular Dynamics Study of Antamanide in Solution. *J. Am. Chem. Soc.* **114** (1992) 2289-2302.

31. P. Schmieder, H. Kessler: Determination of the φ Angle in a Peptide Backbone by NMR Spectroscopy with a Combination of Homonuclear and Heteronuclear Coupling Constants. *Biopolymers* **32** (1992) 435-440.

32. P. Schmieder, M. Kurz, H. Kessler: Determination of Heteronuclear Long-Range Couplings to Heteronuclei in Natural Abundance by Two-Dimensional and Three-Dimensional NMR Spectroscopy. *J. Biomol. NMR* **1** (1991) 403-420.

33. D. F. Mierke, H. Kessler: Combined Use of Homo and Heteronuclear Coupling Constants as Restraints in MD Simulations. *Biopolymers* **32** (1992) 1277-1282.

34. M. Eberstadt, D. F. Mierke, M. Köck, H. Kessler: Peptide Conformation from Coupling Constants: Scalar Couplings as Restraints in MD Simulations. *Helv. Chim. Acta* (1992) in press.

35. H. Kessler, D. F. Mierke, D. Donald, M. Furber: Toward the Understanding of Immunosuppression. *Angew. Chem. Int. Ed. Engl.* **30** (1991) 954-955.

36. G. D. van Duyne, R. F. Standaert, P. A. Karplus, S. L. Schreiber, J. Clardy: Atomic Structure of FKBP-FK 506, an Immunophilin-Immunosuppressant Complex. *Science* **252** (1991) 839-842.

37. D. F. Mierke, P. Schmieder, P. Karuso, H. Kessler: Conformational Analysis of the *cis*- and *trans*-Isomers of FK 506 by NMR and Molecular Dynamics. *Helv. Chim. Acta* **74** (1991) 1027-1047.

Synthesis of Biologically Interesting Glycopeptides
A Problem of Chemical Selectivity

Horst Kunz

Institut für Organische Chemie, Universität Mainz
D - 6500 Mainz, Germany.

Summary: As the carbohydrate parts of natural glycoproteins play key functions in the biological recognition on membranes, e. g. in the regulation of the cell growth or in infectious processes, the synthesis of glycopeptides representing partial structures of glycoproteins is receiving increasing interest. The glycosidic bonds of glycoproteins are potentially acid-sensitive and in some cases also destroyed under moderate basic conditions. Therefore, mild and selective deblocking reactions are crucial in the synthesis of glycopeptides. Applying the 9-fluorenylmethoxycarbonyl (Fmoc) group, removable with the weak base morpholine, and the allylic protecting groups, which are cleavable under neutral conditions via palladium(0)-catalyzed reactions, glycopeptides of increasing complexity were synthesized, deblocked and coupled to carrier proteins. An important effect for practical glycopeptide synthesis has been revealed by the observation that acyl protection within the saccharide portion strongly stabilizes the glycosidic linkages. New methods were developed for the synthesis of β-mannosides, for using the azido group as an anomeric protecting group, and for the introduction of α-fucoside units stable to acids. These tools were applied in the synthesis of partial structures of virus envelope components and tumor-associated antigens.

Carbohydrates of Glycoproteins in Biological Recognition Processes

For a long time, carbohydrates and proteins have been considered separate classes of natural products. However, the results of cell biological, biochemical and immunological research of the past decades have revealed that most of the natural proteins in reality are glycoproteins. The saccharide portions of glycoproteins not only influence the physical properties and the preferred conformation but can protect these biomolecules against proteolytic attack.[1] In numerous selection processes, the carbohydrates of glycoproteins have been found to be important recognition signals, e. g., in the selective transport of serum components through membranes, in the docking of bacteria or viruses to the cells, in cell differentiation and in the regulation of cell growth.[2] Normal cells and tumor cells are quite different in the glycoprotein profiles of their cell membranes. Altered glycoproteins of tumor membranes have been described as tumor-associated antigens, e. g. *O*-glycoproteins of the Tn antigen structure **1**.[3,4] The *O*-glycoproteins contain *O*-glycosidic bonds between the peptide and the carbohydrate parts. Among them the mucin-type glycoproteins (e.g. **1**) are characterized by α-*O*-glycosidic linkages between *N*-acetylgalactosamine and serine or threonine.

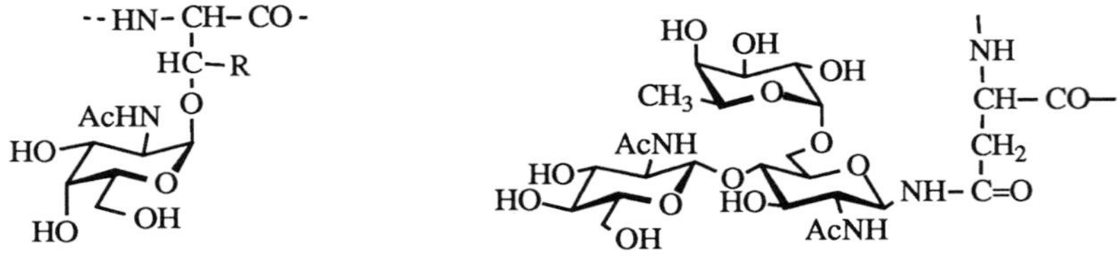

1, R = H or CH$_3$ **2**

Scheme 1

N-Glycoproteins are most frequently found in nature. They almost invariably contain the β-*N*-glycosidic linkage between *N*-acetylglucosamine and asparagine.[1] *N*-Glycoproteins fucosylated in the core region **2** have been suspected to be tumor-associated antigens. This structure is also characteristic for virus envelope components.[5]

In contrast to peptides and proteins, glycopeptides and glycoproteins are not easily accessible by gene-technological methods. The isolation of glycoproteins from biological sources is also difficult. In most cases, only very small amounts can be obtained. Furthermore, their purification suffers from the microheterogeneity of natural glycoconjugates. The investigation of recognition phenomena and of virological and immunological effects demands model compounds of exactly specified structure. In this respect the chemical synthesis of glycopeptides is of interdisciplinary interest.

Limitations in Glycopeptide Synthesis - The Sensitivity of Glycosidic Bonds

Glycopeptides contain numerous functional groups. Their regio- and stereoselective synthesis demands versatile and selective protecting group techniques.[6] This holds true, in particular, for glycopeptides with more complex saccharide parts which themselves are to be constructed in laborious multi-step oligosaccharide syntheses.[7,8] In natural glycoproteins, the peptide and carbohydrate portions are exclusively linked glycosidically. The intersaccharidic and glycosidic bonds, being acetal-like bonds, are more or less acid-sensitive. Strong acids often applied in peptide chemistry can cleave or anomerize these glycosidic bonds.[6] Moreover, *O*-glycosyl serine and threonine derivatives, e. g. **3**, are easily decomposed under basic conditions by a characteristic β-elimination of the carbohydrate.[9]

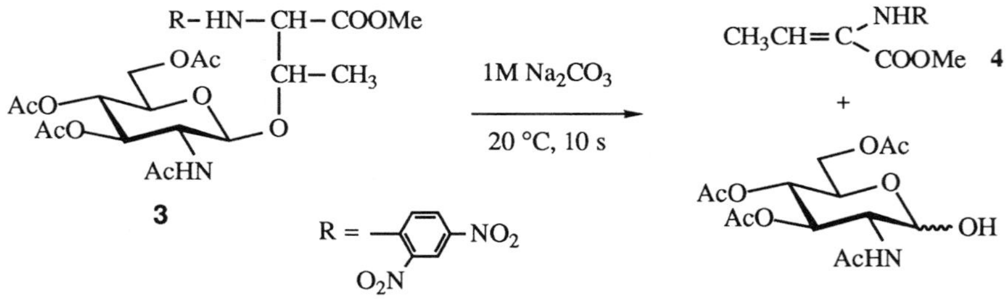

Scheme 2

The high leaving group tendency of the carbohydrate which becomes obvious in this β-elimination stands in a marked contrast to the stability of serine and threonine *tert*-butyl and benzyl ethers well known in peptide chemistry. The interpretation of this leaving group quality (Scheme 2) in terms of the complexing abilities of carbohydrates[6] not only initiated the concept of using the chirality and complexing properties of carbohydrates in asymmetric syntheses,[10,11] but also led to a new methodology in peptide formation: The xyluronic acid amide ester of the *N*-protected glycine **5** did not react with phenylalanine *tert*-butyl ester. However, after addition of lithium bromide under otherwise identical conditions, the dipeptide **7** was formed in high yield (Scheme 3).[12]

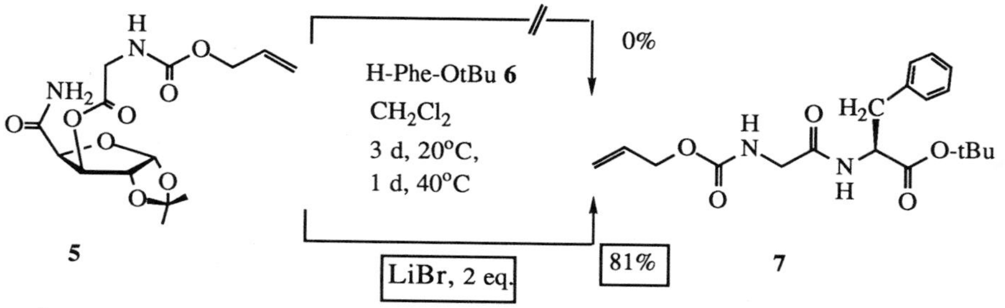

Scheme 3

It is assumed that the complex **8** is formed in the course of the lithium ion-initiated peptide condensation. The complexation of the metal ion not only enhances the leaving group quality of the carbohydrate but also reduces the activation entropy via precoordination of the reacting species.

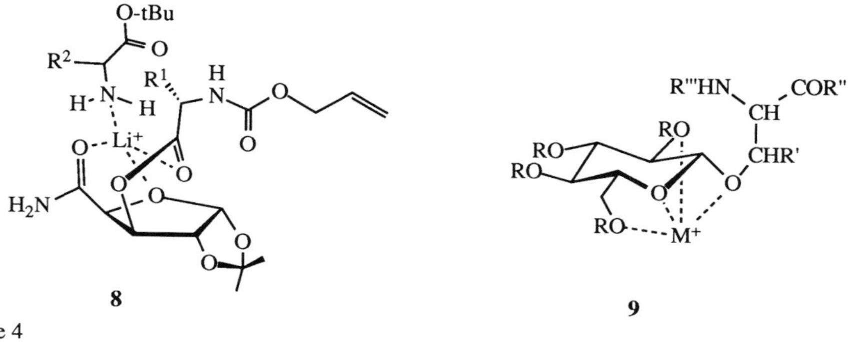

Scheme 4

Whereas this complex formation (**8**) makes possible a new mild peptide synthesis, the coordination of cations to *O*-glycosyl serine or threonine derivatives (**9**) causes an increased base-sensitivity and, thus, constitutes a limiting factor in the glycopeptide synthesis.

The (9-Fluorenyl)methoxycarbonyl (Fmoc) Group in Glycopeptide Synthesis

The Fmoc-group,[13] frequently applied in peptide chemistry, is usually removed via β-elimination carried out with piperidine in dimethylformamide. Since the *O*-glycosyl serine and threonine peptides (**4**) are base-sensitive themselves, we used the weak base morpholine (pKa 8.3) in order to selectively cleave the Fmoc group from the O-glycopeptides, e. g. **10**.[14] Under these very mild conditions, the other protecting groups and the *O*-xylosyl serine linkage in **11** remained absolutely unaffected. Chain extension was now easily realized to give **12**. Removal of the Fmoc group with morpholine again proceeded completely and selectively to yield **13**.[14]

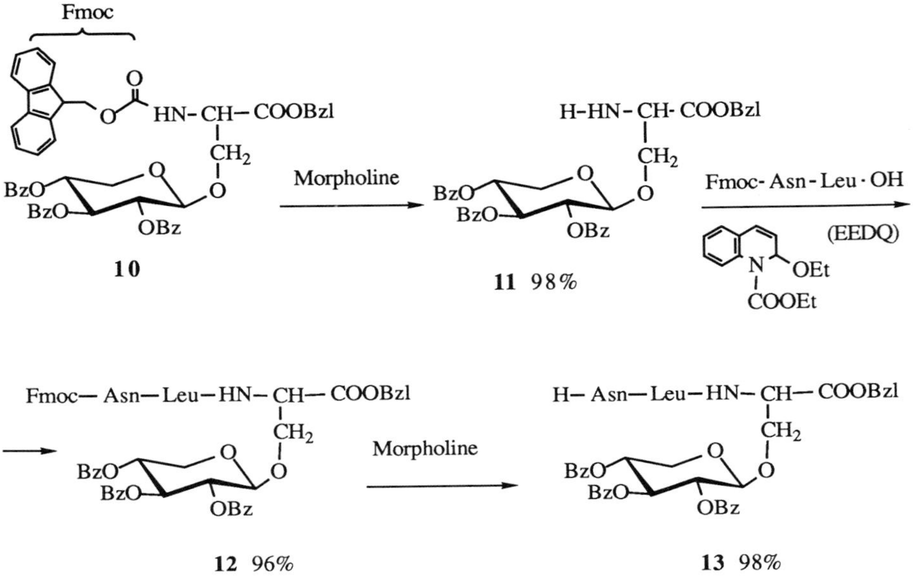

Scheme 5

On the basis of this method, tumor-associated Tn- and T-antigen glycopetides were successfully synthesized, deprotected and coupled to bovine serum albumin.[15] Active esters of glycosylated Fmoc amino acids have also been applied in glycopeptide chemistry.[16] Recently, the Fmoc methodology was successfully adopted in the solid phase synthesis of glycopeptides.[17-19]

The Allyl Ester as the Carboxy Protection

In search of a small and stable blocking group for glycopeptide synthesis, we found the allyl ester to be a very useful carboxy protection. It can be removed either by rhodium (I)-catalyzed isomerization[20] or, more favorably, by palladium(0)-catalyzed allyl transfer.[21] In this sense, the treatment of the xylosyl serine allyl ester derivative **14** with catalytic amounts of the palladium(0) phosphine complex in the presence of morpholine as the weakly basic allyl trapping reagent in tetrahydrofuran at room temperature resulted in a selective and quantitative removal of the otherwise stable allyl ester protection. The other blocking groups and the β-*O*-xylosyl serine linkage of **15** remained completely stable. The xylosyl serine structure constitutes the typical linkage region of connective tissue glycoproteins.

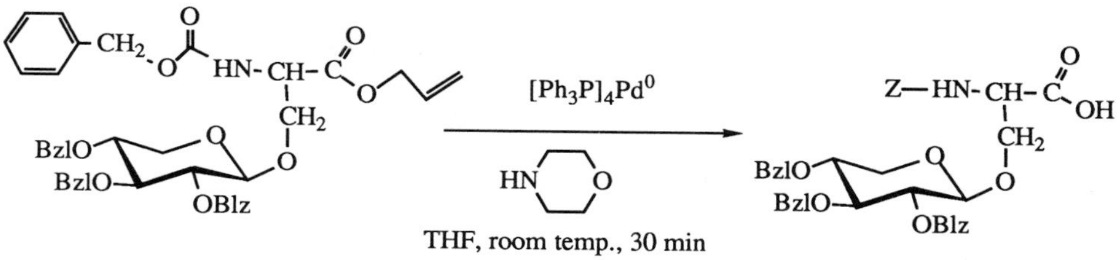

Scheme 6

The allyl ester can advantageously be combined with the Fmoc group as was demonstrated in the selective deblocking of the T antigen component **16**. The Fmoc group was selectively cleaved using morpholine to give **18**, leaving the allyl ester untouched. The allyl ester, on the other hand, was removed selectively by Pd(0)-catalyzed allyl transfer to *N*-methylaniline as the trapping nucleophile which was proven to be too weakly basic for affecting the Fmoc group of **17**.

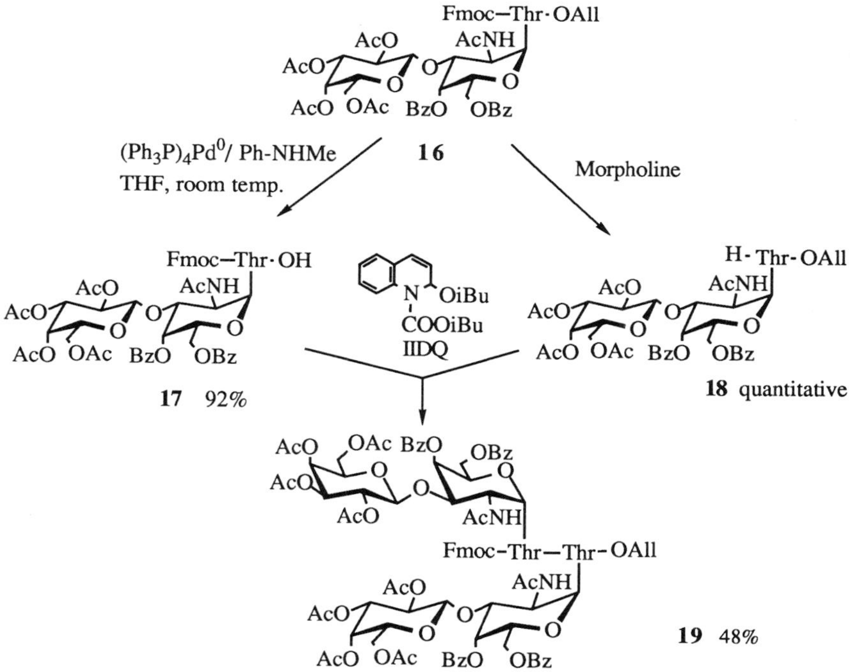

Scheme 7

Condensation of the selectively deblocked compounds **17** and **18** delivered the clustered glycopeptide **19** containing the tumor-associated[3] T antigen saccharide side chains.[22]

The Allylic Anchoring Principle in Solid Phase Synthesis of Peptides and Glycopeptides

The advantageous properties of the allylic protecting group were utilized in the solid phase synthesis by the development of a new allylic anchoring principle.[23] This method allows one to detach the synthesized peptides and glycopeptides from the polymeric support under practically neutral conditions. One particular version of this allylic anchor are amino acid or peptide esters, e. g. **20**, of the polymer-linked 4-hydroxycrotonylaminomethyl structure (HYCRAM™; HYCRAM is a Trade Mark of the Orpegen Company, Heidelberg, Germany).[24] Using the Pd(0)-catalyzed cleavage and morpholine as the trapping nucleophile, the sensitive glycopeptide **21** was released from the polymer-linked allyl ester **20** in high yield and without destroying the O-glycosidic bond or the numerous other protecting groups.[17]

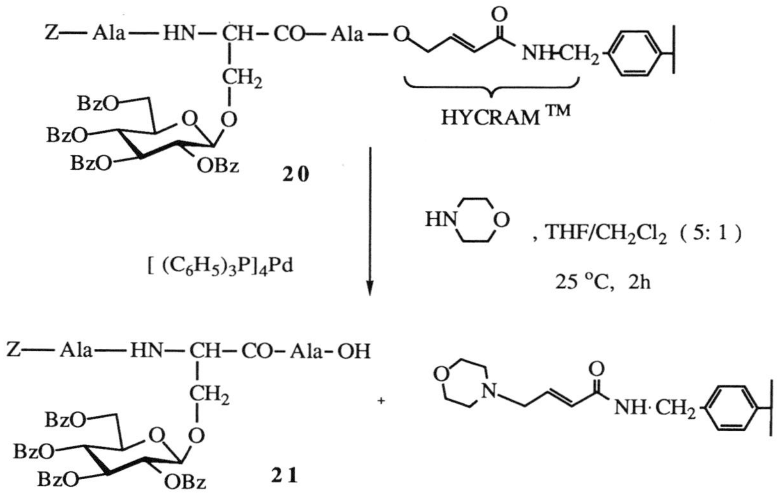

Scheme 8

An improved modification of the allylic HYCRAM anchor contains a β-alanine insert between the hydroxycrotonyl and the aminomethyl group (βHYCRAM). On the basis of the βHYCRAM anchor, the resin-linked octapeptide **22** carrying a type 1 lactosamine side chain β-N-glycosidically coupled to asparagine was constructed. The stepwise solid phase synthesis was performed with Fmoc protected amino acids.[25] Side chain protection of the tert-butyl-type was applied. Using the Pd(0)-catalyzed allyl transfer reaction carried out in DMSO-containing solvents under exclusion of oxygen and light and with N-methyl aniline as the allyl acceptor, the glycopeptide **23** was detached from the resin without destruction of the tert-butyl-type protecting groups and the N- and O-glycosidic bonds. As glycopeptides selectively deblocked in only one functional group can be purified more efficiently than fully deblocked ones, pure **23** was obtained on a gram-scale (yield: 60%) by preparative HPLC.[25] The glycopeptide **23** constitutes a N-glycosylated peptide T which is a partial sequence of gp 120 of HIV-1 and has been reported to inhibit the docking of HIV-1 to the CD$_4$ receptor of lymphocytes.[26]

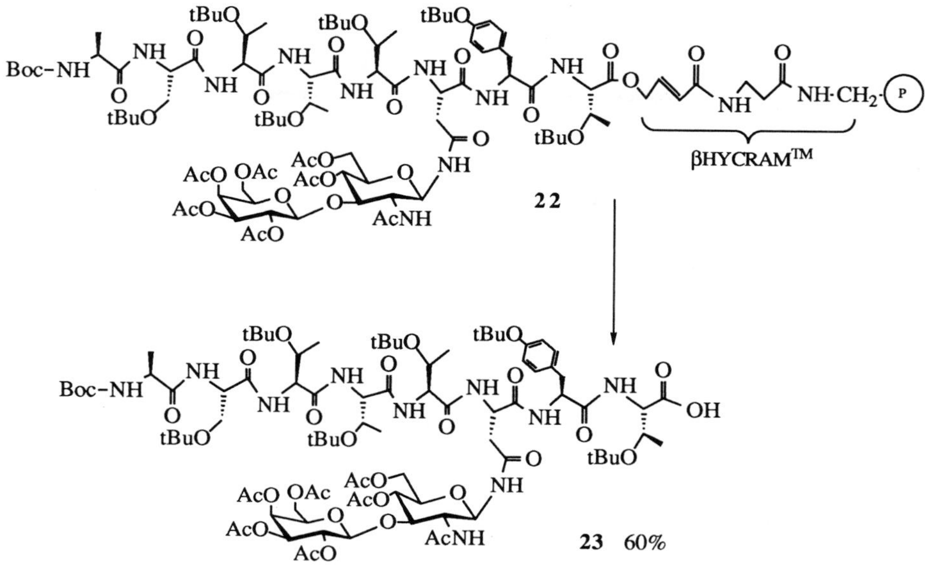

22

23 60%

Scheme 9

The Allyloxycarbonyl (Aloc) Group

The allyloxycarbonyl (Aloc) group was described more than 30 years ago.[27] However, it was not considered useful for peptide synthesis because of the nonselective or harsh conditions required for its removal (hydrogenolysis or acidolysis). The application of the Pd(0)-catalyzed cleavage now makes the Aloc group a very valuable tool in the synthesis of peptides and glycopeptides.[28] For example, the Aloc group was removed from the fucosyl chitobiose asparagine conjugate **24** by Pd(0)-catalyzed allyl transfer to dimedone (5,5-dimethyl-cyclohexane-1,3-dione) in tetrahydrofuran at room temperature with high yield and complete selectivity.[29] The selectively deblocked carboxy component **25** was now accessible to chain extension.

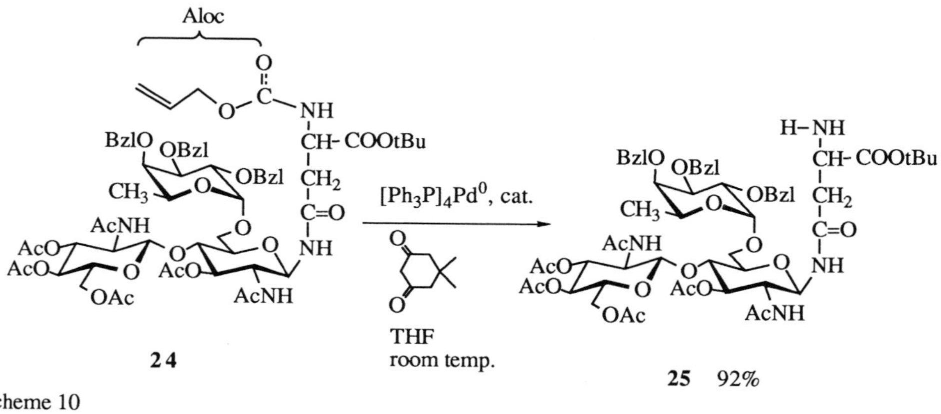

24

$[Ph_3P]_4Pd^0$, cat.

THF
room temp.

25 92%

Scheme 10

In contrast, attempts to remove the *tert*-butyl ester of **24** by treatment with trifluoroacetic acid resulted in the complete cleavage of the fucoside bond. Since the sensitivity of the fucoside linkage was traced
back to the ether-type protection of this monosaccharide moiety, a new synthesis was developed for the trisaccharide, aiming at a product which exclusively carries *O*-acetyl protection.

The Indirect Protection of Glycoside Bonds Exhibited by Acyl Protecting Grou s within the Saccharide Portion

The revised strategy for the construction of the fucosyl chitobiose structure was based on two major
concepts:[28]
a) the use of the anomeric azido group as the protected form and precursor of the anomeric amine required
    for the formation of *N*-glycosyl asparagine, and
b) the exchange of the ether-type protection, at first needed for the formation of the α-fucoside linkage, for
    an acetyl protection.

According to this strategy, the selectively deblocked chitobiosyl azide **26**[28,29] was reacted with the 4-
methoxybenzyl (Mpm) protected fucosyl bromide **27**[28,30] under in situ-anomerization conditions[31] to
give the corresponding trisaccharide azide **28**. Treatment of **28** with ceric ammonium nitrate resulted in selective removal of the Mpm protecting groups[32] without affecting the azido group. Subsequent *O*-acetyla-
tion and hydrogenolysis furnished the desired fucosyl chitobiosylamine **29**.

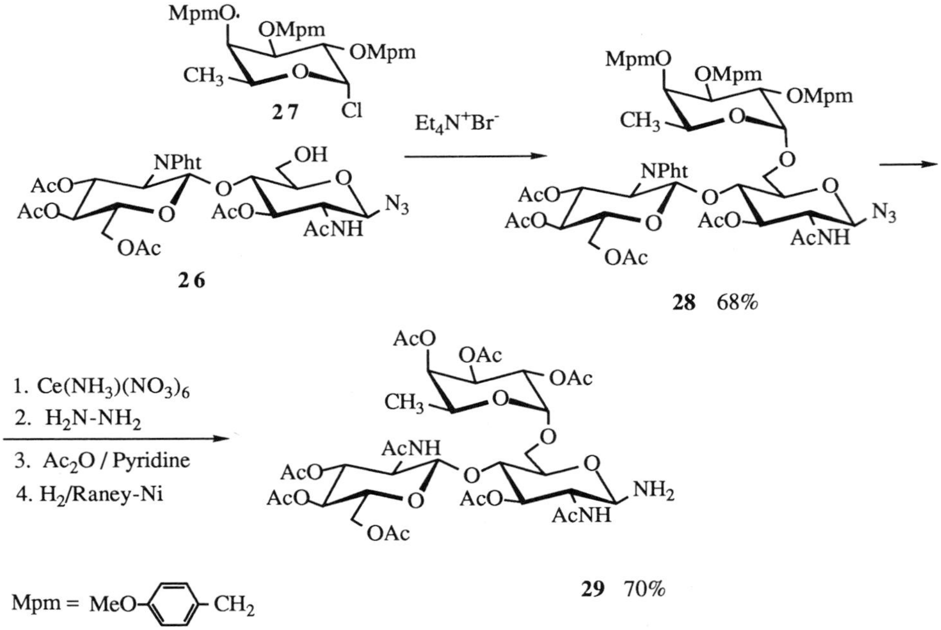

Scheme 11

Condensation of **29** with Aloc aspartic acid α-*tert*-butyl ester gave the conjugate **30**. Removal of the
Aloc group from **30** via Pd(0)-catalyzed allyl transfer to N,N'-dimethylbarbituric acid as the allyl trapping

reagent[33] proceeded with complete selectivity. Subsequent terminal chain extension with Aloc alanylleucine promoted by a water soluble carbodiimide in the presence of 1-hydroxy-benzotriazole delivered the trisaccharide tripeptide **31** almost quantitatively.[28] The treatment of **31** with trifluoroacetic acid in dichloromethane actually resulted in a completely selective cleavage of the *tert*-butyl ester. The fucoside bond of **32** remained absolutely unaffected.[28] The opposite behaviour of the *O*-benzyl protected conjugate **24** on the one hand and the *O*-acetyl protected glycopeptide on the other on treatment with acid discloses a marked influence of the protecting group pattern within the carbohydrate portion on the stability of the glycosidic bonds towards acids.

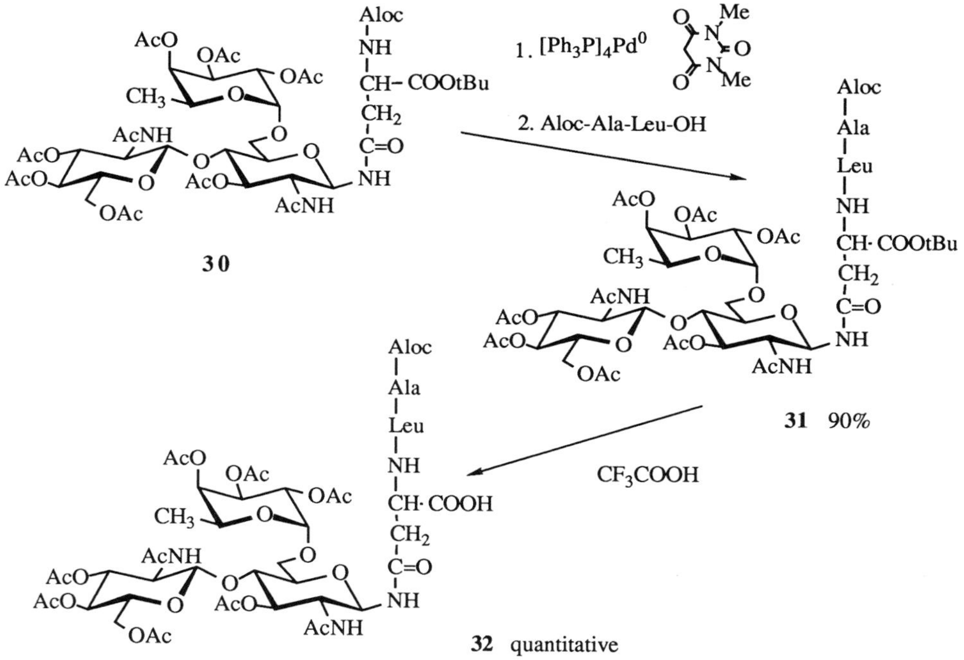

Scheme 12

The acetyl protecting groups exhibit inductive effects and are protonated on their carbonyl oxygens. This results in a Coulomb repulsion of further protons approaching the glycosidic regions and, thus, indirectly protects the intersaccharidic bonds. Chain extension of **32** yielded a glycohexapeptide partial structure of an envelope glycoprotein of a leukemia virus, which after deblocking was coupled to bovine serum albumin to give a synthetic glycoprotein antigen.[28]

Synthesis of β–Mannosyl Chitobiose Asparagine Conjugates

The construction of β-mannoside bonds is a particularly difficult problem in oligosaccharide chemistry since both the anomeric effect and the neighboring-group participation strongly prefer the formation of α-mannosides .[7] As the β-mannoside chitobiose conjugate constitutes a crucial structure in the core region of *N*-glycoproteins, a stereoselective formation of this bond is highly desirable. Earlier β-mannoside syntheses carried out by the use of insoluble silver silicate catalysts were only applicable to reactive glycosyl acceptors.[34] Another concept consisted of the transformation of β-glucosides to β-mannosides via an oxidation-reduction sequence at the 2-position.[35] Both procedures, however, could not prevent the accompanying formation of the undesired anomers or epimers, respectively. The new strategy developed includes the initial stereoselective formation of the corresponding β-glucosides which subsequently are epimerized at C-2 via S_N2-type inversion of configuration. Since the S_N2 reaction at C-2 of glycopyranose is difficult because of the repulsive interaction of the lone pair of the ring oxygen with the approaching nucleophile, the nucleophile was presented intramolecularly as a carbamoyl group in 3-position. In this way, the β-glucosyl glucosamine derivative **33** was stereoselectively converted to the desired β-mannosyl conjugate **34** in high yield.[36]

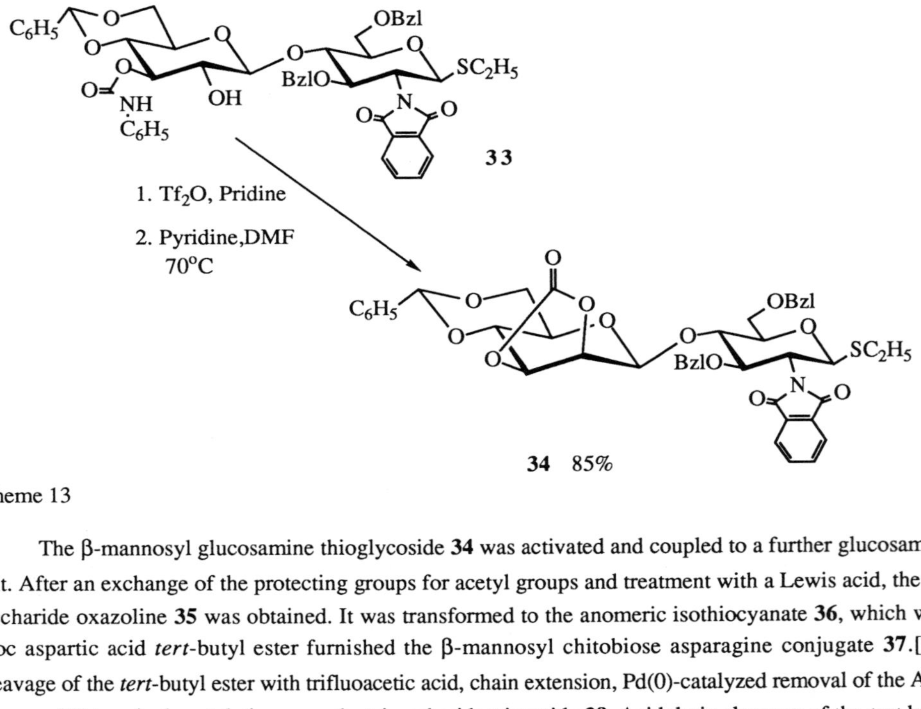

Scheme 13

The β-mannosyl glucosamine thioglycoside **34** was activated and coupled to a further glucosamine unit. After an exchange of the protecting groups for acetyl groups and treatment with a Lewis acid, the trisaccharide oxazoline **35** was obtained. It was transformed to the anomeric isothiocyanate **36**, which with Aloc aspartic acid *tert*-butyl ester furnished the β-mannosyl chitobiose asparagine conjugate **37**.[37] Cleavage of the *tert*-butyl ester with trifluoacetic acid, chain extension, Pd(0)-catalyzed removal of the Aloc group and N-terminal acetylation gave the trisaccharide tripeptide **38**. Acidolytic cleavage of the *tert*-butyl type protection again without affecting the saccharide bonds (see the preceding chapter) and removal of the O-acetyl protecting groups using hydrazine/methanol.[14] furnished the core region N-glycopeptide **39**, which was coupled to bovine serum albumin to give the synthetic glycoprotein useful for immunological evaluation (Scheme 14).

The construction of the glycoconjugate **40** in condensed form shows the solution of major problems in the synthesis of glycopeptides. It includes a new strategy in the stereoselective synthesis of oligosaccharides of demanding structure. The efficient regioselective and chemoselective deblocking of the complex carbohydrate amino acid and peptide conjugates was realized by using the Pd(0)-catalyzed cleavage of the Aloc group. The indirect protecting effect of the O-acyl groups within the carbohydrate on the intersaccharide bonds was demonstrated and utilized for the application of the acid-labile *tert*-butyl ester in glycopeptide chemistry.

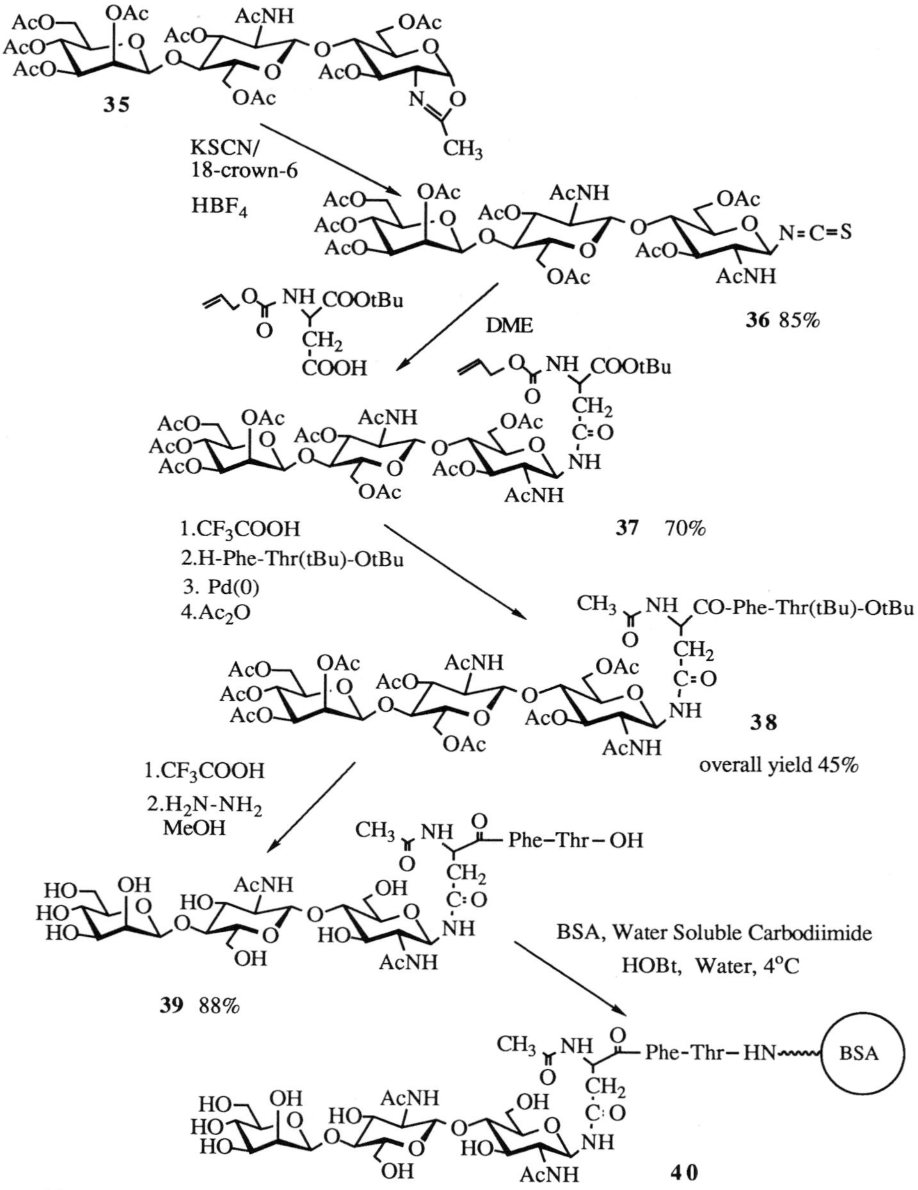

35

KSCN/
18-crown-6

HBF₄

36 85%

DME

37 70%

1. CF₃COOH
2. H-Phe-Thr(tBu)-OtBu
3. Pd(0)
4. Ac₂O

CH₃ ... NH ... CO-Phe-Thr(tBu)-OtBu

38

overall yield 45%

1. CF₃COOH
2. H₂N-NH₂
MeOH

CH₃ ... NH ... Phe–Thr–OH

39 88%

BSA, Water Soluble Carbodiimide
HOBt, Water, 4°C

CH₃ ... NH ... Phe-Thr–HN〜〜 BSA

40

Scheme 14

   The mild removal of the acetyl protecting groups from the saccharide portion was achieved by methanolysis catalyzed by the weak base hydrazine. The binding of the synthetic glycopeptide to a carrier protein, a precondition for immunological applications, was realized without insertion of artificial spacer units.
These methodical tools together with recent developments in the solid phase synthesis and the elaborated
use of the Fmoc group in glycopeptide chemistry now make available glycopeptides which are exactly spe-

cified partial structures of biologically interesting glycoproteins, e. g. tumor-associated antigens, virus coat structures or recognition components.

References

1. Review: J. Montreuil in "Comprehensive Biochemsitry", Vol. 19 B II (A. Neuberger, L. L. M. van Deenen, Eds.), Elsevier, Amsterdam, 1982, p. 1.
2. For a survey, see: R. J. Ivatt, Ed., *The Biology of Glycoproteins*, Plenum Press, New York, 1984.
3. G. F. Springer, *Science* **224**, 1198-1206 (1984).
4. T. Feizi, *Nature* **314**, 53-57 (1985).
5. M. Schlüter, D. Linder and R. Geyer, *Carbohydr. Res.* **138**, 305-312 (1985).
6. For review, see: H. Kunz, *Angew. Chem. Int. Ed. Engl.* **26**, 294-308 (1987).
7. H. Paulsen, *Angew. Chem. Int. Ed. Engl.* **21**, 155-224 (1982).
8. R. R. Schmidt, *Angew. Chem. Int. Ed. Engl.* **25**, 212-235 (1986).
9. J. R. Vercellotti, R. Fernandez, C. J. Chang, *Carbohydr. Res.* **5**, 97-106 (1967).
10. H. Kunz in D. Schinzer (ed.), Selectivity in Lewis acid-promoted-reactions, Kluwer Acad. Publ., Dordrecht, 1989, p. 189-202.
11. H: Kunz, B. Müller, W. Pfrengle, K. Rück, W. Stähle in *ACS Symp. Ser.* (R. Guiliano, ed.) **494**, 131-146 (1992).
12. H. Kunz, R. Kullmann, *Tetrahedron Lett.* **33**, 6115-6118 (1992).
13. L. A. Carpino, *Acc. Chem. Res.* **20**, 401-407 (1987).
14. P. Schultheiss-Reimann and H. Kunz, *Angew. Chem. Int. Ed. Engl.* **22**, 62-63 (1983).
15. H. Kunz and S. Birnbach, *Angew. Chem. Int. Ed. Engl.* **25**, 360-362 (1986).
16. J. M. Lacombe, A. A. Pavia, *J. Org. Chem.* **48**, 2557 (1983).
17. H. Kunz, B. Dombo and W. Kosch, in *Peptides 1988* (G. Jung and E. Bayer, Eds), W. de Gruyter, Berlin, 1989, p. 154-156.
18. B. Lüning, T. Norberg and T. Tejbrant, *Glycoconjugate J.* **6**, 5-19 (1989).
19. H. Paulsen, G. Merz, S. Peters and U. Weichert, *Liebigs Ann. Chem.* **1990**, 1165; S. Peters, T. Bielefeldt, M. Meldal, K. Bock and H. Paulsen, *J. Chem. Soc. Perkin Trans. I* **1992**, 1163-1171.
20. H. Waldmann, H. Kunz, *Liebigs Ann.Chem.* **1983**, 1712-1725.
21. H. Kunz and H. Waldmann, *Angew. Chem. Int. Ed. Engl.* **23**, 71-72 (1984).
22. M. Ciommer, H. Kunz, *Synlett* **1991**, 593-595.
23. H. Kunz and B. Dombo, *Ger. Pat. Appl.* P 3720269.3 (19th June 1987).
24. H. Kunz and B. Dombo, *Angew. Chem. Int. Ed. Engl.* **27**, 711-713 (1988).
25. H. Kunz, W. Kosch and J. März, in *Peptides - Chemistry and Biology* (J. A. Smith and J. E. Rivier, Eds), ESCOM, Leiden, 1992,. p. 502-504.
26. D. E. Brenneman, J.M. Buzy, M. R. Ruff and C. B. Pert, *Drug Develop. Res.* **15**, 371-379 (1988).
27. C. M. Stevens, R. Watanabe, *J. Am. Chem. Soc.* **72**, 725 (1950).
28. H. Kunz and C. Unverzagt, *Angew. Chem. Int. Ed. Engl.* **27**, 1697-1699 (1988); see, also *ibid.* **23**, 436-437 (1984).
29. C. Unverzagt, H. Kunz, *J. Prakt. Chemie*, **334**, 570-578 (1992).
30. H. Kunz, C. Unverzagt, *J. Prakt. Chemie*, **334**, 579-583 (1992).
31. R. U. Lemieux and H. Driguez, *J. Am. Chem. Soc.* **97**, 4063-4069 (1975).
32. R. Johannsson and B. Samuelsson, *J. Chem. Soc. Perkin Trans I* **1984**, 2371-2374.
33. H. Kunz, J. März, *Angew. Chem. Int. Ed. Engl.* **27**, (1988).
34. H. Paulsen, O. Lockhoff, *Chem. Ber.* **144**, 3102-3114 (1981).
35. C. Augé, C. D. Warren, R. W. Jeanloz, *Carbohydr. Res.* **82**, 71-95 (1980).
36. H. Kunz, W. Günther, *Angew. Chem. Int. Ed. Engl.* **27**, 1086-1087 (1988).
37. W. Günther, H. Kunz, *Angew. Chem. Int. Ed. Engl.* **29**, 1068-1069 (1990); *Carbohydr. Res.* **228**, 217-241 (1992).

Studies Directed towards the Total Synthesis of Vancomycin and Related Antibiotics

A.V. Rama Rao

Indian Institute of Chemical Technology

Hyderabad 500 007, India

Summary: An exceptionally simple approach for the synthesis of isodityrosine-derived cyclic peptides such as K-13 and diphenyl ether cross-linked amino acids present in vancomycin is described. The new methodology involves the displacement of one or two bromine atoms of 2,6-dibromobenzoquinone with tyrosine or substituted tyrosine units, providing the aryloxybenzoquinone. Subsequent manipulations of the benzoquinone skeleton to the corresponding aryl amino acid relied on the Pd-catalyzed cross-coupling reaction of aryl triflates with allyl- or vinyltributyltin and the Sharpless asymmetric dihydroxylation reaction, yielding either the isodityrosine unit of K-13 or diphenyl ether cross-linked aryl amino acids present in vancomycin.

Introduction

The current level of interest in the synthesis of cyclic peptides and glycopeptides containing oxidatively coupled aromatic nuclei is indicated by the increasing number of publications in this area [1]. By virtue of their impressive biological profiles coupled with extremely complex structural arrangements, naturally occurring peptides represented by K-13 (**1**) [2], OF-4949 (I-IV) (**2**) [3], bouvardins (**3**) [4], vancomycin (**4**) [5], ristocetin (**5**) [6], teicoplanin (**6**) [7], etc., have now formed an important area of research. Isodityrosine molecules (K-13, OF-4949) are enzymatically formed by the coupling of two (*S*)-tyrosines whereas coupled fragments of 4-hydroxyphenylglycine and substituted tyrosines provide the partial structure of the glycopeptide vancomycin. K-13, which was isolated from *Micromonospora halophytica*, posseses inhibitory activity against angiotensin I converting enzyme [2]. OF-4949 (I-IV) originates from *Penicillium rugulosum* and possesses inhibitory activity against aminopeptidase B [3]. Isolation of vancomycin from the fermentation broth of the *actinomycete Nocardia orientalis* marked the beginning of a new era of unique glycopeptide antibiotics possessing valuable biological activity, particularly in the treatment of staphylococcal infections. The clinically useful vancomycin has been partly responsible for our present understanding of the molecular basis of drug action [8].

Not surprisingly therefore, several synthetic chemists became interested in advancing new concepts required for the syntheses of these important antibiotics. On examination of the structures of these cyclic peptides and glycopeptides, it became obvious that they had several structural elements in common and the knowledge gained from the syntheses of simple molecules such as K-13 (**1**) and OF4949 (**2**) could in principle be applied to the design of synthetic protocols for complex glycopeptides such as vancomycin (**4**).

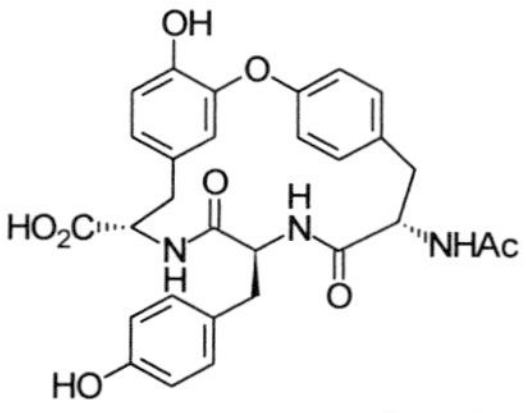

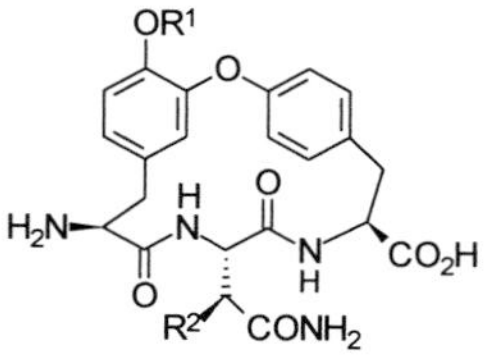

R^1	R^2	
CH$_3$	OH	OF 4949-I
H	OH	OF 4949-II
CH$_3$	H	OF 4949-III
H	H	OF 4949-IV

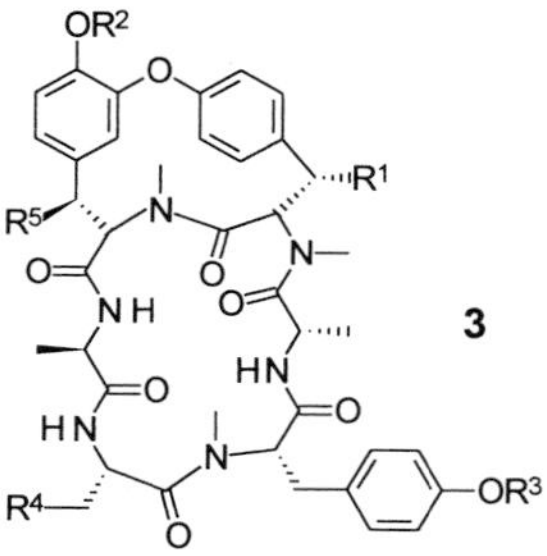

R^1	R^2	R^3	R^4	R^5	
OH	H	Me	H	H	Bouvaridin
H	H	Me	H	H	Deoxybouvaridin (RA-V)
H	H	Me	OH	H	RA-I
H	Me	H	H	H	RA-II
H	Me	Me	OH	H	RA-III
H	Me	Me	H	OH	RA-IV
OH	Me	Me	H	H	O-Methylbouvaridin (RA-VI)
H	Me	Me	H	H	O-Methyldeoxybouvaridin (RA-VII)

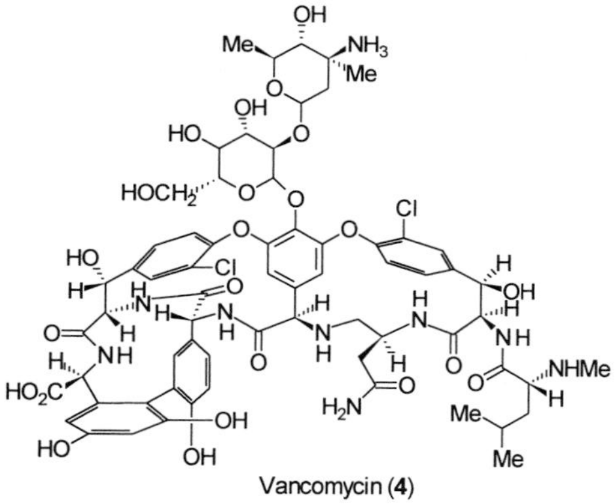

Vancomycin (**4**)

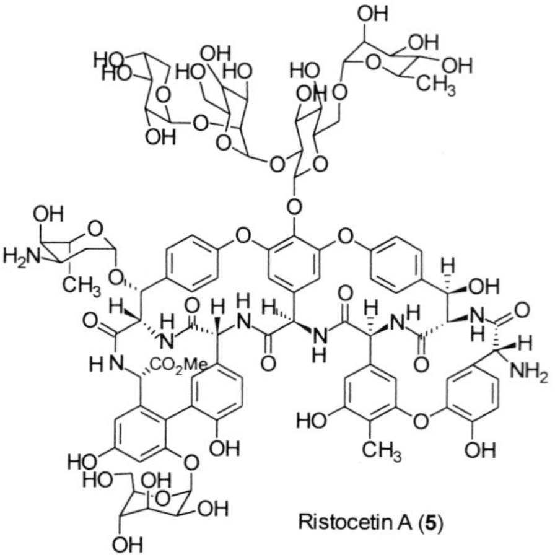

Ristocetin A (**5**)

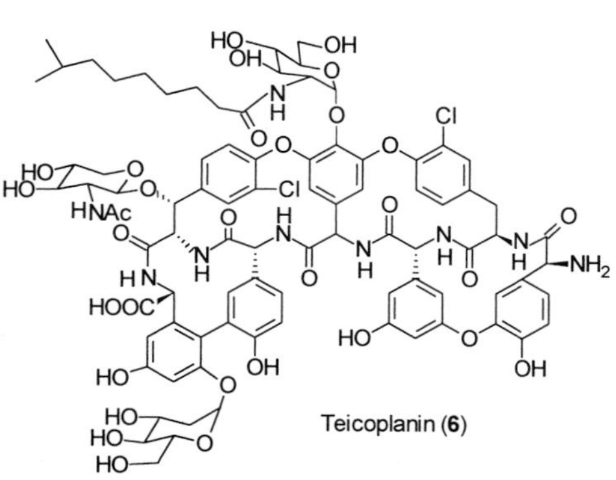

Teicoplanin (**6**)

Synthetic strategies

On the methodological front, the Ullmann reaction [9] and thallium (III) nitrate (TTN) oxidative coupling [10] approaches have formed the basis of assembling aromatic nuclei through ether linkages. Other strategies, although not as versatile, have also been developed to construct diphenyl ether linkages by using aryl iodonium salts [11] and arene-Mn(CO)$_3$ or arene-Ru complexes [12]. It is appropriate to briefly discuss the merits and demerits of these approaches in order to critically evaluate the difficulties likely to arise while dealing with more complex glycopeptide syntheses.

Needless to say, the Ullmann reaction has traditionally been a rigorous method to oxidatively couple aromatic nuclei. The high temperature and long reaction period have been the major bottlenecks in using this reaction. It would be foolish to believe that the delicate functionalities and stereochemical centers present in substrates undergoing the coupling reaction would remain totally immune to such harsh conditions. In fact, the Ullmann reaction between the two tyrosine precursors **7** and **8** provided the isodityrosine derivative **9** in the annoyingly low yield of 1.5% [2] (Scheme 1).

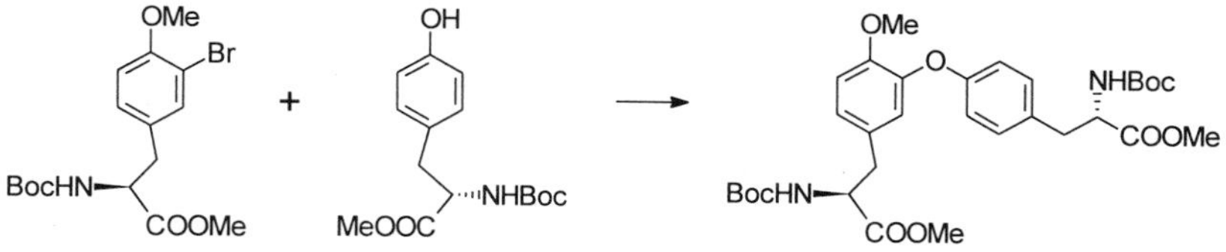

Scheme 1

In a modified version of the Ullmann reaction, based upon the approach of Schmidt et al. [13], the coupling component **10** containing a rigid but suitable functional group for derivatizing the amino acid side chain was employed with an improvement in the yield of the coupled product **11**. Elaboration of **11** by the use of asymmetric homogeneous reduction procedures completed the synthesis of OF-4949 (Scheme 2).

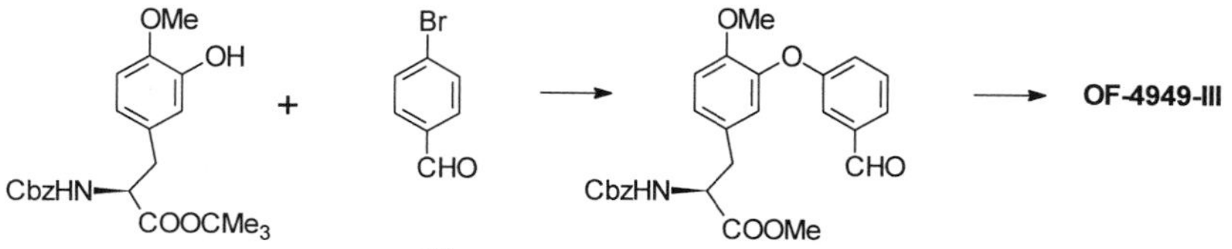

Scheme 2

With the same basic idea, Evans' group [14] prepared the oxidatively coupled cinnamic acid (**12**, 93%) which was then used to elaborate the amino acid side chains on both aromatic nuclei in a stepwise fashion involving the author's own approach to diastereoselective electrophilic azidation of chiral imide enolates. The formation of the common intermediate (**13**) permitted the total synthesis [14] of K-13 (**1**) and OF-4949 (III) (**2**) (Scheme 3).

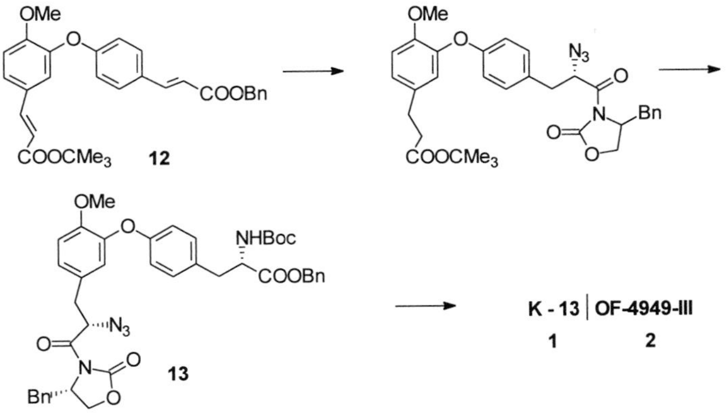

Scheme 3

The synthesis of K-13 employing the optimal Ullmann coupling partner of 3-bromo-4-nitro-benzaldehyde (**14**) was studied in this laboratory [15]. Due to the presence of an electron withdrawing nitro group, the electrophilicity of the adjacent carbon was enhanced, thereby permitting the coupling with a tyrosine derivative under moderate conditions. Subsequent elaboration of the aldehyde into the chiral side chain followed by coupling with another tyrosine derivative provided **15** which was macro-cyclized to generate **16**. Final conversion of N-Boc to hydroxyl and deprotection furnished K-13 (**1**) (Scheme 4).

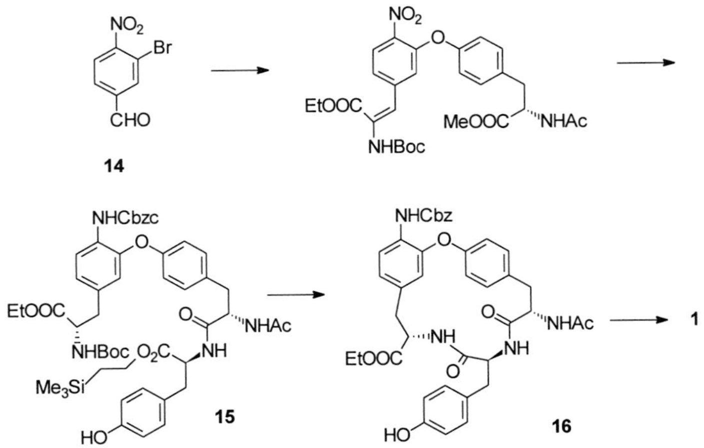

Scheme 4

The TTN oxidative phenolic coupling, pioneered by Yamamura et al., found its judicious utilization in the synthesis of K-13 and OF-4949 (III). This biomimetic phenolic oxidation is truly a versatile method for effecting both intra as well as intermolecular coupling reactions. In TTN oxidation, the o,o'-dihalophenol was employed in order to control the oxidation potential and the regioselectivity. For example, in the synthesis of OF-4949 (III), the derived tripeptide **17** was oxidatively coupled using TTN and reduced with zinc to give the advanced intermediate **18**, from which removal of halogens and deprotection led to OF-4949 (III) (**2**) [16] (Scheme 5).

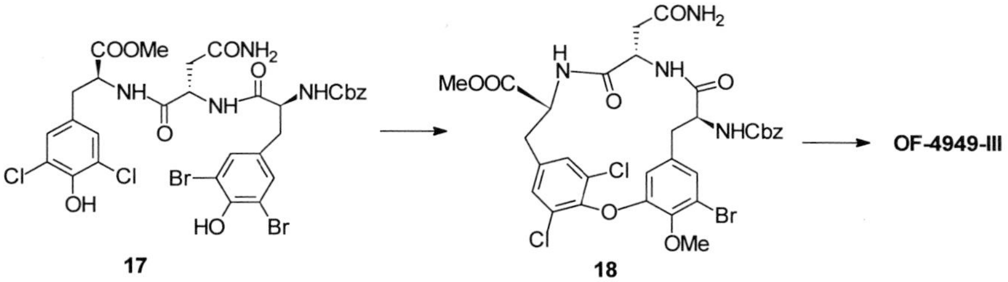

Scheme 5

In an analogous fashion, Yamamura et al. designed the synthetic protocol for K-13. Interestingly, the TTN oxidation of the tripeptide intermediate **19** underwent a different mode of cyclization giving an undesired product **20** for which a mechanistic consideration based on stereochemical strain in the transition state was forwarded. This consideration suggested the replacement of the bromine with iodine, resulting in the preparation of the tripeptide **21**. TTN oxidation and zinc reduction of **21** brought forth the desired cyclization, but the product **22** was isolated in only 15% yield. This latter intermediate **22** was successfully converted into K-13 in 24% yield [17] (Scheme 6).

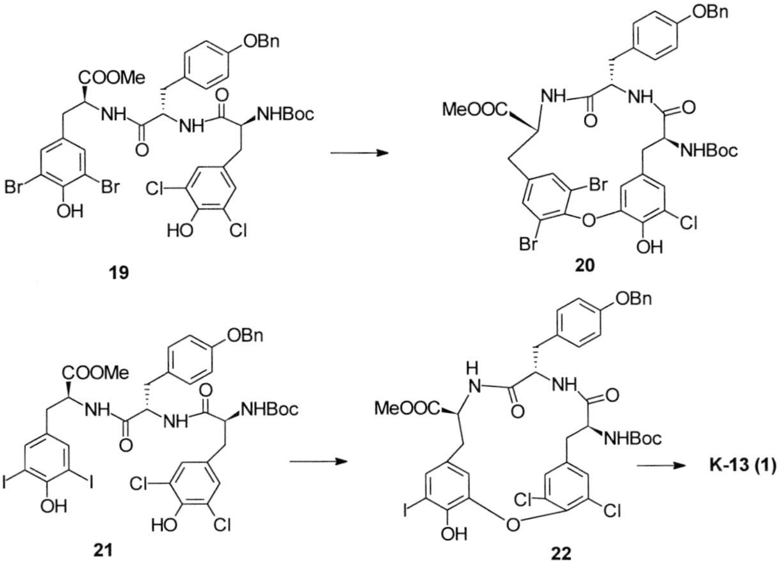

Scheme 6

The TTN oxidation was also judged to be a promising method for the synthesis of the active center of vancomycin [18]. For instance, the Evans' group adopted TTN oxidation for constructing cyclic peptides by stepwise C,D and D,E ring closures for the model of vancomycin subunits [19]. The substrate **23** was oxidatively coupled with TTN and reduced with CrCl$_2$ to afford the cyclic compound **24**. Further elaboration and ring closure formed the model C,D,E diphenyl ether segment **25** of vancomycin (Scheme 7).

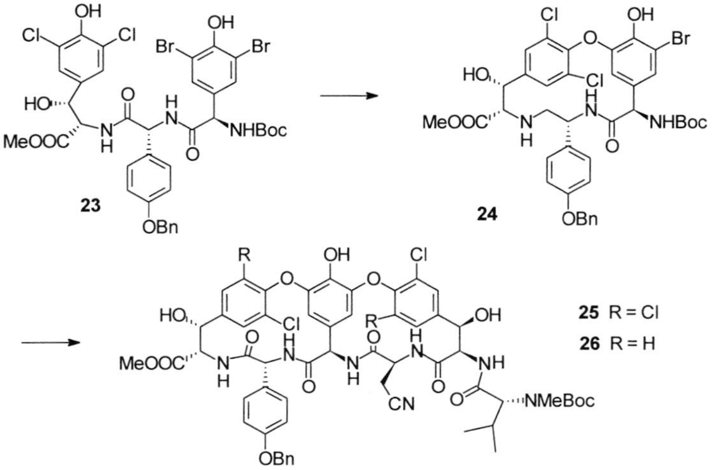

Scheme 7

In principle, **25** could be transformed into the actual vancomycin subunit **26** by effecting selective removal of a chlorine atom from each of the aryl amino acids. However, this endeavour may turn out to be a herculean task. The modalities of the TTN oxidation method where ortho positions of the phenolic units are substituted with halogens or alkyl groups may not be particularly suitable for the syntheses of glycopeptides such as vancomycin and ristocetin.

Expedient approaches towards isodityrosine and diphenyl ether cross-linked amino acids

Despite many advances, we believe that intricate problems of synthetic chemistry still remain to be addressed. For instance, no methodology has yet succeeded in directly coupling two protected aryl amino acids required for the active center of glycopeptides. Another important aspect needing attention is the development of an exceptionally potent methodology to construct oxidatively coupled aromatic nuclei under seemingly mild conditions, ensuring beyond doubt the stereochemical features of reacting molecules and offering high yields of the product. With these aspects in mind, we have developed conceptually varied processes as tactical devices to form diphenyl ether linkages suitable for the syntheses of isodityrosine derived cyclic peptides, such as K-13 and diphenyl ether cross-linked amino acids present in glycopeptides.

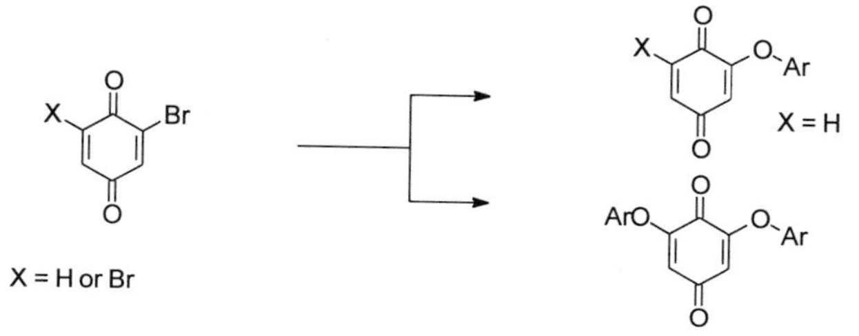

Basically, the approach features the displacement of bromine atoms of bromobenzoquinones with phenolic derivatives providing mono- or di-aryloxybenzoquinones in good yields. Subsequent manipulations of the benzoquinone skeleton to the corresponding aryl amino acid relied on the Pd-catalyzed cross-coupling reaction [20] of aryl triflates with alkenyltributyltin and the Sharpless asymmetric dihydroxylation reaction [21].

In our preliminary investigations, 2-bromobenzoquinone (27) was coupled with p-cresol in the presence of KF (3.0 equi.) to afford the 2-aryloxybenzoquinone derivative 28a in 85% yield. We observed that K$_2$CO$_3$ could also be used instead of KF in the coupling reaction with marginal yield improvement. We further demonstrated that benzyl N-Boc-L-tyrosinate containing a chiral amino acid side chain and labile functional groups could also be used as a suitable coupling partner in the above reaction, giving 28d-e in 80% yield. The efficacy of this reaction was substantiated by a number of displacement reactions carried out with a variety of phenoxides (Scheme 8). It should be pointed out that the conditions for this displacement were mild and the chiral amino acid groups were tolerated in the phenoxide without any degree of racemization.

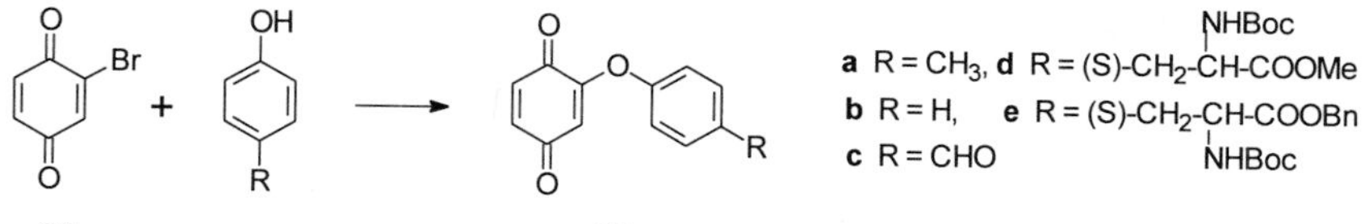

Scheme 8

Inspired by the success of this reaction, we directed our efforts towards the synthesis of K-13. Compound 28e was chosen as the starting point for this exploration. Subsequent reduction of 28e with dithionite provided the hydroquinone 29 in almost quantitative yield. Its direct trifluorosulfonylation to 30 indicated the higher reactivity of the 1-hydroxyl group, probably due to hydrogen bonding (as steric factors suggest that the 4-hydroxyl group should be the preferred site of reaction). Therefore to obtain the 4-O-triflate derivative, we first reacted hydroquinone 29 with TBS-Cl and then methylated with DMS/K$_2$CO$_3$ in acetone to generate 32 in 70% overall yield. Successive silylation and trifluoro-methanesulfonylation of 32 then gave the requisite 4-O-triflate derivative 33 (Scheme 9).

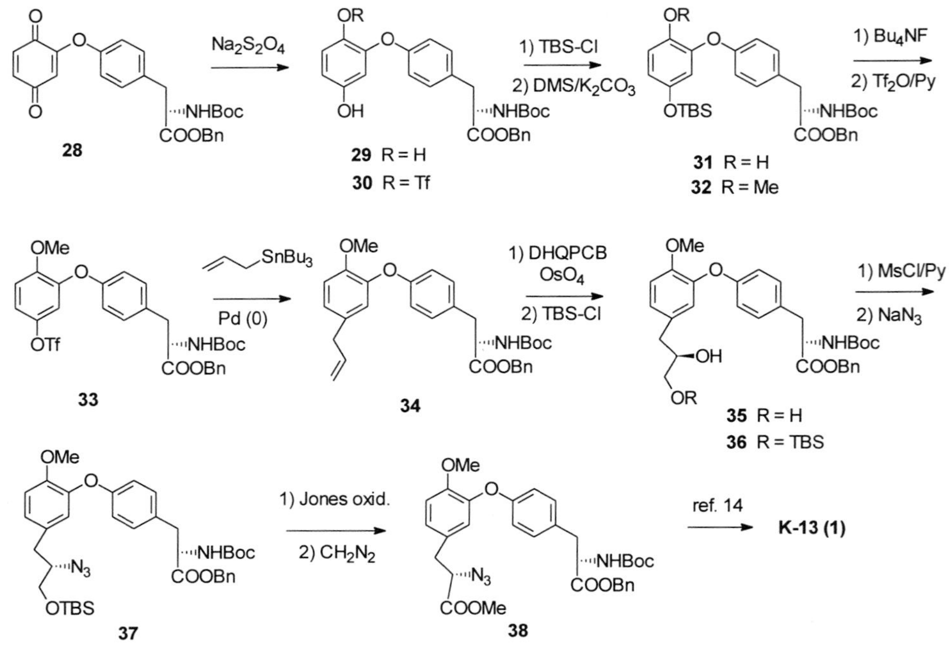

Scheme 9

Our next concern was to refunctionalize the benzoquinone skeleton into the corresponding aryl alanine derivative in optically pure form. Reaction of **33** with allyl tributyltin in the presence of LiCl and Pd(PPh₃)₄ in refluxing dioxane gave the allyl derivative **34**. This C-C bond cross-coupling reaction was found to be highly selective, providing **34** in 80% yield; that the functional groups present in the tyrosine segment remained unaffected was particularly gratifying. Subsequent Sharpless asymmetric dihydroxylation reaction [21] of **34** with dihydroquinidine-p-chlorobenzoate as a chiral ligand gave the diol **35** which was first silylated at the primary hydroxyl group. Then, by the Mosher ester method, the diastereomeric excess of 62% was determined using ^{19}F-NMR spectroscopy [22].

Finally, **36** was transformed into the azido derivative **37** via the mesylate. Jones oxidation and subsequent esterification provided the key intermediate **38** (60% overall yield) whose spectroscopic analysis was comparable with the reported data. This key intermediate was previously converted into K-13 by the Evans' group [14].

We next focused our attention on the displacement reactions of 2,6-dibromobenzoquinone (**39**) [23] with phenoxide in order to obtain 2,6-diaryloxybenzoquinones. Developing the technique of manipulating the benzoquinone skeleton to the corresponding aryl glycine derivative, based on concepts already discussed, may not be a difficult proposition.

Substitution of bromine atoms in **39** with 2 equivalents of various tyrosine derivatives in the presence of 6 equi. KF or K₂CO₃ furnished the requisite diaryloxybenzoquinones **40** in good yield. The stereochemical integrity of the product was not affected during the reaction and was carefully analyzed by spectroscopic analysis. Careful examination of the reaction by TLC suggested that the substitution reaction probably occurred in a stepwise fashion. This provided us the opportunity to attempt selective substitution of one bromine atom of **39** with 1 equivalents of tyrosine derivative, leading to the formation of monobromomonoaryloxybenzoquinone **41** in 81 % yield. Not surprisingly, therefore compound **41** with one equivalent of p-cresol furnished the diaryloxy derivative (**42**, 76 %). This stepwise substitution reaction was particularly relevant for the synthesis of the vancomycin antibiotic because the C and E rings present in the active center of the molecule contained stereogenically different hydroxytyrosine derivatives coupled through ether linkages. A number of phenol derivatives were introduced by the stepwise approach, generating diaryloxybenzoquinones containing different aryl ether substitutions on benzoquinone.

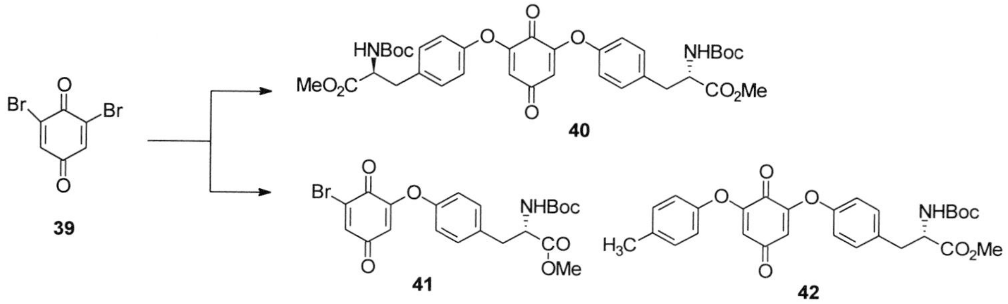

The ability of our approach to introduce aryl ethers in a stepwise fashion was exploited in the synthesis of the model C,D,E diphenyl ether fragment of vancomycin. Compounds **40** and **42** represent the first successful examples of direct coupling of two protected aryl amino acids.

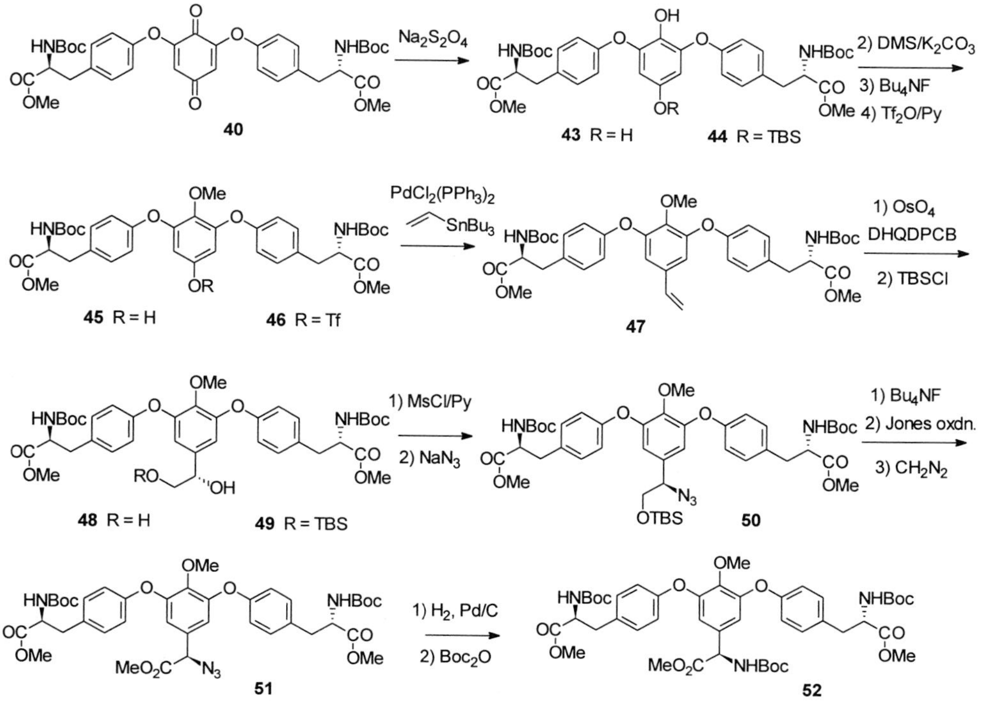

Scheme 10

The less hindered 4-OH group in **43**, obtained from **40**, was selectively silylated to afford **44**. Successive methylation and removal of the TBS ether group generated **45** with a free 4-OH group (overall 52% yield). **45** was converted into the triflate derivative **46**. Initial attempts to introduce a vinyl group with vinyltributyltin in the presence of Pd(PPh $_3$)$_4$ were not satisfactory; however, with PdCl$_2$(PPh$_3$)$_2$ as catalyst and DMF as a solvent, the reaction proceeded smoothly to give the styrene derivative **47** in 86% yield. The catalytic asymmetric dihydroxylation of **47** with dihydroquinine-p-chlorobenzoate [21] as chiral ligand gave the diol (**48**, 91%) which was converted into the TBS ether **49**. The diastereomeric excess (50%) was determined by the ^{19}F NMR spectra of the Mosher ester. With dihydroquinine-9-O-(9'-phenanthryl) ether as a chiral ligand [24], the dihydroxylation of **47** provided the diol **48** with improved d.e. of 80% (Scheme 10).

Treatment of **49** with mesyl chloride followed by nucleophilic displacement reaction with LiN$_3$ gave the azido derivative **50** (64%). Removal of the TBS group followed by Jones oxidation and esterification furnished the azido ester **51** (51% overall yield). Finally, catalytic reduction of **51** reduced the azido group followed by protection gave the N-Boc derivative **52** in 61% overall yield.

Studies directed towards the total synthesis of vancomycin

Our efforts in this field have been rewarded with the development of a general methodology for elaboration of oxidatively coupled aromatic amino acids. The foregoing retrosynthetic analysis (Scheme 11) of vancomycin (**4**) now requires the following considerations. Of the five aromatic amino acids present in **4**, three are locked up with aromatic ether linkages (C,D,E segment): the successful synthesis of K-13 and model vancomycin C,D,E segment appeared to confirm the likely utility of this strategy for the actual C,D,E, ring system **53** in which the β-hydroxytyrosine intermediates are present. The remaining two aryl amino acids are present as a characteristic C-terminal biphenyl moiety **54**, common in all glycopeptides of this class. Consequently, the development of a successful strategy to synthesize it became our first goal in the effort towards achieving the total synthesis of vancomycin.

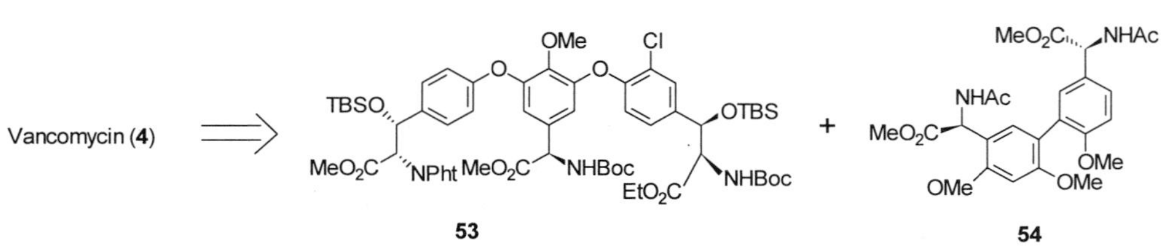

Scheme 11

A close look at the biphenyl segment **54** indicated that the biphenyl linkage is formed between the 3-position of (R)-4-hydroxyphenylglycine and the 2-position of (S)-3,5-dihydroxyphenylglycine. The former is available and forms the starting point of our synthetic endeavour. The other segment was synthesized starting from the 3,5-dihydroxybenzoic acid (**55**) which was subjected to a sequence of reactions as indicated in Scheme 12 to obtain 3,5-dimethoxy-2-bromobenzoic acid (**56**) [25]. Esterification of **56** with *N*-Boc-*O,N*-isopropylidene-4-hydroxy-(*R*)-phenylglycinol (**57**) [26] gave **58** which was then subjected to a palladium-catalyzed intramolecular coupling reaction [27] to obtain the biphenyl derivative **59**. Our next concern was to generate the amino acid moiety from **60**, from which opening of the isopropylidene group and oxidation of the primary hydroxyl group were carried out, thus yielding **61**. In order to generate the second amino acid functionality our own strategy [28] of a diastereoselective Strecker synthesis using phenylglycinol as chiral auxiliary was successfully employed. Thus, the reaction of the derived aldehyde **62** with (*R*)-phenylglycinol followed by treatment with trimethylsilyl cyanide gave the (*S*)-aminonitrile **63** which on hydrolysis gave **64**. Oxidative cleavage with Pb(OAc)$_4$ furnished the required biphenyl moiety **54** of vancomycin (**4**) [29].

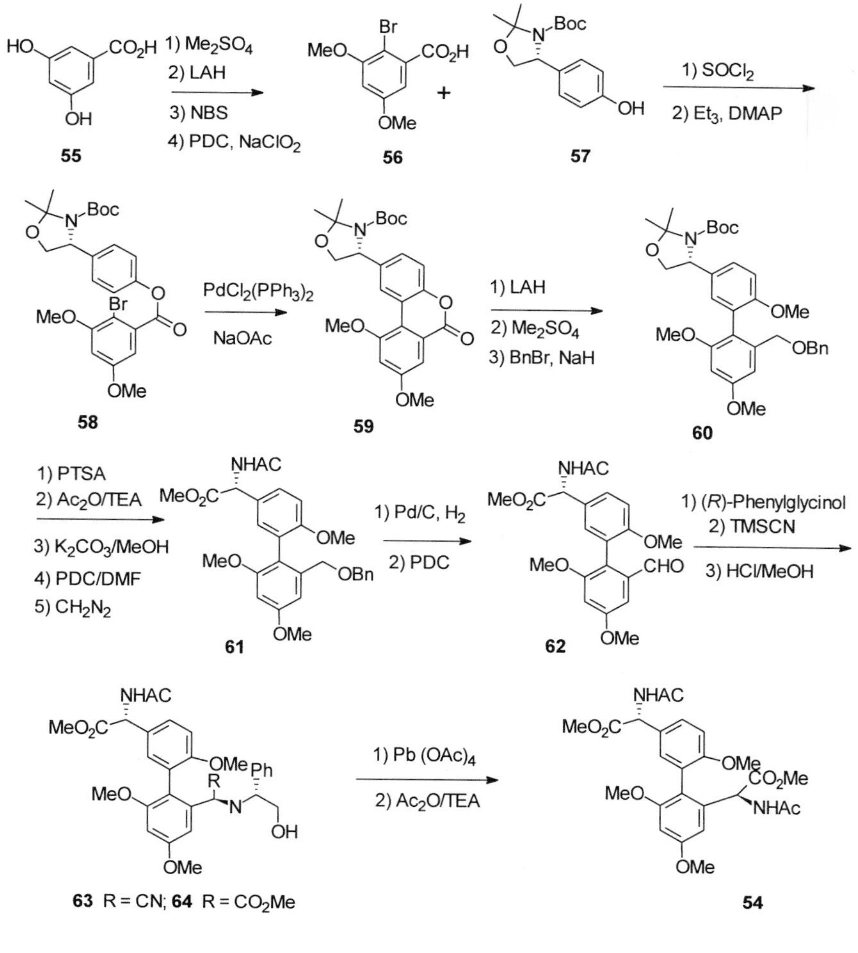

Scheme 12

Synthesis of the C,D,E diphenyl ether fragment **53** of vancomycin was then undertaken by extrapolating the synthetic protocol developed in our laboratory. The preparation of the unusual β-hydroxytyrosine derivatives **65** and **66**, which are coupled through ether linkages with the centrally located 4-hydroxyphenylglycine, have yet to be investigated in this context.

The route selected for the intermediate **65** involved a straightforward transformation of *N*-phthalido-tyrosine derivative **67** into the corresponding methyl *N*-phthalido-3-chloro-4-acetyl-(*S*)-tyrosinate **(68)**. Benzylic bromination of **68** with NBS produced an almost 1:1 diastereomeric mixture of bromides **69**. The separation of diastereomers was not required because subsequent hydrolysis with aqueous silver nitrate provided the β-hydroxytyrosinate derivatives **70** and **71** in a more respectable ratio of 9:1. The high degree of diastereoselectivity during the hydrolysis could be attributed to the preferential attack of the nucleophile (OH) on the carbocation from the sterically favoured α-face [30]. The pure isomer **70** was transformed into **65** as shown in Scheme 13.

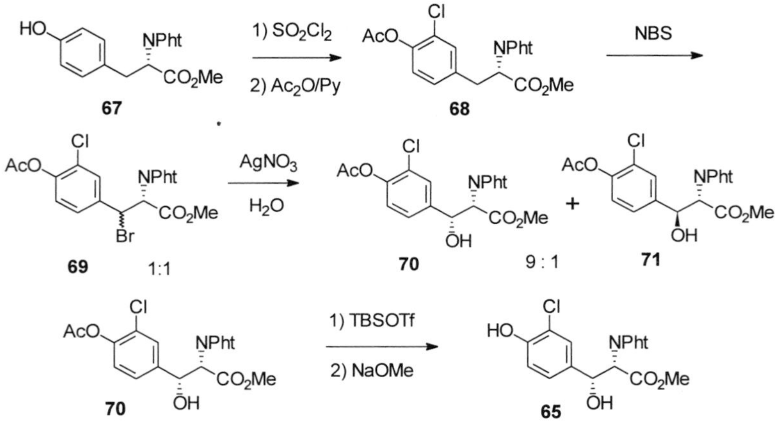

Scheme 13

For the planned synthesis of the second hydroxytyrosinate **66**, the readily available 3-chloro-4-benzyloxybenzaldehyde **(72)** was chosen. Wittig olefination and Sharpless asymmetric dihydroxylation with dihydroquinine-p-chlorobenzoate as a chiral ligand gave the diol **73** with 98% d.e. (chiral HPLC). The advantage of greater reactivity of the hydroxyl group [31] at C-2 was then exploited to prepare the corresponding 2-azido derivative **74** via the tosylate intermediate. Sequential reactions as shown in Scheme 14 furnished **66**.

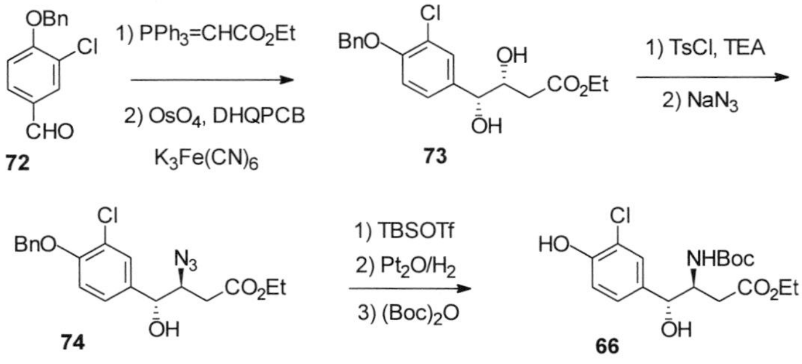

Scheme 14

Total Synthesis of Vancomycin 275

Having obtained both key intermediates **65** and **66**, all that remained was the stepwise introduction of these phenoxides onto the 2,6-dibromobenzoquinone skeleton under conditions detailed earlier to obtain **75**. Further elaboration to get the styrene derivative **77** as described in Scheme 15 brought us to a point where a crucial derivatization of the amino acid was required. In this endeavour, Sharpless asymmetric dihydroxylation with the newly developed dihydroquininephenanthryl ether ligand gave the diol which was protected as its TBS ether; the diastereomeric excess of 85% was then determined by the Mosher ester method. Conversion of the TBS-ether into the corresponding azido derivative via the mesylate was followed by reduction, *N*-Boc protection and selective de-silylation to yield **79**. Oxidation of the primary hydroxyl group in **79** with PDC-DMF and esterification gave **53**, which constituted the C,D,E fragment of vancomycin.

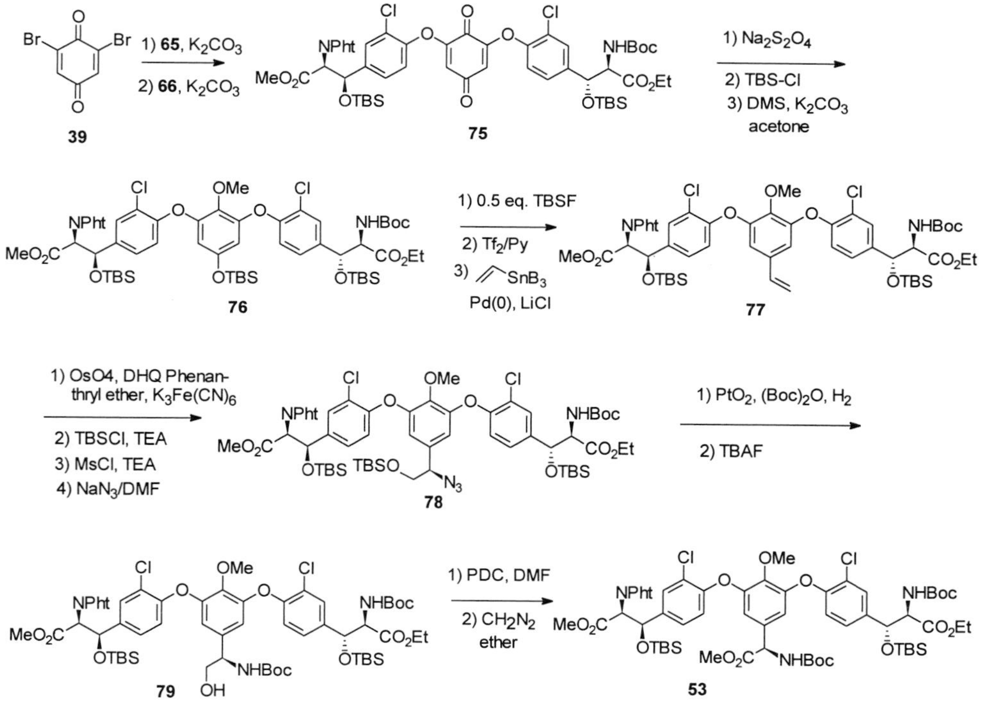

Scheme 15

Acknowledgements: I am indebted to all my colleagues, Dr. M. K. Gurjar, V. B. Khare, V. Kaiwar, A. Bhaskar Reddy, K. Laxma Reddy, S. P. Joshi, Dr. T. K. Chakraborty and A. Srinivasa Rao, for their dedication and devotion in carrying out the work on vancomycin and related antibiotics. I am especially thankful to my senior colleague, Dr. Gurjar, for his help in preparing this manuscript.

276   A. V. Rama Rao

References

1. a) R. M. Williams, Synthesis of optically active α-amino acids, Pergamon Press, London, 1989; b) F. Parenti and B. Cavalleri, *Drugs of the Future*, **15** (1990) 57 and the references cited therein; c) S. Yamamura and S. Nishiyama, *Studies in Natural Products Chemistry*, Atta-Ur-Rahman (Ed), Elsevier Science Publishers, Amsterdam, **l0** (1990) 629.

2. a) H. Kase, M. Kaneko and K. Yamada, *J. Antibiot.* **40** (1987) 450; b) T. Yasuzawa, K. Shirahat and H. Sano, *J. Antibiot.* **40** (1989) 455.

3. S. Tamai, M. Kaneda and S. Nakamura; *J. Antibiot.* **35** (1982) 1130; b) M. Kaneda, S. Tamai, S. Nakamura, T. Hirata, Y. Kushi and T. Suga,.*J. Antibiot.* **35** (1982) 1137; c) S. Nishiyama, K. Nakamura, S. Suzuki and S. Yamamura, *Tetrahedron Lett.* **27** (1986) 4481; d) S. Sano, M. Veno, K. Katayama, T. Nakamura and A. Obayashi, *J. Antibiot.* **39** (1986) 1697; e) S. Sano, K. Ikai, Y. Yoshikawa, T. Nakamura and A. Obayashi, *J. Antibiot.* **40** (1987) 519.

4. S. D. Jolad, J. J. Hoffman, S. J. Torrance, R. M. Weidhopf, J. R. Cole, S. K. Arora R. B. Bates, R. L. Gargiulo and G. R. Kriek, *J. Am. Chem. Soc.* **99** (1977) 8040.

5. M. H. McCormick, W. M. Stark, G. E. Pittenger, R. C. Pittenger and J. M. McGuire, *Antibiot. Ann.* **1955/56**, 606; b) G. M. Sheldrick, P. G. Jones, O. Kennard, D. H. Williams and G. A. Smith, *Nature*, London, **271** (1978) 223; c) M. P. Williamson and D. H. Williams, *J. Am. Chem. Soc.* **103** (1981) 6580; d) C. M. Harris, H. Kopecka and T. M. Harris, *ibid* **105** (1983) 6915.

6. W. E. Grundt, A. C. Sinclair, R. J. Teraiult, A. W. Goldstein, C. J. Ricker, H. B. Warren, T. J. Oliver and J. C. Sylvetster, *Antibiot. Ann.* **1956-57**, 687.

7. A. Borghi, G. Coronelli, L. Fanivolo, G. Allievi, R. Palloanza and G. G. Gallo, *J. Antibiot.* **37** (1984) 615.

8. D. H. Williams; *Acc. Chem. Res.*, **17** (1984) 364.

9. M. Tomito, K. Fujitani and Y. Aoyogi, *Chem. Pharm. Bull.* **13** (1965) 1341.

10. H. Noda, M. Niwa and S. Yamamura, *Tetrahedron Lett.* **22** (1981) 3247; b) S. Nishiyama, K. Nakamura, Y. Suzuki and S. Yamamura, *Tetrahedron Lett.* **27** (1986) 4481.

11. M. J. Crimmin and A. G. Brown, *Tetrahedron Lett.* **31** (1990) 2017.

12. a) A. J. Pearson and J. G. Park, *J. Org. Chem.* **57** (1992) 1744; b) A. J. Pearson, J. G. Park and P. Y. Zhu, *J. Org. Chem.*, **57** (1992) 3583.

13. U. Schmidt, O. Weller, A. Holder and A. Lieberknecht, *Tetrahedron Lett.* **29** (1988) 3227.

14. D. A. Evans, A. Jonathan and A. Ellman, *J. Am. Chem. Soc.* **111** (1989) 1063.

15. A. V. Rama Rao, T. K. Chakraborty, K. Laxma Reddy and A Srinivasa Rao, *Tetrahedron Lett.*, **33** (1992) 4799.

16. S. Nishiyama, Y. Suzuki and S. Yamamura, *Tetrahedron Lett.* **29** (1988) 559.

17. S. Nishiyama, Y. Suzuki and S. Yamamura, *Tetrahedron Lett.* **30** (1989) 379.

18. S. Yamamura, S. Nishiyama and Y. Suzuki, *Tetrahedron Lett.* **30** (1989) 6043.

19. D. A. Evans, A. Jonathan and K. M. Devries, *J. Am. Chcem. Soc.* **111** (1989) 8912.

20. J. K. Stille and M. A. Echavarren, *J. Am. Chem. Soc.* **109** (1987) 5478.

21. H. L. Kwong, C. Sorato, Y. Ogino, H. Chen and K. B. Sharpless, *Tetrahedron Lett.* **31** (1990) 2999.

22. J. A. Dale, D. L. Dull and H. S. Mosher, *J. Org. Chem.* **34** (1969) 2543.

23. H. E. Ungnade and K. T. Zilch, *J. Org. Chem.* **16** (1951) 64.

24. W. Amberg, M. Beller, H. Chen, J-. Hartung, Y. Kawanami, D. Ludben, E. Manoury, Y. Agino, T. Shibata, T. Ukita and K. B. Sharpless, *J. Org. Chem.* **56** (1991) 4595.

25. a) C. A. Townsend, S. G. Davis, S. B. Chrifitenson, J. C. Link and C. P. Lewis, *J. Am. Chem. Soc.* **103** (1981) 6885; b) E. Dalcanale and F. Montanari, *J. Org. Chem.* **51** (1986) 567.

26. It was necessary to start with the reduced form in order to carry out selective manipulation of the ester function on ring A.

27. a) G. Bringmann, R. Walter and R. Weirich, *Angew. Chem. Int. Ed. Eng.* **29** (1990) 977; b) G. Bringmann, J. R. Jansen and H. F. Rink, *Angew. Chem., Int. Ed. Eng.* **25** (1986) 913; c) D.E. Ames and A. Opalko, *Tetrahedron* **40** (1984) 1919.

28. T. K. Chakraborty, G. V. Reddy and K. A. Hussain, *Tetrahedron Lett.* **52** (1991) 7597.

29. A. V. Rama Rao, T. K. Chakraborty and S. P. Joshi, *Tetrahedron Lett.* **33** (1992) 4045.

30. C. J. Easton, C. A. Hutton, E. W. Tan and E. R. T. Tiekink, *Tetrahedron Lett.* **31** (1990) 705.

31. P. R. Fleming and K. B. Sharpless, *J. Org. Chem.* **56** (1991) 2869.

Synthetic Studies toward Enediyne Antitumor Antibiotics

Minoru Isobe and Toshio Nishikawa

Laboratory of Organic Chemistry, School of Agriculture, Nagoya University
Chikusa, Nagoya 464-01, Japan

Summary: New method to introduce acetylenic bond into acylimminium or oxonium centers as a model study directed toward an antitumor antibiotic, dynemicin A. A short routes to bicyclo-[7.3.1]-tridecenediyne system, common skeleton for this class molecule, are described.

Introduction

The synthesis of the enediyne class of molecules possessing DNA cleaving activity has posed an irresistible challenge to synthetic methodologies in organic chemistry. In recent years, the discovery of esperamicin, calicheamicin, and dynemicin containing bicyclo[7,3,1]tridecendiyne carbocyclic skeleton has dictated that new synthetic methodologies be explored in order to construct their complicated structures. Indeed, the aim of contemporary synthetic science is to establish the theory and methodology necessary to access to such molecules. This article illustrates our approach to problems of contemporary synthesis by providing a survey of recent work from this laboratory. The bond forming reactions involving acetylenes are discussed for the construction of the enediyne class compounds. The enediyne antitumor antibiotic, dynemicin A (**1**), was isolated from fermentation broth of *Micromonospora sp.* [1] The enediyne moieties was shown to be responsible for DNA cleaving activity. Since the structural elucidation in 1989, this compound as well as other enediyne natural products, esperamicin and calicheamicin, have attracted attention for chemical synthesis.

Dynemicin A (1)

Coupling of Acetylene to Imine

Our efforts on the synthesis of **1** initially focused on the formation of a bond between an acetylene and a carbon bearing an aniline nitrogen atom. Retrosynthesis of **1** prompted us to consider the simplified model **2** containing the identical functional groups. The further simplified compound **3** was designed to explore acetylenic bond formation between **4** and **5**. Related acetylenic chemistry was reported in 1987 by Ichikawa, Isobe *et al.* in the C-glycosidation of an oxonium cation intermediate.[2]   Mukaiyama simultaneously reported acetal carbon bond formation.[3] Lundkvist *et al.* reported that AlCl₃ catalyzed the formation of a cationic intermediate from an α-alkoxy cyclic amide to convert into acetylene adduct.[4]

2   **3**   **4**   **5**

These reports suggested that a silyl (or tin) acetylene reagent **4** (M = Si, Sn)   would show good nucleophilicity toward the cationic center. The amide ether **5** was considered to be a good system which stabilized the possible intermediate cation (such as **9** *vide infra*). It was easy to prepare from *p*-methoxyaniline **6** and cinnamaldehyde in two steps; first, imine formation (**8**) and then, acylative ethanol addition in 66 % overall yield.[5]

6   **7**   **8**   **5**

The acyliminium intermediate **9** was generated from the amide ether **5** by treatment with AlCl₃. Several acetylenes were examined with various Lewis acids at low temperatures. The conjugated cation **9** was so designed that the intermediate would be stabilized to give a higher yield of the product, **10**. Tin acetylene with TiCl₄ in fact afforded the desired 1,2-product.[6]

5   **9**   **10**

A short route to Bicyclo [7.3.1]-tridecenediyne System

Our synthetic plan is shown in the following scheme. Retrosynthetic disconnection of the propargylic and acetylenic bonds in simplified **1** afforded the aminoalcohol **11** and then the aminoaldehyde **12**. This precursor was analyzed as the enonealcohol **13**, from which the aromatic and aliphatic moieties were separated into **14**.[7] The key step in this scheme would be the ring closure between **11** and **12**.

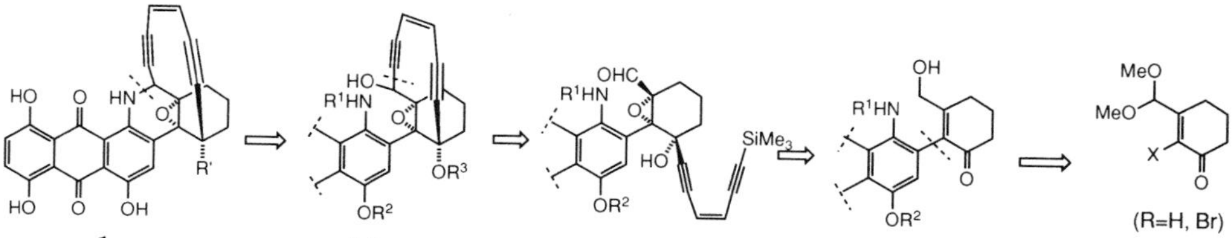

We studied a model system having no anthraquinone part in order to focus on the enediyne ring formation. Compound **14** was designed without the aromatic part according to the above retrosynthetic route.[8]

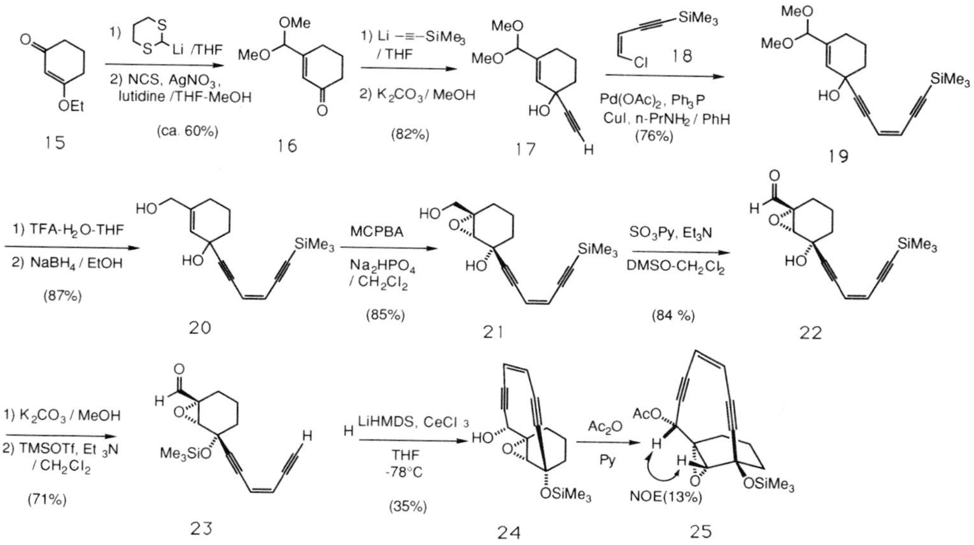

This synthesis started from the alkoxyenone **15**, which was homologated with 1,3-dithiane to **16**.[9] Lithium trimethylsilyl acetylide was added to **16** and the product was desilylated to afford the propargyl alcohol **17**. Coupling of **17** with the (Z)-vinyl chloride **18** [10] under Sonogashira's palladium condition [11] afforded the acyclic enediyne **19**. Acid hydrolysis of the acetal was followed by sodium borohydride reduction of the aldehyde to give the diol **20**. Epoxidation of the allylic alcohol **20** with MCPBA at 5°C afforded the *syn* epoxide **21**. The epoxyalcohol **21** was oxidized with SO₃-Py activated DMSO [12] to yield the epoxyaldehyde **22** in high yield (84%). Sequential desilylation from the silylacetylene and silylation of the *tert*-alcohol in the aldehyde **22** afforded the cyclization precursor **23**.

For the cyclization, the lithium acetylide of **23** in the presence of CeCl₃ [13] was the most effective to afford **24**. The reaction without CeCl₃ gave a mixture of **24** (in very low yield) and bimolecular products. Stereochemistry of the alcohol **24** was assigned from the positive NOE between the two protons as shown in **25**, the corresponding acetate. In this stage, we obtained the mimic compound of dynemicin A in our hands.

Common Skeleton Synthesis

Two cyclic enediyne antibiotics and our synthetic compound are compared. The product **25** posseses the bicyclo[7.3.1]-tridecendiyne system, which was regarded as a new esperamicin/ calicheamicin enediyne analogue that contained a trigger of dynemicin A. The epoxide **25** was converted to the allylic alcohol **26** (R=H) by heating in the presence of acid, and the product was isolated as its acetate (**26**, R=Ac) in 40% overall yield. Thus, **26** is an important analogue of bicyclo[7.3.1]-tridecadiendiyne system as the esperamicin aglycone.

25   26   Dynemicin A   Esperamicin aglycon

Tricyclic Model with Bicyclo[7.3.1]-tridecendiyne System

The above studies provided a promising method toward the bicyclo [7.3.1]-tridecendiyne system of dynemicin A and related compounds. For the selective opening of the epoxide ring, the aromatic ring was necessary to ensure the stabilization of the benzylic cation.

Dynemicin A (**1**)   **27a** R= H, **27b** R= Me   **24**

From our preliminary studies on **27a**, we had found substantial difficulties obtaining the precursor aldehydes (corresponding to **32** or **33** R= H, respectively) in sufficient amounts for the next cyclization because of the unstability of the epoxy aldehyde under the oxidation conditions. Therefore, we switched the target molecule to the homologous compound **27b** through the ketones **32** and **33** (R= Me). [14] The starting quinoline aldehyde **28**, prepared in about 50% yield from commercially available lepidine after SeO₂ oxidation, was reacted with MeMgBr, and the resulting alcohol was silylated to provide **29**. Addition of the magnesium acetylide to **29** with concomitant carbamoylation under Yamaguchi conditions [15] afforded the adduct **30** [m/z 457 (M⁺)] in high yield as a diastereomixture (ca. 1:1), which was not separated. Selective removal of the *O*-silyl group from **30** under acidic conditions (TFA/ MeOH) yielded the allylic alcohol **31** [m/z 343 (M⁺)]. Epoxidation of **31** with MCPBA was followed by SO₃-Py/DMSO oxidation [16] and subsequent desilylation gave the single product **32** as crystals [mp 116 °C]. High stereoselectivity in the epoxidation might be due to the steric hindrance of the acetylenic substituent. Stereochemistry of **32** was temporarily assigned from the small coupling constant between Ha and Hb. Its terminal acetylene was coupled with (Z)-vinylchloride under Sonogashira's condition [10] to give the enediyne **33**. The attempted cyclization of **33** with LiF, KF or CsF [17] did not give the cyclized product **34**. *In situ* generation of the acetylide anion from the silylacetylene **33** with CsF and 18-crown-6 fortunately gave **34** (**27b** R'= COOEt) in 16% yield. The stereochemistry of newly generated asymmetric center in **34** was almost homogeneous (>95%), and was determined from NOE data (23% enhancement) between the methyl group and epoxide proton. The synthesis of the bicyclo[7.3.1]-tridecadiyne system with an epoxide ring in it was concluded.

The cycloaromatization of **34** was achieved under acidic conditions to cleave the epoxide ring and to diminish the strain to the macro ring system. We first expected that the epoxide opening of **34** might afford **36** *via* **35** through Bergman cycloaromatization. But the product which was obtained, in fact, from **34** with *p*-toluenesulfonic acid in the presence of cyclohexa-1,4-diene in THF solvent was the methyl ketone **38**, instead.

This aromatized compound **37** from the 1,2-glycol **34** was obtained *via* pinacol-pinacolone rearrangement of the hypothetical intermediate **35**.[18] The same product **37** could alternatively be considered as another Bergman cycloaromatization through **38**, a pinacol-pinacolone type rearrangement product with concomitant epoxide opening. The rearrangement in the latter case, however, has to be associated with a ring-contraction process for the strained 10-membered enediyne system into 9-membered one in **38**. The following mechanism was supported by the fact that **34** (R= Ac) provided the aromatized compound **36**, instead of the further rearrangement product **37**.

Introduction of Acetylenes to Sugars at Anomeric Position

C-Glycosidation is considered a key reaction for introducing acetylenic groups to sugars. A generalization of alkynylation to *D*-glucals can be readily achieved in the presence of Lewis acid as catalyst. This method might provide new methods toward the enediyne class of compounds.

The first example of reaction between tri-*O*-acetyl-D-glucal **39** and bistrimethylsilyl acetylene as shown above gave the glycosidation product **40** in quantitative yield. This reaction can also be done with tributylstannyl- and trimethylsilyl-acetylene in high yields.[19] The more readily available silylacetylenes proved to be sufficiently reactive for the C-glycosidation. The reaction mechanism, as illustrated in Scheme 1, involves elimination of the 3-acetoxy group in **41** (= **39**), and subsequent formation of the oxonium intermediate **42** to which pi-electrons of the silylacetylene should participate. A possible cationic charge would develop on the beta carbon in **43**, which is highly stabilized by the silicon atom.[20] Departure of the trimethylsilyl group results in the formation of the product **44**. Four more examples of C-glycosidation with Me₃Si-C≡C-R on tri-*O*-acetyl-D-glucal **39** in the presence of Lewis acid at low temperatures in dichloromethane for ca. 1 hr are shown in Table (entry 2-5). This glycosidation was completely stereoselective to obtain single stereoisomers, **40**, **44a**, **44b**, **44c** and **44d**. Partial hydrogenation of product **40** (**44**, R= SiMe₃) afforded **45** (R= SiMe₃), of which the NOE between H-5 and the olefinic H showed the alpha orientation. All others are assigned to have the alpha configuration at the C-1 position. The thiophenyl derivative **44a** was obtained with a weaker catalyst BF₃•OEt₂ instead of SnCl₄ due to the additional stabilization effect by the sulfur atom for the cation **45** (R= SPh). Three products, **40**, **44b**, **44d**, containing the trimethylsilylacetylenic moiety on the other end, would potentially be reactive with another *D*-glucal molecule under the given conditions. In fact, the enediyne type product **44d** [FAB-MS m/z 361 (M+1)], in entry 5, was further reactive to allow us to isolate the twice reacted product (in 79% yield).[21]

Similar C-glycosidation to 2-acetoxy-D-glucal **46** with the silyl acetylenes proceeded with tin tetrachloride at 0 ~ -20°C as illustrated in Scheme 2 to produce the acetylenic products **47**, most of which were unstable to acid, to base and even to silica gel for purification. These products were converted, after aqueous work-up, into the α,β−unsaturated ketones **48**, which were sufficiently stable to measure their nmr spectra (R= SiMe₃ α−H, δ 6.3 ppm & β−H 7.1 ppm), but were too unstable to be isolated by silica gel chromatography. Usually the products (**48**) were treated with a reducing agent such as sodium borohydride in the presence of ceric trichloride to give the allylic alcohols **49** (R'= Ac). Reduction of **38** with lithium aluminum hydride at -40°C also provided the alcohols, **49a-e** (R'= H, overall yields being indicated in Table).[21]

On the other hand, reduction of **48** with LiAlH4 at higher temperatures such as 0°C yielded the corresponding *trans* olefinic compounds **50** (R= Cl, R= -C≡C-SiMe3 in 36, 37% yield, respectively) together with the non-hydroalumination products (**49d,e** in ca. 30%, respectively). This hydroalumination proceeded due to the *cisoid* orientation of the acetylene and alkoxyaluminum at the neighboring positions. This assignment is suggested from the mechanism shown in Scheme 3.

C-Glycosidation of Silyl Acetylenes to D-glucals 39 and 46.[21]

Entry	1	2	3	4	5
Silyl acetylenes	SiMe3 ‖ SiMe3	SiMe3 ‖ SPh	SiMe3 ‖ ‖ SiMe3	SiMe3 ‖ ⟋Cl	SiMe3 ‖ ⟋SiMe3
Lewis Acid	SnCl4	BF3•OEt2	SnCl4	SnCl4	SnCl4
react. temp	-20°C	0°C	-20°C	0°C	-78°C
product from **39**					
yield	99 %	84 %	96 %	81 %	79 %
product from **46**					
yield	50 %	72%	39 %	41 %	48 %

Stereochemistry of 2-hydroxy group is alpha by the hydride attack from the beta side in **A** to result in the formation of the intermediate **B**, which allows the hydroalumination (**C**), resulting in **D** after aqueous work-up. In fact, one of the product (**50**, R= -C≡C-SiMe3) showed NOE between H-5 and H-1' as well as H-6 and H-1 to prove the stereochemistry of C-glycosidation being alpha.

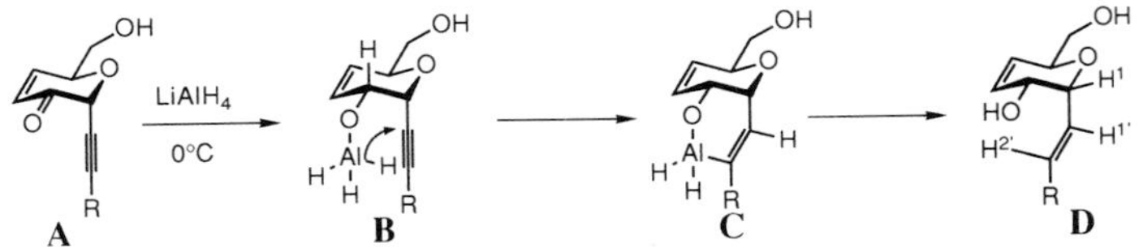

The current C-glycosidation with silyl acetylenes has to be conducted under acidic conditions, which means a limitation on acid-sensitive functional groups such as acetals and ketals. To overcome this limitation, a two-step preparation through the vinyl chloride **44c** was employed, which coupled with some acetylenes with acetal or ketal in the presence of palladium and copper under basic conditions [10] to yield **51a,b** (olefinic H, J= 7.7 Hz). [21]

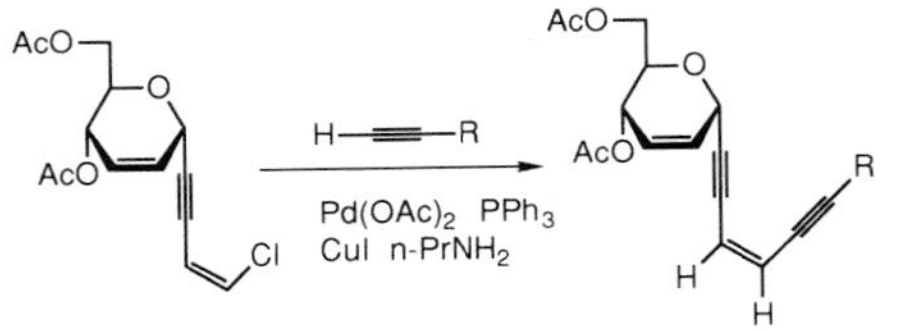

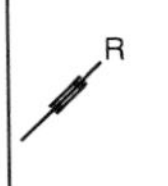

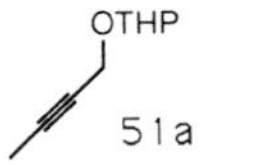

		OTHP	OEt
R		51a	51b
yields		63 %	93 %

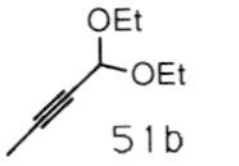

The current method will be used for the synthesis of oxygenated carbon compounds in optically active form. These studies are to be continued for further application to the following cyclization leading to oxabicyclo[7.3.1]tridecendiyne system.

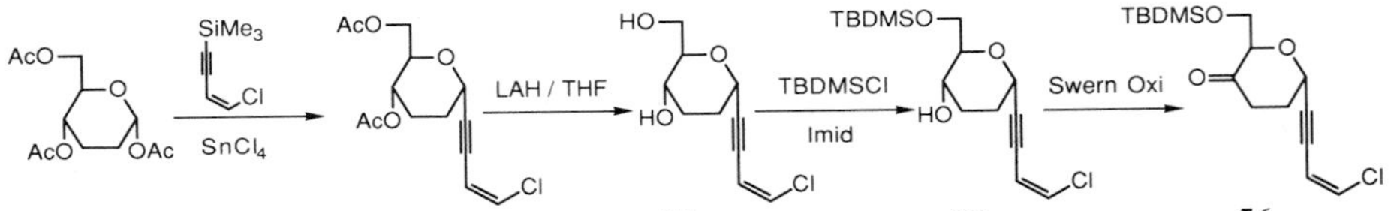

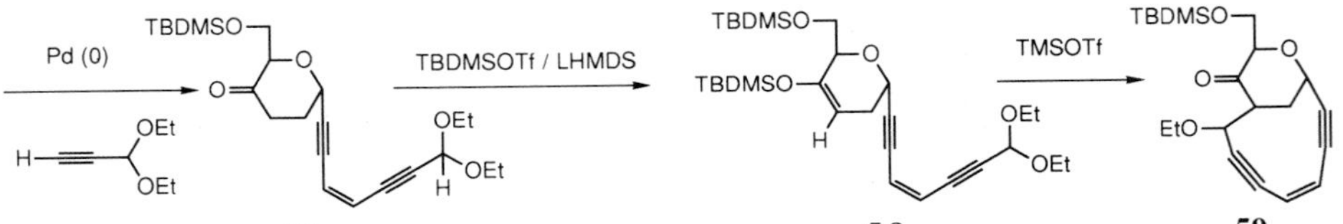

The reaction sequence shown above has been anticipated with dideoxysugar **52** by commencing the C-glycosidation with the *cis*-vinyl silyl acetylene to **53**. The above sequence involves an introduction of an acetal function indirectly through coupling of the propynyl acetal with paladdium to **57**. The manipulation of the hydroxyl groups, on the other hand, to the 4-keto form **56** and then to silyl vinyl ether **58**, which is the precursor of the cyclization to **59**. This cyclization, however, was still not the major product and should be improved for better yields. The essential scheme shown above involves potential utility of the C-glycosidation method.

The above methodologies are focused on the formation of acetylenic bonds with acetylides as well as tin or silyl acetylenes toward oxonium or imminium species.

References

1. Konishi, M.; Ohkuma, H.; Matsumoto, K.; Tsuno, T.; Kamei, H.; Miyaki, T.; Oki, T.; Kawaguchi, H.; VanDuyne, G. D.; Clardy, J. *J. Antibiot.* **1989**, *42*, 1449; *J. Am. Chem. Soc.* **1990**, *112*, 3715.

2. (a) Jr. J. A. Porco, F. J. Stout, J. Clardy, and S. L. Schreiber, *J. Am. Chem. Soc.*, **112**, 7410 (1990). (b) K. C Nicolaou, C. K. Hwang, A. L. Smith, and S. V. Wendeborn, *ibid.*, **112**, 7416 (1990). (c) P. A. Wender and C. K. Zercher, *ibid.*, **113**, 2311(1991).

3. (a) M. Hayashi, A. Inubushi, T. Mukaiyama, *Chemistry Lett.*, 1975 (1987). (b) W. S. Johnson, R. Elliott, J. D. Elliott, *J. Am. Chem. Soc.*, **105**, 2904 (1983).

4. (a) Lundkvist, J. R. M., Ringdahl, B. and Hacksell, U.; *J. Med. Chem.*, **1989**, *32*, 863. (b) Lundkvist, J. R. M., Wistrand, M. L-G. and Hacksell, U.; *Tetrahedron Lett.*, **1990**, *31*, 719.

5. Hiemstra, H., Fortgens, H. P., Stegenga, S. and Speckamp, W. N.; *Tetrahedron Lett.*, **1985**, *26*, 3151.

6. T. Nishikawa, M. Isobe and the late T. Goto; *Synlett*, 99-101 (1991).

7. Furukawa, T.; Horiguchi.; Y.; Kuwajima, I. *32nd Symposium on the Chemistry of Natural Products.* Symposium Papers 411(Chiba, **1990**).

8. T. Nishikawa, M. Isobe and T. Goto; *Synlett*, 393-394 (1991).

9. M. L.; Schlessinger, R. H. *Synthetic Commun.* **1976**, *6*, 555.

10. Kende, A. S.; Smith, C. A. *Tetrahedron Lett.* **1988**, *29*, 4217.

11. Sonogashira, K.; Tohda, Y.; Hagihara, N. *Tetrahedron Lett.* **1975**, 4467.

12. Parikh, J. R.; Doering, W. V. E. *J. Am. Chem. Soc.* **1967**, *89*, 5505.

13. (a) Imamoto, T.; Sugiura, Y.; Takiyama, N. *Tetrahedron Lett.* **1984**, *25*, 4233. (b) Recent application to synthesis of NCS-chr. analogue: Myers, A. G.; Harrington, P. M.; Kuo, E.Y. *J. Am. Chem. Soc.* **1991**, *113*, 694.

14. T. Nishikawa, A. Ino, M. Isobe and T. Goto; *Chemistry Lett.*, 1271-1274 (1991).

15. R. Yamaguchi, Y. Nakazono, and M. Kawanishi, *Tetrahedron Lett.*, **24**, 1801 (1983).

16. J. R. Parikh and W. V. E. Doering, *J. Am. Chem. Soc.*, **89**, 5505 (1967).

17. E. Nakamura and I. Kuwajima, *Angew. Chem., Int. Ed. Engl.*, **15**, 498 (1976); I. Kuwajima, E. Nakamura, and K. Hashimoto, *Tetrahedron*, **39**, 975 (1983); A. B. Holmes, C. L. D. Jenning-White, A. H. Schulthess, B. Akinde, and D. R. M. Walton, *Chem. Commun.*, **1979**, 840.

18. K. C. Nicolaou, A. L. Smith, S. V. Wendeborn, and C.-K. Hwang, *J. Am. Chem. Soc.*, **113**, 3106 (1991).

19. (a) M. Hayashi, A. Inubushi, T. Mukaiyama, *Chemistry Lett.*, 1975 (1987).   (b) W. S. Johnson, R. Elliott, J. D. Elliott, *J. Am. Chem. Soc.,* **105**, 2904 (1983).   (c) (a) Y. Ichikawa, M. Isobe, M. Konobe and T. Goto; *Carbohydrate Research,* **171**, 193 (1987).

20. (a) A. Hosomi and H. Sakurai; *Tetrahedron Lett.,* 1295 (1976). (b) M. Isobe, Y. Ichikawa and T. Goto; *Tetrahedron Lett.,* **27**, 963 (1986). (c) M. Isobe, Y. Ichikawa, D. Bai, H. Masaki and T. Goto; *Tetrahedron,* **43,** 4767 (1987).   (d) S. Danishefsky and J. F. Kerwin, Jr.; *J. Org. Chem.,* **47**, 3805 (1982).   (e) I. Fleming, *Comprehensive Organic Chemistry,* Chapter 13 Organo Silicon Chemistry, Pergamon Press (1979).

21. T. Tsukiyama, M. Isobe; *Tetrahedron Lett.,* **33**, in press (1992).

Molecular Recognition in Neocarzinostatin Complex and Synthesis of New DNA Cleaving Molecules

Masahiro Hirama

Department of Chemistry, Faculty of Science, Tohoku University, Sendai 980, Japan

Summary: We designed and synthesized enantiomerically pure functional 10-membered ring core analogs of neocarzinostatin and hybrids with DNA binding groups. They show striking and selective DNA cleaving abilities as well as antibiotic and cytotoxic activities. We also discuss the three dimensional structure of neocarzinostatin complex determined by 2D NMR analysis and distance geometry calculations to learn how neocarzinostatin apoprotein binds and stabilizes the chromophore.

1. Introduction

Neocarzinostatin (NCS) is a potent antitumor antibiotic isolated from the culture of *Streptomyces carzinostaticus* by Ishida et al.[1] NCS is a complex composed of an unstable chromophore (**1**)[2] and an apoprotein (**2**) of 11,000 daltons.[3] The chromophore (**1**) is very labile to heat, light, and pH above 6, but it is stabilized substantially when bound to the apoprotein (**2**).[2] Thus, the apoprotein (**2**) serves as a stabilizer and a carrier of the unstable chromophore (**1**).[3] However, their binding structure and mechanism of stabilization are yet unknown, so these issues are very interesting in terms of molecular recognition and protein transport. Another intriguing issue is the mode of action. Goldberg and Edo elucidated that the

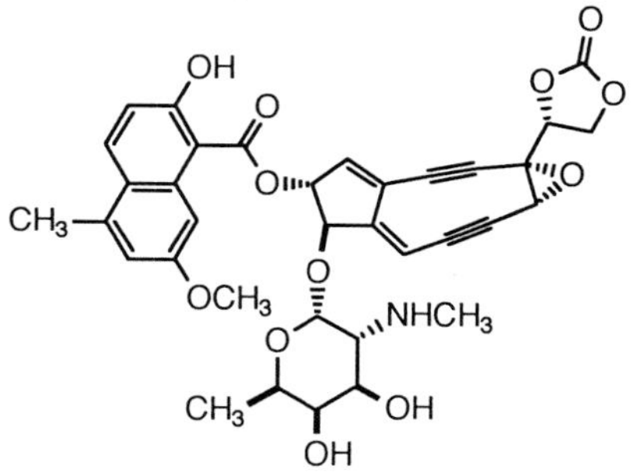

1

									10									20

Ala-Ala- Pro- Thr- Ala- Thr- Val- Thr- Pro- Ser- Ser- Gly- Leu- Ser- Asp- Gly- Thr- Val- Val- Lys-
30   40
Val-Ala- Gly- Ala- Gly- Leu- Gln- Ala- Gly- Thr- Ala- Tyr- Asp- Val- Gly- Gln- Cys- Ala- Trp- Val-
50   60
Asp-Thr-Gly- Val- Leu- Ala- Cys- Asp- Pro- Ala- Asn- Phe- Ser- Ser- Val- Thr- Ala- Asp- Ala- Asn-
70   80
Gly-Ser- Ala- Ser- Thr- Ser- Leu- Thr- Val- Arg- Arg- Ser- Phe- Glu- Gly- Phe- Leu- Phe- Asp- Gly-
90   100
Thr-Arg-Trp- Gly- Thr- Val- Asn- Cys- Thr- Thr- Ala- Ala- Cys- Gln- Val- Gly- Leu- Ser- Asp- Ala-
110
Ala-Gly-Asp- Gly- Pro- Glu- Gly- Val- Ala- Ile- Ser- Phe- Asn

2

chromophore (**1**) is fully responsible for the biological activity of NCS, and that a thiol activates **1** to generate a carbon radical, which abstracts hydrogen from DNA and leads to DNA scission under aerobic conditions.[2] Myers proposed a chemical mechanism for the action of the chromophore (**1**): thiol addition to C12 and a concomitant opening of the epoxide initiates the cascade of reactions that leads to the generation of the dehydroindacene diradical through Masamune-Bergman type cyclization[4] of an enynecumulene intermediate.(Fig. 1).[5]

Our mechanistic and synthetic studies refer to the following questions and research plans:
(1) Is the highly strained 9-membered ring essential for the thiol-triggering aromatization ?
(2) Since there is no precedent for an enynecumulene system, synthesis of such systems is necessary to explore their chemical behavior.
(3) How does the apoprotein (**2**) recognize and stabilize the chromophore (**1**) ?
(4) We intend to design and synthesize new DNA cleaving molecules inspired by **1**, aiming for low molecular weight, sequence-specific DNA cleavers.

This paper will focus the first half on the above synthetic work. The three dimensional structure of the NCS complex and the apoprotein-chromophore interactions will then be discussed.

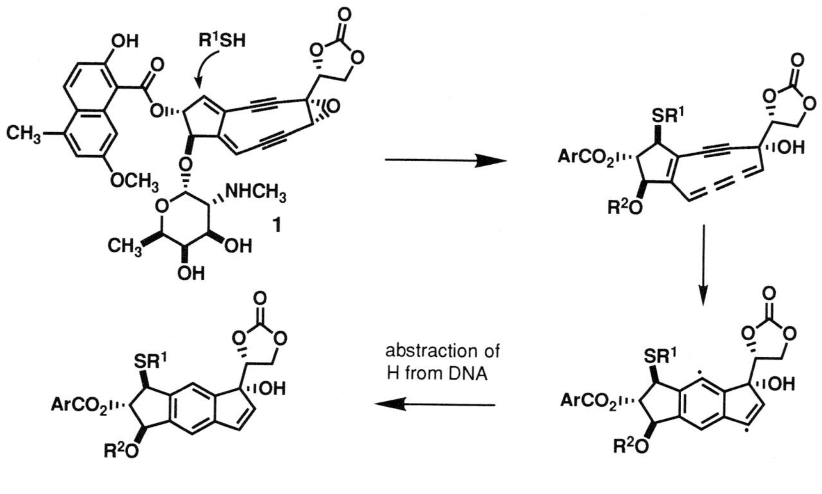

Fig. 1. Myers' proposed mechanism for action of NCS-chromophore (**1**)[5]

2. Synthesis and Chemical Properties of Acyclic Enynecumulenes

To answer the first question, we designed and synthesized an acyclic cross-conjugated diene-diyne system (**3**) in a straightforward manner, and treated it with thiol in acetonitrile (Fig. 2).

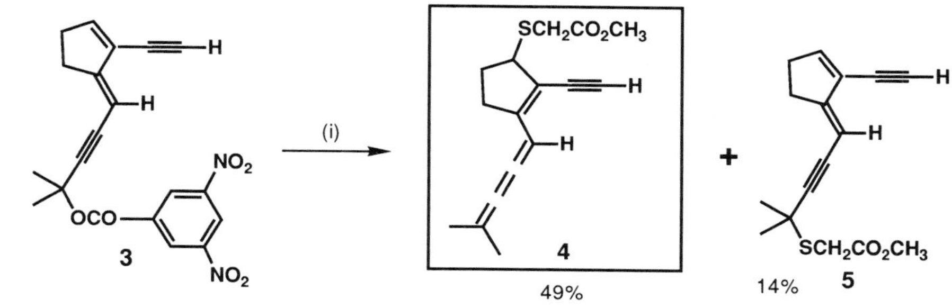

Fig. 2. Reagents and conditions: (i) HSCH$_2$CO$_2$CH$_3$ (1.5 equiv), Et$_3$N (1 equiv), CH$_3$CN, 25°C, 2 h.

Under carefully controlled conditions, we isolated a vinylogous S$_N$2' reaction product, enynecumulene (**4**), in moderate yield, although, interestingly, a formally direct substitution product (**5**) was formed exclusively when a large excess of thiol or DMSO was used as solvent. The ^{13}C-NMR spectrum strongly supported the cumulene structure for **4**, which was stable at room temperature in the absence of air.[6]

The next question was whether **4** would undergo the Masamune-Bergman type cyclization.[4] Heating at 80°C in 1,4-cyclohexadiene produced a styrene derivative (**6**). When the reaction was carried out in cyclohexadiene-d$_8$, two deuteriums were incorporated at the positions indicated in Fig. 3, which clearly indicates that this system can undergo the Masamune-Bergman type cyclization[4] leading to the generation of

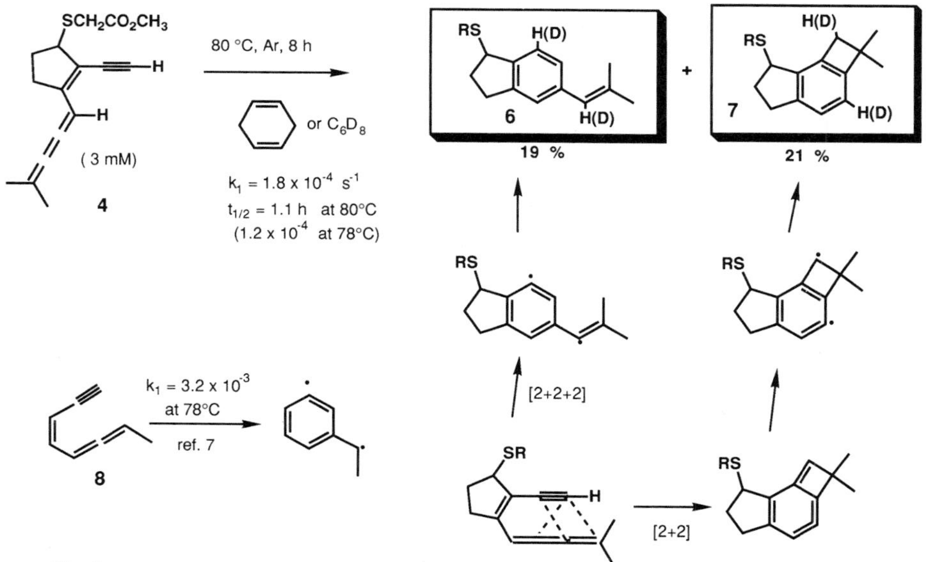

Fig. 3.

σ-diradicals. In addition, benzocyclobutane (**7**) formation turned out to be another major reaction pathway in this conformationally flexible system. The first-order disappearance rate is approximately 30 times slower than that of the related enynallene system.[6,7]

3. Synthesis, Cycloaromatization, and DNA Cleaving and Biological Activities of the Second Generation of NCS Chromophore Analogs

We have synthesized the 10-membered ring analogs (**9-11**)[8-10] of the NCS chromophore and demonstrated that these analogs can also undergo the thiol triggering aromatization. However, they do not exhibit DNA cutting or cytotoxic activities, possibly because they lack a hydrophilic or DNA binding group. Therefore, we designed the second generation models (**12-15**) to improve binding.[11-13] Our basic strategy for designing new DNA cleaving molecules was to combine the DNA cleaving unit with some DNA binding group.

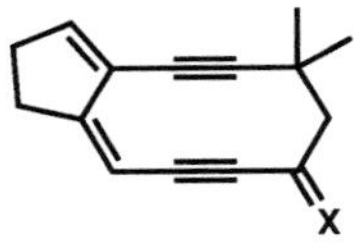

9, X=O
10, X=H, OAc

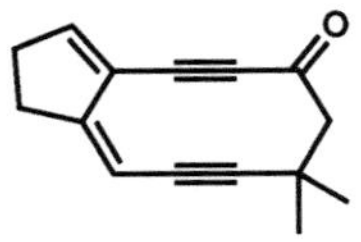

11

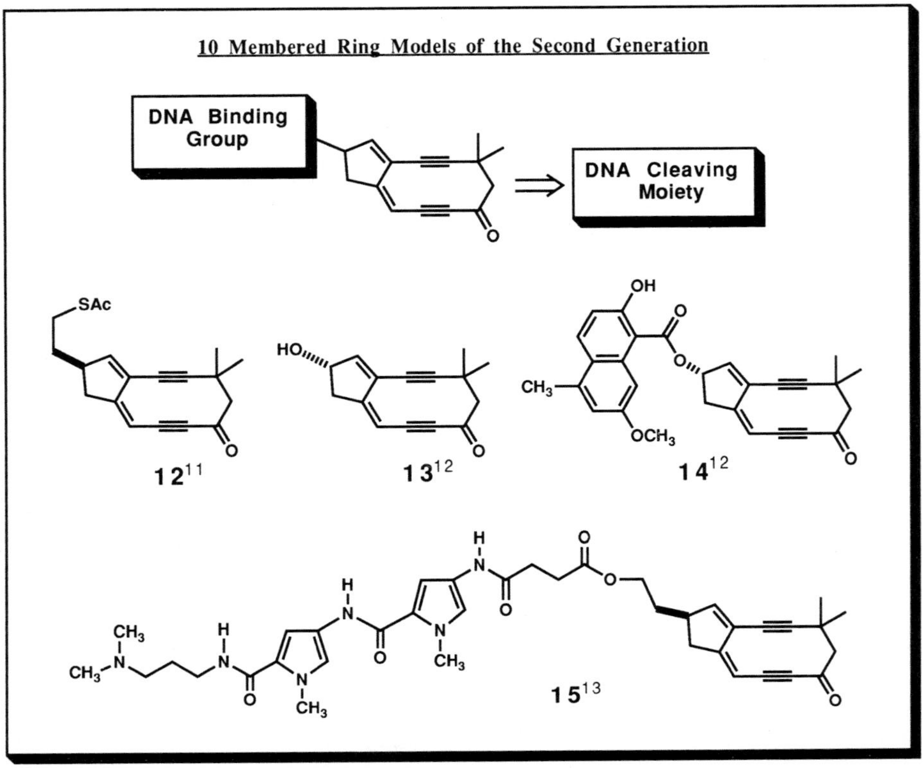

10 Membered Ring Models of the Second Generation

DNA Binding Group ⟹ **DNA Cleaving Moiety**

12[11]   **13**[12]   **14**[12]

15[13]

3.1. A New Model Equipped with an Intramolecular Trigger

First, we synthesized a new model (**12**) equipped with an intramolecular nucleophile, inspired by the related potent antitumor antibiotics, calicheamicin and esperamicin,[14] as summarized in Fig. 4.[11] Palladium mediated cyclization of **16** according to Stille's procedure[15] gave the 10-membered ring compound (**17**). Direct cyclization in the related system without the stannyl group didn't work well, and instead, resulted in the formation of dimeric oxidative coupling products.[8] The final oxidation and concomitant dehydration proceeded smoothly by Swern oxidation using excess reagents.[8,11]

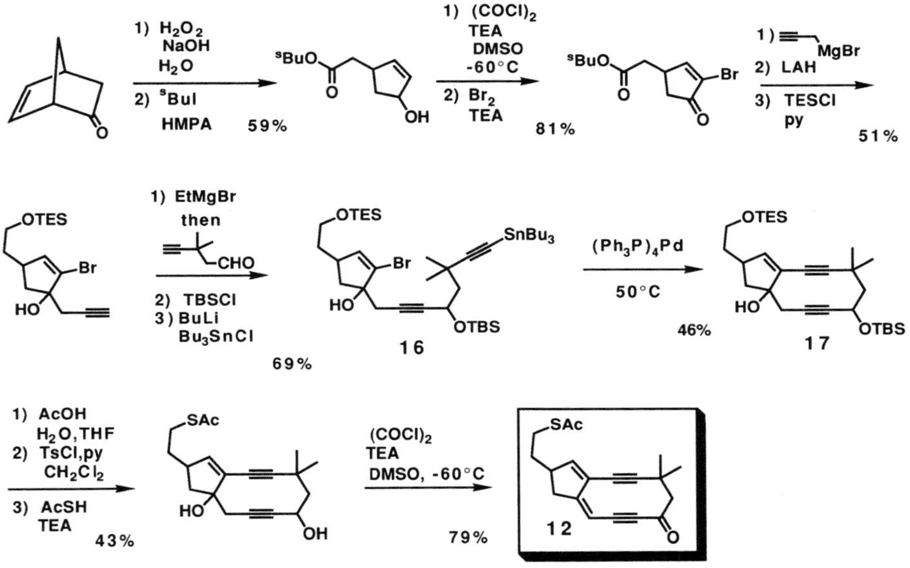

Fig. 4[11]

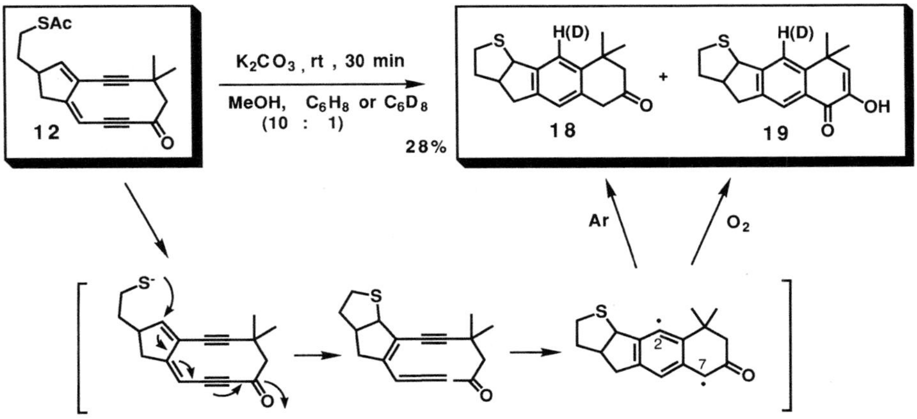

Fig. 5. Cycloaromatization of **12** under weakly alkaline conditions.[11]

Cycloaromatization of **12** was then tested. Methanolysis using potassium carbonate in the presence of 1,4-cyclohexadiene at room temperature was complete within 30 minutes in an argon atmosphere to give the reductive cycloaromatization product (**18**) through the intramolecular addition of thiolate (Fig. 5).[11] Deuterium incorporation was observed at C-2 in the presence of cyclohexadiene-d8. In air and in the presence of cyclohexadiene-d8, the corresponding diketone (**19**) with deuterium at C-2 was also formed. These results clearly support the putative reaction cascade shown in Fig. 5. The reactive C-2 σ radical could abstract carbon-bound deuterium, while C-7 π radical, stabilized by resonance, might be trapped by a thiol hydrogen or an oxygen molecule.[11] Fortunately, our intramolecular trigger model (**12**) showed an appreciable activity toward *Micrococcus luteus* without addition of external thiol.

3.2. New Functional Analogs

We hoped the naphthoate conjugate **14** might exhibit a T-selective DNA cleavage similar to that of NCS chromophore (**1**).[2] Furthermore, the NCS apoprotein (**2**) would serve as a carrier of this lipophilic molecule (**14**).[3] The naphthoate moiety is believed to be responsible for T-specific DNA scission by **1**, possibly through its specific recognition of a T residue by intercalation.[16] Our recent NMR analysis and distance geometry calculations of the three dimensional structure of the NCS complex (*vide infra*) have shown that the naphthoate moiety is responsible for the specific binding of the chromophore.[17] Enantiomerically pure alcohol (**13**) and naphthoate conjugate (**14**) were synthesized as shown in Fig. 6.[12] The key cyclization step was also accomplished as before.

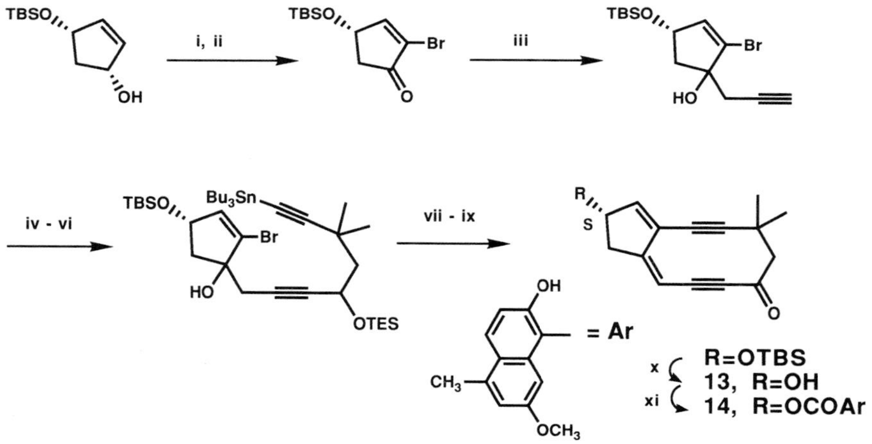

Fig. 6.[12] (i) (COCl)$_2$, DMSO, Et$_3$N, -60°C; (ii) Br$_2$, CH$_2$Cl$_2$, 0°C (85%); (iii) HCCCH$_2$MgBr, ether, -88°C (91%); (iv) BuLi, THF-HMPA, -78°C, and then 3,3-dimethylpent-4-yn-1-al; (v) TESCl, py; (vi) BuLi, THF, -78°C, and then Bu$_3$SnCl (50%); (vii) (Ph$_3$P)$_4$Pd, THF, 60°C, 89 h, (49%); (viii) THF-H$_2$O-AcOH (8 : 1 : 4); (ix) (COCl)$_2$, DMSO, Et$_3$N, -60°C; (x) THF-H$_2$O-AcOH (1 : 1 : 1) (38%); (xi) Me$_2$HN$^+$(CH$_2$)$_3$N=C=NEt Cl$^-$, ArCO$_2$H, CH$_2$Cl$_2$ (54%).

The thiol-triggering cycloaromatization of **13** and **14** proceeded smoothly at room temperature in ethanol in the presence of 1 equivalent of triethylamine (Fig. 7).[12]

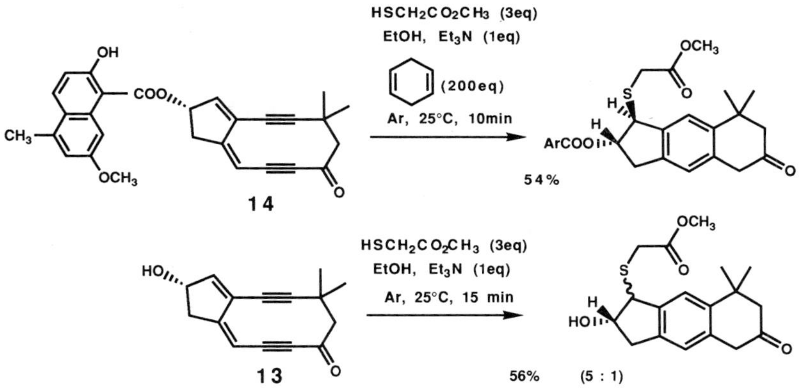

Fig. 7. Thiol triggering cycloaromatization of **13** and **14**.[12]

3.3. DNA Cutting and Biological Activities of 13 and 14

The alcohol (**13**) clearly showed antibiotic activity and cytotoxic activity against *B. licheniformis* and K562 leukemia cell, respectively. On the other hand, the activities of naphthoate (**14**) alone were quite sluggish, but in the presence of NCS apoprotein (**2**), the mixture showed the distinctive activities. This is likely to result from the complex formation and transport function of the apoprotein (**2**).[12] Furthermore, the DNA cutting ability of these molecules was confirmed on supercoiled pBR322 DNA and the site specificity was examined on the 5'-end-^{32}P-labeled pBR322 DNA. Remarkable G specificity was found for alcohol **13**. This is surprising because **13** cannot be an intercalator nor does it possess an apparent DNA binding moiety. On the other hand, combination of the naphthoate **14** and NCS apoprotein (**2**) appears to show some sequence selectivity, namely G of 5'-G̲GT and T of 5'-GCT̲, possibly due to the intercalation of the naphthoate group.[12] This T selectivity is reminiscent of the selectivity of the NCS chromophore (**1**). We are now exploring whether **13** can recognize a guanine base and cleave DNA by a carbon radical, or if a simple alkylation mechanism is working. It will be discussed in the near future.

3.4. A New Hybrid Molecule Containing a Minor Groove Binder

The antibiotic netropsin is known to be a strong minor groove binder to the AT-rich regions of DNA through hydrogen bonding and van der Waals interaction.[18] So, we anticipated that the hybrid netropsin analog (**15**) would function as an effective AT-specific DNA cleaver. Synthesis of the netropsin conjugate has been achieved as summarized in Fig. 8.[13] One key step is the asymmetric hydrolysis of acetate (**20**) by lipoprotein lipase Amano III. Another key step is the successful synthesis of the versatile intermediate (**21**) containing an active ester. It enabled us to couple with various amine derivatives, including the netropsin analog (**22**), simply by mixing in DMF. The compound (**22**) is readily available according to Shibuya's procedure.[19]

This netropsin conjugate (**15**) undergoes the thiol-triggering cycloaromatization without addition of external amine as expected (Fig. 9).[13]

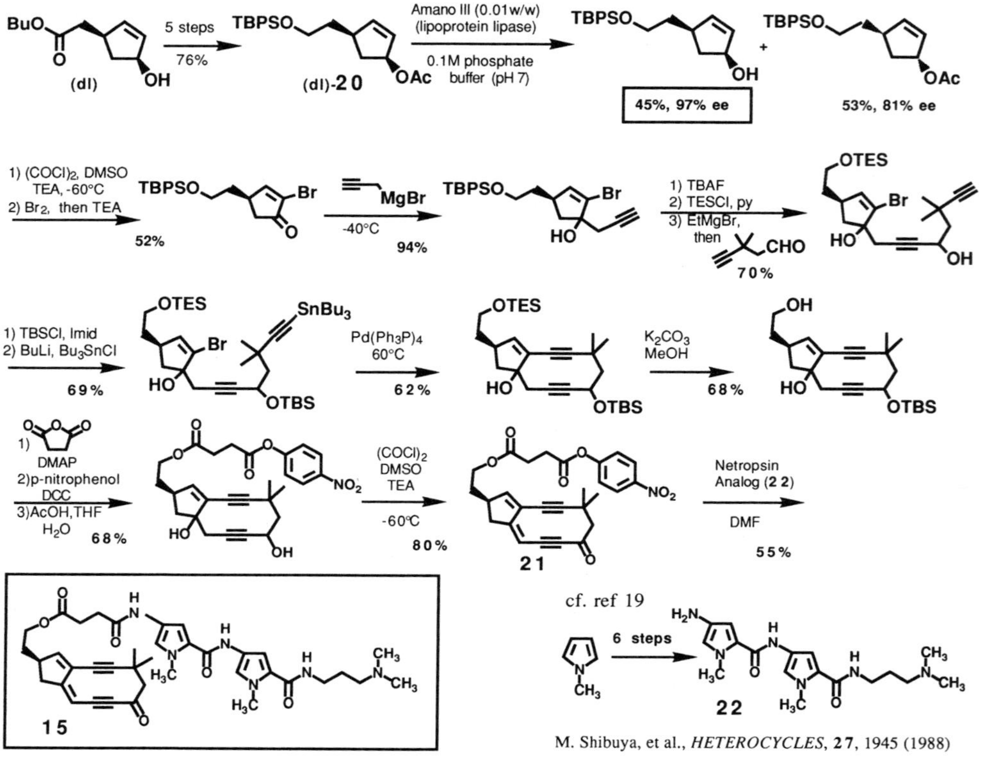

Fig. 8. Synthesis of Hybrid **15**[13]

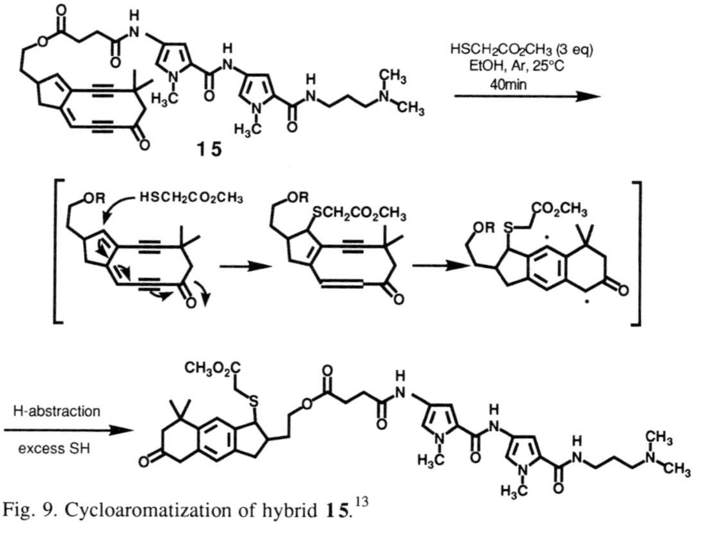

M. Shibuya, et al., *HETEROCYCLES*, **27**, 1945 (1988)

Fig. 9. Cycloaromatization of hybrid **15**.[13]

We compared the DNA cleaving abilities of synthetic molecules **12**, **13**, and **15** on supercoiled pBR322 DNA. Agarose gel electrophoretic patterns clearly showed that netropsin conjugate (**15**) cleaved supercoiled DNA-Form I to nicked circular Form II and further to linear Form III most effectively.[13]

4. Apoprotein-Chromophore Binding Structure and Stabilizing Interactions

Although the crystal structure of the apoprotein (**2**) at low resolution[20] and its solution structure[21] have been studied, the precise binding structure of the NCS complex has not yet been revealed. So, we recently carried out a computer modeling study[22] and the 2D NMR analysis of the solution tertiary structure of the NCS complex.[17,23]

4.1. Three Dimensional Structure of Apoprotein (2)

The DADAS90 distance geometry calculations using the MolSkop System®, based on the constraints obtained by 2D NMR analysis of the NCS complex, revealed that the apoprotein consists of three antiparallel β-sheets without helix secondary structure and that the motif is like a hand holding a ball (Fig. 10).[17,22] We could easily find a pocket indicated by the arrow in Fig. 10. The first feature of the pocket is a deep valley centered by the three Gly35, Gly96, and Gly107, where the side chains of the residues adjacent to these glycines point down to expose the main chains. The second feature is a disulfide bridge of Cys37 and Cys47 making a hydrophobic ridge at the bottom.[17,22] The third feature is the pocket rim. On the right side is a big hydrophobic wall of Phe78, Leu45, and Try39 side chains. On the other side is a big hydrophilic wall of two Ser98, Asp33, and Ser54. In addition, phe52 is placed on the back side wall, and Gln94 is in the front.

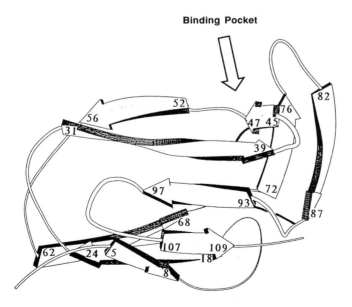

Fig. 10. Motif of NCS apoprotein tertiary structure[17,22,23]

4.2. Three Dimensional Structure of NCS Complex and Essential Functionalities for Specific Binding

Since all of those residues discussed above showed the intermolecular NOEs with chromophore (1), the pocket must be a chromophore binding pocket.[17,22,23] A complete structure of the NCS complex satisfying both the intra- and intermolecular NOEs has been computed by DADAS90 and DGEOM programs.[23] The naphthoate moiety of 1 gets into the deep bottom of the pocket, while the N-methylfucosamine moiety is facing outward.[17] The naphthoate and the carbocyclic core moieties are surrounded by the side chains of the residues at the rim region, Phe52 and Phe78, Leu45, Try39, Gln94, Ser98, and Asp33, as shown in Fig. 11.[23] This binding structure indicates that the naphthoate moiety must be essential for its specific and strong binding to NCS apoprotein (2),[17] and the several interactions between the functional groups on the naphthoate ring and the side chains of residues appear to play an important role for it.[23]

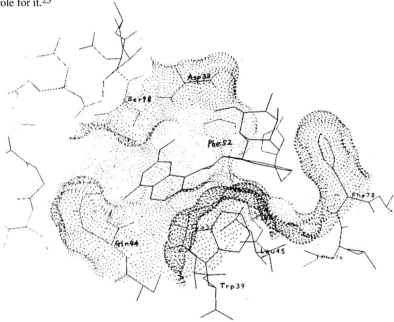

Fig. 11. Illustration of binding structure of NCS complex. Dots indicate molecular surface of the residues forming the binding pocket.[23]

In order to confirm the importance of the naphthoate moiety of 1 for the specific binding, we synthesized the stable simple analogs listed in Fig. 12 and carried out their binding experiments to the NCS apoprotein (2).[24] On this occasion, we needed an efficient synthetic way to get the naphthoic acid (23). The reported syntheses are elegant but too laborious. So, we have developed a simple route to 23 as summarized in Fig. 13.[25] The association constants of the model compounds obtained by fluorescence quenching titration (Fig.12) support that the naphthoate moiety is most essential for apoprotein binding.[17,22,23]

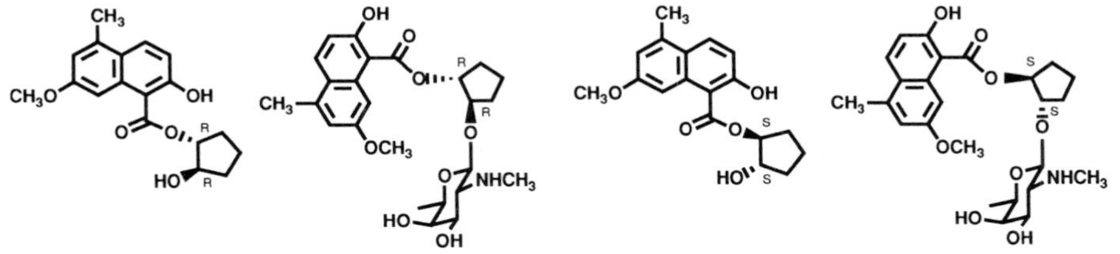

Ka = 1.0 x 10^4 Ka = 1.9 x 10^4 Ka = 1.1 x 10^4 Ka = 1.7 x 10^5

Cf. Ka of NCS-Chr ≥ 1.4 x 10^5 [K. Edo, et al. *J. Antibiot.*, **41**, 554 (1988)]

≥ 3 x 10^6 [Hirama et al., unpublished result]

in 5% MeOH / 0.1 M acetate buffer (pH 4.0)

≥ 5 x 10^7 [I. H. Goldberg et al., *Biochem.*, **19**, 4773 (1980)]

in 20% MeOH / 20 mM Tris buffer (pH 8.0)

Fig. 12. Binding constants of simple model compounds to apoprotein (**2**).[24]

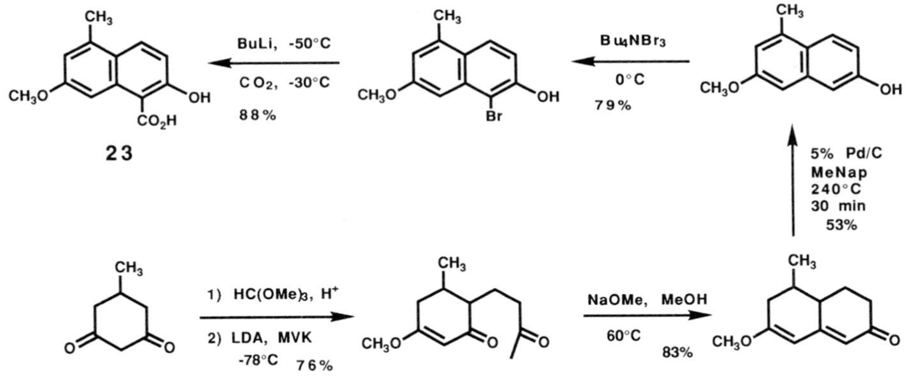

Fig. 13. A simple six-step synthesis of 23.[25]

4.3. Stabilization Interactions

Based on the above binding structure, we can suggest several stabilization interactions. First, amino sugar moiety of **1** has an induced conformation in which the protonated methylamino group is covering the C12 active center of **1** from nucleophile attack.[17] Second, the core of **1** is sitting on the disulfide ridge, and the epoxide is facing down to the hydrophobic bottom, prevented from acid-catalyzed epoxide activation (Fig. 11). Third, as fluorescence of naphthoate group is quenched by interactions with apoprotein (**2**), the chromophore (**1**) can elude photodecomposition caused by photoexcitation of the naphthoate.[26]

Acknowledgment: I wish to thank my talented and enthusiastic coworkers, Dr. K. Fujiwara, T. Tanaka, M. Tokuda, T. Gomibuchi, K. Takahashi, H. Sakai, and Dr. T. Suzuki, and express my sincere thanks to Prof. Y. Sugiura and M. Uesugi (Kyoto University; DNA cleavage), Dr. M. Ishiguro and Dr. S. Imajo (Suntory Co. Ltd.; DGEOM calculation), K. Fujita (JEOL Co. Ltd.; MolScop® calculation), and Dr. H. Komatsu (POLA Co. Ltd.; biological activity) for their great help and good collaborations. Thanks are also due to Professors N. Ishida (Emeritus Prof. of Tohoku Univ.), M. Mizugaki (Tohoku Univ.), and K. Edo (Fukushima Medical College) for valuable discussions. Financial support for the work described in this paper is greatfully acknowledged: Ministry of Education, Science, and Culture, Japan, Japan Society for Promotion of Science, Mitsubishi Science Foundation, CIBA GEIGY Foundation, Shorai Science Foundation, TERUMO Life Science Foundation, Takeda Science Foundation, and Fujisawa Medicinal Resources Foundation.

References

1. N. Ishida, K. Miyazaki, K. Kumagai, M. Rikimaru: *J. Antibiot.* **18** (1965) 68.

2. M. A. Napier, B. Holmquisy, D. J. Strydom, I. H. Goldberg: *Biochem. Biophys. Res. Commun.* **89** (1979) 635; L. S. Kappen, M. A. Napier, I. H. Goldberg: *Proc. Natl. Acad. Sci. U.S.A.* **77** (1980) 1970; Y. Koide, F. Ishii, K. Hasuda, Y. Koyama, K. Edo, S. Katamine, F. Kitame, N.I shida: *J. Antibiot.* **33** (1980) 342; K. Edo, M. Mizugaki, Y. Koide, H. Seto, K. Furihata, N. Otake, N. Ishida: *Tetrahedron Lett* **26** (1985) 331.

3. K. Kuromizu, S. Tsunasawa, H. Maeda, O. Abe, F. Sakiyama: *Arch. Biochem. Biophys.* **246** (1986) 199; L. S. Kappen, I. H. Goldberg: *Biochemistry* **19** (1980) 4786.

4. N. Darby, C. U. Kim, J. A. Salaün, K. W. Shelton, S. Takeda, S. Masamune: *Chem. Commun.* (1971) 1516; R. R. Jones, R. G. Bergma: *J. Am. Chem. Soc.* **94** (1972) 660; Review: K.Fujiwara, M.Hirama: *Gendai Kagaku (Chemistry Today)*, (1990, No.7) 14.

5. A. G. Myers: *Tetrahedron Lett.* **28** (1987) 4493; A. G. Myers, P. J. Proteau, T. M. Handel: *J. Am. Chem. Soc.* **110** (1988) 7212.

6. K. Fujiwara, H. Sakai, M. Hirama: *J. Org. Chem.* **56** (1991) 1688.

7. A. G. Myers, E. Y. Kuo, N. S. Finny: *J. Am. Chem. Soc.* **111** (1989) 8057; R. Nagata, H. Yamanaka, E. Murahashi, I. Saito: *Tetrahedron Lett.* **31** (1990) 2907.

8. M. Hirama, K. Fujiwara, K. Shigematsu, Y. Fukazawa: *J. Am. Chem. Soc.* **111** (1989) 4120.

9. K. Fujiwara, A. Kurisaki, M. Hirama: *Tetrahedron Lett* **31** (1990) 4329.

10 T. Tanaka, K. Fujiwara, M. Hirama: *Tetrahedron Lett* **31** (1990) 5947.

11. M. Tokuda, K. Fujiwara, M. Hirama: *Synlett* (1991) 651.

12. M. Hirama, T. Gomibuchi, K. Fujiwara, Y. Sugiura, M. Uesugi: *J. Am. Chem. Soc.* **113** (1991) 9851; M. Hirama: *J. Synth. Org. Chem. Jpn.* **49** (1991) 1032.

13. M. Tokuda, K. Fujiwara, T. Gomibuchi, M. Hirama, M. Uesugi, Y. Sugiura: submitted for publication.

14. Review: K. C. Nicolaou, W.-M. Dai: *Angew. Chem. Int. Ed. Engl.* **30** (1991) 1387.

15. J. K. Stille, J. H. Simpson: *J. Am. Chem. Soc.* **109** (1987) 2138.

16. A. Galat, I. H. Goldberg: *Nucleic Acids Res.* **18** (1990) 2093; H. Sugiyama, T. Fujiwara, H. Kawabata, N. Yoda, N. Hirayama, I. Saito: *J. Am. Chem. Soc.* **114** (1992) 5573.

17. T. Tanaka, M. Hirama, M. Ueno, S. Imajo, M. Ishiguro, M. Mizugaki, K. Edo, H. Komatsu: *Tetrahedron Lett* **32** (1991) 3175.

18. M. L. Kopka, C. Yoon, D. Goodsell, P. Pjura, R. E. Dickerson: *Proc. Natl. Acad. Sci. USA* **82** (1985) 1376.

19. E. Nishiwaki, S. Tanaka, H. Lee, M. Shibuya: *Heterocycles* **27** (1988) 1945.

20. L. C. Sieker, L. H. Jensen, T. S. A. Samy: *Biochem. Biophys. Res. Commun.* **68** (1976) 358.

21. C. D Study: K. Saito, Y. Sato, K. Edo, Y. A. Murai, Y. Koide, N. Ishida, M. Mizugaki: *Chem. Pharm. Bull. Jpn.* **37** (1989) 3078. NMR Studies: M. L. Remerowski, S. J. Glaser, L. C. Sieker, T. S. A. Samy, G. P. Drobny: *Biochem.* **29** (1990) 8401; E. Adjadj, J. Mispelter, E. Quiniou, J.-L. Dimicoli, V. Favaudon, J.-M. Lhoste: *Eur. J. Biochem.* **190** (1990) 263; Tertiary structure of apoprotein: X.Gao: *J. Mol. Biol.* **225** (1992) 125.

22. M. Ishiguro, S. Imajo, M. Hirama: *J. Med. Chem.* **34** (1991) 2366.

23. Refined three dimensional structure of NCS complex has been recently determined: T. Tanaka, M. Hirama, K. Fujita, S. Imajo, M. Ishiguro, manuscript in preparation.

24. K. Takahashi, T. Tanaka, M. Hirama: manuscript in preparation.

25. K. Takahashi, T. Suzuki, M. Hirama: *Tetrahedron Lett.* **33** (1992) 4603.

26. T. Gomibuchi, K. Fujiwara, T. Nehira, M. Hirama: unpublished results.

New Synthetic Strategies for the Synthesis of the "Enediyne" Antibiotics Esperamicin/Calicheamicin and Dynemicin-A

Troels Skrydstrup and David S. Grierson

Institut de Chimie des Substances Naturelles, CNRS, 91198 Gif-sur-Yvette, France

Summary: The [2,3]-Wittig rearrangement of 13-membered macrocyclic ethers has been explored as a means to access highly strained 10-membered bicyclic enediynes related to the core structure of calicheamicin/esperamicin and dynemicin-A. In particular, the ring contraction of a bicyclic ether containing a 3-oxygen-substituted hexa-1,5-diyne unit in the 13-membered bridging ring was shown to be efficient. Besides providing a new triggering mechanism, the base-induced elimination of the O-mesylate derivative of this [2,3]-Wittig product generated a reactive enediyne that spontaneously cycloaromatized to a 1,4-diyl in which one of the phenyl radical centers was quenched internally (radical translocation).

The "enediyne" antibiotics calicheamicin γ_1^I (1), esperamicin-A_1 (2) and dynemicin-A (3), isolated during the mid 1980's from fermentation (culture) extracts of soil bacteria of widely diverse origin, are amongst the most potent anti-tumor agents actually known displaying *in vitro* and/or *in vivo* activity at ng/ml levels (IC_{50}'s) against a number of tumor systems (B16 melanoma, Moser human carcinoma, HCT-116 carcinoma, and normal and vincristine resistant leukemia) (Scheme 1).[1-3] The anti-tumor properties of these molecules derives from their capacity to cleave double stranded DNA by a totally unprecedented and very efficient mechanism. Calicheamicin and esperamicin cleave DNA by a sequence of events involving [1,2,4] initial site selective complexation with DNA and activation through nucleophilic attack on the novel trisulfide moiety, liberating a thiolate anion which reacts at C-1 of the conjugated enone system (Michael addition) giving **4**. This alters the conformation of the cyclohexanone ring, promoting Bergman type cycloaromatization [5] of the highly strained 3-ene-1,5-diyne bridge to a 1,4-phenylene diradical **5** which abstracts hydrogen from the ribose backbone of duplex DNA causing single and/or double strand breaks. In dynemicin-A, the anthraquinone component provides both the recognition element (inter-chelation) and the triggering device (bioreduction) for activating the molecule through epoxide ring opening.[3,6] Reaction of intermediate **6** with hydroxide ion alters the conformation of the D/E rings in such a way that the distance between the acetylene carbons C_{23} and C_{28} is diminished and the activation energy for Bergman cyclization is lowered to a point where cycloaromatization to **7** becomes spontaneous at physiological temperature.

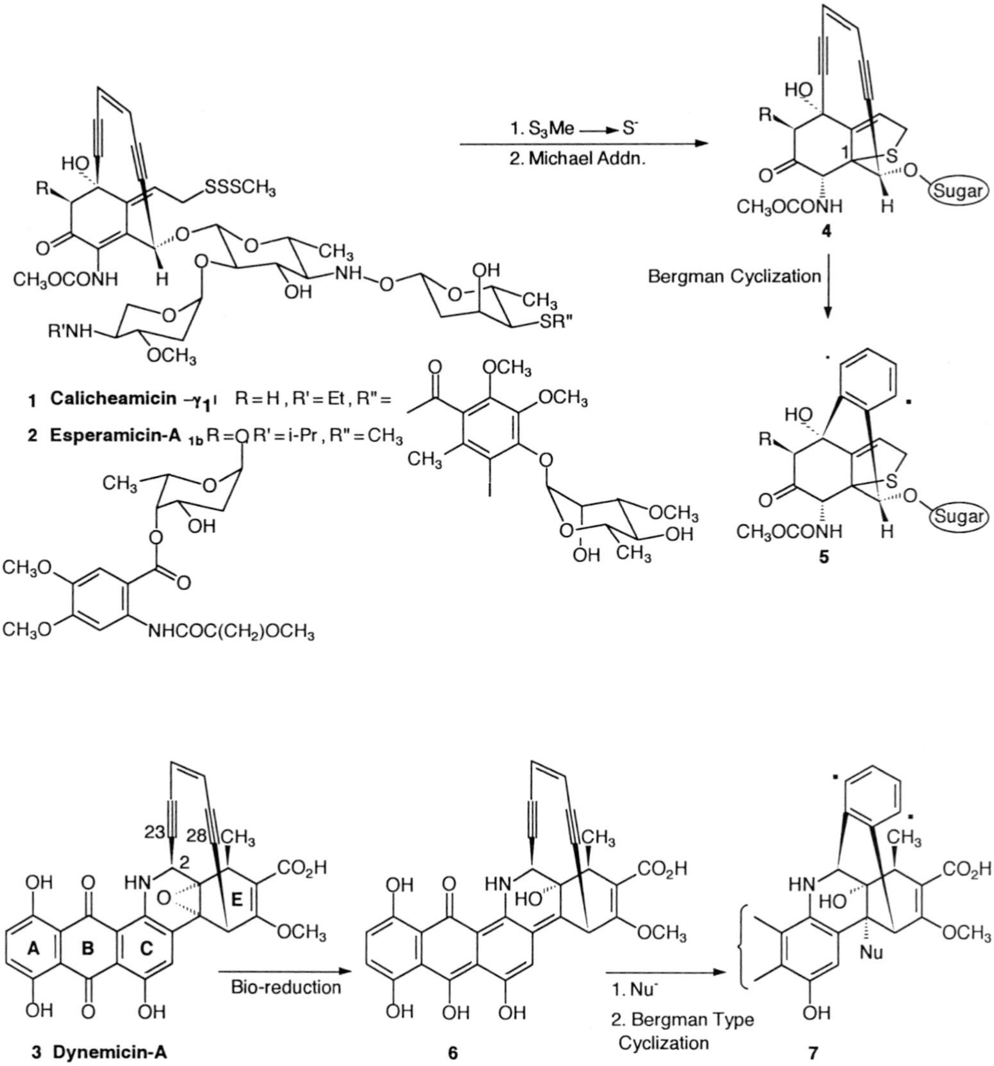

Scheme 1

From the synthetic viewpoint, the construction of the central core structure of these antibiotics can be looked upon in simple terms to involve two sequential condensations of the dianion **8** a of 3-ene-1,5-diyne with a highly functionalized, conformationally rigid keto-aldehyde such as **9** (Scheme 2). Indeed, the essential features of this strategy were incorporated into the elegant work by Danishefsky and Nicolaou *et al* in their respective syntheses of calicheamicinone.[7,8] Isobe and co-workers have also considered the route involving coupling of an anthraquinone to an intermediate related to **11** in their work on the synthesis of dynemicin-A.[9] At the time when we initiated our project on the synthesis of the enediyne antibiotics, methodology for the efficient coupling of the enediyne bridge to the six membered ring platform was lacking. In fact, due to the highly strained nature of the enediyne system in the core structure of these antibiotics, the second intramolecular condensation step alluded to in **8 + 9** still remains a problem.

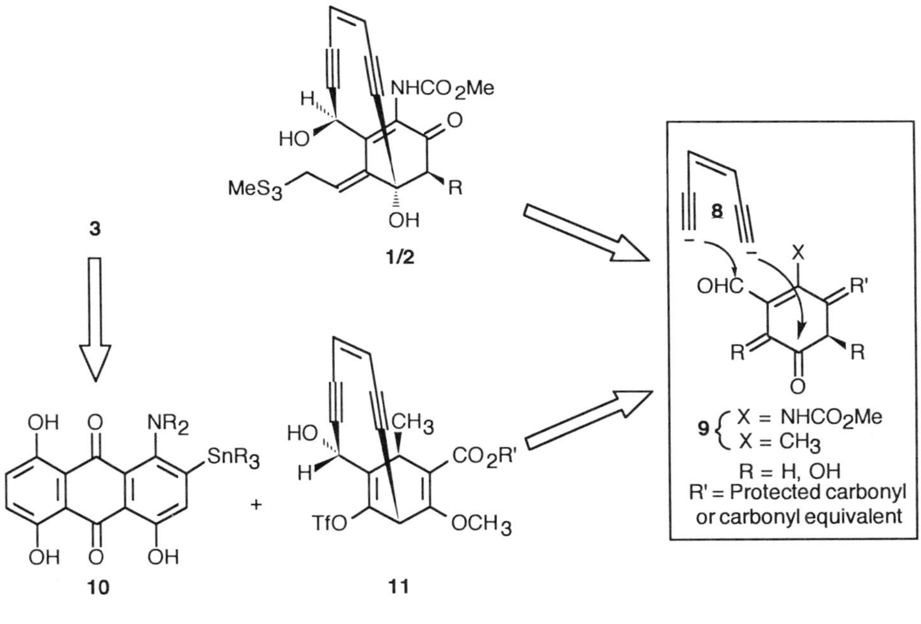

Scheme 2

Sensitive to this issue, our approach has been to study the construction of the strained 3-ene-1,5-diyne bridge through a [2,3]-Wittig rearrangement of an unstrained 13-membered precursor **12** (Scheme 3).[10,11] In the case where X = CH_2, this would result in formation of compound **13** with an exocyclic double bond. In principle, hydrolysis of the ketal function in **13** would be accompanied by double bond isomerization, leading to compound **15** which is stable to cycloaromatization at room temperature. Alternatively, the vinyl urethane intermediate **16** could be accessed directly from **12** by a new innovative version of the [2,3]-Wittig reaction in which (X = $NHCO_2CH_3$). Compounds **15** and **16** are advanced intermediates in our projected syntheses of dynemicin-A and calicheamicin/esperamicin, respectively.

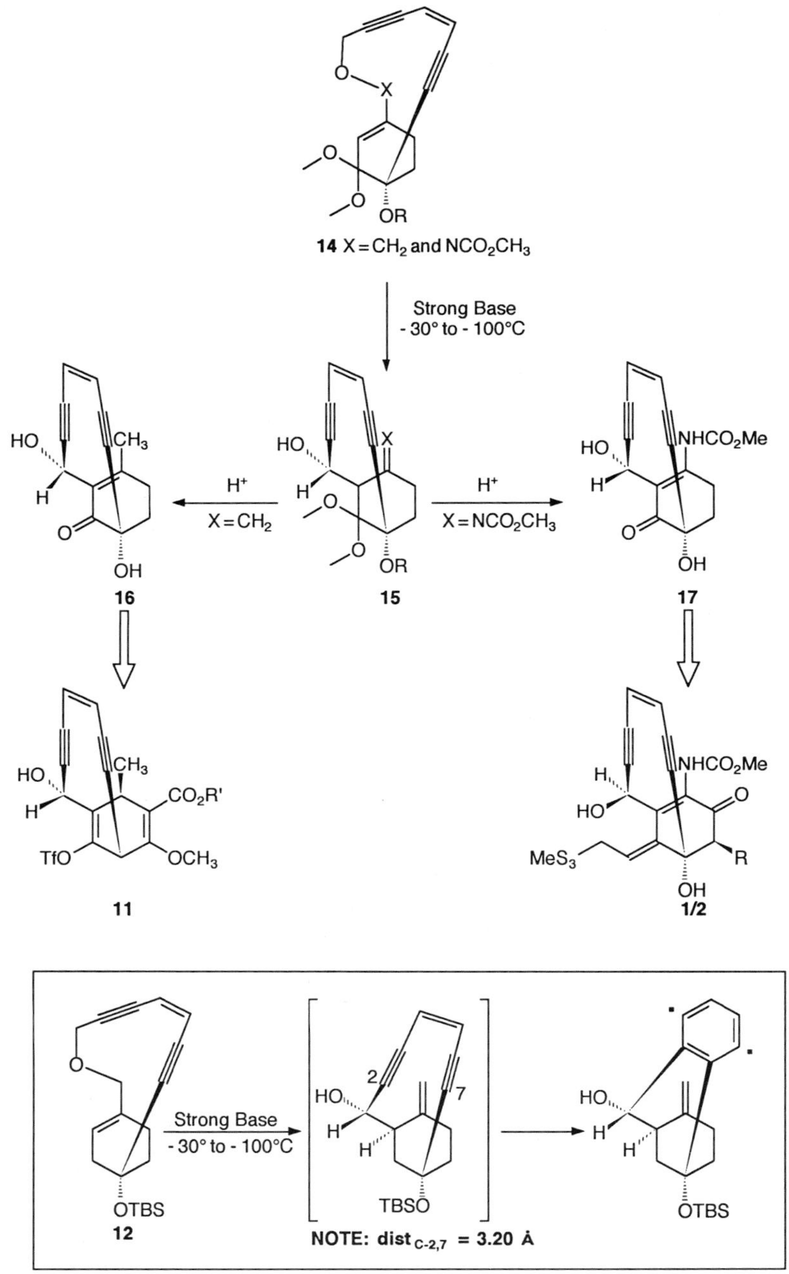

14 X = CH₂ and NCO₂CH₃

Strong Base
- 30° to - 100°C

16   **15**   **17**

H⁺
X = CH₂

H⁺
X = NCO₂CH₃

11   **1/2**

Strong Base
- 30° to - 100°C

12

NOTE: dist $_{C\text{-}2,7}$ = 3.20 Å

Scheme 3

In the first stage of this project, our objective was to prepare the macrocycle **17** and to study its [2,3]-Wittig rearrangement to **18**. Molecular mechanics calculations (MMX; $r_{2,7} = 3.20$ Å), and the results obtained on closely related model systems studied by Magnus et al. [12] strongly suggested that compound **18** should undergo ambient temperature Bergman cyclization to diradical **19**. The cyclohexen-3-one precursor **24** to **17** was prepared in five steps on a multigram scale as depicted in scheme 4. Particularily noteworthy was the observation that whereas the O-benzoate derivative of ketal **23** was readily hydrolyzed to the desired β,γ-unsaturated ketone **24**, the reaction of the corresponding O-acetate derivative of **23** under the same conditions led entirely to formation of the conjugated ketone **27**.

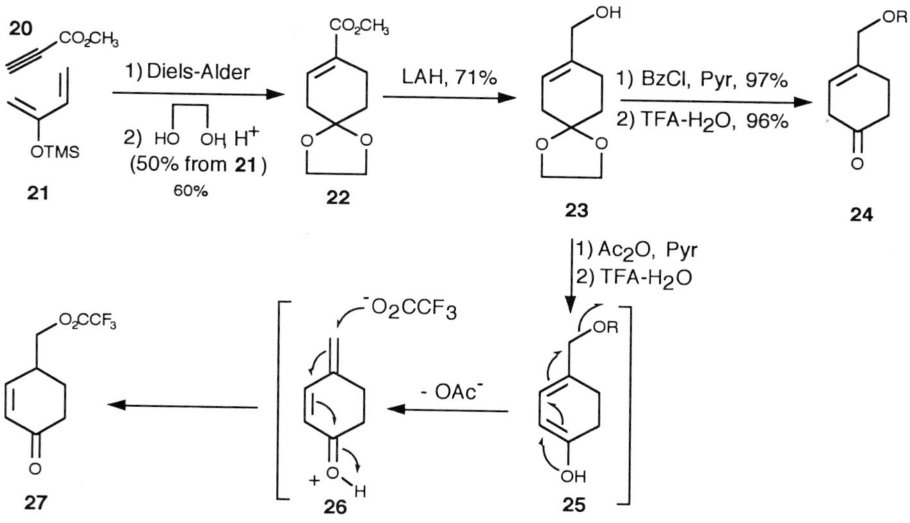

20 CO₂CH₃

1) Diels-Alder

2) HO   OH H⁺
(50% from **21**)
60%

OTMS

21

CO₂CH₃

22

LAH, 71%

OH

1) BzCl, Pyr, 97%
2) TFA-H₂O, 96%

23

OR

24

1) Ac₂O, Pyr
2) TFA-H₂O

O₂CCF₃

27

⁻O₂CCF₃

- OAc⁻

26

25

OR

OH

Scheme 4

Although the reaction of cyclohexenone **24** with the cerium reagent derived from a preformed O-THP protected 6-hydroxymethyl-3-ene-1,5-hexadiyne provided compounds of type **29** in a single step, the yields for the formation of the enediyne derivative and its condensation with **24** were low. No advantage was thus gained over the stepwise approach (Scheme 5) in which **24** was first converted to the monoacetylene substituted intermediate **28**. In **28**, the tertiary alcohol function was protected as its O-TBS ether primarily to reinforce the half chair conformation of the cyclohexene ring in which the acetylene substituent is axial. The other primary hydroxy group was, in contrast, liberated in order to avoid competing π-allyl palladium reactions during the subsequent Pd⁰ (Sonogashira) coupling reactions used to construct the enediyne system.[13] Conversion of compound **29** to bromide **30** was achieved in high overall yield using the standard protocol (1. MsCl-LiBr, 2. TBAF). It was interesting to observe that macrocyclization of this intermediate occurred only after several drops of water were added to the reaction mixture containing NaH in THF. With the desired 13-membered cyclic ether **17** in hand, its reaction to give **18** under standard [2,3] Wittig reaction conditions were examined, i.e. alkyllithium reagents in THF at low temperature (-78° to -100°C). These reactions proved to be visually very impressive. Over a span of less than 10 minutes, the reaction changed color from red-brown to blue to green (on work-up). TLC examination of the crude product mixture revealed the presence of at least forty components, none of which contained an aromatic ring in their structure (NMR). We deduced from these observations that the planar enediyne system reacts as an electron sink promoting a variety of single electron transfer reactions. Interestingly, using the non-nucleophilic base, lithium 2,2,6,6-tetramethylpiperidide (LiTMP) [11], the alternative [1,2]-Wittig rearrangement product was obtained.

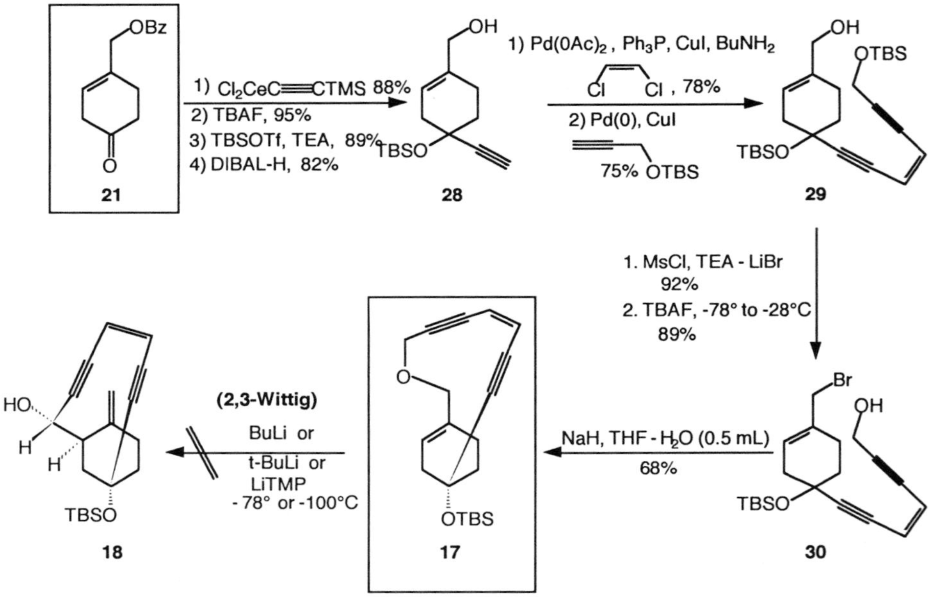

Scheme 5

To determine whether failure to obtain the desired intramolecular Wittig product was due to the reactivity of the enediyne unit rather than excessive strain in achieving the transition state for rearrangement, we embarked upon a study of the corresponding dihydro (1,5-diyne) intermediate **33**. The preparation of **33**, facilitated by the fact that 1,5-hexadiyne is commercially available, was achieved as illustrated in scheme 6. It was satisfying to observe that on reaction with LiTMP in THF at -25°C this dihydro intermediate underwent clean rearrangement to the bicyclic 10-membered ring product **34** (72%). It was apparent also that the C_4 proton in **34** is relatively acidic since in the presence of excess base it isomerizes to allene **35**. An X-ray structure of **33** was eventually obtained which clearly shows that the molecule adopts a conformation in which the C-1 methylene center sits almost directly over C-10 of the allylic double bond waiting for Wittig rearrangement to occur. NOE experiments indicate that this conformation is also preferred in solution. Compound **33** can thus readily reach a transition state on treatment with base where formation of the new C_1-C_{10} bond is synchronized with the removal of the pro-R proton at C-1.[14] This process leads to stereospecific formation of product **34**.

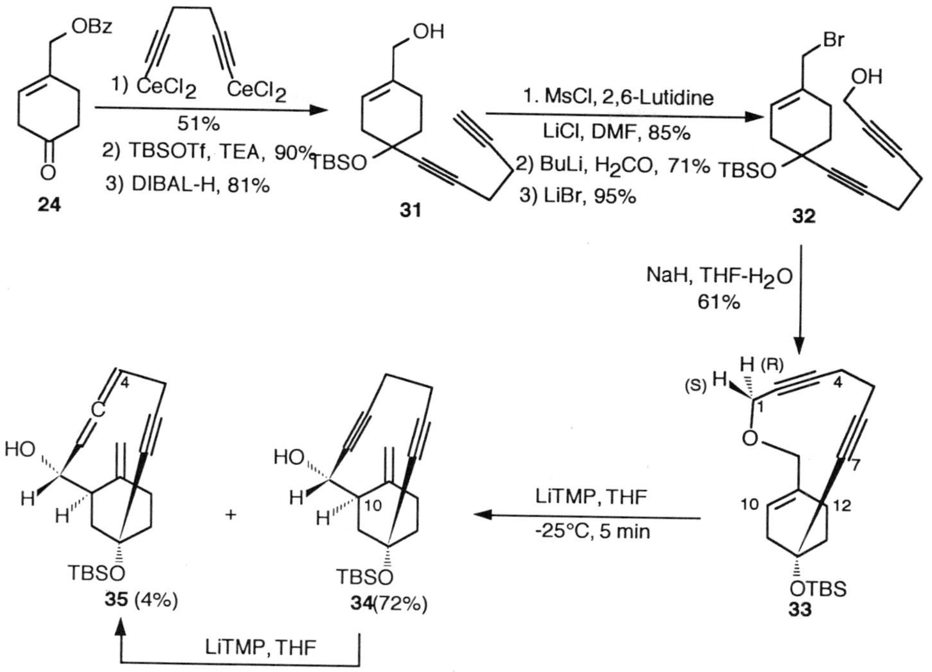

Scheme 6

Having shown that [2,3] Wittig ring contraction of macrocyclic ether **33** to the dihydro (1,5-diyne) compound **34** occurs in good yield, we conjectured that elimination of the elements of HX from the related dihydro derivative **36** , where **X** is a labile functional group under acid, base, or photochemical conditions, would provide a number of novel possibilities for triggering the generation of highly reactive 1,4-benzenoid diradical intermediates (Scheme 7). Our efforts in this direction involved the synthesis of compound **36** (X = OMs) and a study of its base-induced cycloaromatization reactions. Starting with alkyne **37**, the cyclic ether **40** was obtained as a 1:1 epimeric mixture in 6 steps by what are now well established reactions. [2,3]-Wittig rearrangement of **40** proved once again to be efficient, giving the [7.3.1] bicyclic product **41** in 62% yield. In the cycloaromatization experiments, the mixture of compounds **41** was reacted with 4 equivalents of DBU in THF:1,4-cyclohexadiene (CHD) [3:1], without precautions to exclude oxygen. A series of products was formed from which compound **42**, and in particular, compounds **43** and **44** were isolated by column chromatography. Although the formation of compound **42** can be explained by quenching the intermediate diradical species with CHD, we suspected that radical translocation by intramolecular 1,5-hydrogen transfer from the O-methyl to the C-3 center of the intermediate diyl was involved in the formation of **43** and **44**. Indeed, such hydrogen atom transfers are precedented, and more specifically have been reported by Bergman and coworkers in cycloaromatization studies on simple Z -enediynes.[15]

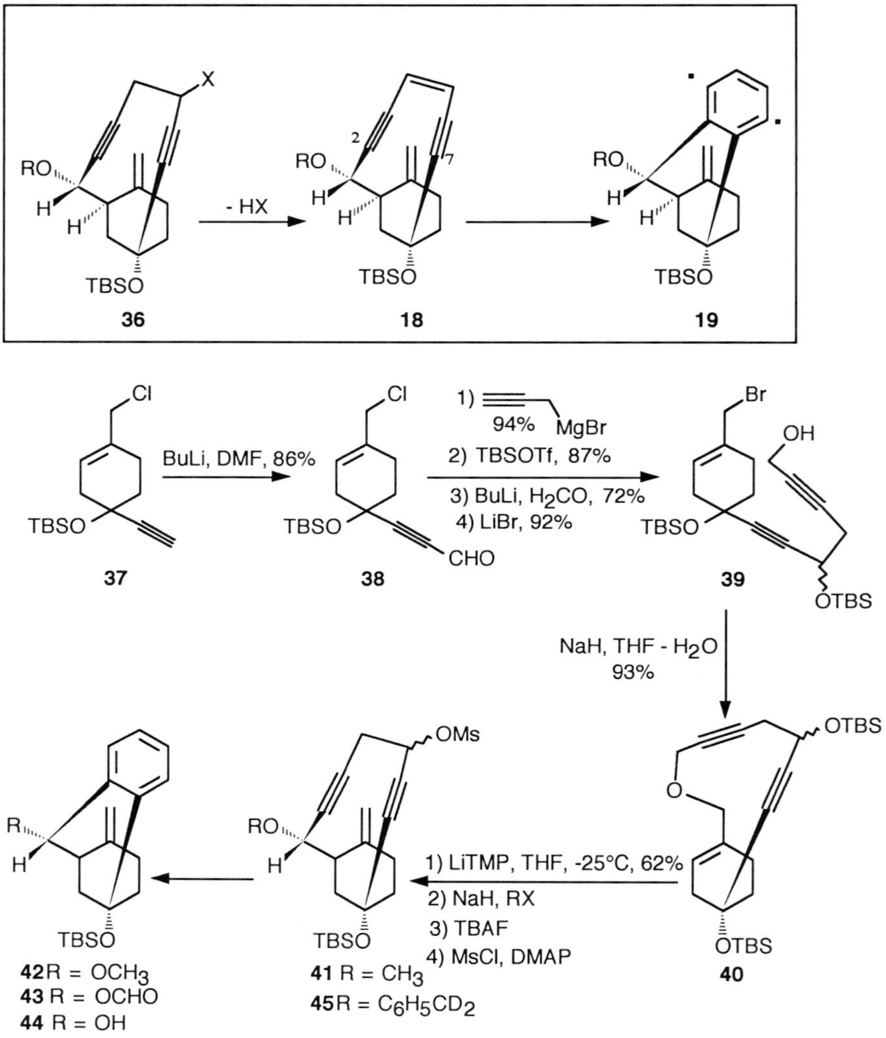

Scheme 7

To confirm this hypothesis, O-mesylate elimination was studied using the deuterium-labelled compound **45** (Scheme 8). As anticipated, deuterium incorporation at C-3 of the phenyl ring was observed in the three major reaction components, benzoate **46**, alcohol **47** and ketone **48**. Several pathways can be proposed for the generation of these products from diyl **49**. One scenario, which agrees with the experimental data, involves radical translocation giving intermediate **50**, which in the presence of oxygen forms the hydroperoxide **51**. Subsequent base-promoted fragmentation (pathway **a,c**) of **51** would afford compounds **46** and **48**, whereas reduction of the hydroperoxide to the corresponding hemiketal followed by hydrolysis would give the debenzylated product **47** (pathway **b**). Recently both Wender [16] and Goldberg [17] have provided experimental evidence that internal 1,5-hydrogen atom transfer of the activated neocarzinostatin systems does occur. Our results demonstrate that simple analogues of calicheamicin also undergo these events, implying that the choice of substituent at the C_1-OH group is important for the generation of a fully effective 1,4-benzenoid diradical.

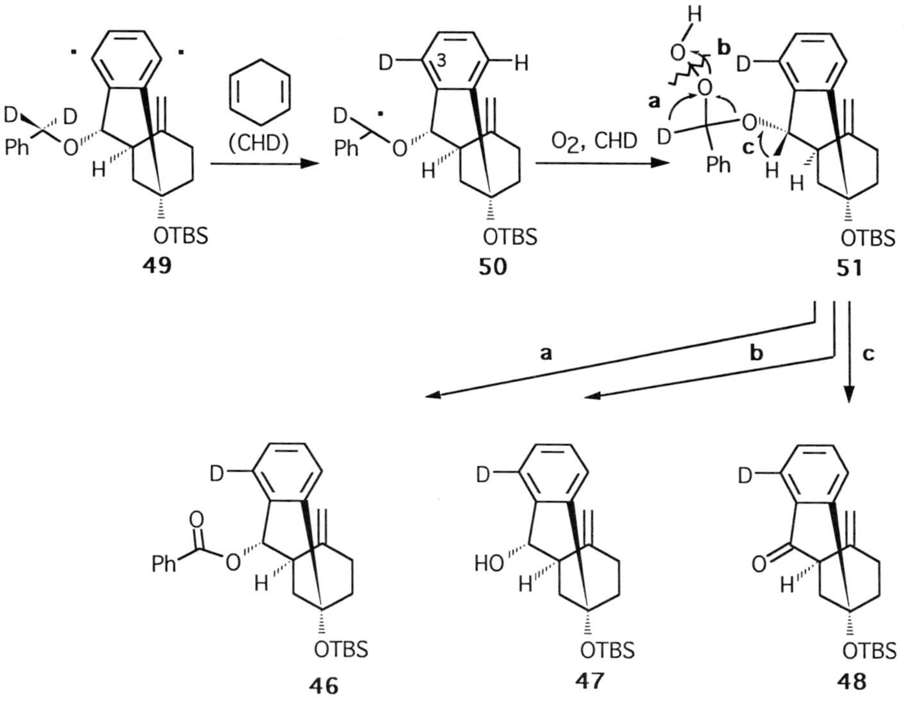

Scheme 8

At present, work on the alternate [2,3]-Wittig approach leading to the vinylogous urethane of type **16** (Figure 3) has reached the point where preparation of the oxime ether **54** has been achieved (Scheme 9). There is literature precedence indicating that N-acylation of **54** to give **55** will be possible.[18] The projected ring contraction of this intermediate to the N-acylenamine **56** (or its 11,12-double bond isomer) will provide an interesting test of the mechanism of the [2,3]-Wittig rearrangement. In view of the concerted nature of this reaction it is entirely feasible that rearrangement to **56** will be observed rather than formation of compound **57** resulting from an E₁cb elimination process. Time and careful experimentation will determine whether this reaction can be controlled in the desired sense.

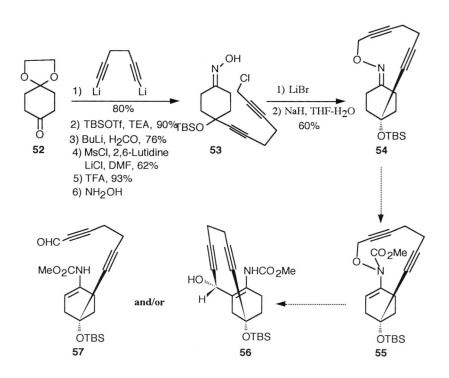

Scheme 9

References

1.   a) M. D. Lee, T. S. Dunne, M. M. Siegel, C. C. Chang, G. O. Morton, D. B. Borders, *J. Am. Chem. Soc.* **109** (1987) 3464-3466; b) M. D. Lee, T. S. Dunne, C. C. Chang, G. A. Ellestad, M. M. Siegel, G. O. Morton, W. J. McGrahren, D. B. Borders, *J. Am. Chem. Soc.* **109** (1987) 3466-3468.

2.   a) J. Golik, J. Clardy, G. Dubay, G. Groenewold, H. Kawaguchi, M. Konishi, B. Krishnan, H. Ohkuma, K. Saitoh, T. W. Doyle, *J. Am. Chem. Soc.* **109** (1987) 3461-3462; b) J. Golik, G.

Dubay, G. Groenewold, H. Kawaguchi, M. Konishi, B. Krishnan, H. Ohkuma, K. Saitoh, T. W. Doyle, *J. Am. Chem. Soc*. **109** (1987) 3462-3464.

3.   M. Konishi, H. Ohkuma, K. Matsumoto, T. Tsuno, H. Kamei, T. Miyaki, T. Oki, H. Kawaguchi, G. D. VanDuyne, J. Clardy, *J. Am. Chem. Soc*. **112** (1990) 3715-3716.

4.   M. D. Lee, G. A. Ellestad, D. B. Borders, *Acc. Chem. Res*. **24** (1991) 235-XXX.

5.   R. G. Bergman, *Acc. Chem. Res*. **6** (1973) 25-31.

6.   a) Y. Sugiura, T. Shiraki, M. Konishi, T. Oki, *Proc. Natl. Acad. Sci*. **87** (1990) 3831-3835; b) Y. Sugiura, T. Arakawa, M. Uesugi, T. Shiraki, H. Ohkuma, M. Konishi, *Biochemistry* **30** (1991) 2989-2992.

7.   J. N. Haseltine, M. Paz Cabal, N. B. Mantlo, N. Iwasawa, D. S. Yamashita, R. S. Coleman, S. J. Danishefsky, G. K. Schulte, *J. Am. Chem. Soc*. **113** (1991) 3850-3866.

8.   A. L. Smith, C.-K. Hwang, E. Pitsinos, G. R. Scarlato, K. C. Nicolaou, *J. Am. Chem. Soc*. **114** (1992) 3134-3136.

9.   T. Nishikawa, M. Isobe, T. Goto, *Synlett* **1991**, 393-395.

10   K. Mikami, T. Nakai, *Synthesis* **1991**, 594-XXX.

11.   J. A. Marshall, J. Lebreton, B. S. DeHoff, T.M. Jenson, *J. Org. Chem*. **52** (1987) 3883-3889.

12.   P. Magnus, P. Carter, J. Elliot, R. Lewis, J. Harling, T. Pitterna, W. E. Bauta, S. Fortt, *J. Am. Chem. Soc*. **114** (1992) 2544-2559.

13   a) K. Sonogashira, Y. Tohda, N. Hagihara, *Tetrahedron Lett*. **1975**, 4467-XXX; b) D. Guillerm, G. Linstrumelle, *Tetrahedron Lett* **26** (1985) 3811-3812.

14.   J. A. Marshall, J. Lebreton, B. S. DeHoff, T.M. Jenson, *J. Org. Chem*. **55** (1990) 1421-1423.

15.   T. P. Lochart, P. B. Comita, R. G. Bergman, *J. Am. Chem. Soc*. **103** (1981) 4082-4090.

16.   P. A. Wender, M. J. Tebbe, *Tetrahedron Lett*. **32** (1991) 4863-4866.

17.   D.-H. Chin, I. H. Goldberg, *J. Am. Chem. Soc*. **114** (1992) 1914-1915.

18.   V. J. Lee, R. B. Woodward, J. Org. Chem. **44** (1979) 2487-2489.

Applications of Allylsilanes in Natural Product Synthesis

Dieter Schinzer

Institut für Organische Chemie der Technischen Universität Braunschweig, Hagenring 30,
W-3300 Braunschweig, Germany

Summary: The synthesis of biologically interesting molecules by the use of intramolecular additions of allyl- and propargylsilanes to enones and dienones is described. The organometallic routes approaching these systems use iron-, copper-, silicon-, and palladium-mediated reactions which proceed in a regio- and diastereoselective way.

Introduction

Almost 15 years ago Sakurai and Hosomi published a pioneering paper on the intermolecular addition of an allylic silane to α,β-unsaturated ketones in the presence of titanium tetrachloride as the Lewis acid. [1-4]

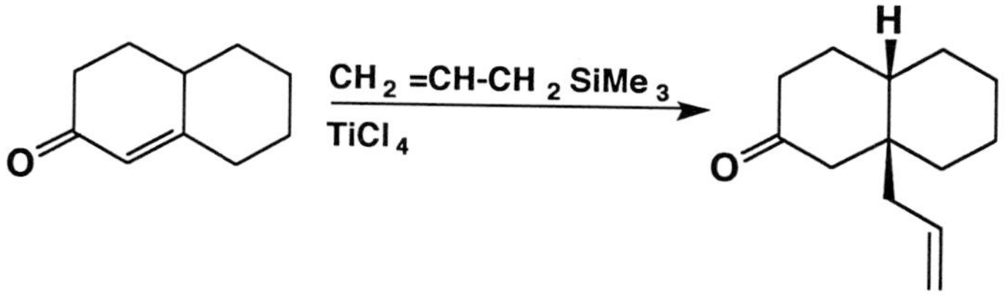

Scheme 1

The reaction was shown to proceed with exclusive production of the cis-fused decalone by 1,4-addition in high chemical yield. Sakurai reactions proceed regiospecifically with a large variety of electrophiles due to the so-called ß-effect. [5-9]

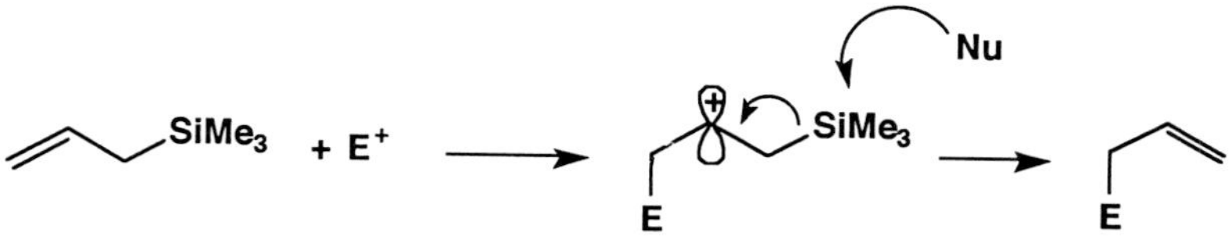

Scheme 2

For almost 10 years, several research groups have focused their interest on intramolecular additions of allyic and propargylic silanes to enones. In early studies a variety of Lewis acids were tested and reaction conditions were optimized to make this reaction a powerful tool to synthesize spiro-annulated and 1,2 fused ring systems. This methodology has found interesting applications in the total synthesis of natural products. [10,11]

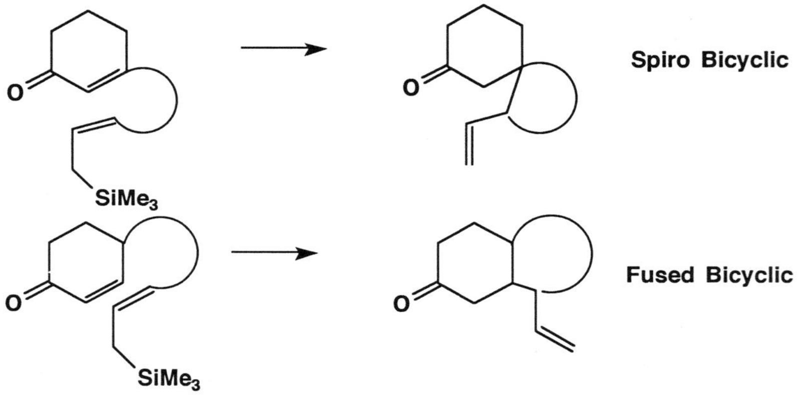

Scheme 3

Stereoselective Approach to Enedinye Systems: Allylsilane-Terminated Cyclization

In this account I will focus first on the synthesis of calicheamicin model systems. [12] Antibiotics of the calicheamicin and esperamicin natural product class [13-15] have had an enormous impact on organic synthesis during the last few years. [16,17] Based mainly on the pioneering work of Nicolaou et al., [18-20] many research groups throughout the world have focused on the synthesis of simplified model compounds that might be powerful reagents for DNA scission via the Bergman cyclization of the enediyne moiety. [21-23]

Our approach to the synthesis of large rings of this type uses iron-, copper-, silicon-, or palladium-mediated reactions; in the latter a highly conjugated allylsilane is used, for the first time under Lewis acid conditions, to construct the bicyclic enediyne unit.

In a convergent approach we first used an iron-mediated coupling [24] of **1** with the lithium salt of trimethylsilylacetylene to yield compounds **2a** and **2b** in a ratio of about 3:1. The observation of the endo product **2b** represents a very rare case of an endo attack of the iron carbonyl complex by a lithium reagent via pre-complexation of the lithium compound with the iron complex. [25]

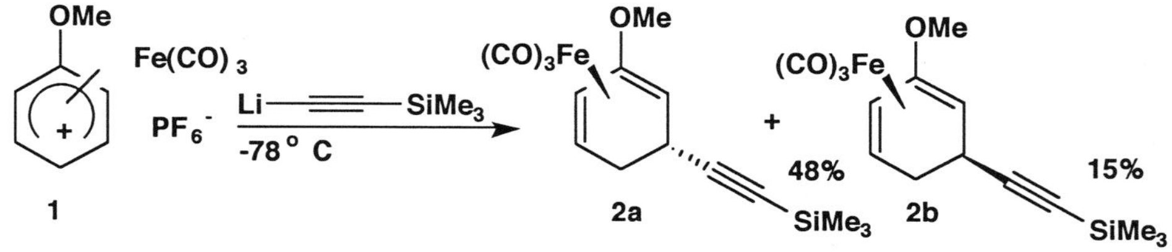

The addition of the softer cuprate reagent of trimethylsilylacetylene afforded almost quantitatively the predicted exo compound **2a** in 98% isolated yield. [26] The relative configuration has been confirmed by x-ray analysis. [27] Desilylation under standard conditions generated the terminal acetylene **3**, which can be used for further couplings. Oxidative demetallation with Fe(NO3)3 · 9 H2O yielded directly the enone **4**.

The functionalized allylsilane **7** has been synthesized in a straightforward way in a one pot procedure starting with a cross-coupling [28] of the allylsilane **5** and tributylstannyl acetylene to afford intermediate **6**, which was directly transformed into compound **7**.

The other building block of the desired enediyne unit was obtained by a Lewis acid-promoted Sakurai reaction with the highly conjugated allylsilane **7**, which proceeded regio- and stereospecifically in the presence of titanium tetrachloride to give compound **8** as a single diastereomer in 68% yield. [29]

Closure of the 11-membered ring was achieved by a palladium-catalyzed reaction in the presence of CuI and BuNH₂. [30] In order to avoid problems arising from the basic reaction conditions, ketone **8** was ketalized by a transketalization in the presence of Amberlyst 15 to yield **9**. [31]

Compound **10** represents the first *trans*-fused enediyne in this series that is stable at room temperature with respect to the Bergman cyclization. [32] The stability of **10** was also predicted by force field calculations using MacroModel and the boat geometry of **10** follows from coupling constants and force field calculations. [32]

The stereoselective reaction sequences presented here demonstrate again the ability of allylsilane chemistry to synthesize highly functionalized large-sized rings under mild conditions in a diastereoselective way. Further investigations, especially of other ring-sizes and biological activities, are currently under way and will be reported in due course.

Stereoselective Approach to Pinguison: Propargylsilane-Terminated Cyclization

In the second part of this chapter, a stereochemical solution to the total synthesis of the natural product pinguison is presented. [33] The compound shows antibiotic properties and has an unusual tricyclic terpene skeleton with four adjacent methyl groups. The retrosynthetic analysis shown in **scheme 4** uses a propargyl silane-terminated cyclization as the key step of the synthesis.

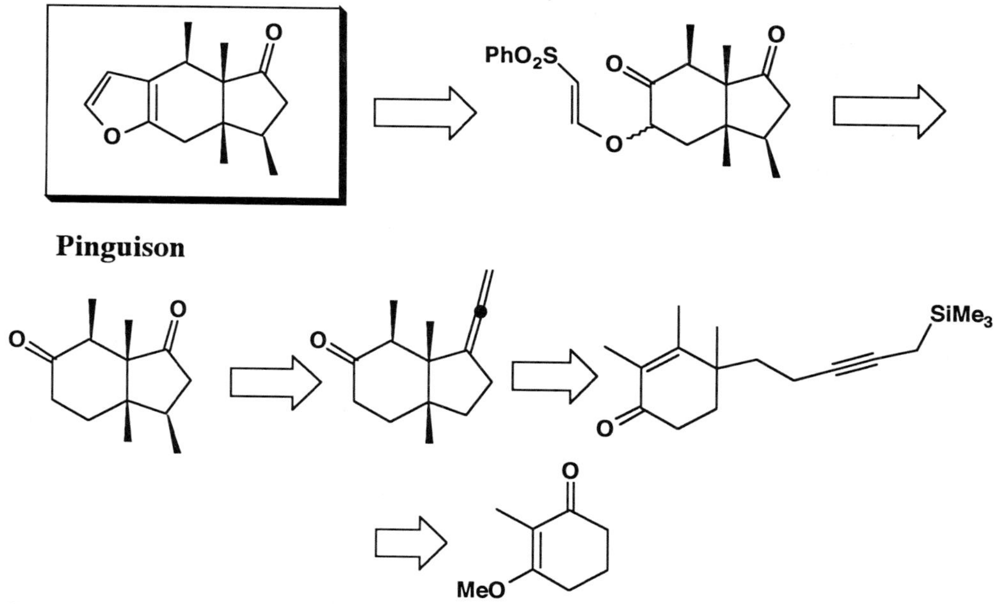

Pinguisone

Scheme 4

The desired starting material **11** for the key cyclization was derived from vinylogous esters by procedures published earlier. The cyclization proceeded smoothly by the use of titanium tetrachloride as the Lewis acid catalyst [34] Two quaternary centers can be set in one step as shown in compound **12** and the third stereogenic center is established during ketalization under thermodynamic conditions. The terminal allene which resulted from the propargylsilane-terminated cyclization was transformed via ozonolysis into the desired ketone **14** which was subsequently transformed to the enone **15**. [35]

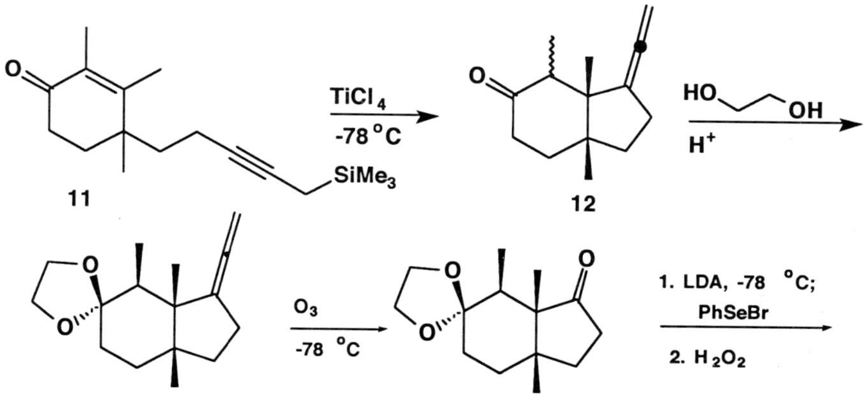

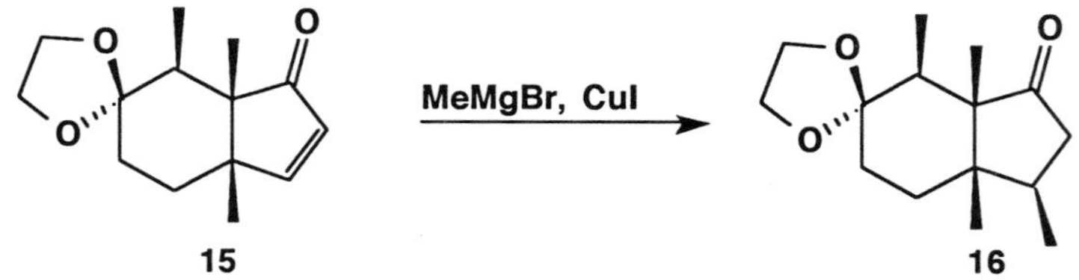
O
O
O
MeMgBr, CuI
O
O
O

15 16

The missing methyl group was added by a copper-catalyzed Grignard addition to yield **16** which established the last stereochemical center by an exo-addition to the bicyclic ketone. The stereochemical outcome has been confirmed by x-ray analysis. [27,35]

Stereoselective Approach to Pinguison: Thexyldimethylsilyl-Mediated Cyclization

In order to approach the natural product, the furan ring has to be annulated starting from the α-hydroxy ketone. Therefore we have focused on a second approach with the hydroxyl group already attached to the cyclization substrate. The key reaction involves a novel vinylogous Dieckmann condensation starting from silyl dienolether **19**. [36]

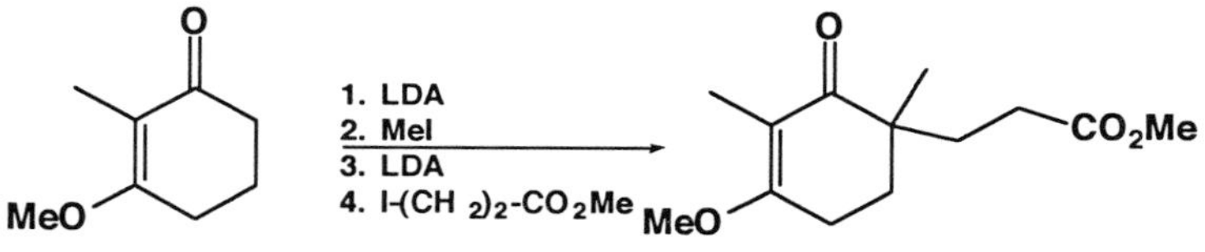
O
MeO
1. LDA
2. MeI
3. LDA
4. I-(CH 2)2-CO2Me
O
CO2Me
MeO

17 18

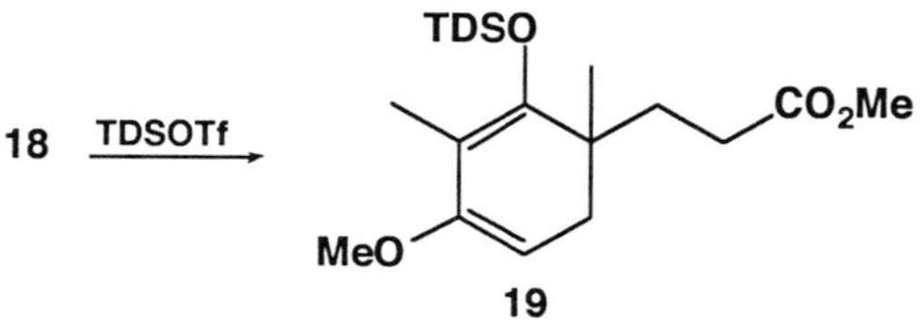
18
TDSOTf
TDSO
CO2Me
MeO

19

A simple alkylation sequence generates the desired vinylogous ester **18** which was cyclized in the presence of thexyldimethylsilyl triflate or titanium tetrachloride to obtain the stereochemically homogeneous compound **20.** [36]

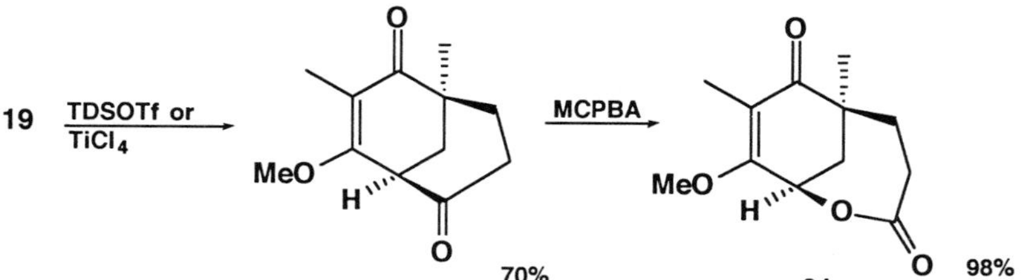
19
TDSOTf or
TiCl4
O
MeO
H
O
70%
MCPBA
O
MeO
H
O
O
98%

20 21

The 1,3 annulated ketone **20** can be lactonized by a standard Baeyer-Villiger protocol to afford the lactone **21** in quantitative yield. [36] A further critical step was the reduction of **21** to the aldehyde **22** which was needed to elaborate the side chain.

The reagent of choice for the selective reduction of this lactone in the presence of a vinylogous ester is diisoamyl borane. [37] The aldehyde **23** was further transformed into the terminal alkyne **24** by treatment with a lithiated diazomethane reagent. [38] The final step to elaborate the desired propargylsilane **25** can be envisioned as an alkylation sequence.

These two strategies demonstrate how flexible silyl-terminated cyclizations can be used in organic synthesis. The first approach shows the striking advantages over classical methods. Even at low temperature two quaternary centers in a row can be set in one step. The second approach uses a novel vinylogous Dieckmann condensation to construct a 1,3-annulated bicyclic ketone as a precursor to attach the critical oxygen atom via a Baeyer-Villiger lactonization. Intermediate **25** then enters into the appoach discussed first.

Stereoselective Total Synthesis of Sesquiterpenoid AE 1: Allylsilane-Terminated Dienone Cyclization

In the final part of this chapter we report he total synthesis of sesquiterpenoid AE 1 **26** which is synthesized by a diastereoselective dienone cyclization with a functionalized allylic silane as the key compound. [39] Sesquiterpenoid AE 1' **26** is a component of the defense secretion of the soldier caste of *Amitermes evuncifer,* a termite species native to several areas of West Africa and particularly to parts of

Nigeria where it is known to reach pest status by inflicting damage on crops and property. [40] The structure has been confirmed by an asymmetric total synthesis using a Diels-Alder approach starting with (-) carvone [40] and (+)-carvone. [41]

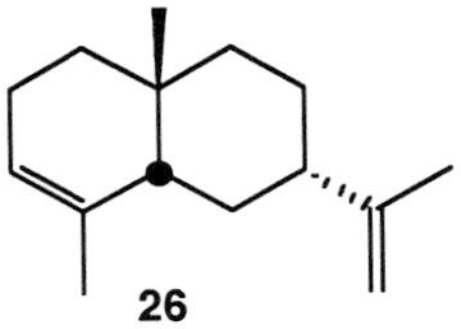

26

Sesquiterpenoid AE 1

Starting from 5-methylresorcinol **27** we obtained **26** in 11 steps employing an allysilane-terminated dienone cyclization as the key step.

Compound **27** was hydrogenated over Raney nickel under basic conditions to yield the 1,3-diketone **28** in 80% yield. The diketone **28** was smoothly transformed into the vinylogous ester **29** with ethanol and triethyl orthoformate under acid catalysis in 99% yield.

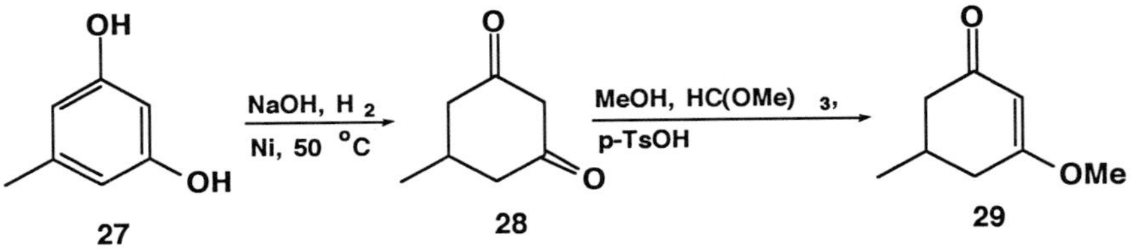

Alkylation of the preformed ester enolate (LDA in THF at low temperature) with 3-methyl-2-butenyl bromide (prenyl bromide) produced the alkylated material **30** again in excellent yield (94%). The stereochemical outcome of the alkylation depends on the reaction conditions, in particular, the reaction temperature. At -78 °C the *trans:cis* ratio was 4:1 (**30a:30b**). Cooling to -105 °C increased the ratio to 6.5:1. Since this is not crucial for the total synthesis (the stereocenter will be destroyed later), we did not optimize this selectivity.

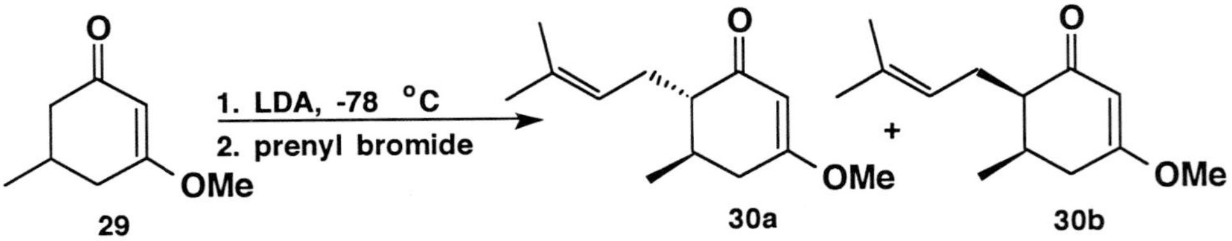

Subsequent chemoselective ozonolysis at -78 °C and reductive workup with zinc and acetic acid led to the aldehyde **31** (78%).

Wittig reaction of **31** with the in situ prepared ß-silylated ylid resulted in the generation of the functionalized allylsilane **32**, which was obtained in 50% yield. [42,43]

Compound **32** was then treated with vinyllithium in ether to yield the key dienone **33** in 73% yield after acid hydrolysis under mild conditions.

Cyclization at -78 °C with 1.1 eq of EtAlCl$_2$ as the Lewis acid catalyst yielded the bicyclic enone **34** diastereoselectively with three defined stereogenic centers as a single diastereomer in 64% yield.

Introduction of the methyl group into compound **34** was accomplished by lithium tetramethylaluminate in the presence of Ni(acac)$_2$ to yield exclusively compound **35** with a ß-methyl group by exo attack from the organometallic reagent (73%). [44]

We elected to examine the selenation of **35** by regioselective deprotonation with LDA at low temperature followed by addition of phenylselenium bromide. To our surprise, this reaction worked very well, and we obtained only the desired compound **36** as a mixture of diastereomers (68% yield).

324   D. Schinzer

Subsequent oxidation of **36** with H₂O₂ at 0 ℃ yielded the enone **37** in 47% yield as a stereochemically homogeneous material.

Subjection of **36** to the procedure developed by Winterfeldt [45] provided **26** after reduction with NaBH₄ in the presence of trifluoroacetic acid in acetonitrile. Compound **26** was identical in all compared respects with that reported for the natural material.

Overall, the conversion of a simple aromatic compound into a terpene with three defined chiral centers again represents an interesting use of intramolecular addition chemistry with allylic silanes.

Further investigations to synthesize the more complex oxygen functionalized natural products of this family are in progress and will be reported in due course.

Conclusions

Allylic and propargylic silanes have been shown to be powerful terminators to solve various problems in organic synthesis. Medium and large rings can be synthesized diastereoselectively in the presence of other functional groups. The methodology described in this account can be used to approach several classes of natural products.

Acknowledgments: I wish to thank my talented coworkers who are mentioned in the references. Financial support from the Deutsche Forschungsgemeinschaft, the Fonds der Chemischen Industrie, and the Volkswagen-foundation is greatly acknowledged.

References

1. Hosomi, A., Sakurai, H. *J. Am. Chem. Soc.* **1977**, 99, 1673.

2. Sakurai, H. *Pure and Appl. Chem.* **1982**, 54, 1.

3. Fleming, I., Dunogues, J. *Organic Reactions* **1989**, 37, 57.

4. Majetich, G. in *Organic Synthesis, Theory and Applications:* T. Hudlicky, Ed., Jai Press, Inc., Greenwich, CT, **1989**, 173.

5. Eaborn, C. *J. Chem. Soc., Chem. Commun.* **1977**, 1255.

6. Chandrasekhar, J., Jorgensen, W. L. *J.Am.Chem.Soc.* **1985**, 107, 1496.

7. Hanstein, W., Traylor, T.G. *Tetrahedron Lett.* **1967**, 4451.

8. Apeloig, Y., Stanger, A. *J. Am. Chem. Soc.* **1985**, 107, 2806.

9. Mayr, H., Pock, R. *Tetrahedron* **1986**, 42, 4211.

10. Schinzer, D. *Synthesis*, **1988**, 263.

11. Majetich, G., Hull, K., Lowery, D., Ringold, C., Defauw, J. in *Selectivities in Lewis Acid-Promoted Reactions*; Schinzer, D., Ed.; Kluwer Academic Publishers Group: Dordrecht, Holland; 1989.

12. Schinzer, D.; Kabbara, J. *Synlett* **1992**, 766.

13. Konishi, H.; Ohkuma, H.; Saitoh, K.; Kawaguchi, H.; Golik, J.; Dubay, G.; Groenewold, G.; Krishnan, B.; Doyle, T. W. *J. Antibiot.* **1985**, 38, 1605.

14. Golik, J.; Clardy, J.; Dubay, G.; Groenewold, G.; Kawaguchi, H.; Konishi, M.; Krishnan, B.; Ohkuma, H.; Saitoh, K.; Doyle, T. W. *J. Am. Chem. Soc.* **1987**, 109, 3461.

15. Golik, J.; Dubay, G.; Groenewold, G.; Kawaguchi, H.; Konishi, M.; Krishnan, B.; Ohkuma, H.; Saitoh, K.; Doyle, T. W. *J. Am. Chem. Soc.* **1987**, 109, 3462.

16. Danishefsky, S. J.; Mantlo, N. B.; Yamashita, D. S. *J. Am. Chem. Soc.* **1988**, 110, 6890.

17. Haseltine, J. N.; Cabal, M. P.; Mantlo, N. B.; Iwasawa, N.; Yamashita, D. S.; Coleman, R. S.; Danishefsky, S. J.; Schulte, G. K. *J. Am. Chem. Soc.* **1991**, 113, 3850.

18. Smith, A. L.; Hwang, C.-K.; Pitsinos, E.; Scarlato, G. R.; Nicolaou, K. C. *J. Am. Chem. Soc.* **1992**, 114, 3134.

19. Nicolaou, K. C.; Ogawa, Y.; Zuccarello, G.; Kataoka, H. *J. Am. Chem. Soc.* **1988**, 110, 7247.

20. Review: Nicolaou, K.C.; Dai, W.-M *Angew.Chem.* **1991**, 103, 1453, *Angew.Chem.Int.Ed.Engl.* **1991**, 30, 1387.

21. Jones, R. R.; Bergman, R. G. *J.Am.Chem.Soc.* **1972**, 94, 660.

22. Zein, N.; Sinha, A. M.; McGahren, W. J.; Ellestrad, G. A. *Science* **1988**, 240, 1198.

23. Zein, N.; Poncin, M.; Nilakantan, R.; Ellestrad, G. A. *Science* **1989**, 244, 697.

24. Birch, A. J.; Haas, M. A. *J. Chem. Soc. C* **1971**, 2465; very recent review: Knölker, H.-J. *Synlett* **1992**, 371.

25. Pearson, A. J. In *Comprehensive Organometallic Chemistry*, Wilkinson, G.; Stone, S. G. A.; Abel, E. W.; Eds.: Pergamon: Oxford, 1982; Vol. 8, Chap. 58, p. 939.

26. Lipshutz, B. H.; Wilhelm, R. S.; Kozlowski, J. A. *Tetrahedron* **1984**, 24, 5005.

27. We express our thanks to Professor Peter G. Jones, Institute of Inorganic and Analytical Chemistry, Technical University of Braunschweig, for the X-ray crystal structure determinations of compounds **3** and **16**.

28. Ratovelomanana, V.; Linstrumelle, G. *Tetrahedron Lett.* **1985**, 25, 6001.

29. Blumenkopf, T. A.; Heathcock, C. H. *J. Am. Chem. Soc.* **1983**, 105, 2354.

30. Kalivretenos, A.; Stille, J. K.; Hegedus, L. S. *J. Org. Chem.* **1991**, 56, 2883.

31. Schinzer, D.; Kabbara, J.; Ringe, K. *Tetrahedron Lett.* **1992**, in print.

32. W. C. Still´s program **MacroModel** (version 3.0) was used. Distance A/B: 2.62 Å; C/D: 3.37 Å. See also: Nicolaou, K. C.; Zuccarello, G.; Ogawa, Y.; Schweiger, E. J.; Kumazawa, T. *J. Am. Chem. Soc.* **1988**, 110, 4866.

33. Benesová, V.; Herout, V.; Sorm F. *Collection Czechoslov. Chem. Commun.* **1969**, 34, 1810.

34. Schinzer, D.; Dettmer, G.; Ruppelt, M.; Sólyom, S.; Steffen, J. *J. Org. Chem.* **1988**, **53**, 3823.

35. Voß, H., PhD thesis, University of Hannover, 1991.

36. Schinzer, D.; Kalesse, M. *Tetrahedron Lett.* **1991**, 4691.

37. Preiß, M., PhD thesis, Technical University of Braunschweig, 1993.

38. Ohira, S.; Okai, K.; Moritani, T. *J. Chem. Soc., Chem. Commun.* **1992**, 721.

39. Schinzer, D.; Feßner, K.; Ruppelt. M. *Liebigs Ann. Chem.* **1992**, 139.

40. Baker, R.; Evans, D. A.; McDowell, P. G. *Tetrahedron Lett.* **1978**, 4073.

41. Caine, D.; Stanhope, B. *Tetrahedron* **1987**, 43, 5545.

42. Seyferth, D.; Wursthorn, K. R.; Mammarella, R. E. *J. Org. Chem.* **1977**, 42, 3104.

43. Fleming, I.; Paterson I. *Synthesis* **1979**, 446.

44. Ashby, E. C.; Heinsohn, G. *J. Org. Chem* . **1974**, 39, 3297; Bagnell, L.; Jeffery, E. A.; Meisters, A. Mole, T. *Aust. J. Chem.* **1975**, 28, 801; T. Mole *ibid* **1975**, 28, 817.

45. Meyer, T., PhD thesis, University of Hannover, 1991.

The Use of Carbohydrate Chemistry in Antibiotic Research

Pierre Sinaÿ

Ecole Normale Supérieure, Laboratoire de Chimie, URA 1110, 24 rue Lhomond, 75231 Paris cedex 05

Summary: Reductive samariation of glycosyl phenyl sulfones in the presence of SmI_2/ THF/ HMPA and a carbonyl derivative selectively afforded a C-glycoside. This reaction complements the reductive lithiation of the anomeric center. The direct radical type condensation between two temporarily connected monosaccharides afforded a novel route to C-disaccharides. This approach is exemplified by a synthesis of C-maltose, using a silicon tether. These methodologies should be useful in the field of antibiotics.

Introduction

A large number of organic molecules displaying antibiotic activity are glycosylated. In many cases 2-deoxy sugars are involved. A classical problem is the sensitivity of such glycosides toward either chemical or biochemical hydrolysis. The replacement of the inter-glycosidic oxygen atom of disaccharides by a methylene group would provide a stable analog. Access to such compounds depends upon an effective synthesis of C-glycosides. We would like to briefly report in this lecture several solutions to this chemical challenge.

The "umpolung" of the anomeric center: A novel route to C-glycosides
1. The anomeric organolithium route

Most of the chemistry of the anomeric center of carbohydrates developed so far relies on its electrophilic character. We have introduced the reductive lithiation of 2-deoxy-D-glycopyranosyl chlorides, phenyl sulfides [1] and phenyl sulfones [2] by lithium naphthalenide (LN). This simple generation of an "umpoled" reactive lithium derivative reverses the characteristic electrophilicity of the anomeric center and opens an entirely novel route to α-C-glycosides. For example, reductive lithiation of phenyl 3,4,6-tri-*O*-benzyl-2-deoxy-1-thio-β-D-*arabino*-hexopyranoside **1** (2 eq., LN, THF, -78°C, 45 min) and reaction with p-anisaldehyde gave a 3:1 diastereoisomeric mixture **2** of α-C-glycosides (65%).

1   **2**

This reaction is characterized by the initial single electron transfer (SET) from LN to the anomeric group X (chloride, thiophenyl or phenylsulfone), to give an anomeric radical. Reduction of this radical was selective and afforded a kinetic α-glycosyl lithium species **3**.

3

The detailed structure of **3** is not known so far, but it nicely reacted with carbonyl derivatives with retention of configuration. The selective synthesis of β-C-glycosides has also been achieved through the generation of a transient reactive β-glycosyl lithium species [3]. 3,4,6-Tri-O-benzyl-2-deoxy-β-D- *arabino*-hexopyranosyl phenyl sulfone **4** was deprotonated with BuLi and quenched with a carbonyl derivative; reductive lithiation of the phenyl sulfonyl group, followed by protonation, gave the β-C-glycoside **5**.

4   **5**

Tributylstannyl lithium treatment of 3,4,6-tri-O-benzyl-2-deoxy-α-D-*arabino*-hexopyranosyl-chloride **6** selectively provided tributyl (3,4,6-tri-O-benzyl-2-deoxy-β-D-*arabino*-hexopyranosyl) stannane **7**. Tin-lithium exchange with butyllithium in THF at -78°C generated the configurationally stable 2-deoxy-β-D-*arabino*-hexopyranosyl-lithium intermediate **8** which reacted with electrophilic compounds with retention of configuration [4].

OBn   OBn   OBn   OBn

Bu$_3$Sn Li   SnBu$_3$ BuLi   Li   E$^+$ E

OBn   OBn   OBn   OBn

BnO   BnO   BnO   BnO

Cl

6   **7**   **8**

2. The anomeric organosamarium route

The successful alkylation of ketones by primary organic iodides or bromides in the presence of samarium (II) iodide has been pioneered by Kagan *et al.* [5]. In order to investigate the potential of this approach in carbohydrate chemistry, the chloride **6** was submitted to reductive samariation in THF/ HMPA in the presence of cyclopentanone [6]. Best trapping results were observed under Barbier conditions when a solution of chloride **6** and cyclopentanone (2eq) in THF was added to a solution of samarium (II) iodide in THF containing HMPA. The course of the reaction is as follows:

OBn

6   SmI$_2$ OBn   •   + ClSm (III) I$_2$ SmI$_2$

THF / HMPA   BnO

OBn

OBn   OH +   BnO   OBn

BnO   BnO   BnO   OH

10 (70%)   **11** (17%)

OBn

OBn   SmI$_2$

BnO

9   cyclopentanone →

OBn   OBn

OBn   +   BnO   O

BnO   BnO   BnO

12 (1%)   **13** (4%)

The stereoselective formation of α-*C*-glycoside **10** was observed in 70% yield. In accordance with previous observations, the condensation with aldehydes, such as n-butanal or isobutyraldehyde, resulted in a lower yield (≈ 25-30%).

A very interesting result was obtained when 2,3,4,6-tetra-*O*-benzyl-β-D-glucopyranosyl phenyl sulfone **14** was condensed with cyclopentanone in the presence of the reducing system SmI$_2$/ THF / HMPA. Three compounds have been isolated : the previously reported glucal **12** (57%), the reduction product **15** (10%), and surprisingly, the β-*C*-glycoside **16** (30%) [7].

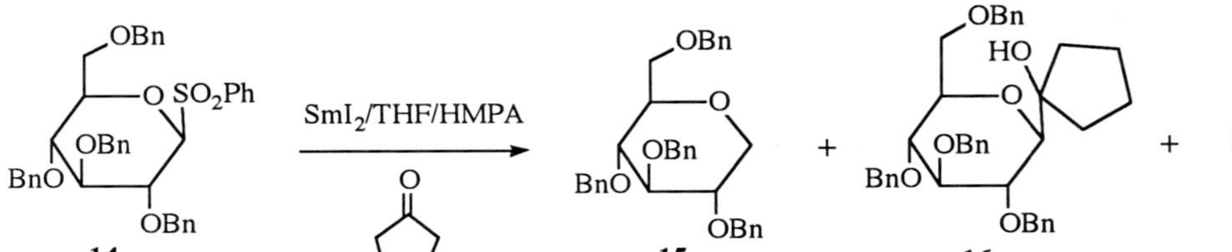

Although the yield of glycoside **16** is not optimized and is low due to the extensive formation of the elimination product **12**, this reaction is remarkable for two reasons :

First, this is a unique example of formation and trapping of a non-stabilized anomeric organometallic species having a leaving group at C-2. Generation of such species has only been achieved previously in the presence of an anion-stabilizing substituent such as a nitro group [8].

Second, the sulfone **14** afforded selectively the β-*C*-glycoside **15**, in contrast with 2-deoxy sugars, with which an α-*C*-glycoside was obtained.

Assuming that *C*-glycoside formation originated from a transient anomeric organosamarium species and that reaction occured with retention of configuration, the reactive intermediate could be the β-glycosyl samarium (III) **17**.

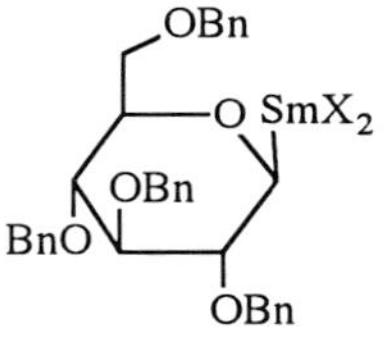

We believe that these novel selective routes to *C*-glycosides are of interest in the construction of antibiotics.

A novel route to *C*- disaccharides

An attractive and straightforward construction of a *C*-disaccharide, the stable analog of an *O*-disaccharide, would rely on the direct condensation between an anomeric radical and an exomethylene sugar.

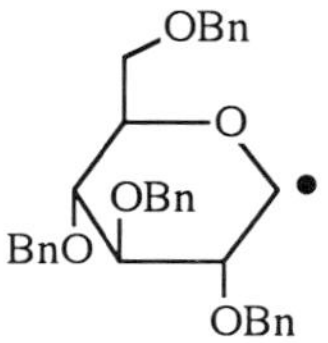

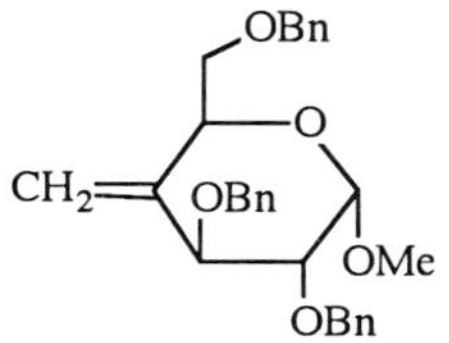

This intermolecular reaction does not occur because a nucleophilic anomeric radical reacts only with an electron deficient double bond. We therefore capitalized on the numerous alcohol groups which obviously offer opportunities for tethering the two sugar units. We now disclose a positive example of this general strategy. The alcohol **20** was easily prepared in two steps from the known orthoester **18** [9]. The selenophenyl group is a suitable precursor of the anomeric radical and the alcohol is ready for the tether.

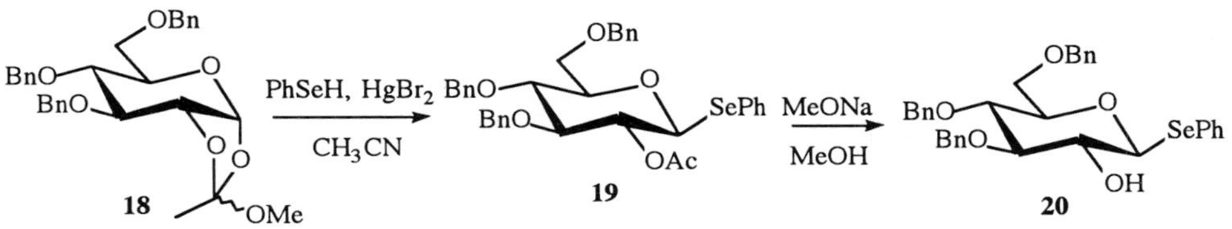

Likewise, the alcohol **22** was prepared from the known exomethylene derivative **21** [10].

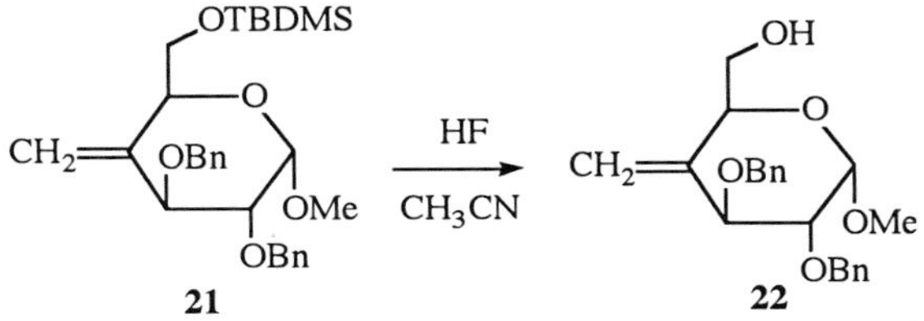

The two alcohols **20** and **22** were then tethered under the mediation of Me$_2$SiCl$_2$, according the following scheme:

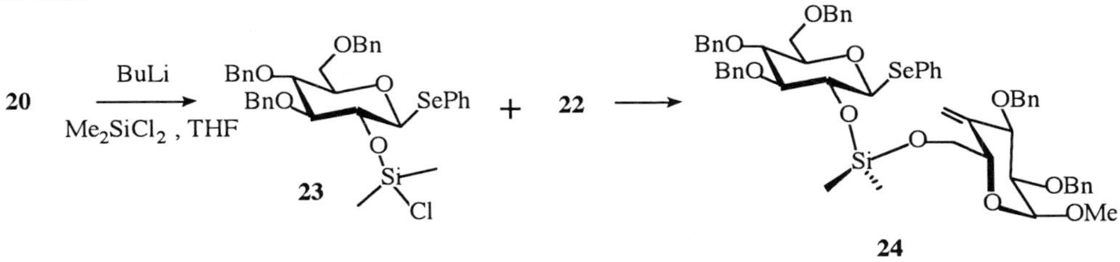

When compound **24** was treated with Bu$_3$SnH in toluene in the presence of AIBN, cyclization took place. Upon removal of the silicon connector in the presence of hydrofluoric acid, a *single C-disaccharide 25 was obtained in 40% yield*, which was fully characterized after acetylation. After removal of the benzyl ether protecting groups, the product was identical to the *C*-maltoside **27** previously prepared by another route by Kishi [11].

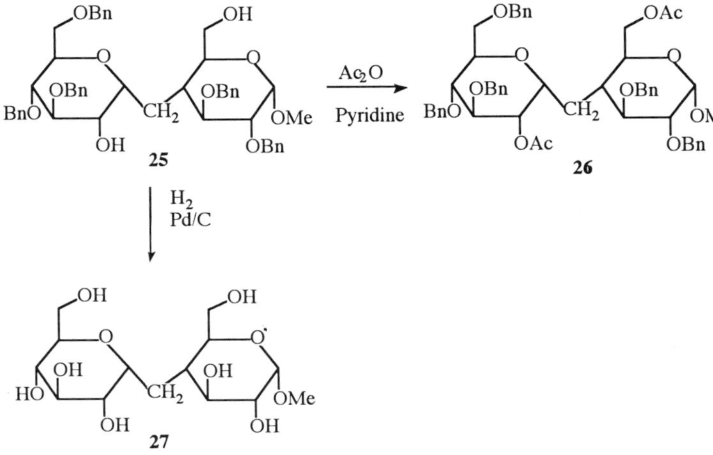

We believe that such a strategy will proved useful for the preparation of modified antibiotics.

Acknowledgements: The author wishes to express his acknowledgements to all participants in these developments, J.-M. Beau, J.-M. Lancelin, L. Morin-Allory and P. Lesimple for their contribution to the anomeric lithiation, J.-M. Mallet, P. de Pouilly, A. Chénedé and B. Vauzeilles for the realization of the anomeric reductive samariation, and Yan-Chao Xin for the synthesis of C-maltose.

References

1.   J.-M Lancelin, L. Morin-Allory, P. Sinaÿ, *J Chem.Soc. Chem. Commun.* (1984) 355-356

2.   J.-M Beau, P. Sinaÿ, *Tetrahedron Lett..* **26** (1985), 6185-88

3.   J.-M Beau, P. Sinaÿ, *Tetrahedron Lett..* **26** (1985), 6189-92

4.   P. Lesimple, J.-M. Beau, P. Sinaÿ, *Carbohydr. Res.* **171** (1987), 289-300

5.   P. Girard, J.-L. Namy, H.B. Kagan, *J. Am. Chem. Soc.* **102** (1980), 2693-2698

6.   P. de Pouilly, B. Vauzeilles, J.-M. Mallet, P. Sinaÿ, *C.R. Acad.Sci.Paris* (1991) **313**, *série* II, 1391-1394

7.   P. de Pouilly, A. Chénedé, J.-M. Mallet, P. Sinaÿ, *Bull.Soc.Chim. Fr.* in press

8.   B. Aebischer, J. H. Bieri, R. Prewo, A. Vasella, *Helv. Chim. Acta* **65** (1982), 2251-2272; B. Aebischer, A. Vasella, Helv. Chim. Acta **66** (1983), 789-794;

9.   P. J. Garregg, L. Maron, *Acta. Chem. Scand.* **33** (1979) 39-41

10. B.F. Molino, J. Cusmano, D. R. Mootoo, R. Faghigh, B. Fraser-Reid, *J. Carbohydr. Chem.* **2** (1987) 479-493.

11. P. G. Goekjian, T.-C. Wu, H.-Y. Kang, Y. Kishi, *J Org. Chem.* **56** (1991) 6422-6434; Y. Wang, S. A. Babirad, Y Kishi, *J. Org. Chem.* **57** (1992) 468-481

Sugar Lactones as Useful Starting Materials

G. W. J. Fleet

Dyson Perrins Laboratory, Oxford University, Oxford OX1 3QY, UK

Summary: (i) Many sugar lactones are widely available and are usually easier to handle in regard to their protecting group chemistry. (ii) Protected seven carbon carbohydrate lactones are readily prepared by the Kiliani synthesis and give flexible intermediates with seven adjacent functional group and five adjacent chiral centres (iii) Lactones are suitable starting materials for the synthesis of highly functionalized carbocycles, nitrogen heterocycles and oxygen heterocycles and in many cases provide the most efficient method for the investigation of biological structure activity relationships.

Syntheses in which carbohydrates are used as starting materials are frequently dominated by protecting group strategies. A major attraction of carbohydrate lactones is the relative ease of protection and deprotection sequences and the absence of anomeric substituents. A number of carbohydrate lactones are commercially available with some, such as gulonolactone, in both enantiomeric forms. Many more lactones are easily prepared from cheap aldoses by well established general methods: (i) by oxidation of an aldose by buffered bromine water to give a lactone with the same number of carbon atoms, (ii) by oxygenation of an alkaline solution of the aldose to give a lactone with one fewer carbon atom, or (iii) by the cyanohydrin extension reaction to give a mixture of the two epimeric sugars with one additional carbon.

Seven carbon lactones with seven adjacent functional groups and five contiguous chiral centres may react with acetone to give a diacetonide with only one of the hydroxyl group unprotected. The Kiliani reaction usually proceeds on unprotected sugars in high yield and with high diastereoselectivity. Protection of the hydroxyl groups prior to the extension normally considerably reduces both the yield and diastereoselectivity [1]. However, the reaction of diacetone mannose (**1**) with cyanide gives a mixture of the two carbohydrate lactones , **2** and **3**, in a ratio of about 3:1 and a yield of about 60% in which only the α-hydroxyl group of the δ-lactones is unprotected [2].

Reaction of **2** with triflic anhydride gives the stable crystalline triflate **4** which can undergo a range of reactions with basic and/or nucleophilic reagents including (A) nucleophilic displacement of triflate, (B) removal of the α-C-H to give the carbanion, and (C) nucleophilic addition to the carbonyl group, followed by ring opening of the lactone to an intermediate hydroxytriflate which then closes to give a tetrahydrofuran by an overall ring contraction. The use of such transformations allows the short and relatively easy synthesis of a number of complex targets.

The chemistry of nucleophilic displacement of triflates α to carbonyl groups can be accompanied by epimerization of either the starting triflate or the product [3]. Thus, both triflates **4** and **5** with sodium azide give the same azide **6** in good yield; what appears to be a simple S_N2 conversion of **5** to **6** in fact involves base-catalyzed isomerization of **5** to **4**, displacement of the triflate in **4** by azide with inversion of configuration, and a second epimerization of the kinetically formed azide to give **6** [4]. It is clear from the efficiency of these reactions that a carbanion α to the lactone carbonyl group may be generated and quenched with a proton source without significant fragmentation of the β-oxygen function.

In contrast to the azide displacements, both **4** and **5** react with iodide to give respectively **7** and **8** in excellent yield, although **4** reacts at a much faster rate [5]; these iodides are not easily interconvertible by further iodide displacement. Further treatment of **7** and **8** with iodide ion affords the deoxygenated material **10**, although the rate of reduction of **7** is faster than reduction of **8**; indeed, treatment of α-triflates of both γ- and δ-lactones with hydrated lithium iodide provides a mild and apparently general method for the deoxygenation of α-hydroxylactones [6]. Again this implies that the intermediate carbanions, such as **9**, do not readily fragment.

The lack of such carbanion fragmentation has been exploited in the development of an approach to the synthesis of very highly substituted cyclopentanes. Thus the iodoaldehyde **11**, readily obtained by hydrolysis of **7** and subsequent periodate cleavage of the resulting diol, gave **12** by a reductive aldol condensation induced by iodide ion; the epimeric iodide **13** give only traces of **12** under the same conditions [7].

13   **14**   **15**

In contrast, **13** with potassium fluoride in acetonitrile undergoes a highly efficient aldol cyclisation to give the bicyclic iodolactone **14** in 81% yield together with 15% of **15** which arises from prior epimerization of the aldehyde function in **13**; however, the iodoaldehyde **11** does not undergo base-induced aldol reactions. The azidoaldehyde **16**, prepared from **6**, can be used for the formation of **17** and **18**. It is thus clear that such an approach should lead to short syntheses of highly functionalized cyclopentanes.

16   **17**   **18**

Polyhydroxylated nitrogen heterocycles provide a powerful group of glycosidase inhibitors [8]. The azidolactone **6** has been used as a common intermediate for the synthesis of a number of potential mannosidase inhibitors such as the alexines **19** and **20** [9], homoDIM (**21**) [10], α-homomannojirimycin (**22**) [11] and the pyrrolizidine **23** [12]. Azidolactones derived from the enantiomers of gulonolactone have been used for the efficient synthesis of pyrrolidines such as **24** [13], piperidines such as deoxymannojirimycin (**25**) [14] and L-deoxyrhamnojirimycin (**26**) [15], and indolizidines such as 6-epicastanospermine (**27**) [16].

19   **20**   **21**

22   **23**   **24**

OH
HO,,,   ,,OH

N
H   CH₂OH

24

OH
HO,,,.   ,OH

N
H   ''''CH₃

25

OH
HO,   ,,,OH
        H
                OH
N

26

It is thus clear that such materials have a very important role in the synthesis of a wide range of complex and highly functionalized nitrogen heterocycles.

The methoxide-induced ring contractions of α-triflates of γ- and δ-lactones provide efficient and easy syntheses of oxetanes and tetrahydrofurans, respectively. Thus, initial addition by methoxide ion to the carbonyl group of the lactone is followed by ring opening to give an open chain hydroxy triflate which then closes to the oxygen heterocycle.

ROH₂C O
        OMe
                =O
(ROHC)ₙ
        OSO₂CF₃

ROH₂C O OMe
                O
(ROHC)ₙ
        OSO₂CF₃

ROH₂C O⁻
                COOMe
(ROHC)ₙ
        OSO₂CF₃

ROH₂C O
(ROHC)ₙ
        COOMe

n = 1 for oxetanes
n = 2 for tetrahydrofurans

This is a highly efficient process for the formation of highly functionalized homochiral oxetanes [17]. However, where there is an oxygen substituent at C-3 of the lactone, the stereochemistry of the carboxylate is always predominantly *trans* to this oxygen functionality, regardless of the stereochemistry of the starting triflate [18]; where there is no such oxygen substituent, the ring contraction occurs with overall inversion of configuration at the carbon bearing the triflate [19]. The product oxetane esters do not equilibrate under the conditions under which they are formed. This difference in behavior can be attributed to the longer lifetime of the open-chain hydroxy triflates with oxygen substituents β-to the triflate. The yields of oxetanes are usually excellent when there are β-oxygen substituents, although if the substituent is too strongly electron-withdrawing, other side reactions compete [20]. This procedure has allowed the synthesis of previously unknown α-chlorooxetanes [21] by the Barton modification of the Hunsdiecker reaction. Such chlorooxetanes are usually quite stable and may be used for the synthesis of oxetane nucleosides by predominantly S_N2 displacements [22] Oxetanocin, a naturally occurring antiviral nucleoside, has been prepared using this strategy [23] although the intermediate chlorooxetanes in this case, which lack β-oxygen substituents, are relatively difficult to handle. Analogues which lack the methylene of the natural product are more readily prepared by this approach since the intermediate chlorooxetanes are more stable [24]. Norepioxetanocin (**33**), which is at least as active against HIV as oxetanocin, was prepared by a short synthesis from the readily available D-lyxonolactone (**27**) [25]; reaction of **27** with benzaldehyde give **28** in which the remaining free hydroxyl group was esterified to give the stable triflate **29**.

(i) PhCHO, H$^+$ (ii) (CF$_3$SO$_2$)$_2$O, pyridine, CH$_2$Cl$_2$
(iii) K$_2$CO$_3$, MeOH   (iv) Pb(OAc)$_4$, N-chlorosuccinimide
(v) adenine, 18-crown-6, DMF   (vi) aq. CF$_3$COOH

Treatment of a methanol solution of **29** with potassium carbonate gave the oxetane **30** in which the ring contraction has occurred with retention of configuration at C-2. Subsequent reaction of **30** with lead tetraacetate and N-chlorosuccinimide gave only the stable chlorooxetane **31** which underwent a predominantly SN2 reaction with adenine in dimethyl formamide to afford the β-nucleoside **32**. Removal of the benzylidene protecting group with aqueous trifluoroacetic acid gave **33**.

Although the ring contractions of α-triflates of δ-lactone derivatives induced by basic methanol occur in high yields and are thus parallel to the ring contractions of γ-lactones to oxetanes, there are also marked differences between the two sets of reactions. In particular, the stereochemistry of carboxylate group in the tetrahydrofuran arises almost exclusively by overall inversion of configuration at C-2 by nucleophilic attack of the ring oxygen; the stereochemistry of the carboxylate is determined by the configuration at the α-position of the lactone. This in contrast to the formation of oxetanes from γ-lactones discussed above in which the stereochemistry of the carboxylate is determined principally by the oxygen substituent at C-3 rather than by that of the leaving group at C-2. Thus the triflate **4** reacts with potassium carbonate in methanol to give the tetrahydrofuran **27** in which all the groups are *cis* to each other; even in the most hindered cases, the contraction takes place with inversion of configuration at C-2; no **28** is formed under these conditions. The iodide **8** under the same conditions gives **28**. Oxetane formation usually requires a triflate leaving group α to the carbonyl function, whereas formation of the tetrahydrofuran ring may happen with a wider range of leaving groups.

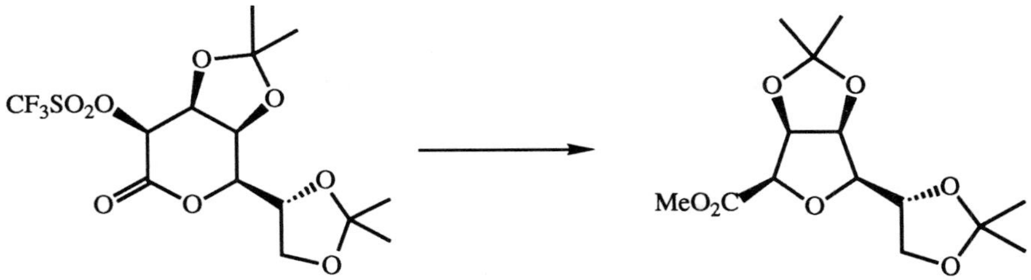

4

27

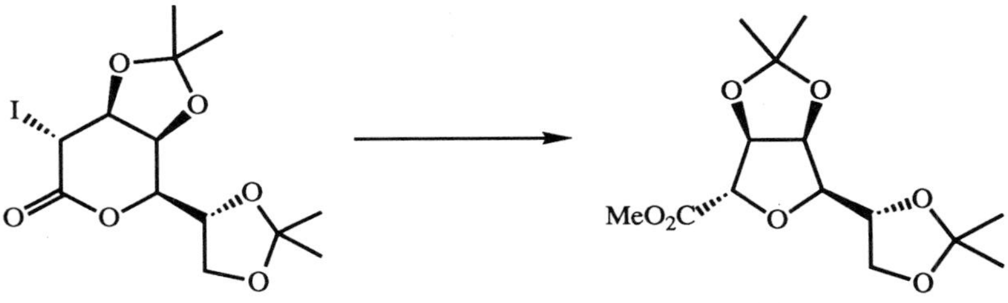

8

28

However these differences in the stereochemical courses of the reactions may arise, it is clearthat the ring contraction of δ-lactone derivatives provides a very powerful strategy for the easy synthesis of complex tetrahydrofurans with carbon substituents at C-2 and C-5 [26]. Such a convenient method for the formation of tetrahydrofurans provides a strategy for the synthesis of precursors to C-nucleosides and polyoxins. One illustration of this in the synthesis of muscarine analogues from the readily available sugar L-rhamnose (**30**). Rhamnose has the correct functionality and stereochemistry at C-2, C-4, C-5 and C-6 for the synthesis of muscarine (**29**) itself by a route that involves displacement of a leaving group at C-2 of a rhamnose derivative by the oxygen at C-5 [27][28].

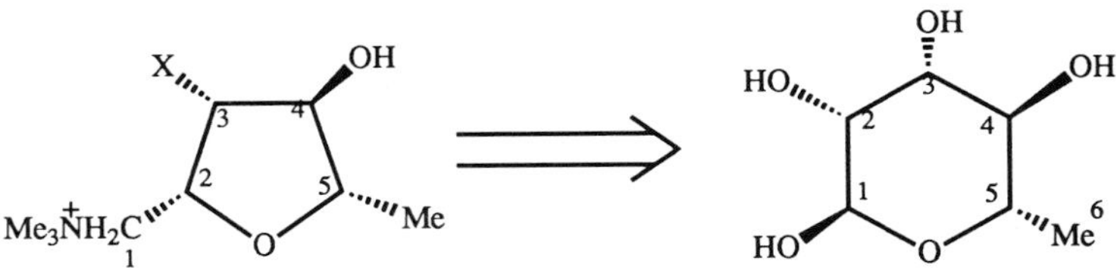

29 muscarine (X = H)
31 hydroxymuscarine (X = OH)

30 L-rhamnose

Although much effort has been invested in the synthesis of analogues of muscarine as a potential strategy for the development of agents for the treatment of Alzheimer's disease, there are no examples of compounds - such as hydroxymuscarine (**31**) - which have a substituent at C-3, the unsubstituted position of the tetrahydrofuran ring. Rhamnose is an ideal starting material for **31** and the sequence does not require the use of any protecting group [29]. Oxidation of **30** with buffered bromine gives the δ-lactone **32** which can be converted to the triflate **33**; treatment with pyridine in methanol induces the ring contraction of **33** to give **34** which on reduction and subsequent tosylation affords **35** which with trimethylamine gives **31**.

(i) bromine water, Ba$_2$CO$_3$ buffer (ii) (CF$_3$SO$_2$)$_2$O, pyridine, THF (iii) pyridine in MeOH (iv) LiAlH$_4$ in THF (v) TsCl in pyridine (vi) Me$_3$N in MeOH

In summary, it is clear that sugar lactones are excellent starting materials for the synthesis of diverse series of very highly functionalized targets. Because of the lack of ambiguity in both the enantiomeric and diaseteromeric products from the reactions, these approaches have significant advantages over the use of asymmetric synthesis in the study of biological structure-reactivity relationships. I am very grateful to my collaborators both in and outside Oxford cited in the references; I hope they have enjoyed working on these problems as much as I have.

References

1. C. J. F.Bichard, A. J. Fairbanks, G. W. J. Fleet, N. G. Ramsden, K. Vogt, O. Doherty, L. Pearce and D. J. Watkin, Acetonides of α-Hydroxy-δ-lactones, *Tetrahedron: Asymm.*, 1991, **2**, 901-912.

2. A. R. Beacham, I. Bruce, I. S. Choi, O. Doherty, A. J. Fairbanks, G. W. J. Fleet, B. M. Skead, J. M. Peach, J. Saunders, and D. J. Watkin, Acetonides of Heptonolactones: Powerful Chirons, *Tetrahedron: Asymm.*, 1991, **2**, 883-900.

3. G. W. J. Fleet, J. C. Dho, J. M. Peach, K. Prout and P. W. Smith, Synthesis of 2R,3S,4S-Dihydroxyproline from D-Ribonolactone; an Approach to the Synthesis of Poly-functionalised D-Amino Acids from Sugar Lactones. X-Ray Crystal Structures of 2-Azido-3,4-O-(R)-benzylidene-2-deoxy-D-ribono-1,5-lactone, 2-Azido-2-deoxy-D-ribono-1,4-lactone, and 2R,3S,4R-3,4-Dihydroxyproline, *J. Chem. Soc., Perkin Trans. 1*, **1987**, 1785-1791.

4. G. W. J. Fleet, I. Bruce, A. Girdhar, M. Haraldsson, J. M. Peach, and D. J. Watkin, Retention and Apparent Inversion during Azide Displacement of α-Triflates of 1,5-Lactones, *Tetrahedron*, 1990, **46**, 19-31.

5. R. P. Elliott, Y. S. Gyoung, G. W. J. Fleet, N. G. Ramsden and C. Smith, Nucleophilic Displacement of 2-O-Trifluoromethanesulphonate Esters of α-Hydroxylactones by Iodide; A Mild Method for the Reduction of α-Hydroxylactones to the Corresponding Deoxylactones by Lithium Iodide Trihydrate, in preparation.

6. R. P. Elliott, G. W. J. Fleet, Y. S. Gyoung, N. G. Ramsden and C. Smith, High Yield Reduction of 2-O-Trifluoromethanesulphonate Esters of α-Hydroxylactones to the Corresponding 2-Deoxylactones by Lithium Iodide Trihydrate, *Tetrahedron Lett.*, 1990, **31**, 3785-3788.

7. R. P. Elliott, G. W. J .Fleet, L. Pearce, C. Smith and D. J. Watkin, A Reductive Aldol Strategy for the Synthesis of Very Highly Substituted Cyclopentanes from Sugar Lactones, *Tetrahedron Lett.*, 1991, **32**, 6227-6231.

8. B.Winchester and G.W.J.Fleet, Amino-sugar Inhibitors: Versatile Tools for Glycobiologists, *Glycobiology*, 1992, **2**, 199-210.

9. S. Choi, I. Bruce, A. J. Fairbanks, G. W. J. Fleet, A. H. Jones, R. J. Nash and L. E. Fellows, Synthesis of Alexines from Heptonolactones,*Tetrahedron Lett.*, 1991, **32**, 5517-5521.

10. P. M. Mysercough, A. J. Fairbanks, A. H. Jones, S.-S.Choi, G. W. J. Fleet, S. S. Al-Daher, I. Cenci di Bello and B. Winchester, Inhibition of some α-Mannosidases by Seven Carbon Sugars: Synthesis of Some Seven Carbon Analogues of Mannofuranose, *Tetrahedron*, 1992, **48**, 10177-10190.

11. I. Bruce, G. W. J. Fleet, I. Cenci di Bello and B. Winchester, Iminoheptitols as Glycosidase Inhibitors: Synthesis of α-Homomannojirimycin, 6-epi-α-Homomannojirimycin and of a Highly Substituted Pipecolic Acid, *Tetrahedron*, 1992, **48**, 10191-10200.

12. A. J. Fairbanks, G. W. J .Fleet, A. H.Jones, I. Bruce, S. Al Daher, I. Cenci di Bello, B. Winchester, Synthesis from a Heptonolactone and Effect on Glycosidases of (1S,2R,6R,7S)-1,2,6,7-Tetrahydroxypyrrolizidine, *Tetrahedron*, 1991, **47**, 131-138.

13. G. W. J. Fleet and J. C. Son, Polyhydroxylated Pyrrolidines from Sugar Lactones: Synthesis of 1,4-Dideoxy-1,4-imino-D-glucitol from D-Galactonolactone and Syntheses of 1,4-Dideoxy-1,4-imino-D-allitol, 1,4-Dideoxy-1,4-imino-D-ribitol and (2S,3R,4S)-3,4-Dihydroxyproline from D-Gulonolactone, *Tetrahedron*, 1988, **44**, 2637.

14. G. W. J. Fleet, N. G. Ramsden and D. R. Witty, Practical synthesis of Deoxymannojirimycin and Mannonolactam from L-Gulonolactone.Synthesis of L-Deoxymannojirimycin and L-Mannonolactam from D-Gulonolactone, *Tetrahedron*, 1989, **45**, 319-326.

15. A. J. Fairbanks, N. C. Carpenter, G. W. J. Fleet, N. G. Ramsden, I. Cenci de Bello, B. G. Winchester, S. S. Al-Daher, and G. Nagahashi, Synthesis of, and Lack of Inhibition of a Rhamnosidase by, Both Enantiomers of Deoxyrhamnojirimycin and Rhamnonolactam: β-Mannosidase Inhibition by δ-Lactams, *Tetrahedron*, 1992, **48**, 3365-3376.

16. G. W. J. Fleet, N. G. Ramsden, R. J. Nash, L. E. Fellows, G. S.Jacob, I. Cenci di Bello, and B. Winchester, Synthesis of the Enantiomers of 6-Epicastanospermine and 1,6-Diepicastanospermine from D- and L-Gulonolactone, *Carbohydr. Res* , 1990, **205**, 269-282.

17. G. W. J. Fleet, G. N. Austin, J. M. Peach, K. Prout, and Jong Chan Son, Chiral Oxetanes from Sugar Lactones: Synthesis of Derivatives of 3,5-Anhydro-1,2-O-isopropylidene- α-D-glucuronic Acid and of 3,5-Anhydro-1,2-O-isopropylidene-ß-L-iduronic Acid, *Tetrahedron Lett.*, 1987, **28**, 4741-4744.

18. D. R. Witty, G. W. J. Fleet, K. Vogt, F. X. Wilson, Y Wang, R. Storer, P. L. Myers, C. J. Wallis, Ring Contraction of 2-O-Trifluoromethanesulphonates of α-Hydroxy-γ-lactones to Oxetane Carboxylic Esters, *Tetrahedron Lett.*, 1990, **31**, 4787-4790.

19. D. R.Witty, G. W. J. Fleet, S. Choi, K. Vogt, F. X. Wilson, Y. Wang, R. Storer, P. L. Myers, C. J. Wallis, Ring Contraction of 3-Deoxy-2-O-trifluoromethanesulphonates of α-Hydroxy-γ-lactones to oxetanes, *Tetrahedron Lett.*, 1990, **31**, 6927-6930.

20. R. P. Elliott, G. W. J. Fleet, K. Vogt, F. X. Wilson, Y. Wang, D. R. Witty, R. Storer, P. L. Myers, C. J. Wallis, Attempted Ring Contraction of α-Triflates of 3-Azido- and 3-Fluoro-γ-Lactones to Oxetanes, *Tetrahedron: Asymm.*, 1990, **1**, 715-718.

21. G. W. J. Fleet, J. C. Son, J. M. Peach and T. A. Hamor, Synthesis and X-Ray Crystal Structure of a Stable α-Chlorooxetane, *Tetrahedron Lett.*, 1988, **29**, 1449-1450.

22. G. W. J. Fleet, J. C. Son, K. Vogt, J. M. Peach and T. A. Hamor, Reaction of Adenine with an α-Chlorooxetane: an Approach to the Synthesis of Oxetane Nucleosides, *Tetrahedron Lett.*, 1988, **29**, 1451-1452,

23. F. X. Wilson, G. W. J. Fleet, K. Vogt, Y. Wang, D. R. Witty, R. Storer, P. L. Myers, C. J. Wallis, Synthesis of oxetanocin, *Tetrahedron Lett.*, 1990, **31**, 6931-6934.

24. F. X. Wilson, G. W. J. Fleet, D. R. Witty, K. Vogt, Y. Wang, R. Storer, P. L. Myers, C. J. Wallis, Synthesis of the Oxetane Nucleosides α- and β-Noroxetanocin, *Tetrahedron: Asymm.*, 1990, **1**, 525-526.

25. Y. Wang, G. W. J. Fleet, R. Storer, P .L. Myers, C. J. Wallis, O. Doherty, D. J. Watkin,K.Vogt, D. R. Witty, F. X. Wilson, J. M. Peach, Synthesis of the Potent Antiviral Oxetane Nucleoside Epinoroxetanocin from D-Lyxonolactone, *Tetrahedron: Asymm.*, 1990, **1**, 527-530.

26. S. S. Choi, P. M. Myerscough, A. J. Fairbanks, B. M. Skead, C. J. F. Bichard, S. J. Mantell, G. W. J. Fleet, J. Saunders and D. Brown, The Ring Contraction of δ-Lactones with Leaving Group α-Substituents: a Strategy for the Synthesis of 2,5-Disubstituted Highly Substituted Homochiral Tetrahydrofurans, *J. Chem. Soc., Chem. Commun.*, **1992**, 1605-1607.

27. S. J. Mantell, G. W. J. Fleet, and D. Brown, (+)-Muscarine from L-Rhamnose, *J. Chem. Soc., Chem. Commun.*, **1991**, 1563.

28. S .J. Mantell, G. W. J. Fleet, and D. Brown, A Practical Synthesis of (+)-Muscarine from L-Rhamnose, *J. Chem. Soc., Perkin Trans. 1*, **1992**, 3023-3028

29. S. J. Mantell, G. W. J. Fleet, and D. Brown, 3R-Hydroxymuscarine from L-Rhamnose without Protection, *Tetrahedron Lett.*, 1992, **33**, 4503-4506.

The Synthesis of Antibiotic Amino Sugars from α-Amino Aldehydes

Janusz Jurczak and Adam Golebiowski

Institute of Organic Chemistry, Polish Academy of Sciences,
01-224 Warszawa, Poland

Summary: New results in the synthesis of valuable antibiotic amino sugars in optically pure form, starting from suitably protected α−amino aldehydes, are presented. The general retrosynthetic approach that proceeds *via* hetero Diels-Alder reaction is shown. The diastereoselectivity control of this reaction gives an easy access to C_4-elongated α-amino acid-derived synthons, which are subsequently transformed into amino sugars. The versatility of this methodology is presented on the previous syntheses of lincosamine and purpurosamines and recent preparations of galantinic, anhydrogalantinic, galantaminic and destomic acids as well as **L**-daunosamine.

Aminoglycoside antibiotics are an important part of clinically used drugs.[1] The name comes from the several amino groups present in their glycosidic moieties. Fig. 1 exhibits some examples of this class of compounds.[2-4]

Another clinically important antibiotic, lincomycin, proved to be the proline amide of lincosamine **4**, a complex amino octose (Fig. 2).[5-6] Structurally related amino sugar, destomic acid **5**, is found in destomycin A, B, C, hygromycin B, SS-56 C, and A-396-I (SS-56 D), (Fig. 2).[7-10]

The synthesis of **amino sugars** present in all these antibiotics has been extensively studied by several research groups. The application of easily available sugars like **D**-glucose or **D**-galactose was the common framework of most of these synthetic approaches.[11]

Recent developments in the **hetero Diels-Alder reaction** and **metalloorganic additions** to α-amino aldehydes [12] make it possible to apply more straightforward retrosynthetic approaches (Scheme 1). Particularly in the case of 6-amino-6-deoxy sugars, this idea is the most efficient approach. In contrast to the traditional approach, the only (if necessary) functionalization is done

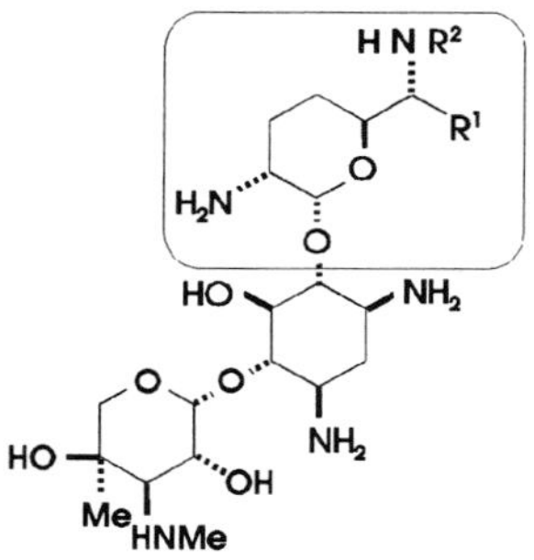

Purpurosamine C (1)
R¹=R²=H

Purpurosamine B (2)
R¹=Me R²=H

6-epi-Purpuros-
amine
B (3)

Gentamicin C₁ₐ
R¹=R²=H
Gentamicin C₂
R¹=Me R²=H

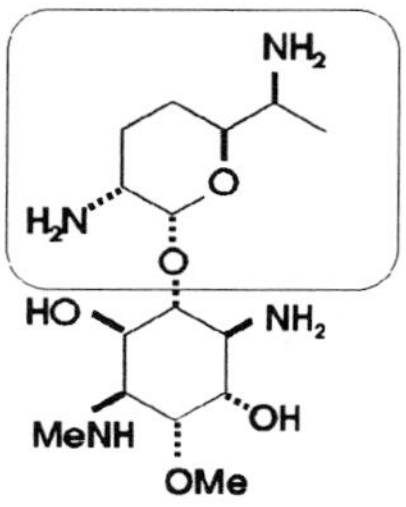

Fortimicin

Figure 1

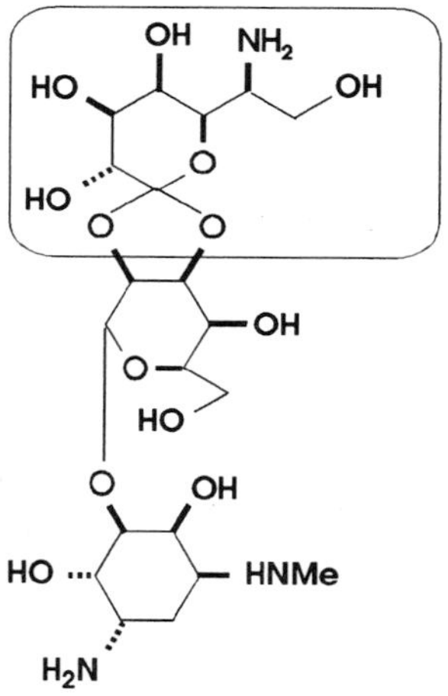

Destomic acid (5)

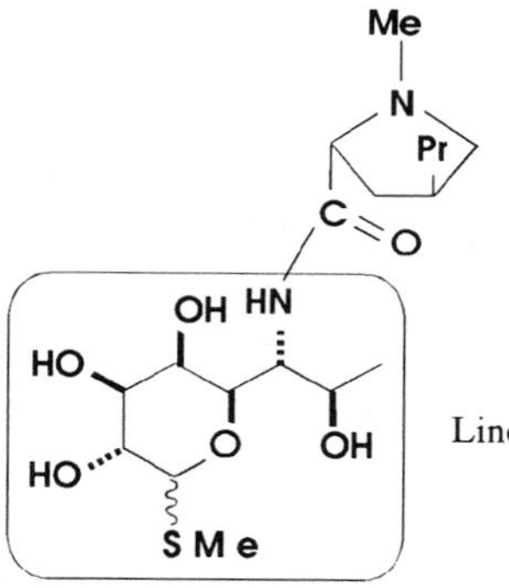

Lincosamine (4)

Lincomycin   Destomycin A

Figure 2

on a pyranose ring-system instead of acyclic moieties (functionalization on the branch-chain part of sugar).

This idea was demonstrated in our recent syntheses of **purpurosamine C (1)**, [13] **B (2)**, [14] **6-epi B (3)**,[15] (Scheme 3) and **lincosamine (4)** (Scheme 2).[16,17] During these syntheses, it was observed that N-monoprotected α-amino aldehydes led to *syn*-adducts, while N,N-diprotected derivatives gave *anti*-isomers as the major product (Scheme 4).[18-21]

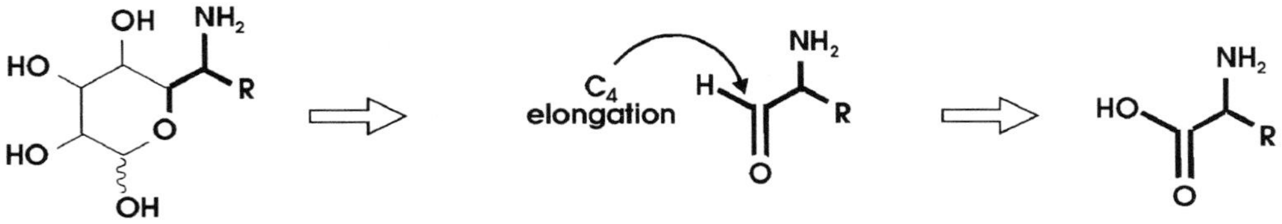

Scheme 1

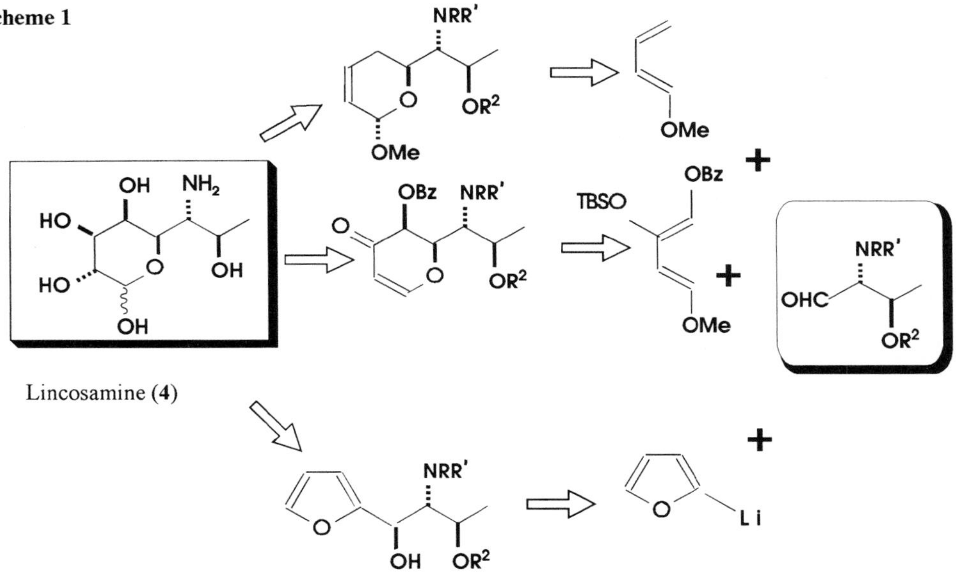

Lincosamine (**4**)

Scheme 2

The explanation of these phenomena was based on chelation versus nonchelation effects in **transition state models** of nucleophilic addition to carbonyl groups (Scheme 5).[22]

For *N*-monoprotected α-amino aldehydes, α-**chelation** model A (assisted by hydrogen bonding-model B) rules the diastereoselectivity of addition leading to *syn*-product. *N,N*-Diprotected α-amino aldehydes undergo steric Felkin-Anh-type addition (Scheme 5, model C). This effect can be amplified by β-chelating interactions (Scheme 5, model D).[16,17]

The methodology we worked out during synthesis of the purpurosamine family and lincosamine was applied recently in the transformation of **L**-serine into **destomic acid (5)**.

6-Amino-6-deoxy-**L**-*glycero*-**D**-*galacto*-heptonic acid (destomic acid (**5**)) is a component of the aminocyclitol antibiotics: destomycin A, B, C, hygromycin B, SS-56C, and A-396-I (SS-56D).[8-10] These antibiotics demonstrate anthelmintic activity and contain a unique structural moiety, an orthoester (glycosylidene) linkage to the hexose fragment. The structure of compound **5** was confirmed by chemical synthesis starting from **D**-galactose, and by X-ray analysis.[23]

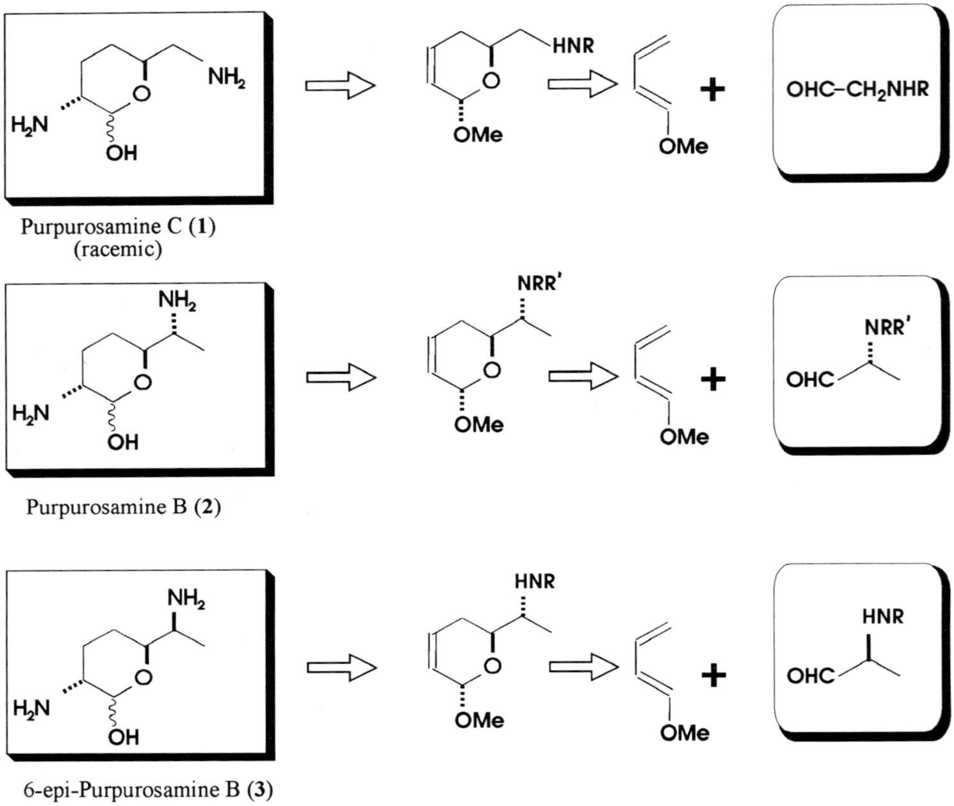

Purpurosamine C (1)
(racemic)

Purpurosamine B (2)

6-epi-Purpurosamine B (3)

Scheme 3

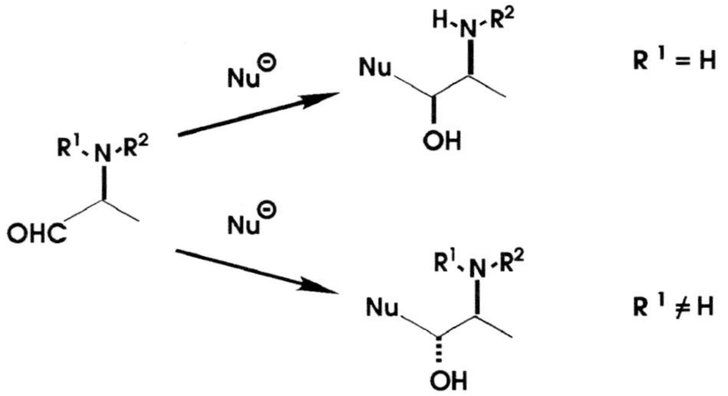

Scheme 4

Retrosynthetic analysis (Scheme 6) of destomic acid led to C$_4$-elongation of the **L**-serine-derived aldehyde **8**.

During the [4+2]cycloaddition step, two stable chiral centers are introduced; the third, the

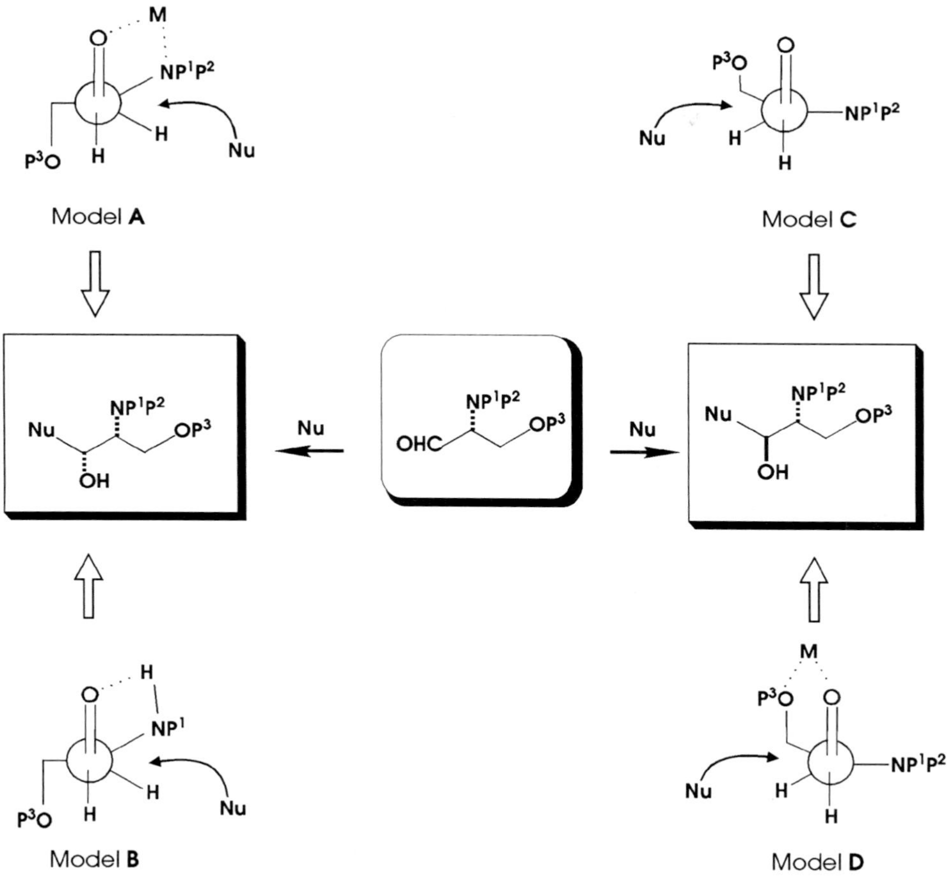

Model A

Model C

Model B

Model D

Scheme 5

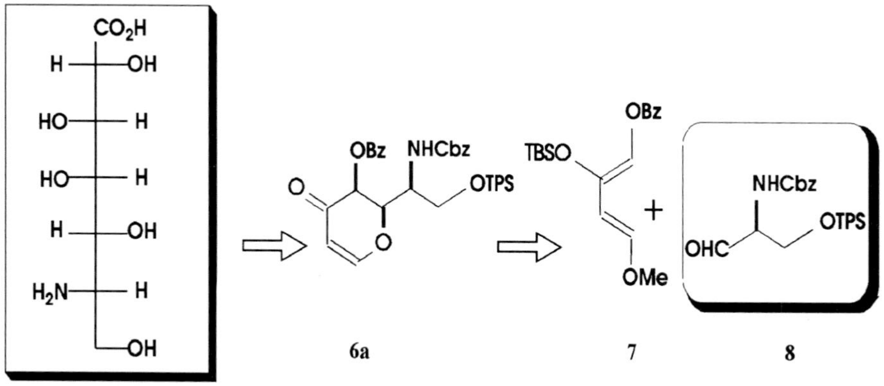

5

6a

7

8

TPS=*tert*-butyl diphenyl silyl

Scheme 6

anomeric center, is lost due to elimination of the methoxy group. The new chiral center of the C-O bond in the ring should be formed with *syn*-stereoselectivity relative to the inducing one (derived from the α-amino aldehyde). This would be expected due to the strong α-chelation that exists in the Lewis-acid-mediated reaction of *N*-monoprotected α-amino aldehydes (Scheme 5, model A). The second, new chiral center, the *O*-benzoate group, should be formed with *cis*-relative stereoselectivity (**6a**, Scheme 7), due to the observed earlier dominance of *exo*-selective [4+2]cycloaddition in Lewis-acid-mediated hetero Diels-Alder reactions. [24]

Indeed, the cyclocondensation reaction of diene **7** with aldehyde **8** in the presence of ZnBr$_2$, followed by treatment with trifluoroacetic acid, led to adduct **6a** as a major product.[25] The diastereomeric proportion was (**6a**):(**6b**):(**6c**):(**6d**) = 87:8:4:1. The Luche-type reduction of the chromatographically pure adduct **6a**, followed by basic debenzoylation, afforded diol **9**. This was subjected to the *cis*-hydroxylation reaction to yield, after isopropylidenation, product **10** (Scheme 8). The transformation of **10** according to the Hashimoto procedure [23] (desilylation, deprotection of

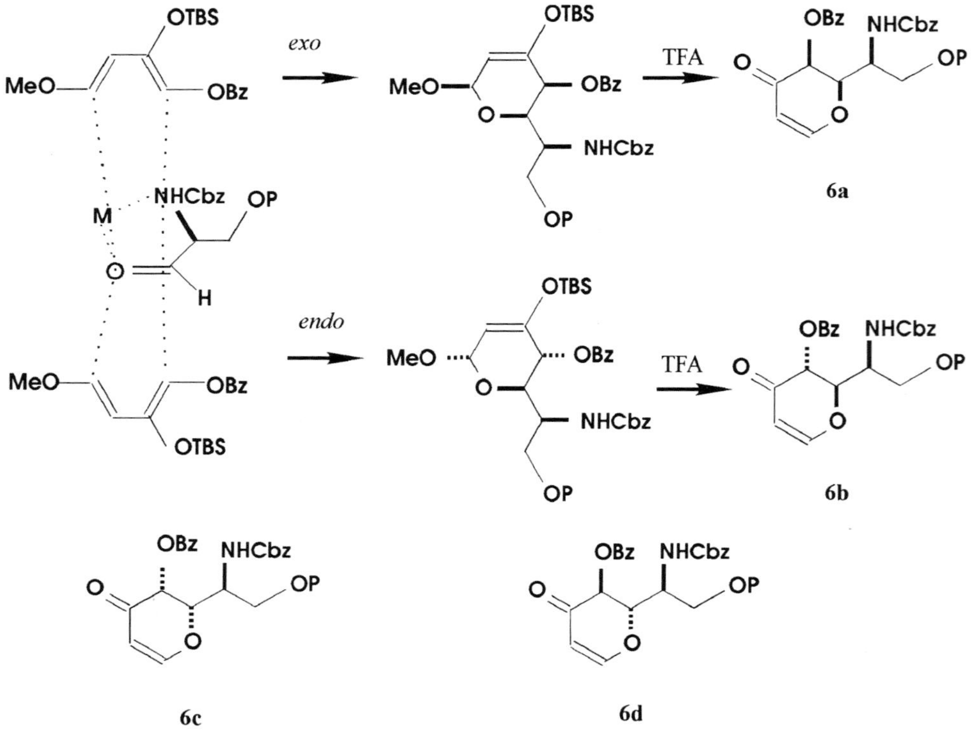

TBS = *tert*-butyl dimethyl silyl

Scheme 7

both acetals, oxidation of the aldehyde function and hydrogenation) afforded destomic acid (**5**).

A structurally different antibiotic, **galantin I**, was isolated from a culture broth of

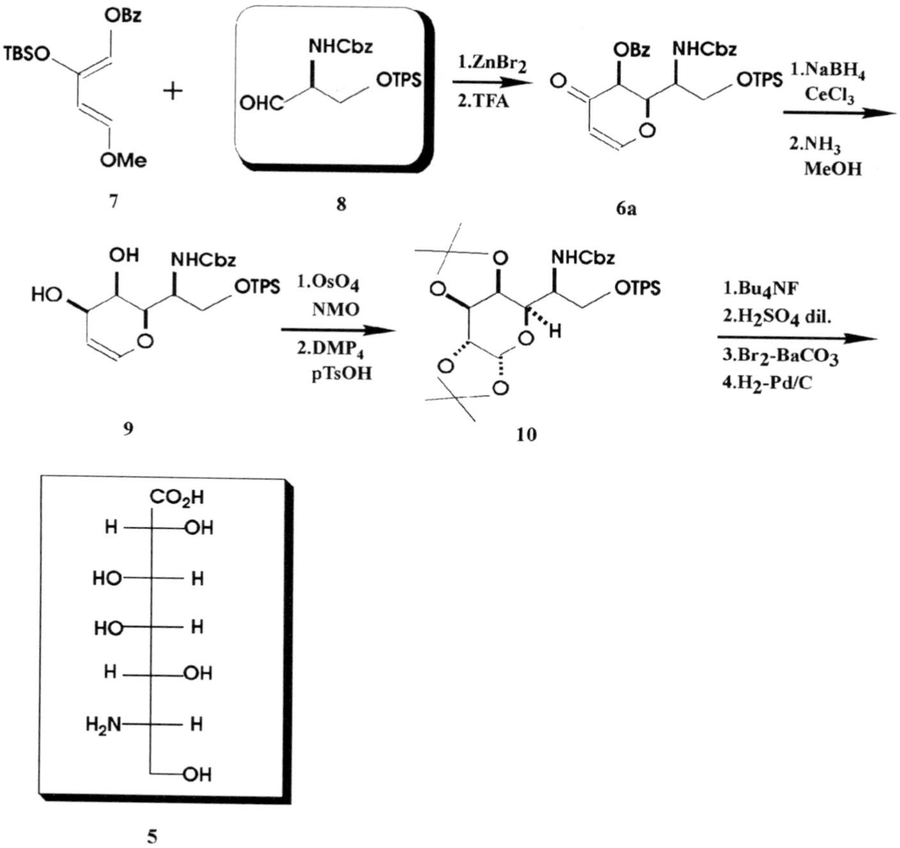

Scheme 8

Bacillus pulvifaciens by Shoji and coworkers.[27] Two fragments of this antibiotic, **galantinic (11)** and **galantaminic (12) acids,** are particularly interesting from the synthetic point of view (Fig. 3).[28,29]

Chemical degradation is one of the oldest and most reliable methods of structural assignment. However some errors occur when the product of degradation undergoes subsequent reactions.

That was the case of galantinic acid, which originally was isolated as a product of a dehydration reaction.[30] Both galantinic (**11**) and **anhydrogalantinic (13) acids** were synthesized by Ohfune and coworkers.[28-30]

Retrosynthetic analysis of anhydrogalantinic acid (**13**) leads to the Diels-Alder reaction of **Danishefsky's diene 16** with serinal **8** as a key transformation (Scheme 9).

In this approach, achievement of high diastereoselectivity of the [4+2]cycloaddition step was

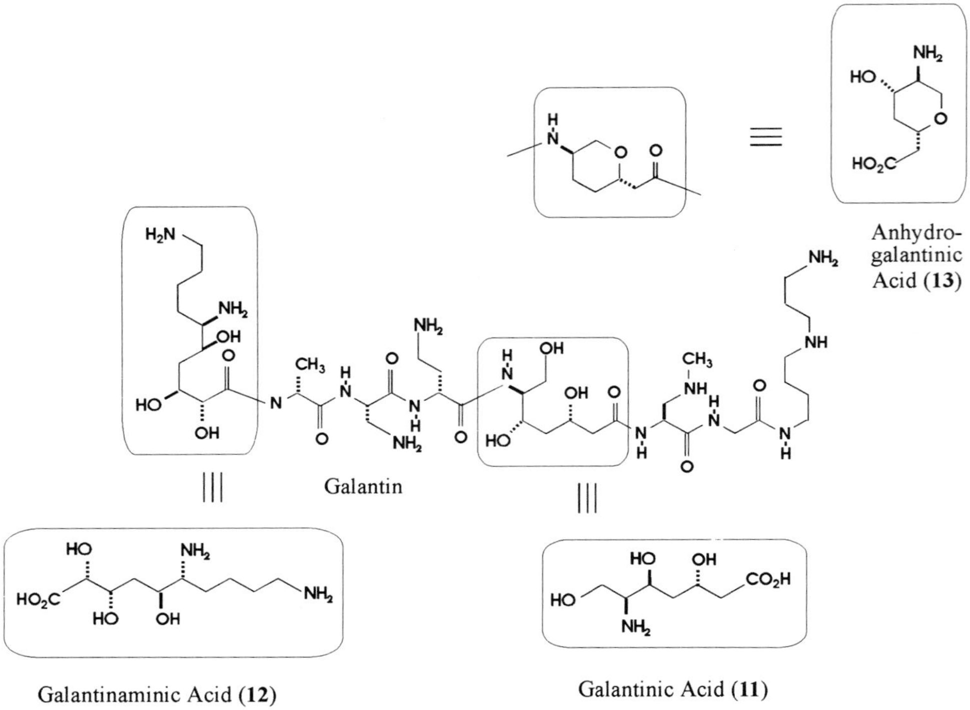

Figure 3

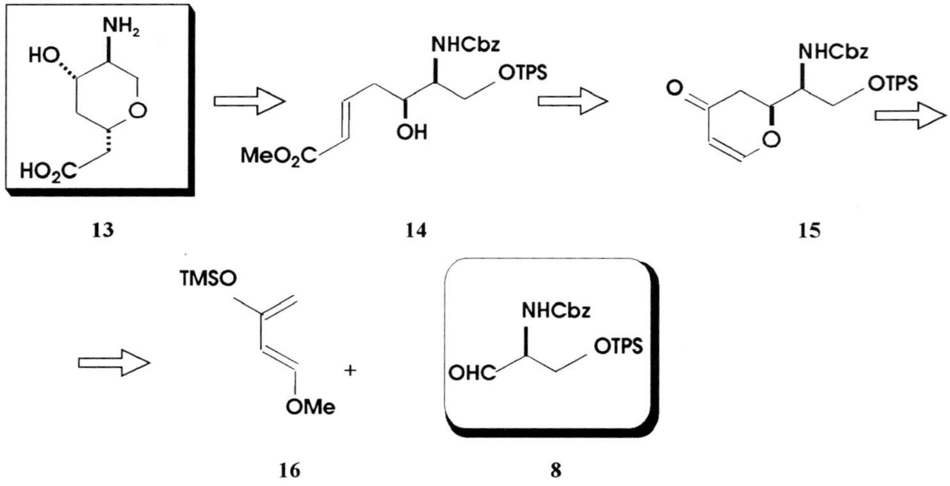

Scheme 9

crucial. According to our earlier results, pyrone 15 was expected to be the major product. Indeed,

N,O-protected serinal **8** (synthesized in a three-step procedure from **L**-serine) afforded, with diene **16**, pyrone **15** as a single product (Scheme 10).[31]

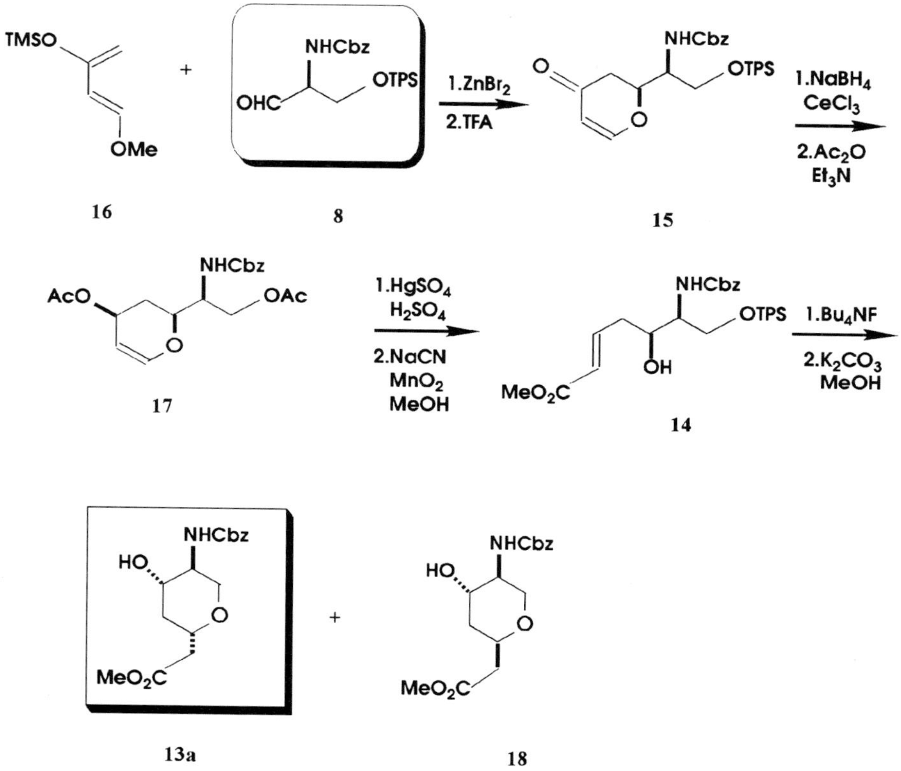

Scheme 10

Luche-type reduction [32] of pyrone **15**, followed by protection of the hydroxy group of the intermediate alcohol, gave acetate **17** in a very high yield. The subsequent dihydropyran-ring opening reaction, [33] followed by Corey's oxidation, [34] afforded ester **14**. After removal of the protecting silyloxy group, the previously described cyclization [28] produced a chromatographically separable 1:1 mixture of (+)-anhydrogalantinic acid derivative (**13a**) and its C-3 epimer (**18**).

The synthesis of galantinic acid (**11**) was achieved using a very similar methodology (Scheme 11). [35]

High pressure [4+2]cycloaddition of diene **19** to **L**-serine-derived aldehyde **8** led, after acidic equilibration, to a 4:1 mixture of *syn* **20a** and *anti* **20b** diasteroisomers. The major *syn*-cycloadduct **20a** was oxidized to α, β-unsaturated lactone **21**, which was subsequently transformed into (+)-galantinic acid (**11**) according to Ohfune's methodology. [28]

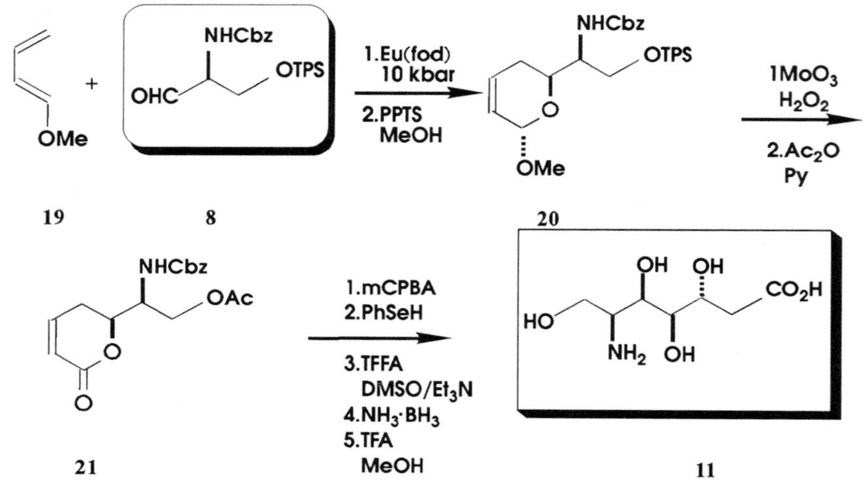

Scheme 11

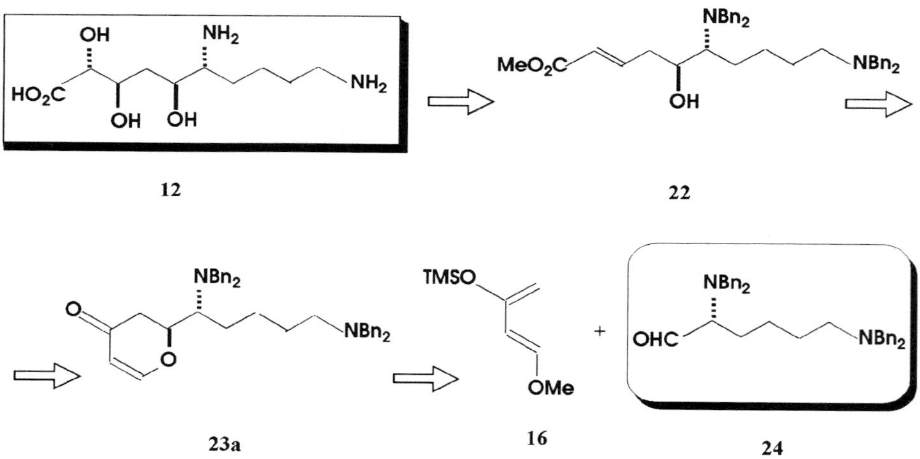

Scheme 12

Retrosynthetic analysis of galantinaminic acid (**12**) shows **D**-lysine-derived aldehyde **24** as a suitable synthon (Scheme 12).

According to our earlier studies, the required *anti*-diastereoselective Diels-Alder reaction can be expected for *N,N*-diprotected amino aldehyde. Indeed, [4+2]cycloaddition of diene **16** to aldehyde **24** led in good yield (80%) to a 9:1 mixture of *anti*-**23a** and *syn*-**23b** cycloadducts (Scheme 13).[36] Chromatographically pure adduct **23a** was transformed similarly as in galantinic acid synthesis into acetate **25** and then into ester **22**. However, the intermediate aldehyde **26** was much less stable than the carbamate-protected serinal-derived product **14**. That fact led to a very poor overall yield of this transformation. Compound **22** was transformed into natural galantaminic acid (**12**) via *cis*-hydroxylation, followed by deprotection steps (hydrogenation on Pd-C, and basic hydrolysis with 0.5 N NaOH). [29]

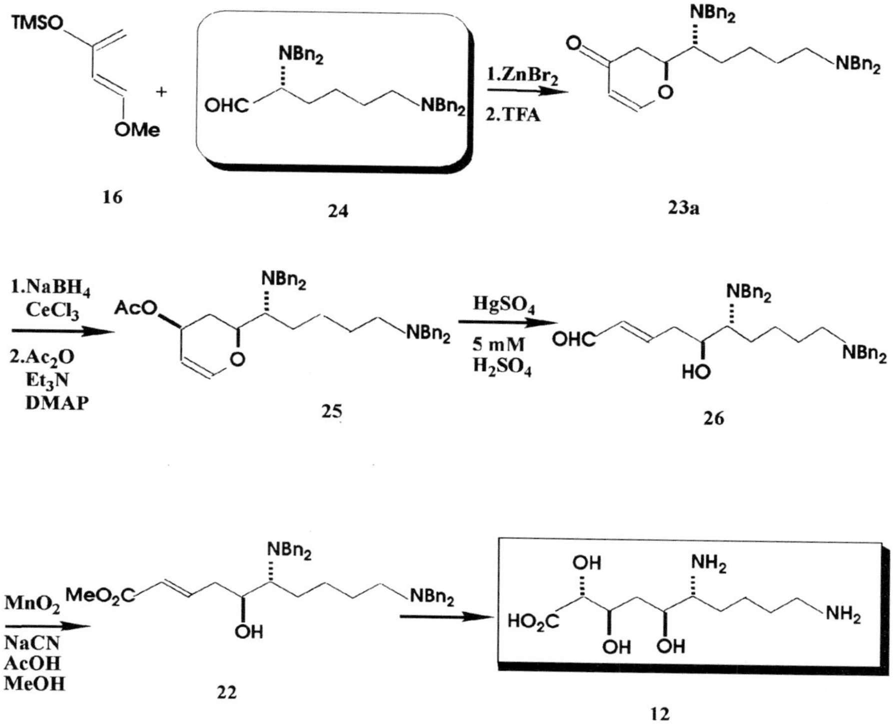

Scheme 13

L-Daunosamine (27), a naturally occurring 2,3,6-trideoxy-3-aminohexose, is an important component of the anthracycline antibiotics daunorubicin (28) and adriamycin (29) (Fig. 4), which exhibit high activity against a wide range of solid tumors and soft tissue sarcomas.[37-39] The importance of the anthracycline antibiotics as antineoplastic agents and the high cost of the microbially produced antibiotics containing L-daunosamine (27) have been the major factors contributing to the great synthetic interest in this sugar. [40,41] Most of the literature methods start from other carbohydrates, and Konradi's and Pedersen's synthesis [42] is the only exception. In Fuganti's approach, [43] D-threonine is used as a masked dihydroxy acid and the amino function is introduced in a separate step. The retrosynthetic analysis shown in Scheme 14 suggests that N,O-dibenzyl-N-tert-butoxycarbonyl-L-homoserinal (33) could serve as a starting material.

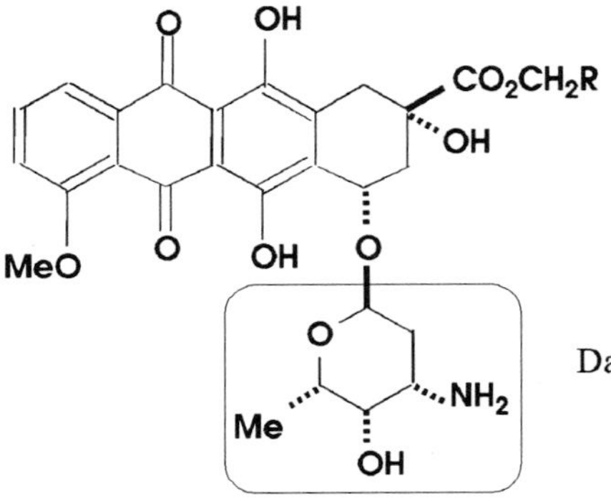

Daunosamine (27)

Daunorubicin 28 : R=H

Adriamycin 29 : R=OH

Figure 4

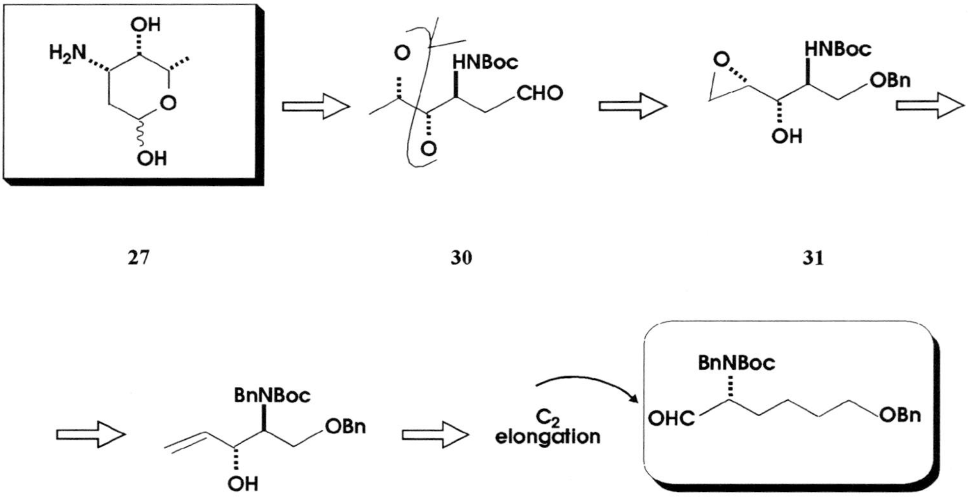

27   30   31

32   33

Scheme 14

The required *anti*-diastereoselectivity of C_2-elongation step was obtained by using the *N,N*-doubly protected α-amino aldehyde (Scheme 5, model C). The preparation of this synthon from L-aspartic acid (**34**) is shown in Scheme 15.

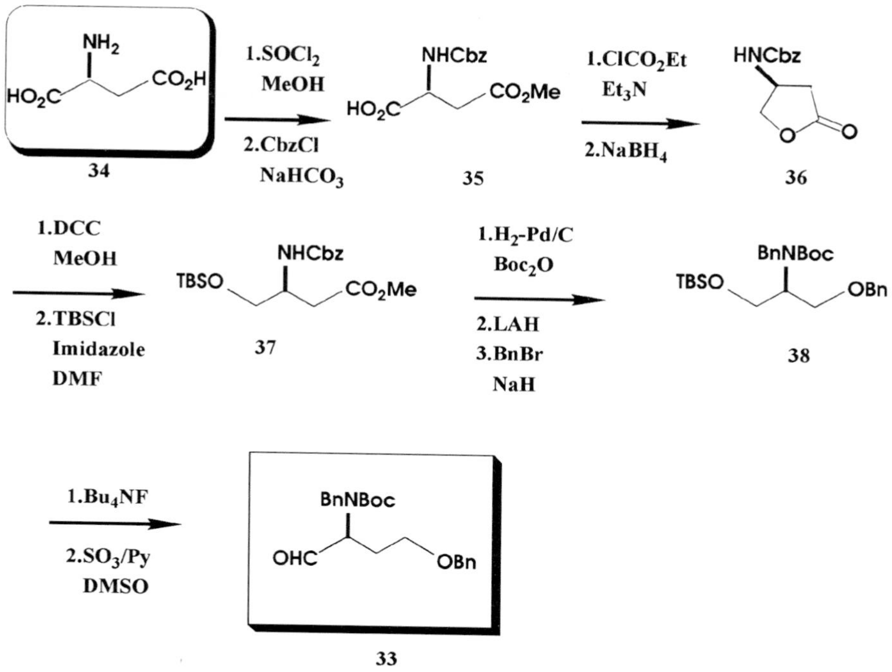

Scheme 15

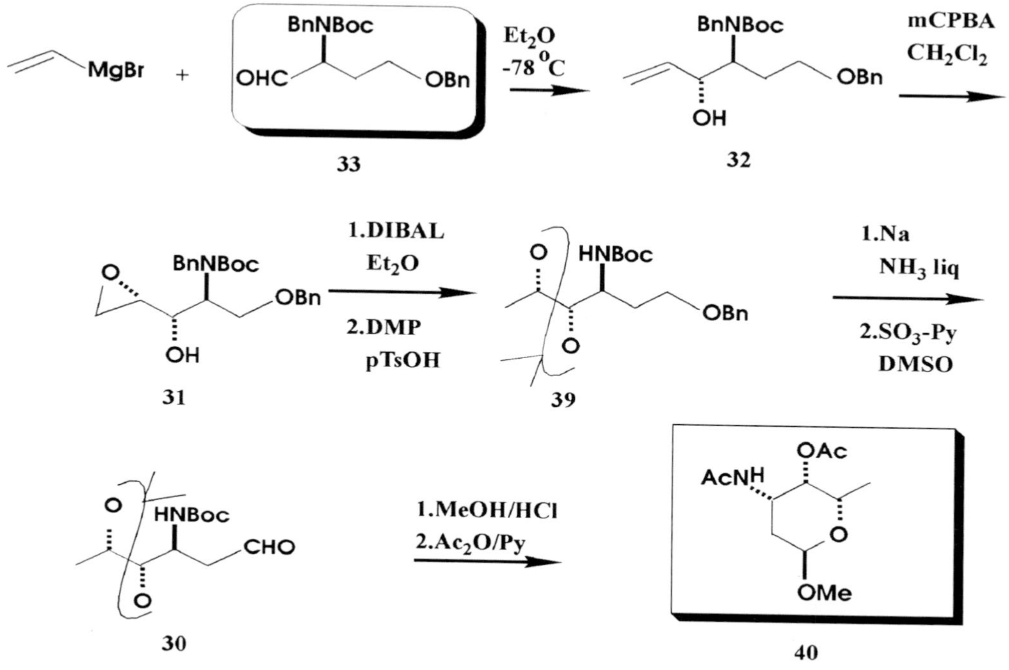

Scheme 16

The starting lactone **36** was obtained from **L**-aspartic acid (**34**) in four steps by the literature method.[44] Transesterification of **36** with methanol, followed by protection of the hydroxy group, led to compound **37**. After the *N*-carbobenzoxy protecting group was transformed into the *N-tert*-butoxycarbonyl one, [45] reduction of the ester group, followed by benzylation, afforded product **38**. Cleavage of the silyl functionality and subsequent oxidation [46] led to the **L**-homoserine derived aldehyde **33**. Vinylmagnesium chloride addition to homoserinal **33** afforded the expected *anti*-adduct **32** with excellent diastereoselectivity (Scheme 16).

Epoxidation of chromatographically pure allylic alcohol **32** gave, as a single product, the epoxide **31** of *syn*-configuration.[47] Reductive ring-opening of the epoxide **31**, followed by protection of the resulting 1,2-diol system, afforded the isopropylidene derivative **39**. Chromatographically pure compound **39** was treated with sodium in liquid ammonia to give an alcohol which was then subjected to oxidation with the sulphur trioxide - pyridine complex, furnishing aldehyde **30**. Finally, treatment of compound **30** with methanolic hydrochloride, followed by acetylation, led to a mixture of methyl α- and β-glycosides of *N,O*-diacetyl-**L**-daunosamine (**40**) in a 95:5 ratio.[48]

REFERENCES

1. S. Umezawa, *Adv. Carbohydr. Res.* (1974) 111.

2. D. J. Cooper, M. D. Yudis, R. D. Guthrie, and A. M. Prior, *J. Chem. Soc. (C)* (1971) 960.

3. J. P. Rosselet, J. Marquez, E. Meseck, A. Murawski, A. Hamdam, C. Joyner, R. Schmidt, D. Miliore, and H. L. Herzog, *Antimicrob. Agents Chemother.* (1963) 14.

4. T. Nara, M. Yamamoto, J. Kawamoto, K. Takayama, R. Okachi, S.Takasawa, and T. Sato, *J. Antibiot.* **30** (1977) 533.

5. D. J. Mason and C. DeBoer, *Antimicrob. Agents Chemother.* (1962) 544.

6. A. Golebiowski and J. Jurczak in *Recent Progress in the Chemical Synthesis of Antibiotics,* G. Lucacs, M. Ohno, Eds., Springer-Verlag Berlin, 1990.

7. S. Kondo, E. Akita, and M. Koike, *J. Antibiot. Ser. A* **19** (1966) 139.

8. S. Kondo, K. Iinuma, H. Naganawa, M. Shimura, and Y. Sekizawa, *J. Antibiot.* **28** (1975) 79.

9. M. Shimura, Y. Sekizawa, K. Iinuma, H. Naganawa, and S. Kondo, *Agric. Biol. Chem.* **40** (1976) 611.

10. N. Neuss, K. F. Koch, B. B. Molloy, W. D. Day, L. L. Hickstep, D. E. Dorman, and J. D. Roberts, *Helv. Chim. Acta* **53** (1970) 2314.

11. S. Hanessian, *Total Synthesis of Natural Products: The "Chiron" Approach,* Pergamon Press, Oxford, 1983.

12. For the review see: J. Jurczak and A. Golebiowski, *Chem. Rev.* **89** (1989) 149.

13. A. Golebiowski, U. Jacobsson, M. Chmielewski, and J. Jurczak, *Tetrahedron* **43** (1987) 599.

14. A. Golebiowski, U. Jacobsson, J. Raczko, and J. Jurczak, *J. Org. Chem.* **54** (1989) 3759.

15. A. Golebiowski, U. Jacobsson, and J. Jurczak, *Tetrahedron* **43** (1987) 3063.

16. A. Golebiowski and J. Jurczak, *Tetrahedron* **47** (1991) 1037.

17. A. Golebiowski, and J. Jurczak. *Tetrahedron* **47** (1991) 1045.

18. J. Jurczak, A. Golebiowski, and J. Raczko, *Tetrahedron Lett.* **29** (1988) 5975.

19. J. Jurczak, A. Golebiowski, and J. Raczko, *J. Org. Chem.* **54** (1989) 2495.

20. J. Raczko, A. Golebiowski, J. W. Krajewski, P. Gluzinski, and J. Jurczak, *Tetrahedron Lett.* **31** (1990) 3797.

21. A. Golebiowski, J. Raczko, U. Jacobsson, and J. Jurczak, *Tetrahedron* **47** (1991) 1053.

22.. N. T. Anh and O. Eisenstein *Nouv. J. Chim.* **1** (1977) 61.

23. H. Hashimoto, K. Asano, F. Fuji, and J. Yoshimura, *Carbohydr. Res.* **104** (1982) 87.

24. S. Danishefsky, E. Larson, D. Askin, and N. Kato, *J. Am. Chem. Soc.* **107** (1985) 1246.

25. A. Golebiowski and J. Jurczak, *J.Chem.Soc., Chem. Commun., (1989) 263.*

26. A. Golebiowski, J. Kozak, and J. Jurczak, *J. Org. Chem.* **56** (1991) 7344.

27. J. Shoji, R. Sakazaki, Y. Wakishima, K. Koizumi, M. Mayama, and S. Matsuura, *J. Antibiot.* **28** (1975) 122.

28. Y. Ohfune and N. Kurokawa, *Tetrahedron Lett.* **25** (1984) 1587.

29. K. Hori and Y. Ohfune, *J. Org.Chem.* **53** (1988) 3886.

30. N. Sakai and Y. Ohfune, *Tetrahedron Lett.* **31** (1990) 4151.

31. A. Golebiowski, J. Kozak and J. Jurczak, *Tetrahedron Lett.*, **30** (1989) 7103.

32. J. L. Luche and A. L. Gemal, *J. Am. Chem. Soc.* **101** (1979) 5848.

33. F. Gonzales, S. Lesage, and A.S. Perlin, *Carbohydr. Res.*, **42** (1975) 267.

34. E. J. Corey, N. W. Golman and B. E. Ganem, *J. Am. Chem. Soc.*, **90** (1968) 5616.

35. J. Kozak and J. Jurczak, unpublished results.

36. J. Kozak, A. Golebiowski, and J. Jurczak, unpublished results.

37. A. Dimarco, F. Arcamone, and F. Zuzino, in *Antibiotics III,* J. Corcoran and P. E. Hahn, Eds., Springer-Verlag Heidelberg, 1975, p.102.

38. S. K. J. Carter, *Natl. Cancer Inst.* **55** (1975) 1265.

39. T. Skovsgaard and N. I. Nissen, *Dan. Med. Bull.* **22** (1975) 62.

40. F. M. Hauser, and S. R. Ellenberger, *Chem. Rev.* **86** (1986) 35.

41. I. F. Pelyves, C. Monneret, and P. Herczegh, *Synthetic Aspects of Aminodeoxy Sugars of Antibiotics*, Springer-Verlag, Heidelberg, 1988.

42. A. W. Konradi and S. F. Pedersen, *J. Org. Chem.* **55** (1990) 4506.

43. C. Fuganti, P. Grasseli and G. Pedrocchi-Fantoni, *Tetrahedron Lett.* **22** (1981) 4017.

44. G. J. McGarvey, J. M. Williams, R. N. Hiner, Y. Matsubara, and T. Oh, *J. Am. Chem. Soc.* **108** (1986) 4943.

45. M. Sakaitani, K. Hori, and Y. Ohfune, *Tetrahedron Lett.* **29** (1988) 2983.

46. Y. Hamada and T. Shioiri, *Chem. Pharm. Bull.* **30** (1982) 1921.

47. K. Hori and Y. Ohfune, *J. Org. Chem.*, **53** (1988) 3886.

48. J. Jurczak, J. Kozak, and A.Golebiowski, *Tetrahedron* **48** (1992) 4231.

Flexible, Stereocontrolled Routes to Sugar Mimics via Convenient Intermediates

Hans-Josef Altenbach

Fachbereich 9 - Organische Chemie
Bergische Universität - GH Wuppertal
5600 Wuppertal 1

Summary: Novel entries to carbasugars and cyclitols with known or potential antibiotic and antiviral activity are described 1. by a "chiral-pool"-approach starting from sugar lactones or sugar alcohols and 2. by a synthesis using an enzymatic resolution of an easily accessible C_2-symmetric intermediate as a key step.

Introduction

There are many natural products with antibiotic and antiviral activity which are structurally reminiscent of monosaccharides. Most of them can be regarded as sugar analogues in which the oxygen of a pyranose has been replaced by carbon or nitrogen, respectively, the so-called carba- [1] or azasugars [2]. Their biological activity may even be due to their resemblance to sugars in that they act as sugar "mimics" being substrate or transition state analogues in enzymatic reactions whose normal substrates are carbohydrates.

Sugar analogues

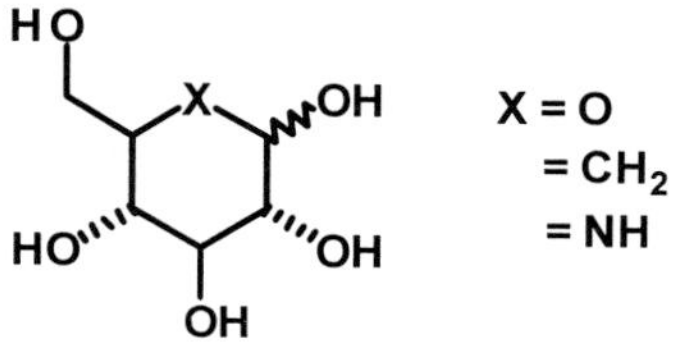

$$X = O$$
$$= CH_2$$
$$= NH$$

Scheme 1

Well known in this respect are many compounds which are inhibitors of the glycosidases whose substrates they most closely resemble [3]. Examples of some naturally occurring and synthetic glycosidase inhibitors are illustrated in Scheme 2.

Glycosidase-Inhibitors

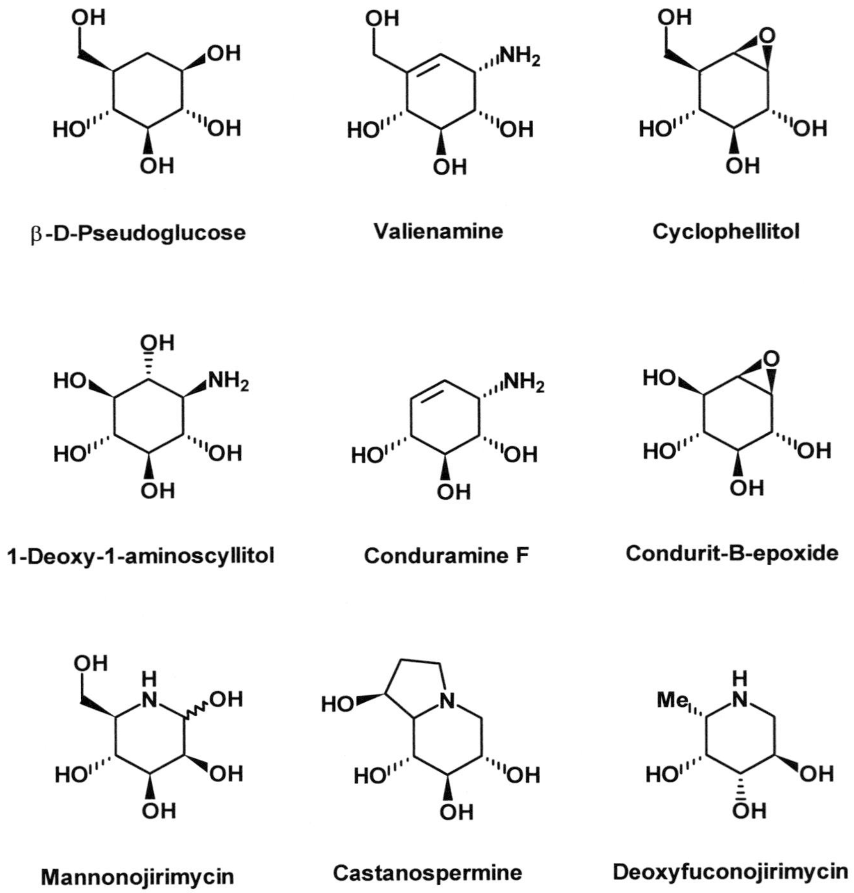

β-D-Pseudoglucose

Valienamine

Cyclophellitol

1-Deoxy-1-aminoscyllitol

Conduramine F

Condurit-B-epoxide

Mannonojirimycin

Castanospermine

Deoxyfuconojirimycin

Scheme 2

One can see that the structural similarity needs not to be too close, because simple cyclitols of the conduritol or inositol type can be effective and specific inhibitors.

Interest in such systems has increased substantially since it became evident that the inhibition of enzymes that are essential in polysaccharide and glycoprotein processing might offer a promising strategy for designing new chemotherapies for treatment of viral diseases, including AIDS. For example cyclophellitol and derivatives of nojirimycin have been studied in that respect.

Consequently, the synthesis of simple sugar analogues, being only of academic interest for a long time, has been recently stimulated. For carbasugars, several different approaches have already been developed, leading even to enantiomerically pure compounds by using starting material from the "chiral pool", mainly naturally occurring carbohydrates [4-8]. Other approaches have used polyoxygenated six-membered carbocyclic natural products such as quebrachitol [9] or quinic acid [10], or asymmetric synthesis using a Diels-Alder-strategy which has been elegantly elaborated especially by Vogel [11] and Ogawa and Suami [1,12] in different ways.

Our first approach started from sugar lactones and used a strategy we had developed years ago as a general method for the conversion of lactones to cycloalkenones of the same ring size [13].

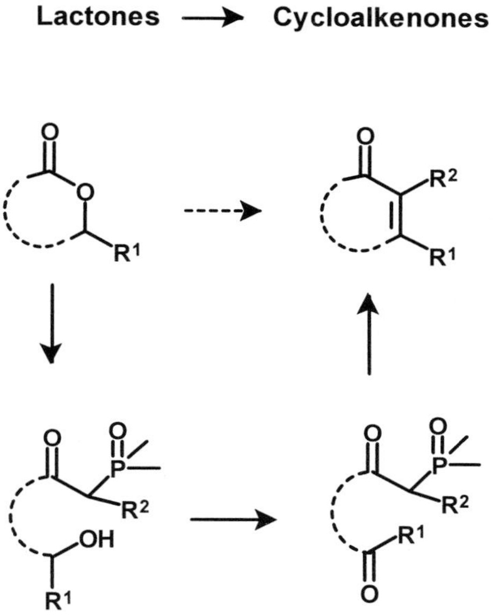

Lactones ⟶ Cycloalkenones

Scheme 3

As a second entry into primarily unbranched cyclitols, we have looked at new ways to synthesize chiral conduritols as key intermediates.

Carbasugars from Sugar Lactones

The method for the conversion of lactones to cycloalkenones consists of reacting the lactone with an alkylphosphonateanion, leading to an adduct which, after ring opening, is oxidized and then cyclized by an intramolecular olefination reaction (Scheme 3).

In the case of six-membered protected sugar lactones such as tetrabenzylgluconolactone (**1**), a problem arose because the adduct existed in the lactol form **2** and exhibited little tendency to open. Consequently, oxidation of the secondary alcohol to the ketone could not be accomplished. As seen in scheme 4, the problem was overcome by reductive opening of the lactol with NaBH$_4$. The resulting diol was then oxidized under specific Swern oxidation conditions (excess of oxidizing reagent in a dilute solution to avoid precipitation of the monoalkoxysulfonium salt intermediate and subsequent treatment with base). We did not isolate the 2,6-dioxo system **4**, but obtained instead a derived cyclized product **5** as an inseparable mixture of diastereomers. This mixture was transformed to the cyclohexenone **6** on treatment with LiBr-DBU in acetonitrile, a basic system which is recommended for

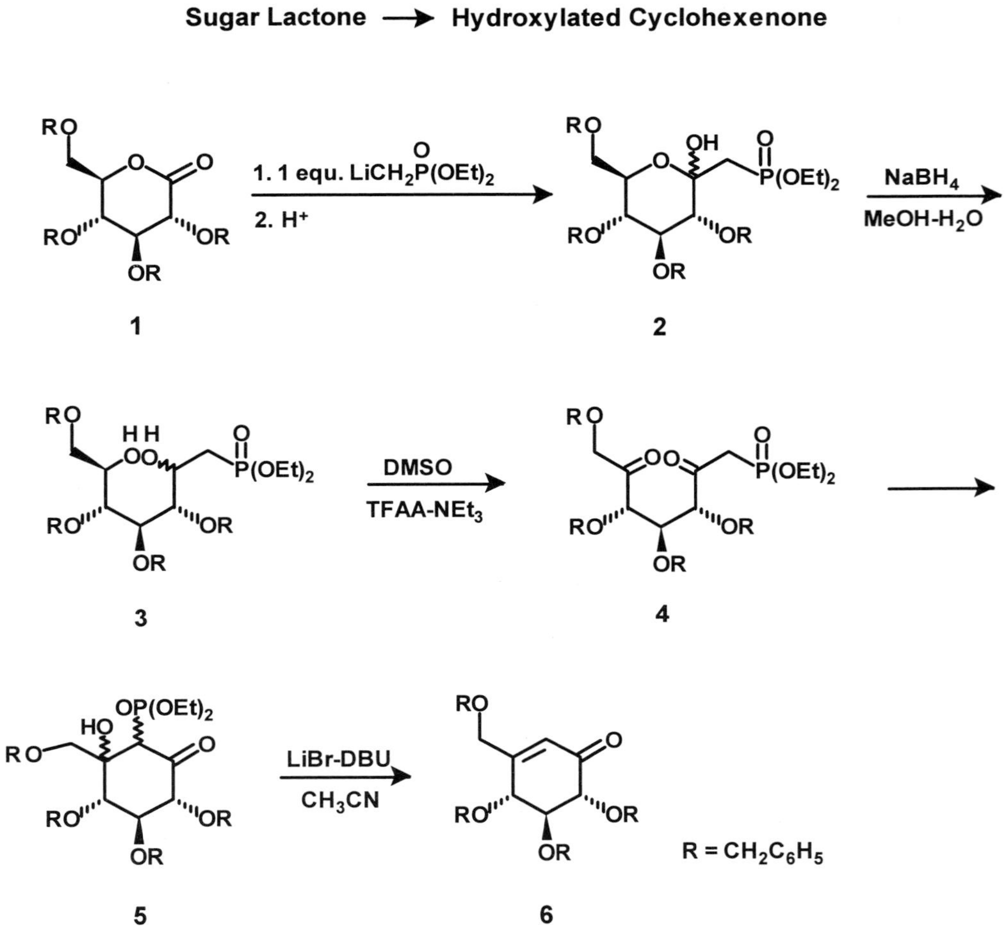

Sugar Lactone ⟶ Hydroxylated Cyclohexenone

Scheme 4

olefination reactions of base sensitive systems [14]. With most other bases, a subsequent β-elimination of benzyl alcohol took place, leading after aromatization to a phenol in which all stereocenters had been lost. Recently Fukase was able to isolate the 2,6-dioxophosphonate system **4** (with a dimethyl- instead of a diethylphosphonate group) under slightly different conditions [15].

Compound **6** can be used as a key intermediate for the synthesis of polyhydroxylated, six membered carbocyclic systems by proper functional group manipulation of the carbonyl group and the double bond. In scheme 5, a highly stereocontrolled synthesis of (+)-valienamine is depicted [16,17]. The cyclohexenone **6** was reduced to the alcohol **7** by L-Selectride, but we were not able to get its epimer **8** by a comparable stereoselective reduction. Under Luche conditions with $NaBH_4/CeCl_3$, we

Synthesis of Valienamine

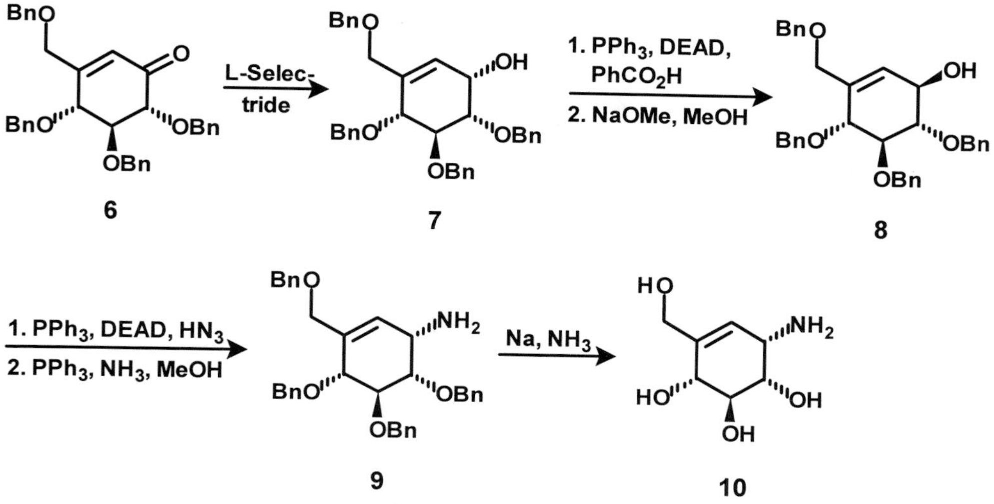

Scheme 5

observed only a mixture of both isomers. Therefore we converted **7** to **8** by a Mitsunobu inversion. The stereocontrolled introduction of the amino function was then performed by another Mitsunobu reaction with HN₃ as the nucleophile. Attempts to introduce the α-NH₂ group more directly by a stereoselective reduction of an imine derivative of the enone **6** failed. The oxime derived from **6** yielded only to mixtures of stereoisomeric unsaturated and saturated amines on reduction with different hydrides.

Compounds **7** and **8** can also be used for the synthesis of carbasugars. Hydrogenation with Ra/Ni was not stereoselective and gave only a mixture of isomers, as had been observed in a very similar system with an unprotected primary hydroxyl group by Paulsen [6], leading finally to the α-D-gluco- or β-L-ido-carbasugar systems. In the case of **13** and **17**, these cycloalkenones with galacto - or manno - like configuration of the stereogenic centers are easily available by the same general strategy starting from the corresponding sugar lactones. These compounds underwent not only a stereoselective reduction of the enone, but also a stereoselective hydrogenation (Schemes 6 and 7), leading to short, fully stereocontrolled syntheses of α-D-pseudogalactopyranose **15** and β-D-pseudomannopyranose **19** [18].

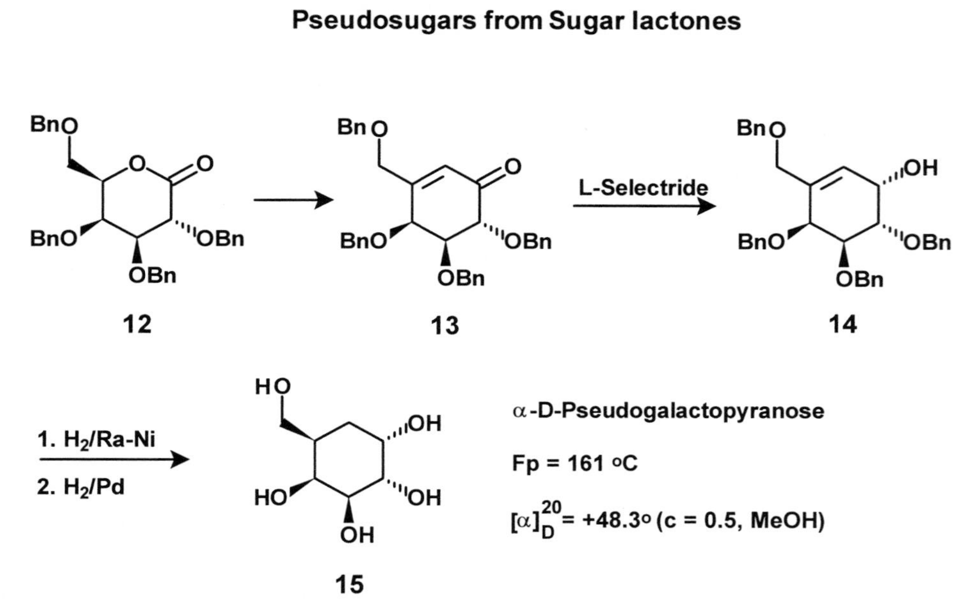

Pseudosugars from Sugar lactones

12   **13**   L-Selectride   **14**

1. H₂/Ra-Ni
2. H₂/Pd

α-D-Pseudogalactopyranose

Fp = 161 °C

$[\alpha]_D^{20} = +48.3°$ (c = 0.5, MeOH)

15

Scheme 6

16   **17**   NaBH₄/CeCl₃
or
L-Selectride   **18**

β-D-Pseudomannopyranose

Fp = 217 °C

$[\alpha]_D^{20} = +11.9°$ (c = 0.46, MeOH)

19

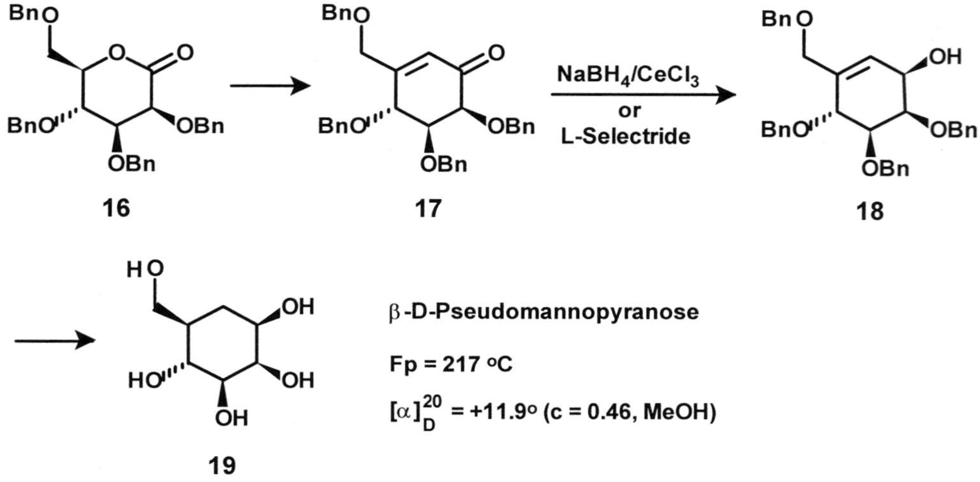

Scheme 7

New synthetic Approaches to Chiral Conduritols

Tetrahydroxycyclohexenes, known as conduritols [19], are natural products and interesting in-termediates in the synthesis of cyclitols. Eight out of ten possible isomers are chiral and there has been a lot of interest recently in the stereocontrolled syntheses of these compounds [20-24], mainly for the preparation of specific hexahydroxycyclohexanes, the inositols, but also for the synthesis of branched cyclitols [25]. In this respect, we envisioned the all trans-cyclohexentetrol, conduritol B (21), as an especially useful intermediate due to its C_2-symmetry, which should facilitate synthetic manipulations by reducing stereochemical problems.

Two different entries to conduritol B were considered (Scheme 8): first, a chiral pool approach starting from an appropriate sugar alcohol 22 and joining its two ends in such a manner that a double bond results; and second, via anti-benzene dioxide 20 which can be regarded as a dianhydro derivative of conduritol B.

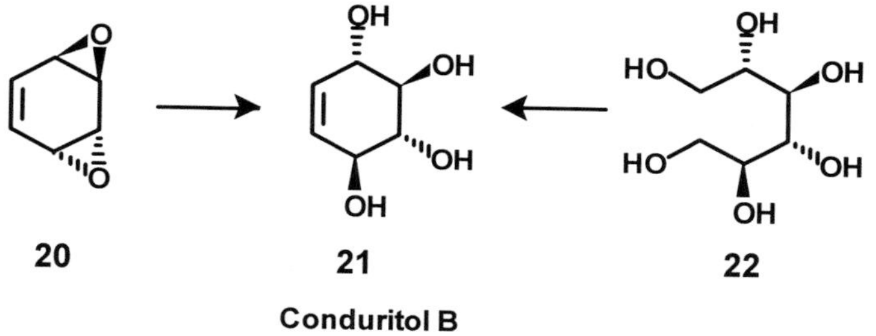

| 20 | 21 | 22 |

Conduritol B

Scheme 8

Synthesis of (+)-Conduritol B and (-)-Conduritol E from D-Mannitol

The retrosynthetic consideration shown in scheme 9 reveals that a Ramberg-Bäcklund rearrangement should lead to the desired cyclohexene systems with retention of the stereogenic centers.

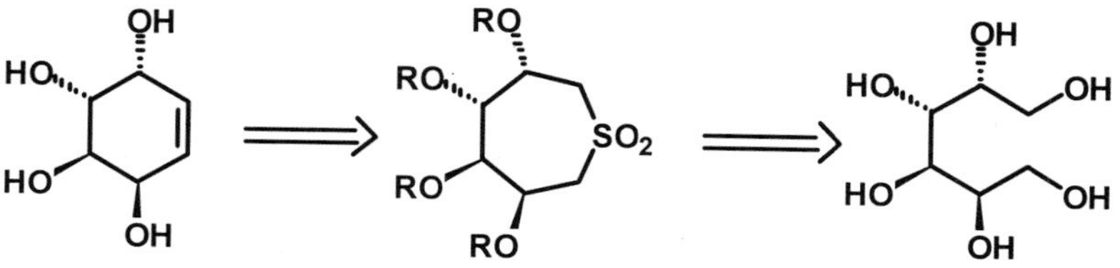

Scheme 9

The preparation of a thiepane ring from the sugar alcohol should be no problem as the primary alcohol functions can be selectively activated. It is even more attractive, however, to start from the known diepoxides 24 and 25 [26] (Scheme 10), which can be readily prepared from the D-mannitol-derived

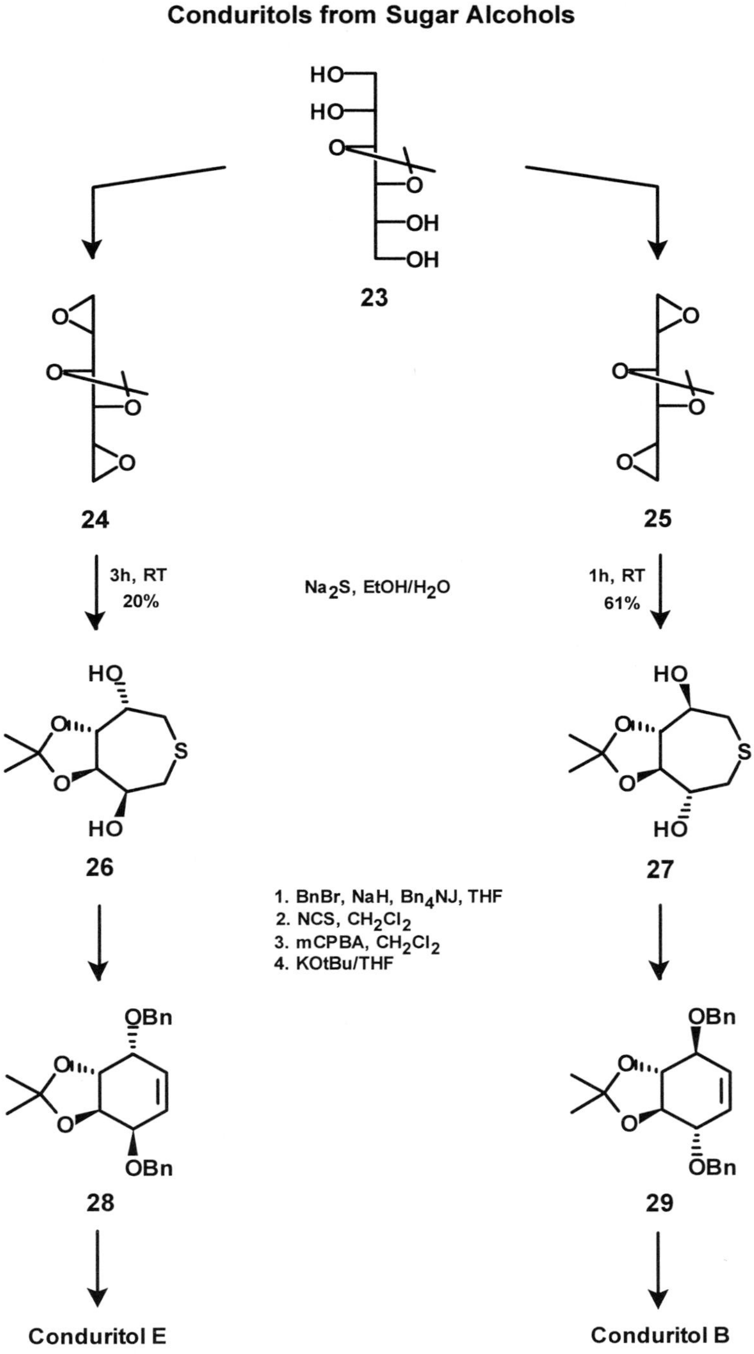

Conduritols from Sugar Alcohols

23

24   **25**

Na$_2$S, EtOH/H$_2$O

3h, RT
20%

1h, RT
61%

26   **27**

1. BnBr, NaH, Bn$_4$NJ, THF
2. NCS, CH$_2$Cl$_2$
3. mCPBA, CH$_2$Cl$_2$
4. KOtBu/THF

28   **29**

Conduritol E   **Conduritol B**

Scheme 10

system **23** as a common precursor depending on whether the primary or the secondary hydroxyl group is transformed to a leaving group. This strategy opens the possibility of preparing both isomers, conduritol B and E, from the same starting material. The synthesis turned out to be straightforward, since reaction with sodium sulfide led to the seven membered ring. En route to conduritol E, however, with compund **24**, the cyclization turned out to be somewhat difficult (only 20% yield) and proceeded more slowly than in the epimeric series with **25**, probably due to the more unfavorable steric situation in the product **26**. After protection of the hydroxyl groups, the Ramberg-Bäcklund rearrangement took place without any problem leading to the corresponding enantiomerically pure derivatives **28** and **29** of conduritol B and E respectively [27]. Whereas these stereoisomers are readily available by such a "chiral-pool-approach" from cheap D-mannitol, the preparation of their enantiomers starts from the much more expensive L-mannitol. For this and other reasons we looked for another, more flexible route to conduritols.

Enantiospecific Synthesis of all Chiral Conduritols via an Enzymatically Resolved C$_2$-Symmetric Intermediate

Anti-benzene dioxide **20** seems to be an ideal precursor for the synthesis of conduritol B as it is known to give regioselective ring opening with nucleophiles at the allylic position, thus establishing the all-trans configuration, and can be prepared in racemic form by a short and very efficient route from p-benzoquinone [28] (Scheme 11).

anti-Benzene dioxide

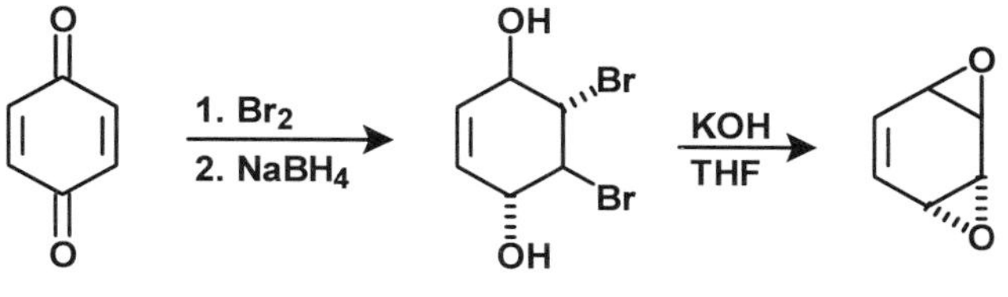

Scheme 11

Even optically active anti-benzene dioxide has been prepared [29], but the described classical resolution procedure via the MTPA-ester of **30** seems to be unsuitable for its preparation on a larger scale. We have now been able to resolve the C$_2$-symmetric intermediate **30** enzymatically [30]. It turned out that lipase catalyzed hydrolysis of the easily derived diacetate **31** was better with respect to yield and selectivity than enzymatic esterification of the diol [31]. The enzymatic hydrolysis of racemic **31** with PPL in a phosphate buffer at pH 7.0 was very slow, but it stopped after 50% conversion (4 days) and proceeded with excellent enantioselectivity for both the remaining diacetate **31** and for the hydrolyzed

Enzymatic Resolution

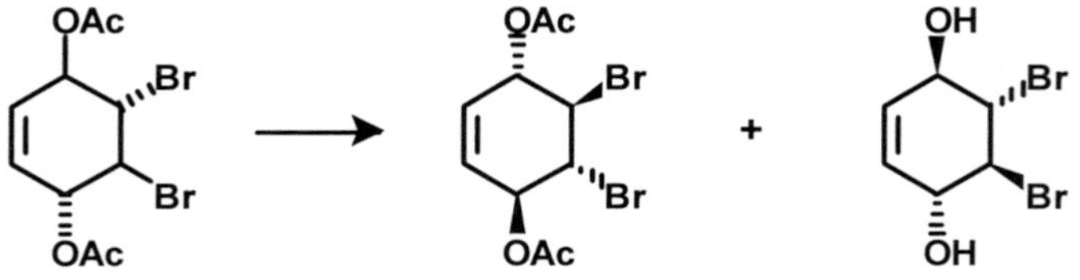

31   (+)-31   (+)-30

racemic   PPL: >99% ee   96% ee
                     PLE: 85% ee

Scheme 12

diol **30**. Due to its much better solubility in nonpolar solvents, **31** was easily separated from **30** by extraction with pentane. Base promoted epoxide ring formation of the pure enantiomers of either **30** or **31** occurred without problems, as described for the racemates.

(+)-anti-Benzene dioxide

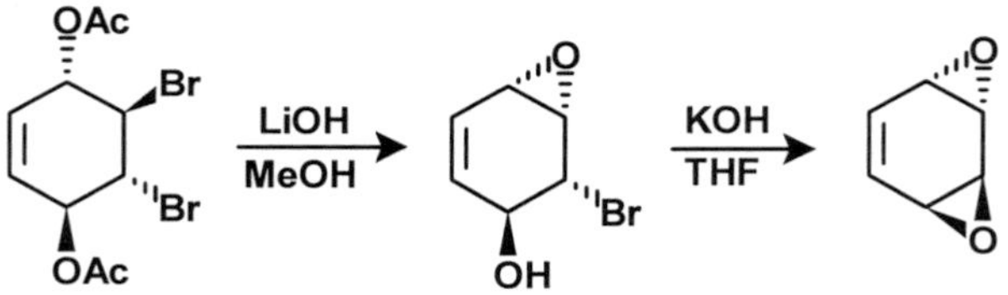

31   32   20

$[\alpha]_D^{20}$: +13°   +150°   +321°

Scheme 13

The direct conversion of (+)- or (-)-diol **30** to enantiomerically pure (+)- or (-)-anti-benzene dioxide **20** was performed with powdered potassium hydroxide in THF. On treating the diacetate **31** with lithium hydroxide in methanol, the formation of only one epoxide ring was achieved. Starting from (+)-**31**, the monoepoxide (+)-**32** was isolated, which may be further transformed to the anti-benzene dioxide (+)-**20** with KOH in THF (Scheme 13). Monoepoxide **32** also constitutes an interesting chiral building block for the synthesis of highly functionalized cyclohexene systems having all four stereogenic centers differentiated.

With compounds **20** and **31** in both enantiomeric forms in hand, enantiospecific routes to all chiral conduritols were devised [32]. (+)-Conduritol B was prepared by reaction of (-)-anti-benzene dioxide **20** with sodium benzyl alcoholate in benzyl alcohol, leading to the exclusive formation of (+)-**33**, which was deprotected to (+)-**21** by reduction with sodium in ammonia (Scheme 14).

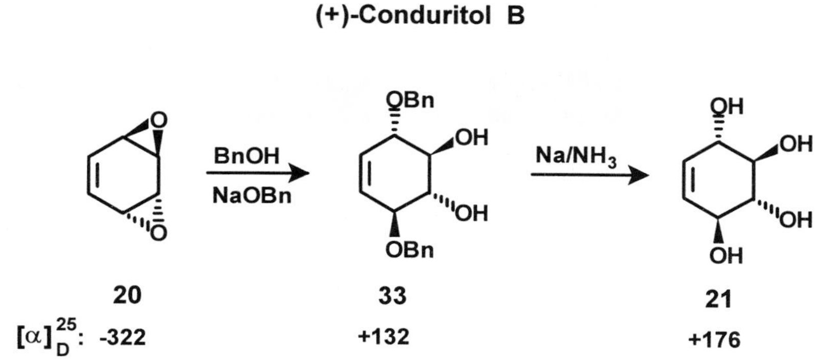

(+)-Conduritol B

20

$[\alpha]_D^{25}$: -322

33

+132

21

+176

Scheme 14

For the synthesis of conduritols C and E, the double bond in **31** was either cis- or trans-hydroxylated to give compounds **34** and **36**, respectively, which were debrominated after acetylation with zinc in acetic acid to the tetraacetates **35** and **37**. Whereas there is no stereochemical problem due to the C_2-symmetry of **31** during the addition to the double bond, the stereoselective formation of the diastereomer **36** in the opening of the intermediate epoxide is remarkable and can be explained by the Fürst-Plattner rule.

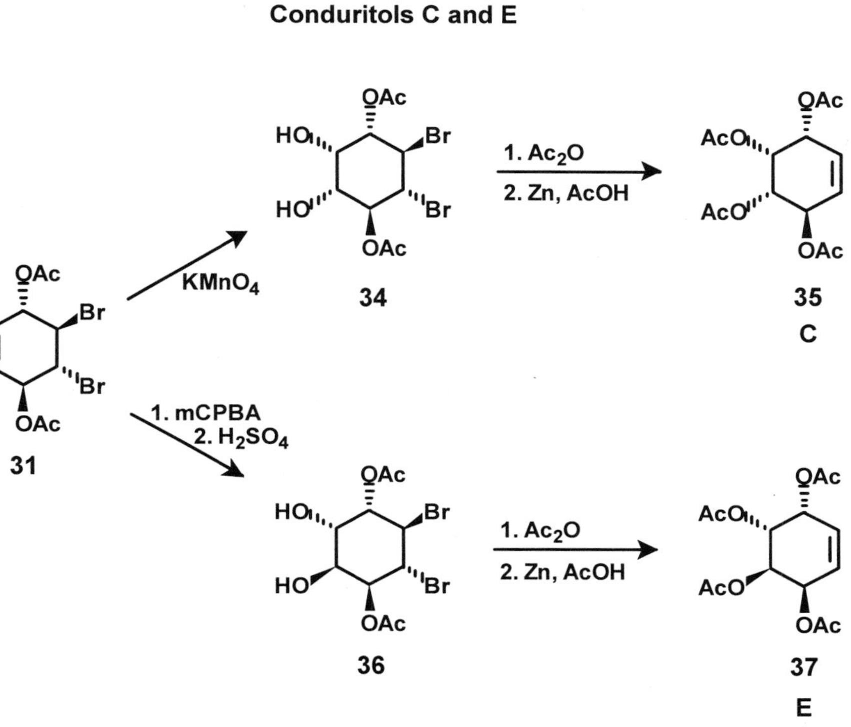

Conduritols C and E

31

34

1. Ac_2O
2. Zn, AcOH

35

C

1. mCPBA
2. H_2SO_4

36

1. Ac_2O
2. Zn, AcOH

37

E

Scheme 15

The last chiral tetrahydroxycyclohexene isomer, conduritol F, was also derived from the key intermediate **30**, but by a more elaborate strategy (Scheme 16). **30** was first transformed to the cyclic carbonate **38** before an epoxide ring closure to **39** was undertaken. Ring opening of the epoxide at the allylic position with benzyl alcohol established the desired configuration of the conduritol F system. The parent system was liberated from **40** with sodium in ammonia. Starting from (+)-**30**, conduritol F in its (-)-form was obtained enantiomerically pure.

Conduritol F

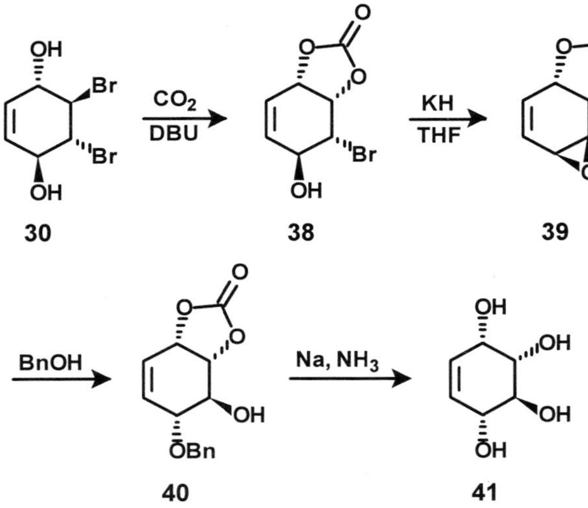

Scheme 16

Thus, enantiospecific routes to all chiral conduritols have been developed from the dibromocyclohexene diol **30** as a chiral building block which is easily available in both enantiomeric forms by an enzymatic resolution procedure.

Furthermore, to demonstrate that specifically substituted inositols can be efficiently prepared, a rather short synthesis of (+)-pinitol is described in scheme 17 starting from **39** [33].

(+)-PINITOL

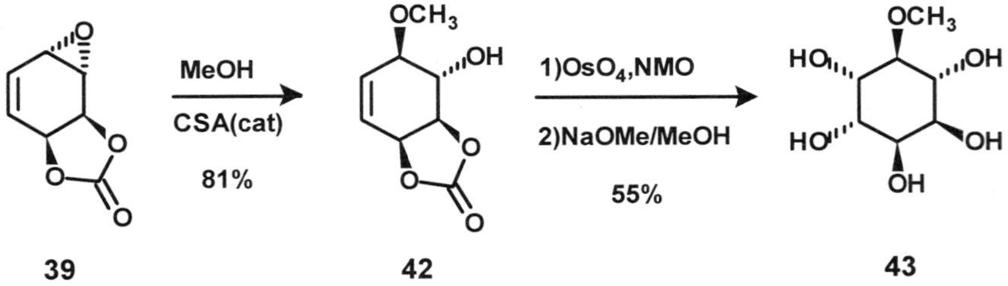

Scheme 17

43 is a *chiro* inositol derivative, where both enantiomers have been found in various plant sources. (+)-**43** has recently found some interest because of its hypoglycemic and antidiabetic activity [34].

Syntheses of other interesting cyclitol systems, including branched ones, will be published in due course.

Acknowledgement: I am very grateful to my coworkers who contributed to the work described here and are named in the references. Furthermore, for financial support, I thank the "Deutsche Forschungsgemeinschaft" and the "Fonds der Chemischen Industrie".

References

1. T. Suami, S. Ogawa, Adv. *Carbohydr. Chem. Biochem.* **48** (1990) 21.

2. H. Paulsen, K. Todt, *Adv. Carbohydr. Chem.* **23** (1968) 115; G. W. J. Fleet, *Chem. Brit.* **1989**, 287.

3. G. Legler, Adv. Carbohydr. *Chem. Biochem.* **48** (1990) 319; M. L. Sinnott, *Chem. Rev.* **90** (1990) 1171.

4. R. Blattner, R. J. Ferrier, *J. Chem. Soc., Chem. Commun.* 1987, 1008; R. J. Ferrier, A. E. Stuz, *Carbohydr. Res.* **205** (1990) 283.

5. D. H. R. Barton, S. Augy-Dorey, J. Camara, P. Dalko, J. M. Delaumény, S. D. Géro, B. Quiclet-Sire, P. Stütz, *Tetrahedron* **46** (1990) 215.

6. H. Paulsen, W. Deyn, *Liebigs Ann. Chem.* **1987**, 125.

7. K. Tadano, H. Maeda, M. Hoshimo, Y. Iimura, T. Suami, *J. Org. Chem.* **52** (1987) 1946.

8. C. S. Wilcox, J. J. Gaudino, *Carbohydr. Res.* **205** (1990) 233.

9. H. Paulsen, W. Röben, F. R. Heiker, *Chem. Ber.* **114** (1981) 3242; T. Akiyama, H. Shima, S. Ozaki, *Tetrahedron Lett.* **32** (1990) 5593; A. P. Kozikowski, A. H. Fauq, G. Powis, D. C. Melder, *J. Am. Chem. Soc.* **112** (1990) 4528; N. Chida, K. Yamada, S. Ogawa, *J. Chem. Soc., Chem. Commun.* **1991**, 588.

10. T. K. M. Shing, Y. Cui, Y. Tang, *Tetrahedron* **48** (1992) 2349.

11. P. Vogel, Bull. Soc. Chim. Belg. **99** (1990) 395; P. Vogel, D. Fattori, F. Gasparini, C. Le Drian, *Synlett* **1990**, 173.

12. S. Ogawa, Y. Iwasawa, T. Suami, *Chem. Lett.* 1984, 355; S. Ogawa, M. Kemura, T. Fujita, *Carbohydr. Res.* **177** (1988) 213.

13. H.-J. Altenbach, W. Holzapfel, G. Smerat, S. Finkler, *Tetrahedron Lett.* **1985**, 6329.

14. M. A. Blanchette, W. Choy, J. T. Davis; A. P. Essenfeld, S. Masamune, W. Roush, T. Sakai, *Tetrahedron Lett.* **1984**, 2183.

15. H. Fukase, S. Horii, *J. Org. Chem.* **57** (1992) 3651.

16. H.-J. Altenbach, W. Holzapfel, H.-J. Dax, unpublished results in these laboratories.

17. For other syntheses of valienamine see: H. Paulsen, F. R. Heiker, *Liebigs Ann. Chem.* **1981**, 2180; R. R. Schmidt, A. Köhn, *Angew. Chem. Int. Ed. Engl.* **26** (1987) 482; M. Yoshikawa, B. C. Cha, Y. Okaichi, Y. Takinami, Y. Yokokawa, I. Kitagawa, *Chem. Pharm. Bull.* **36** (1988) 4236; S. Ogawa, Y. Shibata, T. Nose, T. Suami, *Bull. Chem. Soc. Jap.* **58** (1984) 3387; F. Nicotra, L. Panza, F. Ronchetti, G. Russo, *Gazz. Chim. Ital.* **119** (1989) 577; S. Knapp, A. B. J. Naughton, T. G. Murali Dhar, *Tetrahedron Lett.* **33** (1992) 1025, see also ref. [15] and the contribution of B. M. Trost in this book.

18. H.-J. Altenbach, H.-J. Dax, unpublished results.

19. Review: M. Balci, Y. Sutbeyaz, H. Secan, *Tetrahedron* **46** (1990) 3715.

20. S. Ley, A. J. Redgrave, *Synlett* **1990**, 393.

21. C. Le Drian, J. P. Vionnet, P. Vogel, *Helv. Chim. Acta* **73** (1990), 161.

22. H. A. J. Carless, Tetr. Asymm. **3** (1992) 795; T. Hudlicky, R. Fan, H. Luna, H. Olivo, J. Price, *Pure Appl. Chem.* **64** (1992) 1109; T. Hudlicky, H. Luna, H. F. Olivo, C. Anderson, T. Nugent, J. D. Price, *J. Chem. Soc. Perkin Trans. 1*, **1991**, 2907.

23. C. R. Johnson, P. A. Plé, L. Su, M. J. Heeg, J. P. Adams, *Synlett* **1992**, 388.

24. L. Dumortier, P. Liu, S. Dobbelaere, J. Van der Eycken, M. Vandewalle, *Synlett* **1992**, 243.

25. See f.i.: L. Dumortier, J. Van der Eycken, M. Vandewalle, Synlett **1992,** 245; S. V. Ley, L. L. Yeung, *Synlett* **1992**, 291.

26. Y. Le Merrer, A. Dureault, C. Greck, D. Micas-Languin, C. Gravier, J. Depezay, *Heterocycles* **25** (1987) 541.

27. H.-J. Altenbach, A. Dombert, unpublished results.

28. H.-J. Altenbach, H. Stegelmeier, E. Vogel, *Tetrahedron Lett.* **1978,** 3333.

29. M. Koreeda, M. Yoshihara, *J. Chem. Soc. Chem. Commun.* **1981** 974.

30. H.-J. Altenbach, G. Klein, *Tetrahedron Asymm.*, submitted for publ.

31. The asymmetrization of meso 2,3-diprotected Conduritol A derivatives by lipase catalyzed reactions has been described: see ref. [23, 24].

32. H.-J. Altenbach, D. Bien, G. Klein, *Tetrahedron Asymm.*, submitted for publ.

33. H. J. Altenbach, D. Bien, unpublished results.

34. S. V. Ley, F. Sternfeld, *Tetrahedron* **45** (1989) 3463; T. Hudlicky, J. D. Price, F. Rubis, T. Tsunoda, *J. Am. Chem. Soc.* **112** (1990) 9439.

Transglycosylase Inhibition

Peter Welzel

Fakultät für Chemie der Ruhr-Universität

Postfach 102148, D-4630 Bochum (Germany)

Introduction

Among the different constituents of the bacterial cell wall, the most important for the survival and integrity of the cell is peptidoglycan, cf. **F** in Scheme 1, which shows the peptidoglycan of *E.coli*. [1]
The two successive final reactions in the biosynthesis of cross-linked peptidoglycan (exemplified here for *E.coli*) from the membrane precursor **D** (Scheme 1) are (i) the transglycosylation (**D** ⇒ **E**) that extends the glycan chain and (ii) the transpeptidation (**E** ⇒ **F**) that cross-links the glycan chains through two

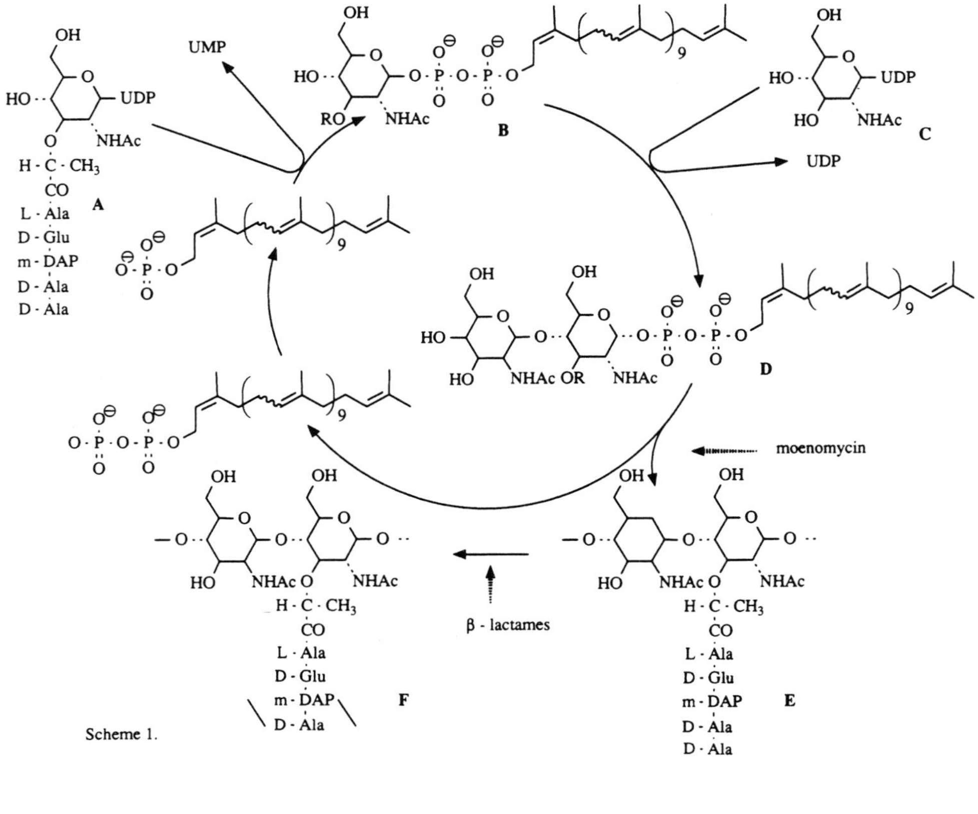

Scheme 1.

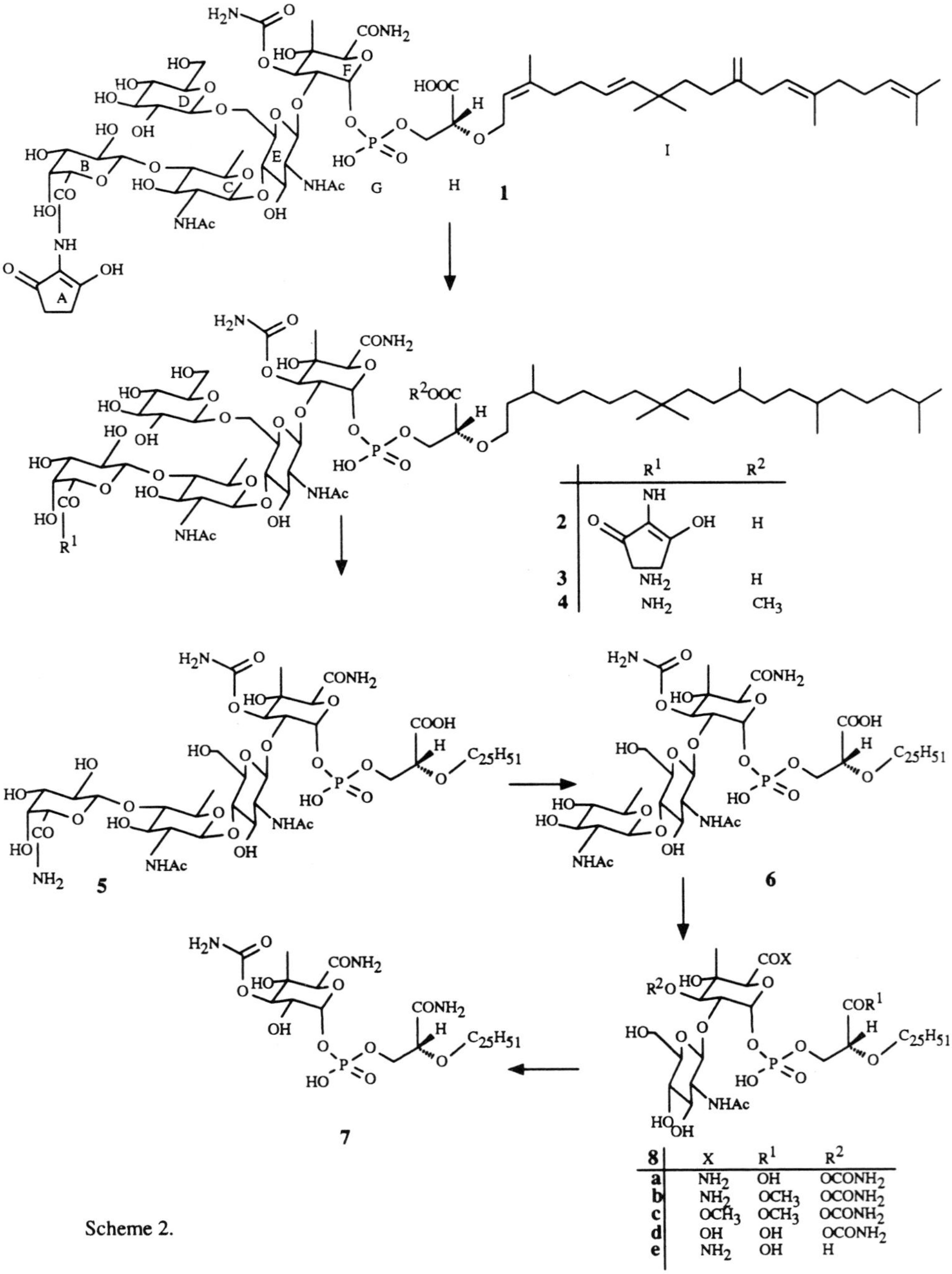

Scheme 2.

peptide units. A number of bifunctional enzymes (penicillin-binding proteins, PBP's) have been identified that catalyze both transglycosylation and transpeptidation. [2] With cell-free systems from *E.coli*, it was demonstrated that the antibiotic moenomycin A selectively inhibits the transglycosylation step by its inhibitory effect on penicillin-binding protein 1b (PBP 1b). An in-vitro assay specific for the polymerization by transglycosylation (using radiolabelled intermediate **D** as substrate) was developed. With the cell-free systems and purified PBP 1b, moenomycin A was inhibitory at concentrations between 10^{-8} and 10^{-7} mol/l. [3] Moenomycin A (the main component of the trade product flavomycin^R), its derivatives, and related antibiotics [4] belong to the rare class of compounds compounds known to efficiently inhibit the transglycosylation reaction.

Stepwise Degradation of Moenomycin A and Structure-Activity Relations.

Moenomycin A has structure **1** as determined by a combination of chemical degradation and spectroscopic methods. [5] With the aim of defining the minimal structural basis of the antibiotic activity a systematic degradation of **1** was performed. [5b,6,5c] Hydrogenation of **1** yielded the decahydro derivative **2**. The tritiated analogue of **2** was used to demonstrate that the antibiotic binds reversibly to PBP 1b. [3] From **2** the enolic unit A was removed oxidatively. Stepwise degradation of the sugar chain (**3** ⇒ **6**, **6** ⇒ **8a**, **8a** ⇒ **7**) was performed using selective periodate cleavage as the key reaction. With the help of protective group chemistry, **3** could also be degraded to tetrasaccharide **5**. With the exception of **7** (and methyl ester **4**), all degradation products were fully active transglycosylase inhibitors. From **8a**, methyl ester **8b**, the diester **8c**, and the uronic acid analogue **8d** were prepared. All these compounds were inactive. Treatment of **8a** with butylamine in methanol gave in a clean reaction to **8e**, again devoid of in vitro activity. In vivo results obtained for *Staph.aureus* parallel those for the *E.coli* cell-free system. From these findings it was concluded that compound **8a** contains all the structural features necessary for full antibiotic activity. There is a striking structural similarity between **8a** and intermediate **D** (Scheme 1) of the peptidoglycan biosynthesis.

Structure-Activity Relations Based on Synthesis

A synthetic route to analogues of **8a** was devised in which suitable E-F and H-I building blocks are coupled to the phosphate group. The procedures summarized in Scheme 3 were used to prepare suitable 2-O-alkylated glyceric acid derivatives. [7] **10** and **11** carry the C_{25} chain of moenomycin and its decahydro derivatives, respectively, whereas the route commencing from D-mannitol (**12**) allows the 2-*O*-alkyl substituent (e.g. **14**) to be chosen arbitrarily.

The disaccharide portions of **8a** analogues were prepared as summarized in Scheme 4. For the construction of the phosphoric acid diester grouping, the Ugi variant [8,9,10] of the phosphite methodology has proven its merits. Thus, treatment of **24** (X = 1,2,4-triazol-1-yl) first with one of the disaccharides **22b**, **22e**, **22f** and then with **11**, followed by oxidation with bis(trimethylsilyl)peroxide [11] provided the phosphoric acid triesters. These were converted into the corresponding phosphoric acid diesters (**25a**, or **25c** or **25e**) by treatment with Zn-Cu couple. Final deprotection in the case of **25a** to give **25b** was achieved by basic hydrolysis. In the case of **25c**, complete removal of all ester groups could not be achieved without also hydrolyzing the uronamide function to give **25b**. For this reason the Troc protected compound **25d** was prepared which on Zn-Cu treatment yielded **25e**. From **25e** the desired analogue **25f** was obtained by mild base hydrolysis.

Understandably, **25b** was antibiotically inactive, since the uronamide grouping in unit F is essential for

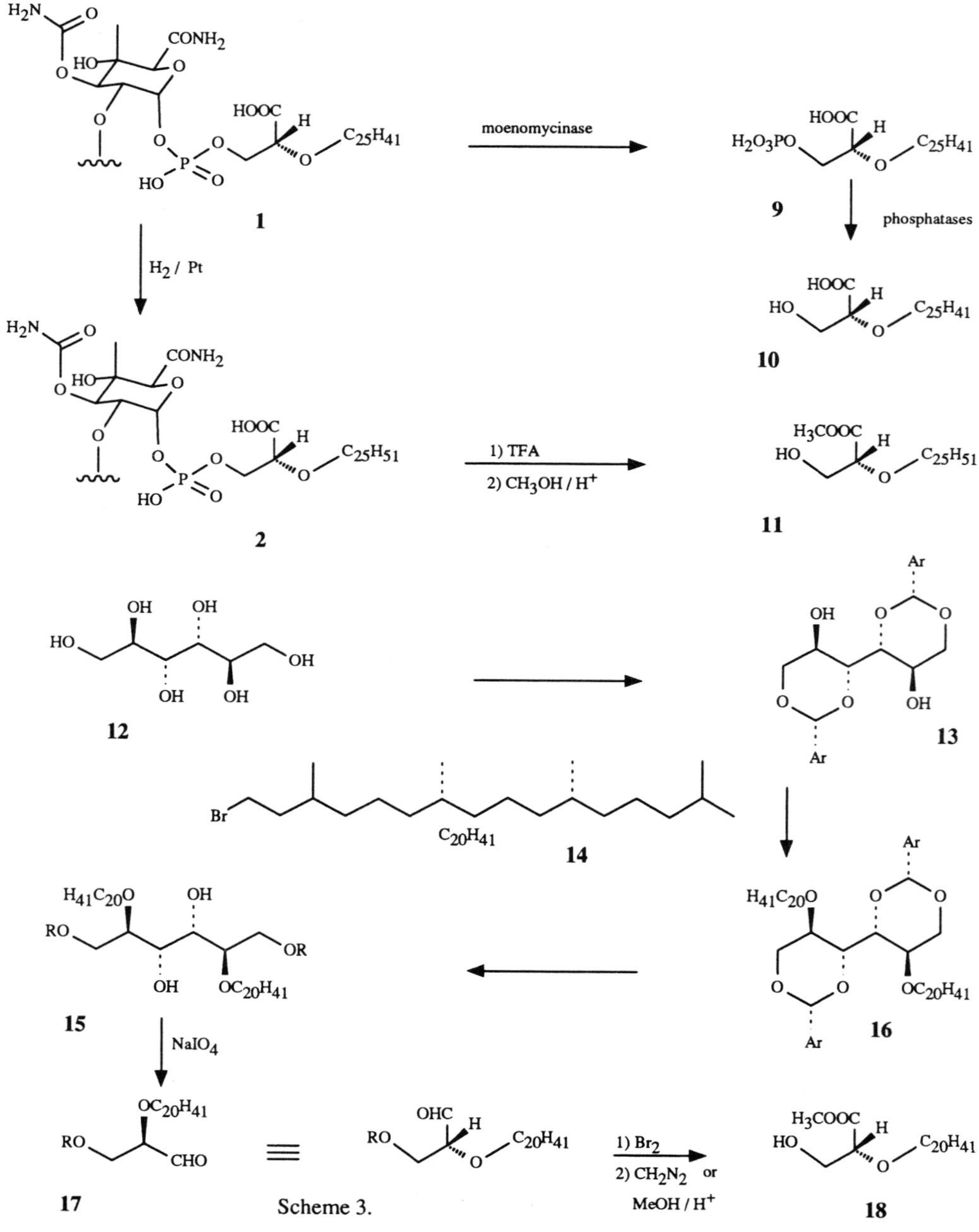

Scheme 3.

antibiotic activity (vide supra). But also **25f** was found to be devoid of antibiotic activity. This result though disappointing, stresses at the same time the high specificity of the interaction of moenomycin (**1**) and degradation products such as **8a** with the binding site at the transglycosylating enzyme that forms the basis of the antibiotic activity. [12]

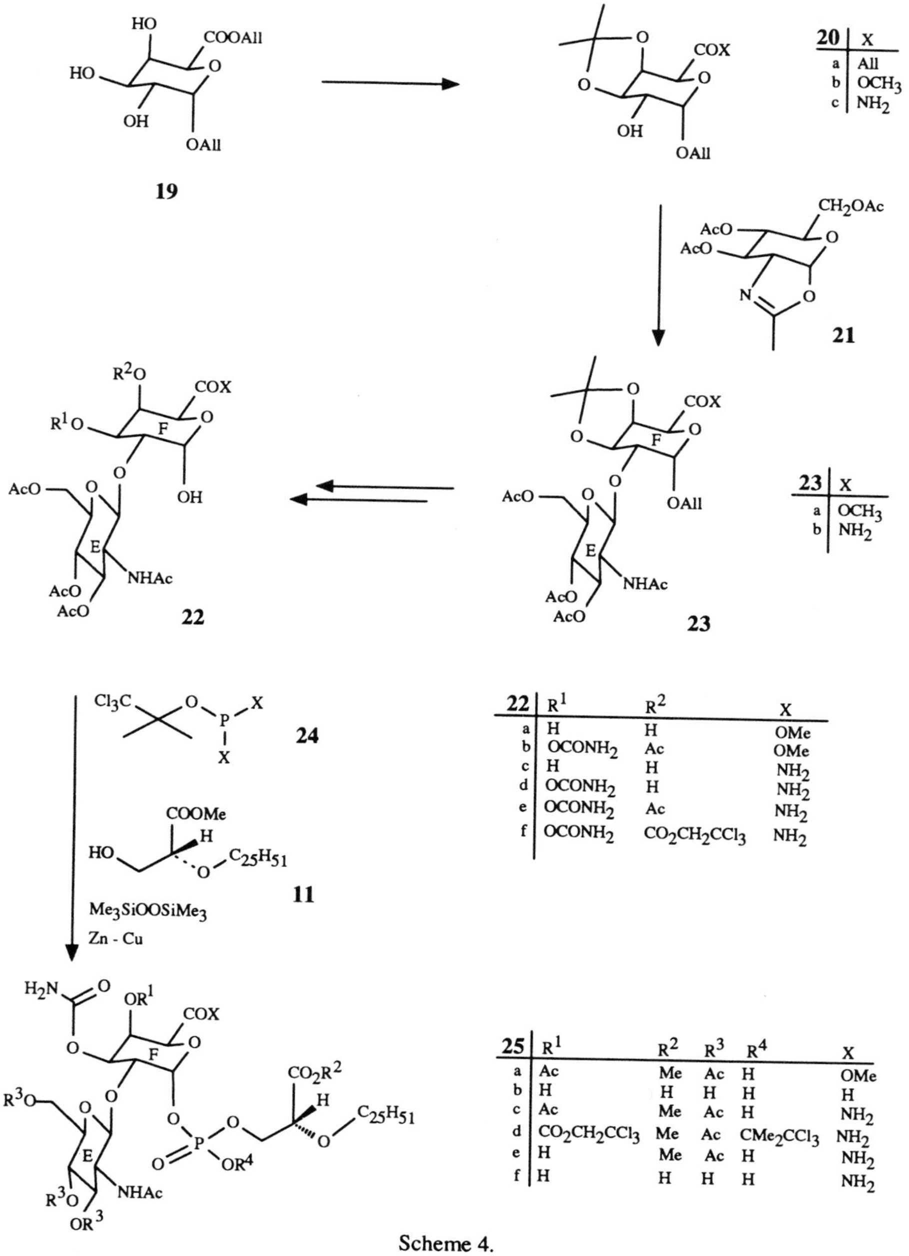

20	X
a	All
b	OCH_3
c	NH_2

23	X
a	OCH_3
b	NH_2

22	R^1	R^2	X
a	H	H	OMe
b	$OCONH_2$	Ac	OMe
c	H	H	NH_2
d	$OCONH_2$	H	NH_2
e	$OCONH_2$	Ac	NH_2
f	$OCONH_2$	$CO_2CH_2CCl_3$	NH_2

25	R^1	R^2	R^3	R^4	X
a	Ac	Me	Ac	H	OMe
b	H	H	H	H	H
c	Ac	Me	Ac	H	NH_2
d	$CO_2CH_2CCl_3$	Me	Ac	CMe_2CCl_3	NH_2
e	H	Me	Ac	H	NH_2
f	H	H	H	H	NH_2

Scheme 4.

Acknowledgements: The work summarized above is the result of the joint efforts of mainly three groups (at the Ruhr-Universität in Bochum, the Université Paris-Sud at Orsay, and the Hoechst AG). It gives me great pleasure to thank all (named in the references) who contributed to what has been achieved in this difficult area. Financial support by the Deutsche Forschungsgemeinschaft, the Fonds der Chemischen Industrie, and the Hoechst AG is gratefully acknowledged.

References

1. N. Sharon "Complex Carbohydrates - Their Chemistry, Biosynthesis, and Functions", Addison Wesley, London 1975.

2. Y. van Heijenoort, M. Gómez, M. F. Derrien, J. Ayala, J. van Heijenoort, *J. Bacteriol.* **1992**, *174*, 3555, and references therein.

3. Review: J. van Heijenoort, Y. van Heijenoort, P. Welzel, in P. Actor; L. Daneo-Moore, M. L. Higgins, M. R. J. Salton, G. D. Shockman (eds) "Antibiotic Inhibition of Bacterial Cell Wall Surface Assembly and Function" American Society for Microbiology, Washington 1988, p. 549.

4. Review: G. Huber in F. E. Hahn (ed.) Antibiotics, Vol. V/1, p. 135, Springer, Berlin 1979.

5. a) P. Welzel, F.-J. Witteler, D. Müller, W.Riemer, *Angew.Chem.* **1981**, *93*, 130; *Angew. Chem. Int. Ed. Engl.* **1981**, *20*, 121; b) P. Welzel, B. Wietfeld, F. Kunisch, Th. Schubert, K. Hobert, H. Duddeck, D. Müller, G. Huber, J. E. Maggio, D. H. Williams, *Tetrahedron* **1983**, *39*, 1583; c) H.-W. Fehlhaber, M. Girg, G. Seibert, K. Hobert, P. Welzel, Y. van Heijenoort, J. van Heijenoort, *Tetrahedron*, **1990**, *46*, 1557.

6. P. Welzel, F. Kunisch, F. Kruggel, H. Stein, J. Scherkenbeck, A. Hiltmann, H. Duddeck, D. Müller, J. E. Maggio, H.-W.Fehlhaber, G. Seibert, Y. van Heijenoort, J. van Heijenoort, *Tetrahedron* **1987**, *43*, 585.

7. a) P. Welzel, F.-J. Witteler, D. Müller, *Tetrahedron Lett.* **1976**, 1665; b) Th. Schubert, F. Kunisch, P. Welzel, Tetrahedron **1983**, *39*, 2211; c) Th. Schubert, K. Hobert, P. Welzel, *Tetrahedron* **1983**, *39*, 2219; d) U. Peters, W. Bankova, P. Welzel, *Tetrahedron* **1987**, *43*, 3803; e) K.-H. Metten, K. Hobert, S. Marzian, U. E. Hackler, U. Heinz, P. Welzel, W. Aretz, D. Böttger, U. Hedtmann, G. Seibert, A. Markus, M. Limbert, Y. van Heijenoort, J. van Heijenoort, *Tetrahedron* **1992**, *48*, 8401.

8. R. G. K. Schneiderwind-Stöcklein, I. Ugi, *Z. Naturforsch.* **1984**, *39b*, 968.

9. J. L. Fourrey, D. J. Shire, *Tetrahedron Lett.* **1981**, *22*, 729.

10. R. L. Letsinger, E. P. Groody, N. Lander, T. Tanaka, *Tetrahedron* **1984**, *40*, 137.

11. L. Wozniak, J. Kowalski, J. Chojnowski, *Tetrahedron Lett* , **1985**, *26*, 4965; Y. Hayakawa, M. Uchiyama, R. Noyori *Tetrahedron Lett.* **1986**, *27*, 4191; and references therein.

12. a) H. Hohgardt, W. Dietrich, H. Kühne, D. Müller, *Tetrahedron* **1988**, 44, 5771; b) U. Möller, K. Hobert, A. Donnerstag, P. Wagner, D. Müller, H.-W. Fehlhaber, A. Markus, P. Welzel, unpublished.

Recent Developments in the Enantiospecific Synthesis of Amaryllidaceae Alkaloids

Yves Chapleur, Françoise Chrétien and Mustapha Khaldi

Laboratoire de Chimie Organique 3, associé au CNRS, Université de Nancy I

BP 239, F 54506 Vandoeuvre-les Nancy (France)

Summary: Several routes toward the synthesis of *enantiomerically pure amaryllidaceae Alkaloids* have been explored starting from either D-*glucose* or D-*gulonolactone*. A short synthetic approach toward the A and C ring assemblies of these *antitumor* compounds, utilizing an *intramolecular aldol reaction*, is described.

Introduction

Some plants of the genus *Amaryllidaceae* produce highly oxygenated alkaloids.[1] Lycoricidine (**1**) and narciclasine (**2**) were isolated, for the first time, about twenty years ago, [2] and exhibit interesting antitumor activities by inhibition of protein synthesis.[3] More recently, Pettit et al. [4] isolated a more complex member of this class of compounds, pancratistatin (**3**).

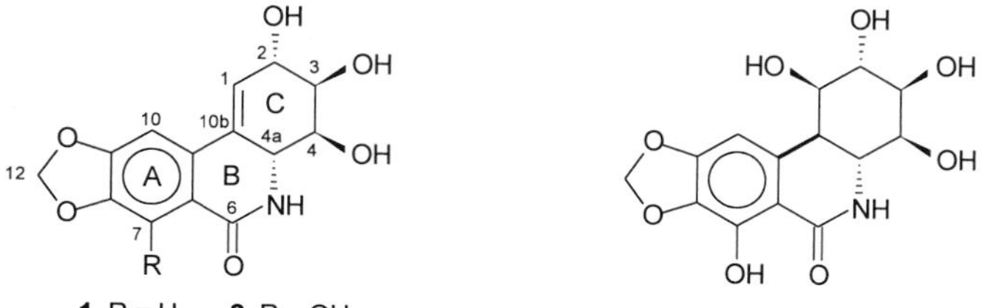

1 R = H   **2** R = OH   **3**

Scheme 1

Their interesting biological properties and intricate structures with four to six contiguous chiral centres have prompted several studies aimed at total synthesis or syntheses of analogues. In a pioneering study starting from natural narciclasine, Krohn and Mondon [5] demonstrated the importance of the C ring of this compound for biological activity. Some synthetic studies directed towards the synthesis of the phenantridinone system of these compounds have been published, [6] but only a few total syntheses have been recorded for lycoricidine [7] and pancratistatin.[8] No synthesis of narciclasine has yet been completed. Some years ago we embarked on a program aimed at the synthesis of narciclasine, pancratistatin and related compounds with a view to obtaining compounds with improved biological activity. In addition, we hoped to get new insights into the biological target and mode of action of these compounds.

Results

We have explored two different routes. The first one is based on the retrosynthetic analysis depicted in Scheme 2.

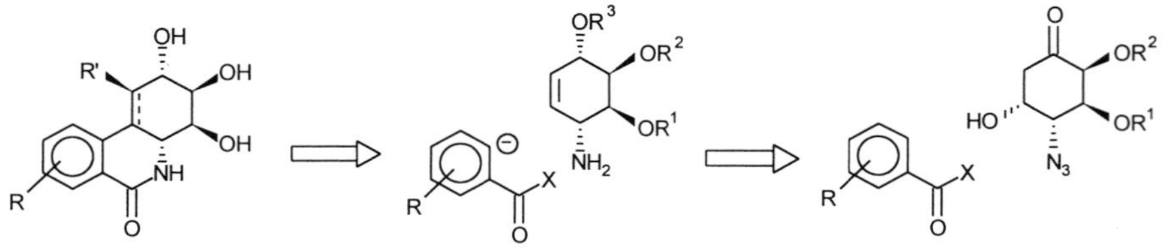

Scheme 2

This convergent synthesis consisted mainly in the coupling of the A and C rings by formation of a C-C bond between a suitable aromatic part and the appropriately functionalized C ring which could be obtained in optically pure form from a carbohydrate precursor. The advantage of this strategy was to allow the synthesis of a large variety of narciclasin analogues modified either on the aromatic ring or on the C ring, with the coupling occurring at a late stage of the synthesis.

Synthesis of the C ring

The synthesis starts from D-glucose and takes advantage of the new carbocyclic ring closure of Ferrier.[9] The key feature of this synthesis was first, the introduction of the amino group in a suitably protected form with the correct stereochemistry and second, the suitable functionalization at C-4 for the subsequent coupling. The preparation of the suitable chiron **8** is summarized in Scheme 3.

Enone **8** was prepared from the known azido alcohol **4** [10] by a standard series of reactions. The intermediate bromo alcohol **5** was transformed into the key intermediate **6,** taking advantage of a new reaction of simultaneous protection/elimination using sodium hydride in DMF. [11] This new reaction overcomes some drawbacks in the formation of 5,6-unsaturated sugars using silver fluoride. The next crucial step was the Ferrier ring closure. Initial attemps to carbocyclize **6** using mercury chloride in wet acetone/hydrochloric acid resulted in extensive aromatization of the expected

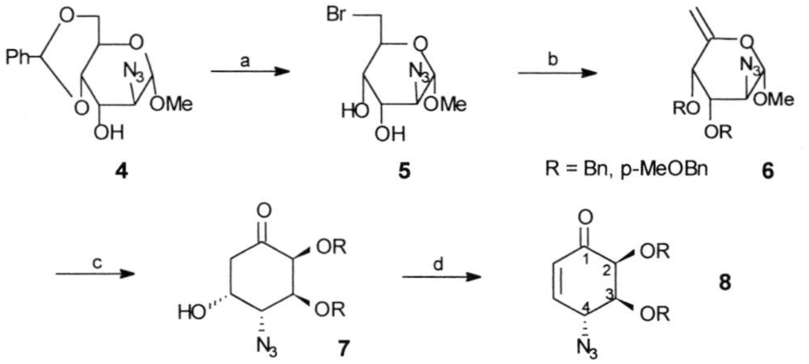

Scheme 3: a) NBS, CaCO$_3$, CCl$_4$; MeONa, MeOH. b) NaH, RBr, DMF. c) HgSO$_4$, acetone/water. d) MsCl, CH$_2$Cl$_2$, pyridine.

cyclohexanone **7** by a series of β-eliminations. It was more convenient to use mercury trifluoroacetate in acetone/water, which cleanly gave **7**. Further dehydration yielded the key enone **8**.

The reactivity of this enone in 1,4-addition reactions with aromatic cuprates was examined next. In a model study, diphenyl copper lithium **9a** cleanly reacted with enone **8** to give the expected cyclohexanone **10a** as a single isomer, resulting from axial attack of the organocopper species. Thus, we examined the preparation and the use of suitably substituted aromatic derivatives which could be

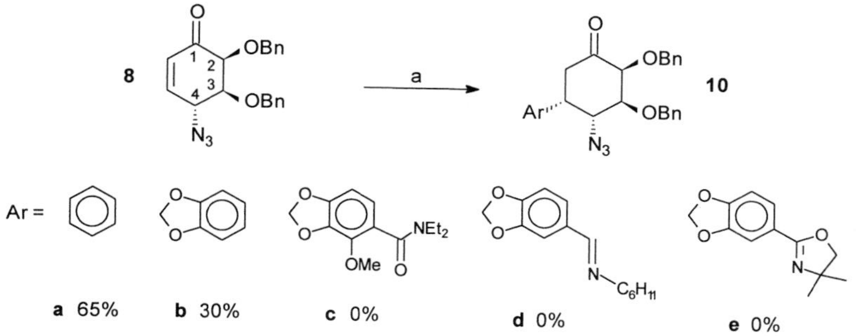

Scheme 4: a) **9** = Ar$_2$CuLi

coupled to **8**. Cuprate **9a** and **9b** gave acceptable results while those derived from **9c**, **9d** and **9e** were unreactive towards 1,4-addition, probably because of the number of complexation sites.

Disappointed by the failure of making the C-C bond in an intermolecular fashion, we planned to used an intramolecular reaction for the B ring closure.

Suitable aromatic acids were prepared and reacted with amino ketone **11** obtained from the azido ketone **7** by hydrogenation. Amide formation using the BOP reagent [12] to activate the carboxylic function cleanly gave **12**. Subsequent elimination and reduction of the enone system under Luche's conditions [13] gave the corresponding allylic alcohol **14** as a single isomer. Mitsunobu [14] inversion gave the inverted benzoate **15**. However, since aminoalcohol **11** was unstable and difficult to handle, we explored an alternate route.

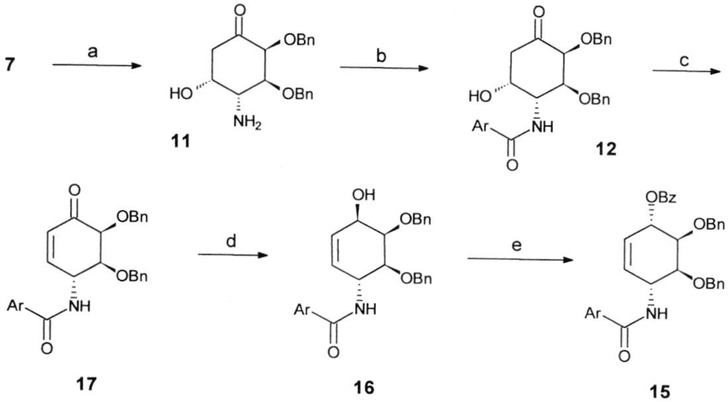

Scheme 5: a) H$_2$, Pd/C 5%. b) BOP, ArCOOH. c) TsCl, pyridine. d) NaBH$_4$, CeCl$_4$, EtOH. e) PPh$_3$ /DEAD, PhCOOH.

This route started from enone **8** which was reduced to **16** under Luche's conditions. Mitsunobu benzoylation with inversion of **16** gave **17** which was subsequently reduced to aminoalcohol **18**. Formation of the amide **19** proceeded as above in good yields.

At this stage, all of the functionalities of narciclasine and lycoricidine, with the correct stereochemistry, were present. It was interesting to check the biological activities of these flexible analogues of narciclasine in which the B ring is open. None of these derivatives was found to be active *in vitro* against L1210 or P388 leukemia below 22.5 M, whatever the substitution of the aromatic ring or the presence of the double bond. This would indicate that these conformationally flexible molecules cannot efficiently bind to the ribosome.

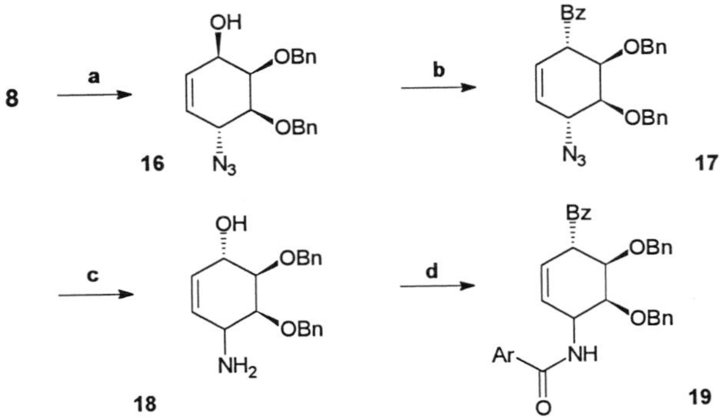

Scheme 6: a) CeCl₃, NaBH₄. b) PPh₃ /DEAD, PhCOOH. c) LiAlH₄. d) BOP, ArCOOH.

Owing to the facile access to unsaturated amides, we thought that the formation of a C-C bond using a Heck reaction could be favored. This reaction has been applied with success to olefins, enones, and unsaturated esters. [15] Thus different haloamides **20**, **22** and **24** were prepared and submitted to various Heck conditions (Pd(PPh₃)₄, proton sponge, DMF, 20°C or Pd(PPh₃)₄, NEt₃, DMF, 60°C or Pd(PPh₃)₄, DMF, 60°C or Pd(OAc)₂, NaHCO₃, tBu₄N⁺HSO₄⁻, DMF).

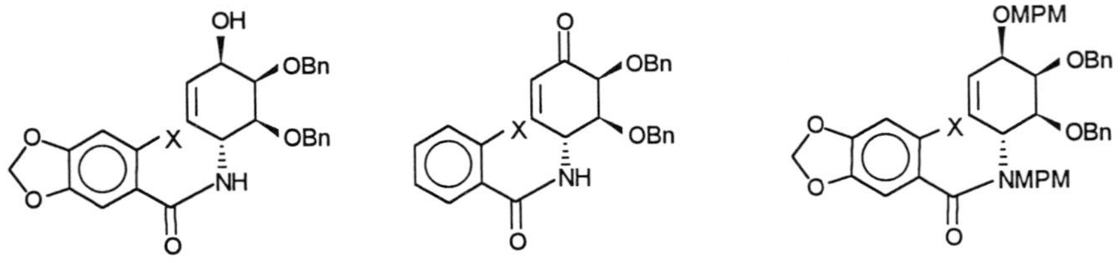

Scheme 7

In all cases, even under forced conditions, only recovery of starting compounds or complete degradation was observed. Recently, after this part of our work had been completed, Chida *et al.* [7e]

published a similar route to lycoricidine. The last step was a Heck reaction using the new conditions described by Grigg. [16] We have reinvestigated our own experiments using the amide **22** which differs only from that of the Japanese workers by the protecting groups at C-3 and C-4 (benzyl instead of MOM ethers). However, in our hands, these conditions failed to give any cyclized compound but led only to recovery of starting compound (54% after four different attempts). No insertion of palladium into the C-Br bond was observed. Further investigations are currently underway to explain these observations.

A 6-*exo* radical ring closure [17] was attempted on bromo derivatives **20** and **24** which were treated with tributyltin hydride to yield only the reduced compounds **21** and **25**. Even with the activated olefin of the model compound **22**, only reduction occured and no cyclization was observed. [18]

At the same time, we became aware of a new method for the coupling of aromatics with olefins proposed by Edström and Livinghouse. [19] We tried to transpose this methodology to our case, first on model compounds. Several aromatic olefins were treated with N-phenyl selenophthalimide (NPSP) in the presence of a Lewis acid such as tin tetrachloride or boron trifluoride in nitromethane. Clearly amides behave differently giving heterocyclization rather than carbon-carbon bond formation. [20]

We found that this drawback could be circumvented using an amine instead of an amide. Thus amines **27** were prepared by reaction of amine **18** with aldehyde **26** followed by reduction of the resulting imine with sodium borohydride. On treatment of **27a** with NPSP and tin tetrachloride, a complex mixture was produced; the main product **28** resulted from the hydroxy selenylation of the double bond. In the absence of an ortho methoxy group on the aromatic ring as in **27b**, only small amounts of the expected cyclized product **29** were isolated.

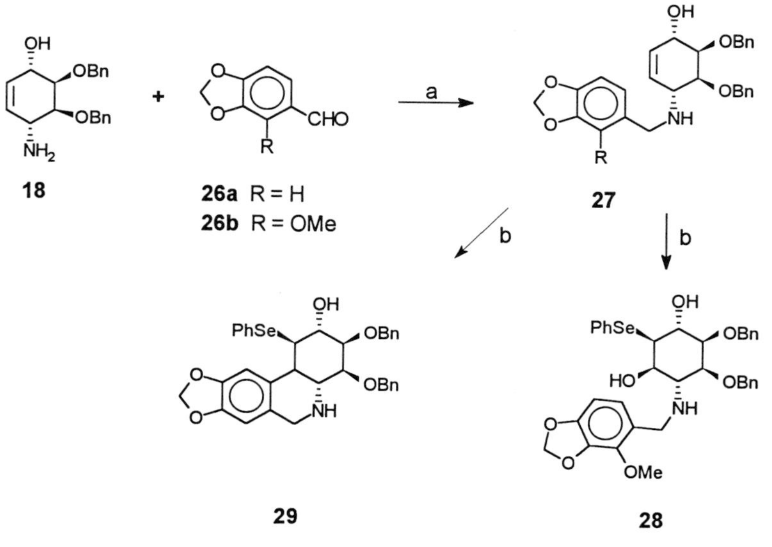

18

26a R = H
26b R = OMe

27

29

28

Scheme 8: a) Benzene, azeotropic reflux, NaBH$_3$CN. b) NPSP, SnCl$_4$.

At the same time, we devised another route, making the assumption that two carbon-carbon bonds would be more easily made than one bond. Thus, we decided to explore the use of benzylic derivatives in which the benzylic carbon could be the future C10b of the alkaloids. This strategy is delineated in Scheme 9.

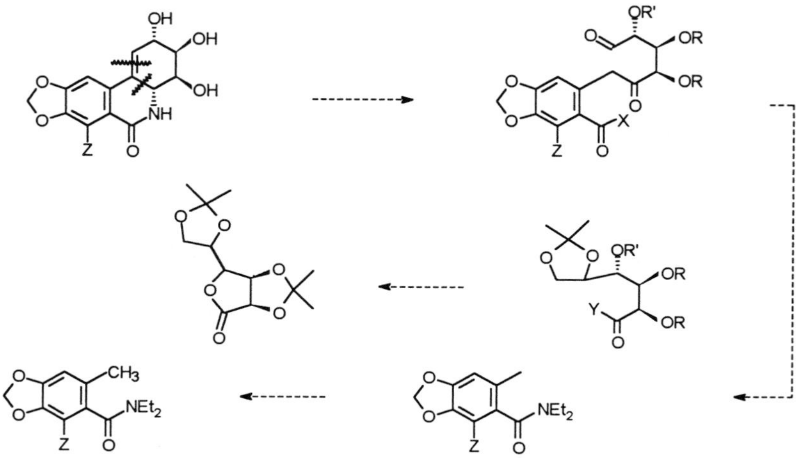

Scheme 9

The key feature is the condensation of a benzylic anion with a suitable lactone followed by intramolecular Knoevenagel reaction between an aldehyde and the benzylic active methylene. The required aldehyde could be obtained from the oxidative cleavage of a 1,2-diol and thus, it would be easy to start from a hexose. D-gulonolactone proved to be the ideal precursor in terms of functionalities and stereochemistry of the three chiral centres already present in this material. After some attempts with different benzylic anions, it was found that tertiary amides gave good results in the formation of such anions.

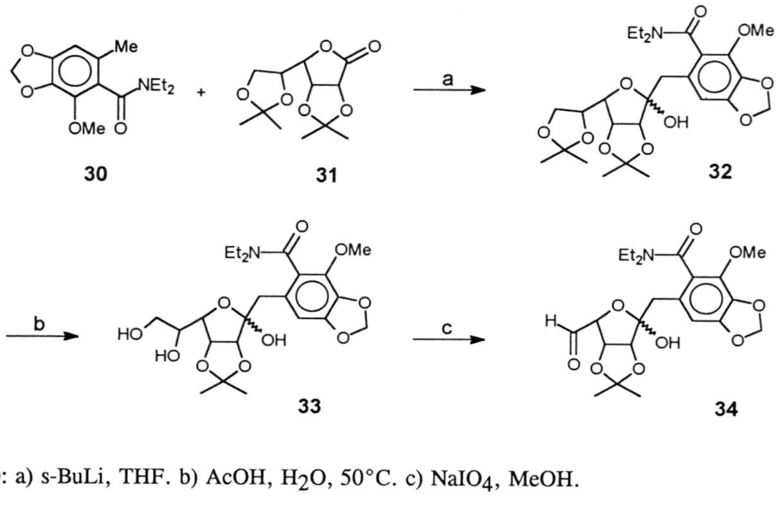

Scheme 10: a) s-BuLi, THF. b) AcOH, H_2O, 50°C. c) $NaIO_4$, MeOH.

Condensation of the preformed anion, from **30** and s-BuLi, with di-O-isopropylidene D-gulonolactone **31** gave the expected alcohol **32** in 65 % yield. This compound was selectively deprotected to provide diol **33** which was cleaved to give the corresponding aldehyde **34**.

After extensive experiments, we found that the intramolecular cyclization proceeded cleanly using sodium carbonate and a catalytic amount of DBU in THF to give cyclohexenone **35** together with the corresponding aldol product **36**. Standard acetylation gave only acetate **37**.

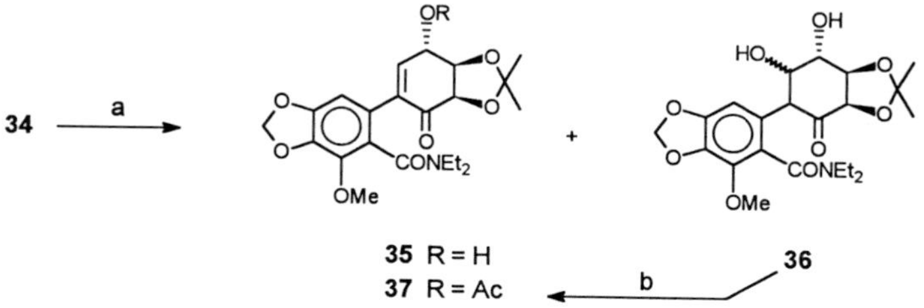

35 R = H
37 R = Ac   **36**

Scheme 11: a) Na₂CO₃, DBU (cat.), THF. b)Ac₂O, DMAP, pyridine.

This sequence was successfully extended to different substituted aromatic rings. Finally, we attempted to introduce the amino group at C5. Cerium chloride/sodium borohydride reduction of the enone **37** gave an 2/1 epimeric mixture of the allylic alcohols **38** and **39** in 73% yield.

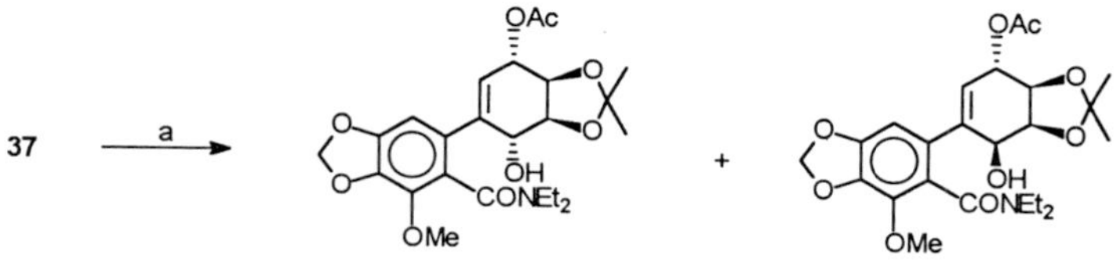

38   **39**

Scheme 12: a) NaBH₄ CeCl₃, MeOH.

After unsuccessful attempts to introduce an azide group by activation with triphenylphosphine/-diethylazodicarboxylate (DEAD) in the presence of Yamada's reagent [(PhO)₂PON₃], the isomer **38** was treated with trifluoromethane sulfonic anhydride in pyridine to give the corresponding triflate **40**. Upon treatment with sodium azide in DMF, a 1/1 mixture of lactone **41** and **42** was isolated.

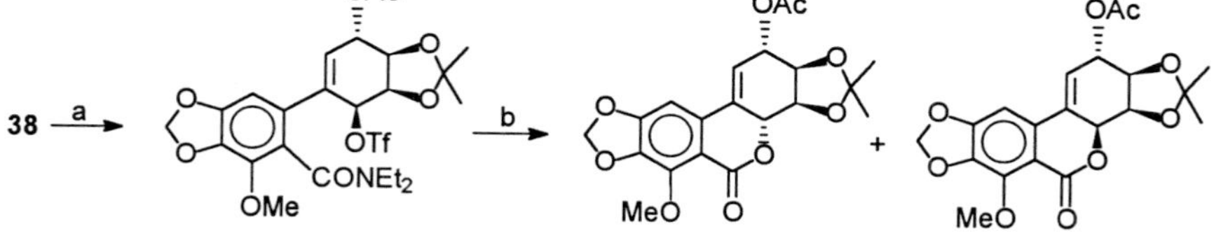

40   **41**   **42**

Scheme 13: a)Tf₂O, CH₂Cl₂, pyridine; b) NaN₃, DMF.

This was explained by a nucleophilic attack of the amide oxygen at position C5 with a probable participation of the acetoxy group at C2 according to Scheme 14.

To overcome this problem, we planned to protect the alcohol **35** with a non-participating group or to reduce the double bond. All attempts to protect this hydroxyl with an acetal group such as MOM, MEM, or BOM gave only poor results.

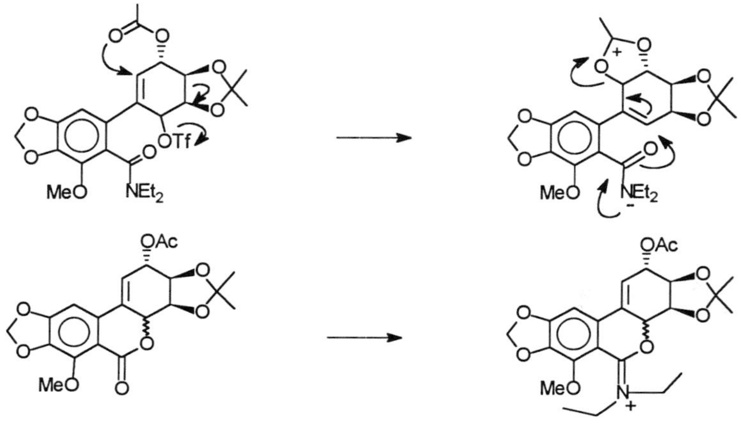

Scheme 14

On the other hand, 1,4 reduction of the enone system of **37** using L-selectride gave the expected isocoumarin **43** in low yield. We turned our attention to the addition on the double bond of an entity suitable for the further elimination to restore the C10b-C1 unsaturation. Thiophenol cleanly added to

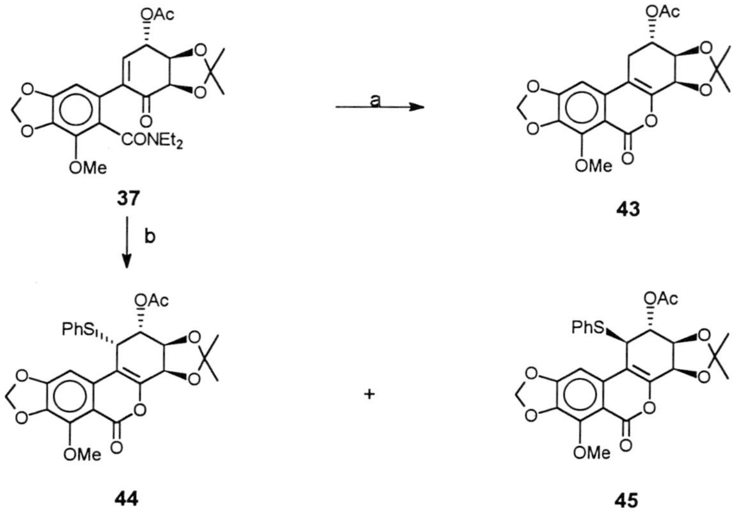

Scheme 15: a) L-selectride, THF, -30°C; b) PhSH, NEt₃, THF, reflux.

the enone **37** in the presence of triethylamine to give a 4/1 mixture of isomers, the major isomer **44** resulting from axial attack of the conjugated system.

The ultimate step was now the formation of the amide function. While this work was in progress, a conceptually similar synthesis of tetra-O-benzyl lycoricidine was described by Kallmerten [7d].

In summary, we have explored two different ways for the preparation of the phenantridinone ring system of Amaryllidaceae alkaloids. The first convergent approach involved the intramolecular formation of a carbon-carbon bond between an olefin and an aromatic ring. This particular reaction was rather difficult to realize but this bond was necessary for the biological activity as proved by the synthesis of some bicyclic analogs of narciclasin and lycoricidin which were devoide of biological activity.

The second linear approach is more efficient, shorter and gives another type of bicyclic analogues. The introduction of the amide function is the last crucial step. Biological testing of the different intermediates could provide new insights in the structure-activity relationship of these alkaloids.

References

1. S. F. Martin "The Amaryllidaceae Alkaloids" in A. Brossi (ed) "The Alkaloids". 1987. pp. 251-376.

2. a) T. Okamoto, Y. Torii, Y. Isogai, *Chem. Pharm. Bull.* **16** (1968) 1860-1864; b) G. Ceriotti, *Nature* **213** (1967) 595-596; c) C. Fuganti, M. Mazza, *J. Chem. Soc. Chem. Commun.* **1972**, 239; d) A. Immirzi, C. Fuganti, *J. Chem. Soc. Chem. Commun.* **1972**, 240.

3. A. Jimenez, A. Santos, G. Alonso, D. Vasquez, *Biochim. Biophys. Acta* **425** (1976) 342-348.

4. G. R. Pettit, V. Gaddamidi, G. M. Cragg, D. L. Herald, Y. Sagawa, *J. Chem. Soc. Chem. Commun.* (1984) 1693-1694.

5. a) A. Mondon, K. Krohn, *Tetrahedron Lett.* (1970) 2123-2126; b) A. Mondon, K. Krohn, *Tetrahedron Lett.* **21** (1972) 2085- 2088; c) A. Mondon, K. Krohn, *Chem. Ber.* **108** (1975) 445-463.

6. a) G. E. Keck, E. Boden, U. Sonnewald, *Tetrahedron Lett.* **22** (1981) 2615-2618; b) G. E. Keck, S. A. Fleming, *Tetrahedron Lett.* (1978) 4763-4766; c) T. Weller, D. Seebach, *Tetrahedron Lett.* **23** (1982) 935-938.

7. a) S. Ohta, S. Kimoto, *Tetrahedron Lett.* (1975) 2279-2282; b) S. Ohta, S. Kimoto, *Chem. Pharm. Bull.* **24** (1976) 2977- 2984; c) H. Paulsen, M. Stubbe, *Tetrahedron Lett.* **23** (1982) 3171-3174; *Liebigs Ann.* (1983) 535; d) B. G. Ugarkar, J. DaRe, E. M. Schubert, *Synthesis* (1987) 715-716; d) R. C. Thompson, J. Kallmerten, *J. Org. Chem.* **55** (1990) 6076-6078; e) N. Chida, M. Ohtsuka, S. Ogawa, *Tetrahedron Lett.* **32** (1991) 4525-4528.

8. S. Danishefsky, J. Y. Lee, *J. Am. Chem. Soc.* **111** (1989) 4829-4837.

9. R. J. Ferrier, *J. Chem. Soc. Perkin Trans. I* (1979) 1455-1458.

10. R. D. Guthrie, D. Murphy, *J. Org. Chem.* **28** (1963) 5288-5296.

11. F. Chrétien, *Synth. Commun.* **19** (1989) 1015-1024.

12. B. Castro, J. R. Dormoy, G. Evin, C. Selve, *Tetrahedron Lett.* (1975) 1219-1223.

13. A. L. Gemal, J. L. Luche, *J. Am. Chem. Soc.* **103** (1981) 5454-5459.

14. O. Mitsunobu, *Synthesis* (1981) 1-28.

15. R. F. Heck, *Organic Reactions* **27** (1982) 345-390.

16. R. Grigg, V. Loganathan, V. Santhakumar, V. Sridharan, A. Teasdale, *Tetrahedron Lett.* **32** (1991) 687-691.

17. a) S. Takano, M. Suzuki, A. Kijima, K. Ogasawara, *Chem. Lett.* (1990) 315-316; b) S. Takano, M. Suzuki, A. Kijima, K. Ogasawara, *Tetrahedron Lett.* **31** (1990) 2315-2318; c) J. H. Rigby, M. Qabar, *J. Am. Chem. Soc.* **113** (1991) 8975-8976.

18. a) K. Shankaran, C. P. Sloan, V. Snieckus, *Tetrahedron Lett.* **26** (1985) 6001-6004; b) V. Snieckus, J. C. Cuevas, C. P. Sloan, H. Liu, D. P. Curran, *J. Am. Chem. Soc.* **112** (1990) 896-898.

19. E. D. Edström, T. Livinghouse, *Tetrahedron Lett.* **27** (1986) 3483-3487; *J. Org. Chem.* **52** (1987) 949.

20. F. Chrétien, Y. Chapleur, *J. Org. Chem.* **53** (1988) 3615-3617.

Steps Toward the *In Vitro* Synthesis of Deoxy Sugars

Werner Klaffke

Institut für Organische Chemie der Universität Hamburg, Martin-Luther-King-Platz 6,
D-2000 Hamburg 13, Fed. Rep. Germany

Abstract: Deoxy sugar nucleosides serving as co-substrates and inhibitors in the various steps of deoxy sugar biosynthesis have been synthesized and an enzymatic approach toward the same target molecules has been studied applying enzymatic routes found in micro-organisms.

1.   Rationale and Aim

The chemical synthesis of complex oligosaccharide structures defines the state-of-the-art in carbohydrate chemistry, which now gives access to gram-quantities of a variety of endogeneous and artificial structures. More and more this technique is accompanied by an alternative preparation, the chemo-enzymatic approach. However, this is still hampered by both, low availability of the particular enzymes found in the sources presently investigated and a rather limited spectrum of glycosyl donors employed in the *in vitro* Leloir pathway [1,2]. Mainly those sugars found in mammalian glycoproteins, e.g. UDP-glucose, UDP-GlcNAc, UDP-galactose and GDP-mannose, have gained the predominant interest. Various groups have responded to this problem and have elaborated valuable protocols for the preparation of nucleoside diphosphate sugars [3]. For deoxy sugars other than GDP-fucose, the biochemical preparation by *in vitro* synthesis still remains unsolved.

On the other hand, deoxy sugars are often linked to secondary metabolites found in microorganisms, which have in many cases gained industrial interest and are used as target structures for further exploitation as antibiotics. The glycosidic part mediates mainly pharmacokinetic parameters, e.g. therapeutic index and bioavailability. Modifications in that side chain often result in better phamacological properties.

Should an enzymatic glycosylation of such target structures be envisaged, the demand for appropriate glycosyl donors has to be realized. We therefore re-visited the biosynthesis of such sugars with the possibility in mind to supply deoxy sugars by those pathways found in bacteria and other micro-organisms. The following will present some initial results concerning the chemical synthesis of sugar phosphates and nucleotides as reference compounds and analogs and the enzymatic synthesis of dTDP-sugars by utilizing steps in bacterial pathways.

2. Biosynthesis of Deoxy Sugars

Following the pioneering work by Strominger *et al.* [4], the biosynthesis of glucose-derived deoxy sugars follows a pathway in which a glucose dehydratase plays a center role as "bottleneck" enzyme. Diversification of the anabolic products follows this step and depends on either the subset of enzymes of the particular organism [5] or on the stereochemistry in the hexose moiety or the base attached to it via the diphosphate linkage [1,6]. Products of this pathway are L-configured sugars, e.g. dTDP-β-L-rhamnose, -2-deoxy-fucose, and 3-deoxy-rhamnose, or D-sugars like mycosamine and perosamine. All products which are observed are formed by an ensemble of enzymes which belong to the galactose epimerase family, not only mediating the epimerization of α-hydroxy carbonyl compounds or the oxidation/reduction sequence of galactose epimerase, but also the mechanistically still unsolved 2-epimerization and finally, the "4→6 H"-shift.

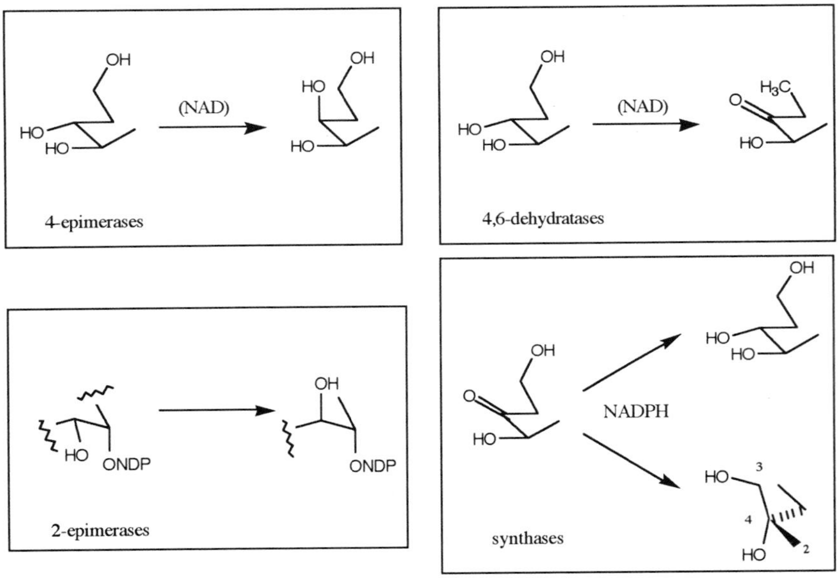

Scheme 1: GalE enzyme family

The latter mechanism has been investigated intensively [7] and by a double labeling experiment, a net 4→6 hydride shift was proven [8]. The 4-OH group is deprotonated and H-4, which was deuterated stereospecifically in this case, is abstracted by cofactor NAD$^+$. From this keto-intermediate, H-5 is abstracted and enolate formation occurs. The allylic 6-OH group is now prone to an S$_N$2 type attack by cofactor bound "D$^-$" and thus C-6 expresses chirality which gave rise to the mentioned hypothesis.

The high chemical potential of a carbonyl group is reflected by the diversity of products arising from the 4-keto-6-deoxy hexose. The most prominent pathway, formation of dTDP-L-rhamnose and other L-configured sugars, is shown in scheme 3. After a set of enolizations and β-facial reprotonations by the action of a 3,5-epimerase the the configurations at C-3 and C-5 are reversed. The last step before the substrate is released from the protein involves a reduction. In the case of 3,6-dideoxy sugars, recent findings by Liu et al.[9] showed this reduction to be an electron shuttle mechanism between a flavonucleotide, a [2Fe 2S] cluster and the substrate.

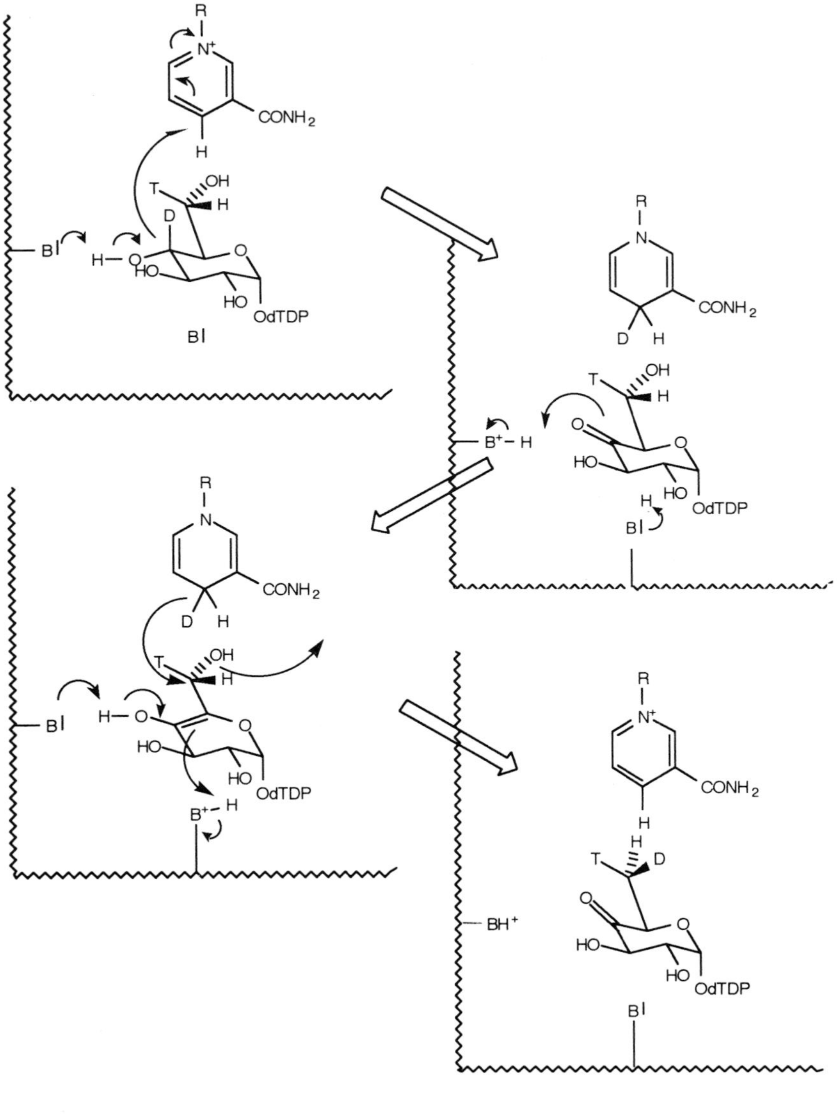

Scheme 2: Mechanism of the dehydratase reaction proposed by Floss et al.

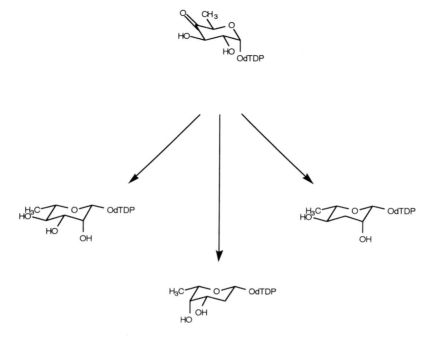

Scheme 3: Conversion of dTDP-4-keto intermediate into L-configured deoxy sugars

3.   Chemical synthesis of reference compounds and analogs

Several substrates and analogs have been envisaged and synthesized, and we focused mainly on a facile access to inhibitors of the dehydratase reaction, such as 4-deoxy compounds. D-galactose was partially esterified to give 1,2,3,6-tetra-O-benzoyl-α-D-galactose (1) [10], which then was converted into its triflate and reacted with sodium iodide to give in 72% yield *gluco*-configured sugar 2. Treatment in refluxing dichloromethyl methyl ether and subsequent reaction with silver diphenylphosphate gave α-phosphate 4 in a moderate yield of 37% after chromatography. Hydrogenation yielded the deblocked phosphate 6 after addition of a molar equivalent of tributyl amine. The ^{1}H-n.m.r. coupling constants $J_{1,2}$ and $J_{H-2,P}$ clearly assign the anomeric substituent to be "α". The same route was applied to obtain 4-amino phosphate sugar 7. However, the reduction still left 15% of a monophenyl phosphate, a fact which was also observed with phenyl phosphates by other authors [11]. The synthesis of 4,6-dideoxy glucosyl phosphate 8 started from D-fucose and followed exactly the same outline as the previous ones.

A major drawback in these syntheses was found to be the glycosylation reaction and the reductive cleavage of the phenyl groups. Therefore the phosphoramidite-method devised by van Boom et al. [12] was adopted. Starting from the same material as above, the anomeric group was cleaved by reaction with DCMME. Subsequent treatment with acetone/water in the presence of silver carbonate gave hexose 9. Reaction of 9 and 10 with chloro 2-cyanoethyl N,N-diisopropylamine phosphoramidite in dichloromethane gave the diastereomeric mixture of α-configured phosphoramidites 11 and 12 in 78 and 87%, respectively, after chromatography. A second product, presumably the less stable β-anomer, which was observed on t.l.c. during the reaction, vanished completely after work-up, probably due to hydrolysis.

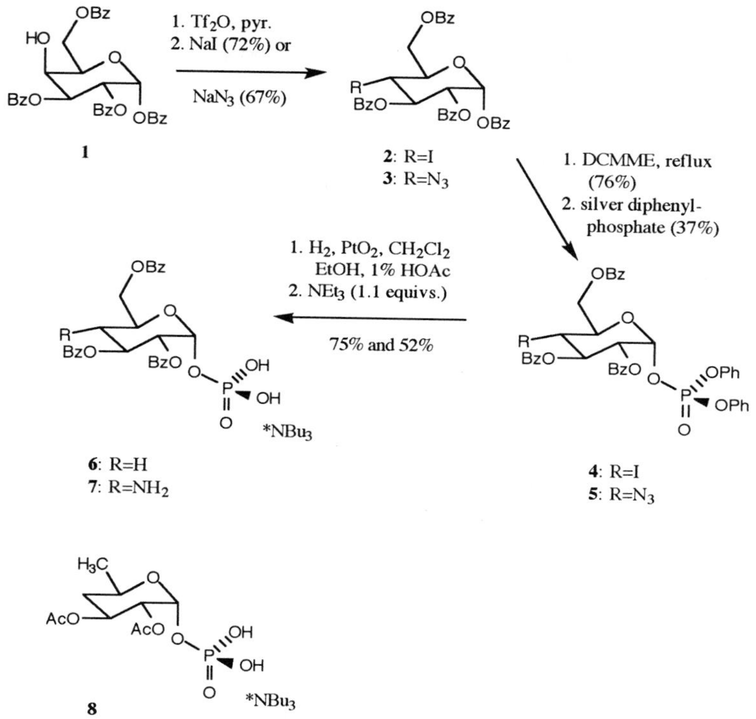

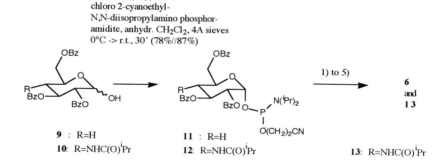

Scheme 4: Synthesis of deoxy sugar phosphates

ethyl diisopropylamine,
chloro 2-cyanoethyl-
N,N-diisopropylamino phosphoramidite, anhydr. CH$_2$Cl$_2$, 4A sieves
0°C -> r.t., 30' (78%//87%)

9 : R=H
10: R=NHC(O)^{i}Pr

11 : R=H
12: R=NHC(O)^{i}Pr

1) to 5)

6
and
13

13: R=NHC(O)^{i}Pr

1) cyanoethanol, acetonitrile, ethyl diisopropylamine
2) tert.-butylhydroperoxide (1.2 equivs), 3) approx. 7N NH$_3$ in MeOH, 2h at 0°C
4) Dowex W50 (py$^+$), 5) 1 equiv. NBu$_3$, MeOH, reflux

Scheme 5: Phosphoramidite route to sugar phosphates

The advantage of this approach becomes evident in the conversion step that yields the benzoylated sugar phosphates. Subsequent treatment of the amidites **11** and **12**, respectively, with 2-cyanoethanol, tert.-butylhydroperoxide and 7N ammonia in methanol resulted in a 60% conversion into the deblocked derivatives **6** and **13**. These were further converted into their respective tributyl ammonium salts and were purified by flash chromatography on silica gel.

Another series of sugar phosphates which gained our synthetic interest are those with the D-*manno*-configuration bearing an amino group at C-3. One of the most prominent members of this group is D-mycosamine, which is β-glycosidically linked to amphotericin B, a polyene macrolide produced by *Streptomyces nodosus*.

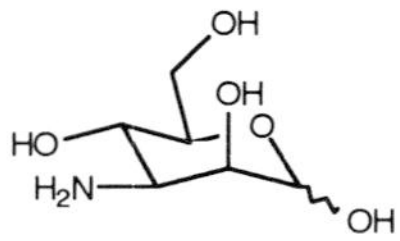

1 4

Scheme 6: D-mycosamine

There are three major concepts regarding how to establish a D-*manno*-3-amino-3-deoxy-configuration: nitromethane cyclization [13], double and single inversions at C-2 and C-3[14] or a [3+2]-cis-oxyamination of a suitable enopyranoside [15]. However, we decided to establish alternatives requiring neither elaborate chromatography techniques nor special equipment. Levoglucosan (**15**) was converted to the 3:2 mixture of *allo*-configured epoxides **16** and **17** [16]. Azide attack and subsequent acetylation yielded 1,6-anhydro-3-azido-3-deoxy-β-D-*gluco*pyranose (**18**) in 75% yield [17]. According to reports in the literature, the acetyl groups of per-O-acetylated levoglucosan are prone to regioselective de-O-acetylation upon treatment by various lipases and esterases.[18] With these results in mind, azide **18** was subjected to several lipases from *Candida cylindracea* and *Pseudomonas sp.* to give the 4-OH monoacetyl derivative **19** in 85-90% yield. Interestingly, the regiochemistry could be reverted, when subtilisin or alcalase (Novo Nordisk) was used. In those cases the conversion into monoacetate **20** was observed in 82% yield.

This, however, was well suited for further synthesis. Inversion by triflate/sodium acetate inversion and acetolysis gave hexopyranoside **21** with the desired D-*manno* configuration. After this synthesis had been carried out, we developed another strategy, which proved to be more effective. Tosylation of levoglucosan **15**, base treatment of the resulting ditosylate and epoxide opening with sulfuric acid in refluxing dioxane resulted in the formation of *gluco*-configured 2-O-tosylate **22** [19]. After acetylation and chromatography, treatment with sodium methoxide gave a mixture of both *manno*-epoxides **23** and **24**, which were equilibrated in 0.1N aq. sodium hydroxide to give, after acetylation, *altro*-configured precursor **25**. From this an azide attack in refluxing DMF and acetolysis again led to 3-azido compound **21** in 66% yield.

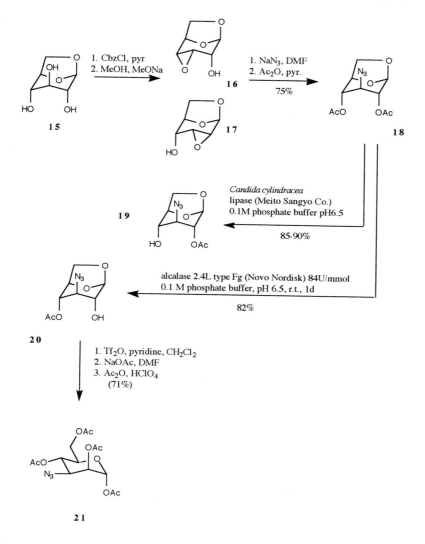

Scheme 7: Synthesis of 3-azido-3-deoxy-D-mannopyranose (route A)

Attempts to convert the azide into a glycosyl phosphate by van Boom´s method failed completely and it was found to be essential to reduce the azide before further treatment. Staudinger reaction with triphenyl phosphine on **21** in dichloromethane containing a defined amount of water delivered the free amine without acetyl migration, which was observed to be the main reaction during hydrogenation. After protection with isobutyric anhydride in pyridine, 3-amino-3-deoxy-D-mannose **26** was obtained, which after chromatography was converted into its tri-acetate and subsequently into α-phosphoramidite **27**. The yields of both steps is in the range of 90% and 69%, respectively. Further conversion into free phosphate **28** followed the procedure discussed above.

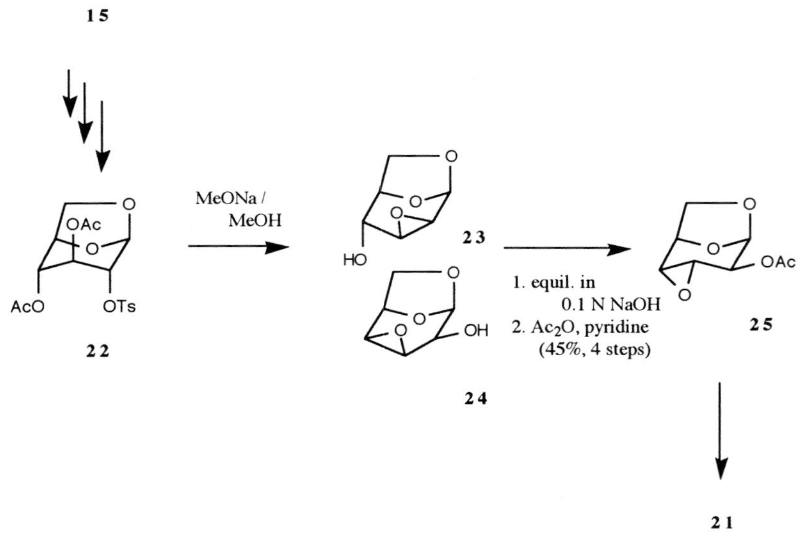

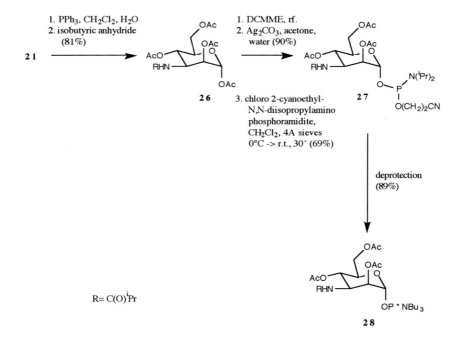

15

22 → (MeONa / MeOH) → **23**, **24** → (1. equil. in 0.1 N NaOH 2. Ac₂O, pyridine (45%, 4 steps)) → **25** → **21**

Scheme 8: Synthesis of 3-azido-3-deoxy-D-mannopyranose (route B)

21 → (1. PPh₃, CH₂Cl₂, H₂O 2. isobutyric anhydride (81%)) → **26** → (1. DCMME, rf. 2. Ag₂CO₃, acetone, water (90%) 3. chloro 2-cyanoethyl-N,N-diisopropylamino phosphoramidite, CH₂Cl₂, 4A sieves 0°C -> r.t., 30′ (69%)) → **27** → (deprotection (89%)) → **28**

R= C(O)ⁱPr

Scheme 9: Synthesis of 3-amino-3-deoxy-D-mannopyranosyl phosphate **28**

By now, there is a complete set of 3-, 4- and 6-deoxy sugars available, which have either been coupled to dTMP by chemical means or which now serve as substrates for enzymic conversions. The coupling to give diphosphate sugars by mixing nucleotide morpholidates with glucosyl phosphates was introduced by Moffatt et al. [20]. We followed this protocol with respect to the nucleotide part and found the yields to be higher when reacted with partially blocked hexosyl phosphates instead of completely de-esterified products [21]. In all cases the coupling yields were between 31 and 43%.

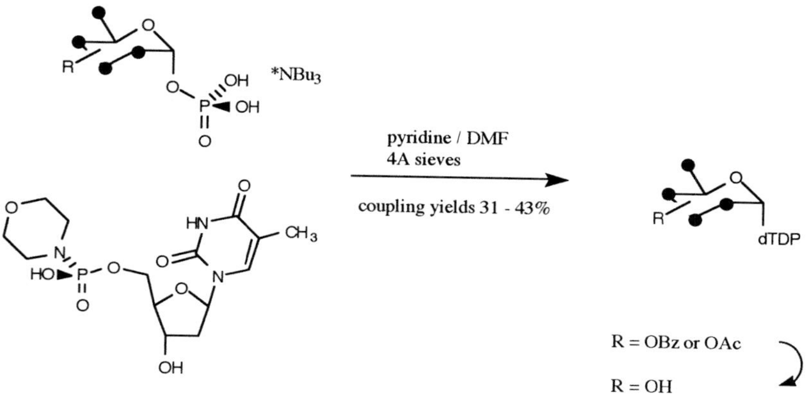

pyridine / DMF
4A sieves

coupling yields 31 - 43%

dTDP

R = OBz or OAc

R = OH

Scheme 10: Coupling scheme for dTDP-sugars (cf. Jeanloz et al. [21])

4.   The enzymatic approach

After the pioneering work of Whitesides et al. [22] who introduced the regeneration cycle for co-factors in a macro-scale synthesis of sugars nucleotides, preparations for almost all members of this family have been disclosed by different groups. For the use in the *E. coli* pathways we had to establish a route toward dTDP sugars, which required several questions to be answered:

- would the hexokinase from yeast accept synthetic modifications in the
    hexose like those discussed above.
- does the phosphoglucose mutase work with such 6-phosphates
- is there an enzymatic way to deoxythymidine triphosphate.

4.1. The hexokinase reaction

Yeast hexokinase [E.C.2.7.1.1] mediates the phosphate group transfer from ATP to the 6-OH group of D-glucose *in vitro* [20]. Modifications by replacing hydroxyl groups with fluorides have been carried out and it was shown that a 2-deoxy-2,2'-difluoroglucose was accepted in the same range as the substrate. Furthermore, the 3- and 4-fluoroglucoses show a 100-fold increase in K_M. However, the former was still accepted and converted into the 6-phosphate by hexokinase [23]. Our experiments with the five deoxy glucoses **29** through **33** meet pretty much the predictions one would make on the basis of these kinetic studies. 3-Deoxyglucose (**29**) and azide **30** are accepted and converted to the respective 6-phosphates in reasonable yields and were obtained after Sephadex G10 chromatography as white powders after lyophilization. Neither of the 4-deoxy-4-azido nor 4-amino compounds (**31**, **32** and **33**, resp.) were phosphorylated and further investigations are presently under way to establish the kinetic data for these and further modified glucoses.

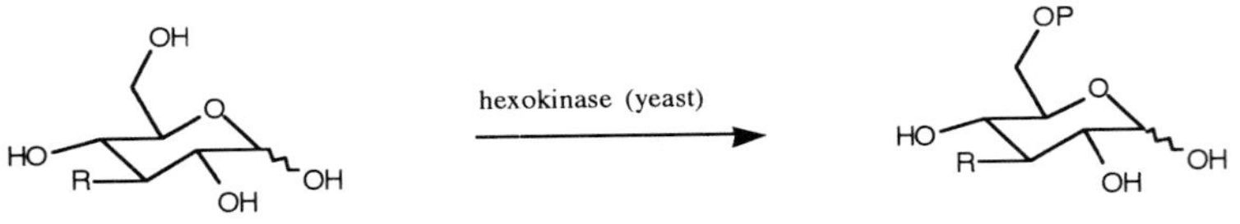

#	R^1	R^2
29	H	OH
30	N$_3$	OH
31	OH	H
32	OH	N$_3$
33	OH	NH$_2$

Scheme 11: Phosphorylation catalyzed by yeast hexokinase

4.2. The cofactor regeneration for the dTDP-pyrophosphorylase reaction

The next question to be answered on the way to an enzymatic synthesis is whether the phosphoglucomutase is working on the artificial substrates phosphorylated by hexokinase. This, however, is particularly difficult to answer, as the glucomutase reaction has to be coupled to the thymidyl phosphate transfer reaction. Therefore a cofactor delivery system has to be established first. Reports in the literature are quite different with regard to the specificity of nucleoside monophosphate kinase [E.C. 2.7.4.4] and pyruvate kinase [E.C. 2.7.1.40] towards deoxythymidine monophosphate and diphosphate respectively. First experiments carried out in our laboratory, however, gave rise to the assumption that both enzymes can work in this regeneration cycle. On the basis of measurements on dTDP and pyruvate kinase [24] and our kinetic data for the nucleotide monophosphate kinase we observed perfect Michaelis-Menten kinetics. With these data in hand, we were able to design the synthesis of dTTP in a 50mg scale. The product was isolated as a white powder after ion-exchange chromatography on Dowex 1X2 and elution with a linear gradient of 0.5 - 1.1 M ammonium bicarbonate and lyophilization.

With both strategic questions being answered, the *in situ* reaction could be designed. After isolating the dTDP-glucose pyrophosphorylase from *E. coli* by a procedure described first by Kornfeld et al. [25], the process was transferred to the enzymatic synthesis of dTDP-glucose analogs **34** and **35**. Despite the fact that a 3-azido compound such as **35** could not be prepared by any of the chemical procedures described above, this enzymatic pathway yields an alternative approach to nucleoside diphosphates of 3-deoxy- and 3-azido sugars, **34** and **35**, in 16% and 17% yield, respectively.

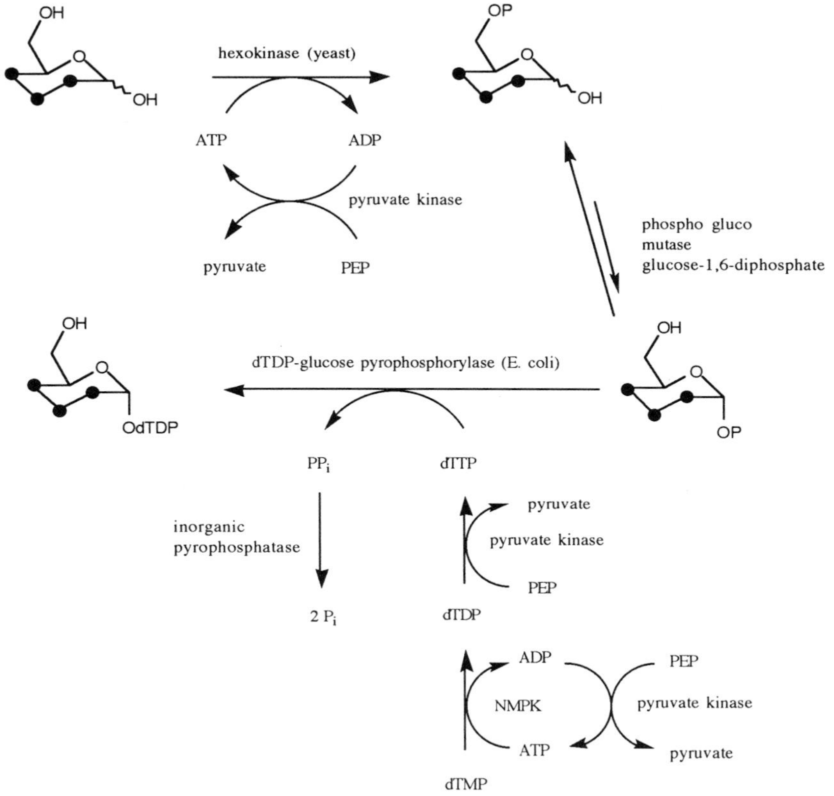

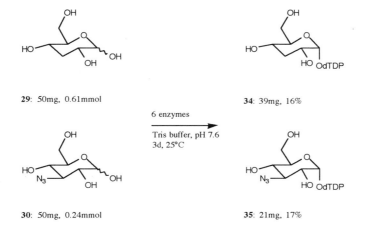

Scheme 12: Enzymatic synthesis of dTDP deoxy sugars with cofactor-regeneration

29: 50mg, 0.61mmol

34: 39mg, 16%

6 enzymes

Tris buffer, pH 7.6
3d, 25°C

30: 50mg, 0.24mmol

35: 21mg, 17%

Scheme 13: dTDP sugars and yields

5.   Isolation of dTDP-Glucose Dehydratase

According to the literature, the dehydratase reaction (cf. scheme 2) can be inhibited competitively by addition of dTDP [5]. We therefore adopted the synthesis of a spacer-modified diphosphate ligand devised by Barker *et al.*. [26] Trifluoroacetylated hexanolamine phosphate was coupled in 56% yield to dTMP-imidazolate to give dTDP-hexanolamine. After basic deprotection, the free amine was coupled to Sepharose 4B to give the immobilized product in 87% coupling yield. The *E. coli* protein was obtained from cells grown in 10 l "LB"-culture medium, harvested in the late log-phase and sonicated at 0°C for 5x5min. After centrifugation, nucleic acids were removed by protamine sulfate-precipitation and the precipitate was spun down. The supernatant was concentrated by filtration through an Amicon cell (cutoff 10000 D) and DEAE-f.p.l.c. yielded a protein fraction of 9.5 U of activity. This, after pooling and reconcentration, gave a raw enzyme preparation which was applied to a 1x5cm column filled with the previously synthesized affinity material. Elution tests were promising: 75% of the enzyme activity was retarded on the column and re-mobilized upon addition of substrate.

Presently several modified affinity materials are under investigation, in order to facilitate the work-up and re-mobilization by applying weaker or sterically more hindered inhibitors on a suitable affinity matrix.

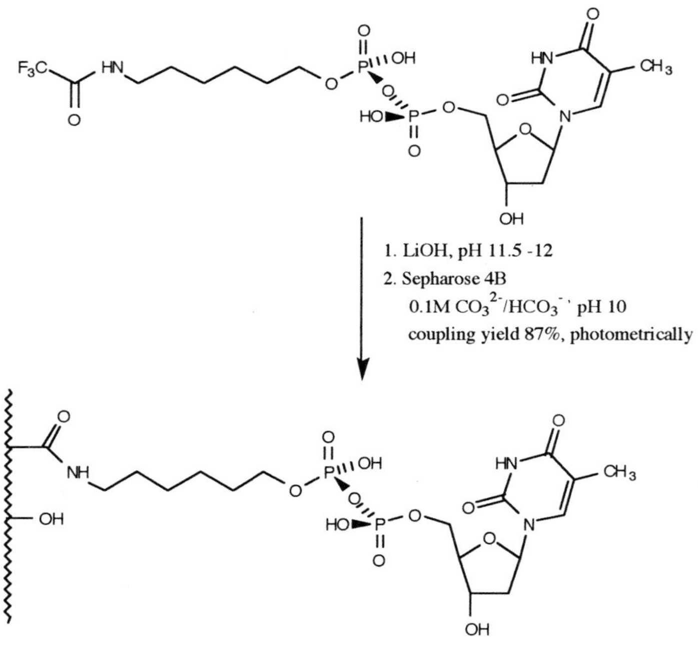

1. LiOH, pH 11.5 -12
2. Sepharose 4B
   0.1M CO_3^{2-}/HCO_3^- pH 10
   coupling yield 87%, photometrically

Scheme 14: Immobilized affinity material for dTDP-Glucose dehydratase purification

6.   Acknowledgements

The author would like to thank the co-workers on the research project: Brigitte Leon for her skillful technical assistance and her extensive help in microbiology, and Andreas Naundorf, who is presently working on the synthesis of affinity materials and the purification of dTDP-glucose dehydratase.

Furthermore, special thanks to Dr. E. Wolfgang Holla (Hoechst AG, Frankfurt) for his co-operation in the enzymatic de-esterification studies and Prof. Joachim Thiem for his support of this project. Financial supports by the *Fonds der Chemischen Industrie* (Liebig-Stipend to W.K.) and by the *Deutsche Forschungsgemeinschaft* are gratefully acknowledged.

7. References

1.   R. Caputto, L. F. Leloir, C. E. Cardini, A. C. Paladini, *J. Biol. Chem.* **184** (1950) 333 - 350.
2.   E. J. Toone, E. S. Simon, M. D. Bednarski, G. M. Whitesides, *Tetrahedron* **45** (1989) 5365 - 5422.
3.   E. J. Toone, G. M. Whitesides, in *Enzymes in Carbohydrate Snthesis* (M.D. Bednarski, E.S. Simon, Eds.), ACS Symposium Series **466** (1991) 1 - 22.
4.   R. Okazaki, T. Okazaki, J. L. Strominger, A. M. Michelson, *J. Biol. Chem.* **237** (1962) 3014-3026.
5.   O. Gabriel. *Methods Enzymol.* **83** (1982) 332- 353.
       L. Glaser, H. Zarkowsky, in *The Enzymes* (P.D. Boyer, Ed.), Vol. 5, p. 465-480 , Academic Press, London, 1971.
6.   J. H. Pazur, E. W. Shuey, *J. Biol. Chem.* **236** (1961) 1780- 1785.
7.   H. Zarkowsky, L. Glaser, *J. Biol. Chem.* **244** (1969) 4750 - 4756.
       S. F. Wang, O. Gabriel, *J. Biol. Chem.* **244** (1969) 3430 - 3437.
       T. E. Barman, *Enzyme Handbook*, Suppl. I, p. 426 ff., Springer Verlag, Heidelberg, 1974.
8.   C. E. Snipes, G. U. Brillinger, L. Sellers, L. Mascaro, H. G. Floss, *J. Biol. Chem.* **252** (1977) 8113 - 8117.
9.   V. P. Miller, H.-w. Liu, *J. Am. Chem. Soc.* **114** (1992) 1880 - 1881;
       H.-w. Liu, XVI Int. Carbohydr. Symp., Abstr. B5, Paris, 1992.
10.   P. J. Garreg, H. Hultberg, *Carbohydr. Res.* **110** (1982) 261 - 266.
11.   H. G. Floss, personal note, 1991.
12.   P. Westerduin, G. H. Veeneman, J. E. Marugg, G. A. van der Marel, J. H. van Boom, *Tetrahedron Lett.* **27** (1986) 1211 - 1214.
13.   H. H. Baer, H. O. L. Fischer, *J. Am. Chem. Soc.* **82** (1969), 3709- 3713.
       H. H. Baer, F. Kienzle, *Can. J. Chem.* **41** (1963) 1606- 1611.
       K. Capek, J. Staněk, Jr., J. Jary, *Coll. Czech. Chem. Commun.* **39** (1974) 1462 - 1478.
14.   J. Staněk, Jr., K. Capek, J. Jary, *Coll. Czech. Chem. Commun.* **39** (1974) 1479- 1487.
15.   I. Dyong, G. Schulte, Q. Lam-Chi, H. Friege, *Carbohydr. Res.* **41** (1979) 257 - 273.
16.   T. Trnka, M. Cerny, M. Budesinsky, J. Pacák, *Coll. Czech. Commun. Chem.* **40** (1975) 3038 - 3045.
17.   B. Leon, S. Liemann, W. Klaffke, *J. Carbohydr. Chem.*, in press.
18.   J. Zemek, S. Kucár, D. Anderle, *Coll. Czech. Chem. Commun.* **52** (1987) 2347 - 2352.
       R. Czuk, B. I. Glänzer, *Z. Naturforsch.* **43b** (1988) 1355 - 1357.
19.   M. Cerny, J. Pacák, J. Staněk, *Coll. Czem. Chem. Commun.* **30** (1965) 1151 - 1156.
20.   G. Moffatt, *Meth. Enzymol.* **8** (1966) 136 - 142.

21. T. Yamazaki, C. D. Warren, A. Herscovics, R. W. Jeanloz, *Can. J. Chem.* **59** (1981) 2247 - 2252.

22. B. L. Hirschbein, F. P. Mazenod, G. M. Whitesides, *J. Org. Chem.* **47** (1982) 3765 - 3766.

23. D. G. Drueckhammer, C.-H. Wong, *J. Org. Chem.* **50** (1985) 3912 - 5913.

24. L. Elling, personal note, 1992.

25. L. Glaser, S. Kornfeld, *J. Biol. Chem.* **236** (1961) 1795 - 1799.

26. R. Barker, K. W. Olsen, J. W. Shaper, R. L. Hill, *J. Biol. Chem.* **247** (1972) 7135 - 7147.

TSAO Derivatives: A Novel Class of HIV-1-Specific Inhibitors

J. Balzarini[1], M.-J. Pérez-Pérez[2], A. San-Félix[2], S. Velázquez [2], M.-J. Camarasa [2], A.-M. Vandamme[1], A. Karlsson[3] and E. De Clercq[1]

[1]*Rega Institute for Medical Research, Katholieke Universiteit Leuven, B-3000 Leuven, Belgium;* [2]*Instituto de Quimica Médica, 28006 Madrid, Spain; and* [3]*Karolinska Institute, S-104 01 Stockholm, Sweden.*

Summary: A number of different structural classes of human immunodeficiency virus type 1 (HIV-1)-specific inhibitors have recently been identified. Among them, TSAO derivatives (prototype compound: TSAO-T [1-[2',5'-bis-*O*-(*tert*-butyldimethylsilyl)-ß-D-ribofurano-syl]thymine]-3'-spiro-5"-(4"-amino-1",2"-oxathiole-2",2"-dioxide) represent a group of nucleoside analogues that are endowed with a potent and highly specific anti-HIV-1 activity. In this paper, the structure-antiviral activity relationship of the TSAO derivatives, their chemical synthesis and molecular mechanism of action, the rapid emergence of TSAO-resistant HIV-1 strains and the biological and molecular characterization of the TSAO-resistant HIV-1 strains are discussed.

Introduction

Recently, several new classes of compounds have been identified as highly specific and potent inhibitors of human immunodeficiency virus type 1 (HIV-1) (Fig. 1). The first compounds shown to be selective inhibitors of HIV-1 were 1-[(2-hydroxyethoxy)methyl]-6-(phenylthio)thymine (HEPT) [1-3] and the tetrahydroimidazo-[4,5,1-*jk*][1,4]-benzodiazepin-2(1*H*)-thione (TIBO) (R82150) [4,5]. Later, the dipyridodiazepinone nevirapine (BI-RG-587) [6,7], the pyridinone derivative L-697,661 [8], the bis(heteroaryl)piperazine (BHAP) derivative U-87201E [9], the 1-[2',5'-bis-*O*-(*tert*-butyldimethylsilyl)-ß-D-ribofuranosyl]-3'-spiro-5"-[4"-amino-1",2"-oxathiole-2",2"-dioxide]thymine derivative TSAO-T [10-14] and the α-anilinophenylacetamide derivatives (α-APA) [15] were also shown to be specific inhibitors of HIV-1 (Fig. 1).

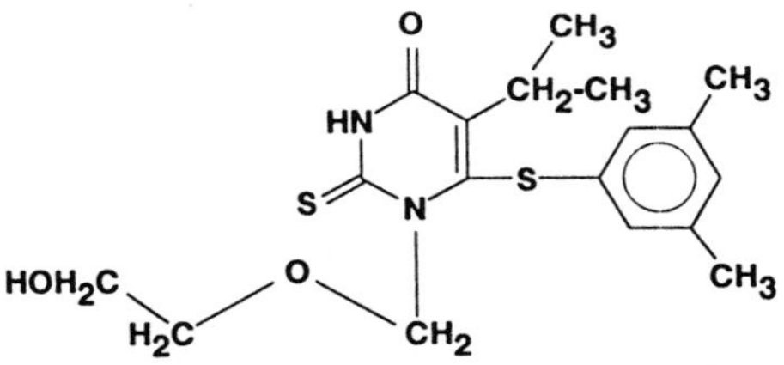

HEPT : I - HEPU - SdM

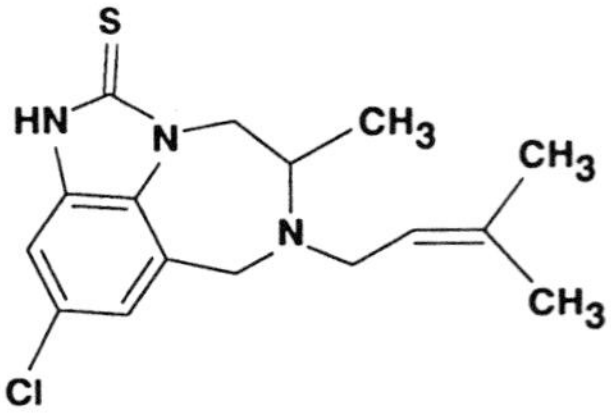

TIBO (R - 82913)

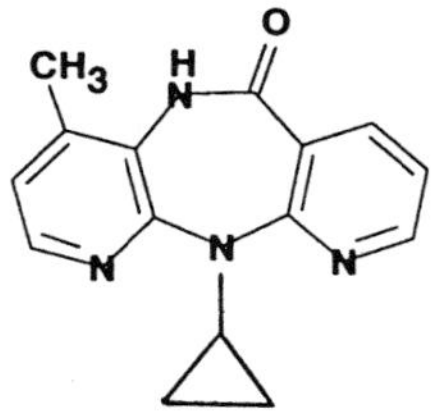

Nevirapine (BI - RG - 587)

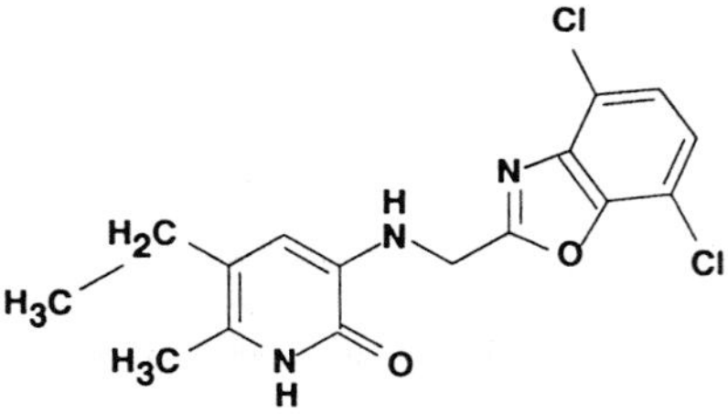

Pyridinone L - 697 661

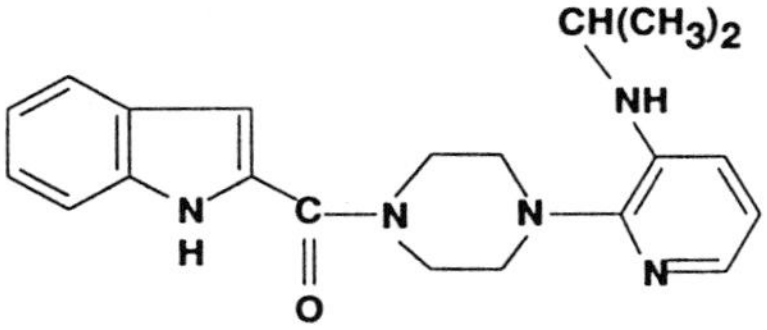

BHAP (U - 88204)

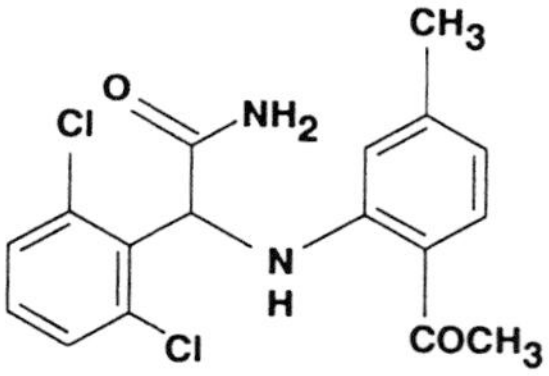

α - APA (R - 89439)

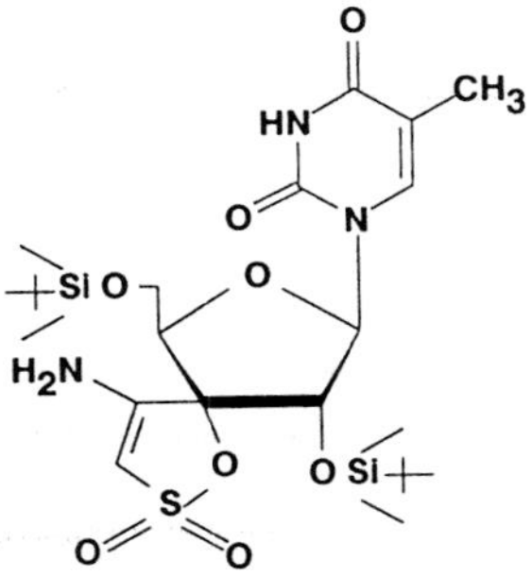

TSAO - T

Fig. 1. Structural formulae of HIV-1-specific RT inhibitors.

The common features of these different compounds are as follows: (*i*) they are highly inhibitory to HIV-1 replication, but do not inhibit the replication of HIV-2, simian immunodeficiency virus (SIV) or other retroviruses (i.e. Moloney murine sarcoma virus, feline immunodeficiency virus), DNA viruses or RNA viruses. (*ii*) They are targeted at HIV-1 reverse transcriptase (RT) at a non-substrate binding site. This peculiar property has first been demonstrated for TIBO [4,16-18] and later confirmed for the other compounds [6,8,9,15,19-23]. (*iii*) They do not inhibit other DNA polymerases (i.e. DNA polymerase α, ß, γ, HSV-1 DNA polymerase, Taq DNA polymerase). (*iv*) They are relatively non-toxic to human cells, and their antiviral selectivity index in cell culture (ratio cytotoxic concentration/effective concentration) is usually very high (up to 100,000) [4-11,15,23]. (*v*) They rapidly select for highly resistant HIV-1 strains *in vitro* in cell culture as well as *in vivo* in patients infected with HIV-1 [24-33].

This overview is focussed on the chemical synthesis of the TSAO derivatives, their structure activity/cytotoxicity relationship, their molecular mechanism of antiviral action, and the emergence and characterization of TSAO-resistant HIV-1 strains. Also, a model of specific interaction of TSAO with HIV-1 reverse transcriptase will be proposed.

Synthesis of TSAO derivatives

Xylo and *ribo* TSAO-T derivatives were prepared according to the following procedure (Scheme 1): treatment of the 3'-ketonucleoside 2 [34] with NaCN and NaHCO$_3$ in a stirred mixture of ethyl ether/water (2:1) gave a mixture of the 3'-cyanohydrin epimers 3 and 4, which were then treated with mesyl chloride in pyridine to give the respective 3'-C-cyano-3'-O-mesyl-ß-D-xylo- and -ribofuranosyl thymine nucleosides 5 and 6.

The *xylo* and *ribo* configuration for 5 and 6 was determined by ^{1}H-NMR, from their $J_{1',2'}$ coupling constant values [12,14] and from the shielding observed for H-2' in compound 5 when compared to the same proton in 6. The stereochemistry of the resulting cyanohydrin depends on the relative steric hindrance of the upper (α) or lower (ß) sides of the furanose ring, which facilitates the approach of the cyanide ion from the less hindered side of the molecule. Thus, a *xylo* configuration for the major cyanohydrin is in agreement with the approach of the cyanide ion from the sterically less hindered α-face of the ulose 2, opposite to the base [35].

Reaction of 5 and 6 with Cs$_2$CO$_3$, in dry acetonitrile at room temperature, gave the *xylo* and *ribo* spiro nucleosides 7 and 8, respectively [12,14]. Deprotection of 7 and 8 with tetrabutylammonium fluoride gave the fully deprotected nucleosides 9 and 10. Reaction of 10 with 1.1 equivalents of *t*-butyldimethylsilyl chloride selectively afforded the 5'-O-silylated 3'-spiro-ribonucleoside 11 (Scheme 1).

The TSAO nucleoside analogues of pyrimidines and purines having a *ribo* configuration were prepared stereoselectively by glycosylation of the appropriate 3-C-cyano-3-O-mesyl sugar intermediate 15 with heterocyclic bases, followed by basic treatment of the cyanomesyl nucleosides obtained, to give exclusively the ß-D-*ribo*-spiro nucleosides [13] (Scheme 2). The *ribo* configuration of the nucleosides was determined by the configuration of the starting cyanohydrin used in the preparation of the sugar intermediate 15, as clearly demonstrated in previous papers [36,37].

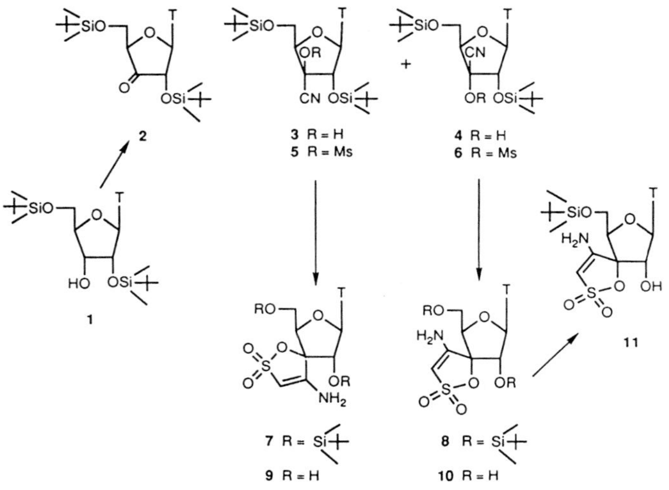

Scheme 1

Thus, reaction of the ribofuranosulose **12** [38] with NaCN and NaHCO$_3$ in a stirred mixture of ethyl ether/water (2:1), afforded exclusively the kinetically controlled *ribo*-cyanohydrin **13** which was transformed without further purification to the 3-*C*-cyano-3-*O*-mesyl derivative **14** by reaction with mesyl chloride in pyridine. The stereochemistry at C-3 of compound **14** was assigned as *ribo* as it has been previously demonstrated in the synthesis of similar α-mesyloxynitriles of pentofuranoses [37]. This configuration is in agreement with the approach of the cyanide ion from the sterically less hindered ß-face of the ulose **12** opposite to the 1,2-*O*-isopropylidene group, as described for other additions of nucleophiles to uloses [39].

Hydrolysis of the 1,2-*O*-isopropylidene group of **14** with aqueous trifluoroacetic acid, followed by acetylation with acetic anhydride/pyridine, gave a (3:2) mixture of the α- and ß-acetyl derivative **15**. Glycosylation of **15** with previously silylated pyrimidine bases [thymine (T), uracil (U), 5-ethyluracil (e^{5}U), or 5-bromouracil (Br^{5}U)] in refluxing acetonitrile, in the presence of trimethylsilyl triflate as condensing reagent [40], afforded the 1-(2'-*O*-acetyl-5'-*O*-benzoyl-3'-*C*-cyano-3'-*O*-mesyl-ß-D-ribofuranosyl)nucleosides **16a-d**. Similarly, glycosylation with the previously silylated purine bases [adenine (A), hypoxanthine (Hx) or xanthine (X)] afforded the N-9-linked-(2'-*O*-acetyl-5'-*O*-benzoyl-3'-*C*-cyano-3'-*O*-mesyl-ß-D-ribofuranosyl) nucleosides of adenine **16e** and hypoxanthine **16f** and the N-7-linked nucleosides of hypoxanthine **16g** and xanthine **16h**. Glycosylation position of **16e,f** and **16g,h** was established as N-9 and N-7, respectively, from their ^{1}H NMR and ^{13}C NMR data [41]. The ß-anomeric configuration was established from the coupling constant value $J_{1',2'} = 6.0$-7.0 Hz, which is in reasonably good agreement with literature data for other ß-D-*ribo*-3'-*C*-branched nucleosides [42,43].

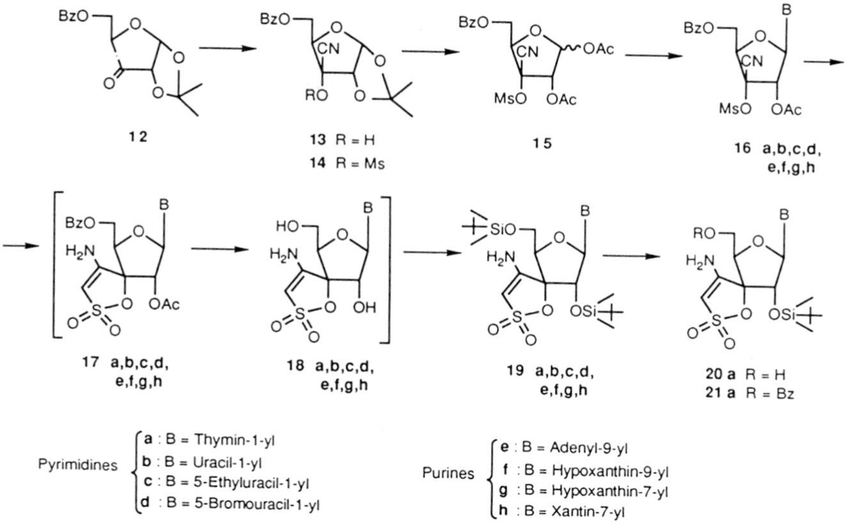

12

13 R = H
14 R = Ms

15

16 a,b,c,d,
    e,f,g,h

17 a,b,c,d,
    e,f,g,h

18 a,b,c,d,
    e,f,g,h

19 a,b,c,d,
    e,f,g,h

20 a R = H
21 a R = Bz

Pyrimidines
a : B = Thymin-1-yl
b : B = Uracil-1-yl
c : B = 5-Ethyluracil-1-yl
d : B = 5-Bromouracil-1-yl

Purines
e : B = Adenyl-9-yl
f : B = Hypoxanthin-9-yl
g : B = Hypoxanthin-7-yl
h : B = Xantin-7-yl

Scheme 2

Treatment of the cyanomesylates **16a-h** with Cs$_2$CO$_3$ in dry acetonitrile at room temperature gave the spiro derivatives **17a-h**. Deprotection of **17a-h** with saturated methanolic ammonia, provided the fully deprotected nucleosides **18a-h**, which, by reaction with an excess of t-butyldimethylsilyl chloride in pyridine, yielded the 2',5'-bis-O-silylated nucleosides **19a-h**.

Selective cleavage of the 5'-O-silylether of **19a** followed by benzoylation with benzoyl chloride/pyridine gave the 5'-O-benzoyl-2'-O-silyl spiro thymine nucleoside **21a**. Structures of the new compounds were assigned on the basis of the corresponding analytical and spectroscopic data [13,41]. Spironucleosides of thymine **19a** and uracil **19b** were transformed to the corresponding derivatives of 5-methylcytosine **22** and cytosine **23**, respectively, by reaction with phosphorous oxychloride and 1,2,4-triazole followed by treatment with an excess of concentrated ammonia (Scheme 3).

Selective N-3-alkylation [44] of spiro pyrimidine nucleosides **19a,b** or N-1-alkylation of spiro purine nucleosides **19f,g** afforded the N-3 (**24** and **25**) or N-1 (**27** and **28**) substituted derivatives. Thus, reaction of the thymine (**19a**), uracil (**19b**) and the N-9- and N-7-hypoxanthine nucleosides (**19f** and **19g**) with methyl iodide in the presence of potassium carbonate gave 3-methyl-3'-spiro pyrimidine nucleosides **24** and **25** and 1-methyl-purine nucleosides **27** and **28**, respectively. Similarly, reaction of **19a** and **19g** with ethyl iodide afforded N-3- and N-1-ethyl nucleosides **26** and **29**, respectively. Their analytical and spectroscopic data were in agreement with the proposed structures [13,41] and with those reported in the literature for other N-3-substituted pyrimidines and N-1-substituted purines [44,45].

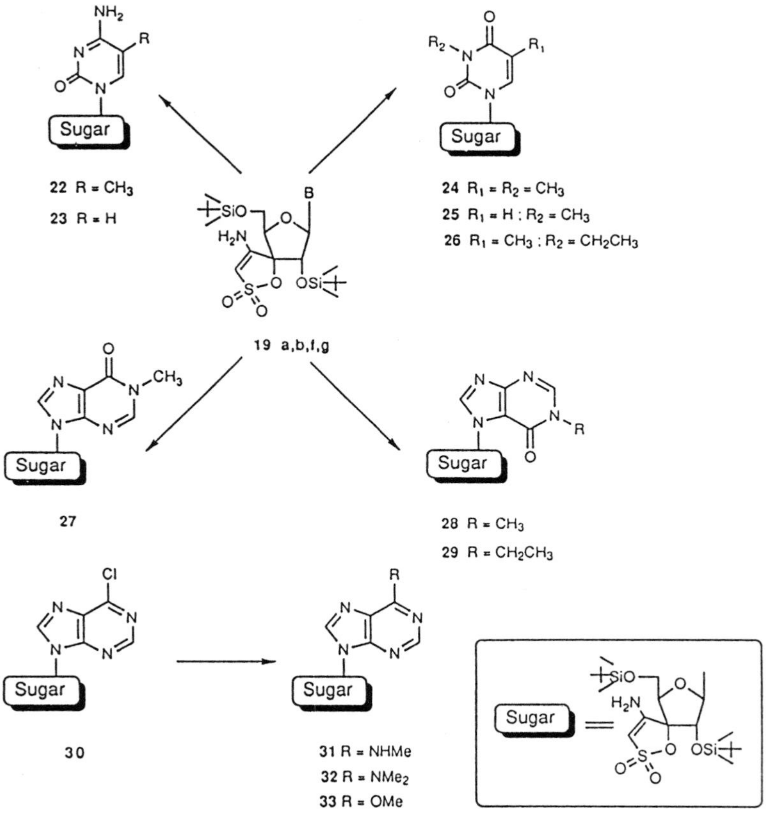

Scheme 3

Finally, N-6 alkylated derivatives of adenine **31** and **32** and hypoxanthine **33** were prepared by reaction of the 6-chloropurine nucleoside **30** (prepared by glycosylation of **15** with 6-chloropurine) with methylamine, dimethylamine and sodium methoxide, respectively [46].

Antiviral and cytostatic activity of TSAO derivatives

A structure-activity relationship analysis including about 50 different TSAO derivatives revealed the structural requirements that the compounds have to fulfil to inhibit HIV-1 replication. Some of the essential features, as applicable to the TSAO-thymine derivatives are presented in Table 1: (*i*) the *tert*-butyldimethylsilyl groups have to be present in both C-2'- and C-5'-position of the ribose moiety. Replacement of the silyl moiety by another lipophilic substituent [i.e. benzoyl (**21a**)], or by *tert*-butyldiphenylsilyl (data not shown) annihilates the antiviral activity of TSAO-T. (*ii*) The presence of the spiro moiety at C-3' (3'-spiro-5"-[4"-amino-1",2"-oxathiole-2",2"-dioxide]) of ribose is obligatory for anti-HIV-1 activity. Replacement of 3'-spiro substituent by OH (1)or H, or C≡N (data not shown) renders the TSAO-T derivative inactive. (*iii*) The 3'-spiro-substituent must be present in the (*R*)-(*ribo*) configuration to confer anti-HIV-1 activity. The (*S*)-(*xylo*) enantiomer is devoid of anti-HIV-1 activity, although it is almost as cytotoxic as its (*R*)-(*ribo*) enantiomer (compare **7** with **8**) (Table 1).

Table 1. Anti-HIV-1 activity of TSAO-T analogues with modifications in the sugar part

Compound[a]	R_2'	R_3'	R_5'	Sugar configuration	EC_{50}[b] (μg/ml)	CC_{50}[c] (μg/ml)	S.I.[d]
10	OH	Spiro	OH	*Ribo*	> 100	> 100	-
11	OH	Spiro	O[Si]	*Ribo*	> 20	45	< 2
20a	O[Si]	Spiro	OH	*Ribo*	> 40	106	< 2.5
1	O[Si]	OH	[OSi]	*Ribo*	> 1.6	5.7	< 3.6
21a	O[Si]	Spiro	OBz	*Ribo*	> 8	20	< 2.5
8	O[Si]	Spiro	[OSi]	*Ribo*	0.034	7.7	227
7	O[Si]	Spiro	[OSi]	*Xylo*	> 4	18	< 4.5

[a]Spiro, 3'-spiro-5"-(4"-amino-1",2"-oxathiole-2",2"-dioxide); O[Si], (*tert*-butyldimethylsilyl), Obz, *O*-benzoyl.
[b]50% effective concentration, or compound concentration required to inhibit HIV-1(III$_B$)-induced cytopathicity in MT-4 cells by 50%.
[c]50% cytotoxic concentration, or compound concentration required to reduce MT-4 cell viability by 50%.
[d]S.I., selectivity index, or ratio of CC_{50} to EC_{50}.

The prototype compound TSAO-T inhibits HIV-1 replication at a 50% effective concentration (EC_{50}) of 0.034 μg/ml, that is at a concentration that is 227-fold lower than its 50% cytotoxic concentration (CC_{50}) (i.e. 7.7 μg/ml) in MT-4 cells (Table 1). The thymine moiety of TSAO-T can be replaced by a number of other pyrimidines (Table 2) and purines (Table 3) without significant loss of antiviral activity. In fact, all natural pyrimidine bases (i.e. uracil, cytosine) as well as 5-substituted pyrimidine bases (i.e. 5-ethyluracil, 5-bromouracil, 5-methylcytosine) show a similar antiviral potency (EC_{50}: 0.034-0.114 μg/ml) and cytotoxicity (CC_{50}: 3.2-17.7 μg/ml) as the prototype compound TSAO-T, except for TSAO-Br^{5}U, which is 7-fold less antivirally effective and 3- to 4-fold more cytotoxic than TSAO-T (Table 2), and TSAO-C, which is 13-fold less antivirally effective than TSAO-T and also markedly (9 30-fold) less toxic (CC_{50}: 200 μg/ml) (Table 2).

Table 2. Anti-HIV-1 activity of TSAO-T analogues with modifications in the pyrimidine part

Compound[a]	EC$_{50}$[b] (μg/ml)	CC$_{50}$[c] (μg/ml)	S.I.[d]
TSAO-T	0.034	7.7	227
TSAO-U	0.114	8.3	73
TSAO-e^{5}U	0.038	3.2	82
TSAO-Br^{5}U	0.206	2.35	11
TSAO-C	0.439	$\geq$ 200	$\geq$ 456
TSAO-m^{5}C	0.072	17.7	246
TSAO-m^{3}T	0.034	139	4,088
TSAO-e^{3}T	0.073	73	1,000

[a]TSAO, t-butyldimethylsilyl-3'-spiro-5"-(4"-amino-1",2"-oxathiole-2",2"-dioxide); T, thymine; U, uracil; e^{5}U, 5-ethyluracil; Br^{5}U, 5-bromouracil; C, cytosine; m^{5}C, 5-methylcytosine; m^{3}T, 3-methylthymine; e^{3}T, 3-ethylthymine. [b,c,d]As for footnotes to Table 1.

Introduction of an alkyl moiety at N-3 (as in TSAO-m^{3}T and TSAO-e^{3}T) decreased the cytotoxicity by 10- to 20-fold without markedly affecting the anti-HIV-1 activity. As a consequence, the selectivity index of these test compounds increased to 1,000-4,088 as compared to 227 for the prototype compound TSAO-T (Table 2).

The TSAO-purine derivatives are in general 3- to 5-fold less effective than the most active TSAO-pyrimidine derivatives (Table 3). Irrespective of the nature of the purine base (adenine, hypoxanthine or xanthine) and of the way by which the purine is linked to the sugar part (through N^7 or N^9 of the purine ring), all TSAO-purines are endowed with similar antiviral and cytostatic potency.

Table 3. Anti-HIV-1 activity of TSAO-purine analogues in MT-4 cells

Compound[a]	EC$_{50}$[b] (μg/ml)	CC$_{50}$[c] (μg/ml)	S.I.[d]
TSAO-A	0.162	7.3	45
TSAO-m^{6}A	0.086	8.5	99
TSAO-dm^{6}A	0.270	10	37
TSAO-Hx	0.092	8.7	90
TSAO-m^{6}Hx	0.160	13	81
7-TSAO-Hx	0.090	8.5	94
TSAO-m^{1}Hx	0.180	> 100	> 555
7-TSAO-m^{1}Hx	0.150	89	593
7-TSAO-e^{1}Hx	0.550	> 100	> 180
7-TSAO-X	0.230	7.7	33

[a]TSAO, t-butyldimethylsilyl-3'-spiro-5"-(4"-amino-1",2"-oxathiole-2",2"-dioxide); A, adenine; m^{6}A, 6-methyladenine; dm^{6}A, 6-dimethyladenine; Hx, hypoxanthine; m^{6}Hx, 6-methoxypurine; 7-TSAO-Hx, hypoxanthine linked *via* N^7 to TSAO; m^{1}Hx, 1-methylhypoxanthine; 7-TSAO-m^{1}Hx, 1-methylhypoxanthine linked *via* N^7 to TSAO; 7-TSAO-e^{1}Hx, 1-ethylhypoxanthine linked *via* N^7 to TSAO; 7-TSAO-X, xanthine linked *via* N^7 to TSAO. [b,c,d]As for footnotes to Table 1.

However, as also observed with the TSAO-pyrimidine derivatives, introduction of an alkyl moiety at N-1 of the purine ring significantly decreases the cytotoxicity of the test compounds, thereby increasing the selectivity index from 45 (for TSAO-A) to > 555 (for TSAO-m^{1}Hx) (Table 3).

The presence of a methylamino (TSAO-m^{6}A) or dimethylamino (TSAO-dm^{6}A) function at C-6 of purine, or a methoxy group at C-6 of purine (TSAO-m^{6}Hx) does not alter the antiviral efficiency, suggesting that hydrogen bonding of the purine base with its target site seems not to be required for antiviral activity (Table 3). Also, the fact that the structural requirements for antiviral activity of the TSAO derivatives more stringently depend on the sugar part than the base part indicates that the sugar part plays a principal role in the interaction of these molecules with their antiviral target.

Mechanism of antiviral action of TSAO nucleoside derivatives

The inhibitory effect of TSAO-T on HIV-1 reverse transcriptase (RT) has been examined in the presence of various artificial homopolymeric templates. TSAO-T proved inhibitory to RT at an IC$_{50}$ of 17 μM when poly(C).(dG) was used as the template/primer and dGTP as the natural substrate [21,22]. No marked inhibitory effect was noted when poly(A).(dT), poly(U).(dA), poly(I).(dC) or poly(dC).(dG) was used in the presence of dTTP, dATP, dCTP or dGTP, respectively. In contrast, other HIV-1-specific compounds [i.e. TIBO (R82150), nevirapine, HEPT] were inhibitory to HIV-1 RT in the presence of various template/primers, although to a lesser extent than when poly(C).(dG) template was used as the template/primer [21,22].

When the kinetics of TSAO-T inhibition of HIV-1 RT in the presence of poly(C).(dG) and dGTP was examined (Lineweaver-Burk plots), non-competitive inhibition of RT with respect to both varying concentrations of substrate (dGTP) and template/primer [poly(C).(dG)] was noted [21,22]. This type of kinetics is indicative of a specific interaction of these inhibitors at an HIV-1 RT site that is distinct from the substrate-binding site of the enzyme and that is not present on HIV-2 RT or any other retroviral RT. The data are also in keeping with our findings that TSAO-T does not act as a DNA chain terminator and is not incorporated into the growing DNA chain. As has also been shown for TIBO (R82150) [16], TSAO-T appears to be a reversible inhibitor of HIV-1 RT [21,22].

TSAO-T is inhibitory to HIV-1 RT only at a concentration that is higher by several orders of magnitude than the concentration needed to exhibit antiviral activity in cell culture. This is most likely due to the circumstances under which the inhibition of HIV-1 RT by TSAO is investigated. In fact, other HIV-1-specific inhibitors such as TIBO, nevirapine and HEPT also exhibit anti-HIV-1 activity in cell culture at concentrations that are markedly lower than those required for inhibition of reverse transcriptase. As a rule, TSAO derivatives (TSAO-m^{3}T, TSAO-e^{3}T, TSAO-m^{1}Hx, etc.) are inactive against those HIV-1 strains that contain a single amino acid mutation in the RT enzyme (see below) suggesting that HIV-1 RT is the principal molecular target for TSAO derivatives.

Selection and biological properties of HIV-1 strains resistant to TSAO-e^{3}T and TSAO-m^{1}Hx

When HIV-1(III$_B$)-infected CEM cells were subjected to 2 to 3 subcultivations in the presence of 2 to 3 times the EC$_{50}$ of TSAO-e^{3}T or TSAO-m^{1}Hx, virus replication resumed and the virus that was recovered proved insensitive to the test compounds [30-33].

When evaluated against the TSAO-e^{3}T- and TSAO-m^{1}Hx-resistant HIV-1 strains, all TSAO-purine and TSAO-pyrimidine derivatives examined proved inactive at subtoxic concentrations (Table 4). Only TSAO-T still had antiviral activity at a concentration of 2.0 μg/ml, that is 3- to 4-fold lower than its CC$_{50}$. Also, a TSAO-T-resistant HIV-1 strain (designated HIV-1/TSAO-T, and selected in the presence of increasing concentrations of TSAO-T) remained sensitive to TSAO-T at 2 μg/ml, while it was insensitive to the other TSAO derivatives (data not shown). These observations suggest that TSAO-T may differ in its mode of action from the other TSAO-T congeners.

The other HIV-1-specific inhibitors (i.e. TIBO R82913, nevirapine, pyridinone L697,661) were almost equally inhibitory to the HIV-1/TSAO-e^{3}T and HIV-1/TSAO-m^{1}Hx strains as the wild-type HIV-1(III$_B$) [31,32] (Table 4). This is noteworthy, as earlier reports indicated that nevirapine- and pyridinone-resistant HIV-1 strains are cross-resistant to other HIV-1-specific inhibitors (i.e. TIBO R82150, TIBO R82913, HEPT derivatives, nevirapine and pyridinones). TSAO-resistant HIV-1 strains remain fully sensitive to 2',3'-dideoxynucleoside analogues (i.e. AZT, DDC, DDI) and acyclic nucleoside phosphonates (i.e. PMEA, FPMPA). This is in agreement with earlier reports that HIV-1 strains which became resistant to the HIV-1-specific RT inhibitors (i.e. nevirapine and pyridinone) do not loose their sensitivity to the inhibitory effects of the nucleoside analogues [24-26].

Table 4. Inhibitory effects of test compounds on HIV-1(III$_B$) mutant strains in CEM cells

Compound[a]	EC$_{50}$[b] (μg/ml)		
	HIV-1(III$_B$)	HIV-1/TSAO-e^{3}T	HIV-1/TSAO-m^{1}Hx
TSAO-T	0.03	2.0	2.0
TSAO-m^{3}T	0.03	35	≥ 50
TSAO-e^{3}T	0.03	> 50	> 50
TSAO-U	0.08	> 5	-
TSAO-A	0.07	> 5	-
TSAO-m^{1}Hx	0.06	-	> 50
7-TSAO-m^{1}Hx	0.05	-	> 20
TIBO R82913	0.02	0.05	0.10
Nevirapine	0.03	0.02	0.01
Pyridinone L697,661	0.007	0.02	0.035
AZT	0.003	0.003	0.003
DDC	0.05	0.03	0.02
DDI	5.0	15	4.3
PMEA	7.0	7.5	8.0
(S)-FPMPA	3.0	1.8	1.5

[a]AZT, 3'-azido-3'-deoxythymidine; DDC, 2',3'-dideoxycytidine; DDI, 2',3'-dideoxyinosine; PMEA, 9-(2-phosphonylmethoxyethyl)adenine; (S)-FPMPA, (S)-9-(3-fluoro-2-phosphonylmethoxypropyl)ade-nine.
[b]50% effective concentration, or compound concentration required to inhibit HIV-1-induced cytopathicity in CEM cells by 50%.

Characterisation of TSAO-resistant HIV-1 strains

HIV-1 strains with reduced sensitivity against different TSAO derivatives were subject of DNA sequence analysis. MT-4 cells (3 x 10^5 cells/ml) were grown in RPMI-1640 medium and infected with the different HIV-1 strains at 200 $CCID_{50}$. After incubation for 3 days at 37°C the cells were centrifuged and washed twice with phosphate-buffered saline. Total cellular DNA was extracted from cell pellets by cell lysis and proteinase K digestion. To 10^6 cells, 100 μl extraction buffer was added containing 2 mM $MgCl_2$, 50 mM KCl, 10 mM Tris-HCl, 10 μg proteinase K, 0.5% Tween-20 and 0.5% NP-40. The cell extracts were incubated at 56°C for 1 h and subsequently at 95°C for 10 min. DNA was extracted with phenol/chloroform and stored at -20°C before PCR analysis. DNA from 5 x 10^4 HIV-1-infected MT-4 cells were used for PCR amplification of the proviral reverse transcriptase gene.

The PCR reactions were performed in 100 μl with 3 mM $MgCl_2$, 50 mM KCl, 10 mM Tris-HCl, 200 μM dNTPs, 2.5 U AmpliTaq DNA polymerase and 0.15 μM of each primer. Oligonucleotides were chosen to give a 727 bp fragment covering the amino acids 50-270 (sense primer 5'-CCTGAAAATCCATACAATACTCCAGTATTTG-3' and reverse complement primer 5'-AGTGCTTTGGTTCCTCTAAGGAGTTTAC-3'). The amplification was performed in a Perkin-Elmer Cetus DNA Thermal Cycler with 35 cycles of DNA synthesis (denaturation at 95°C for 1 min, annealing at 55°C for 30 sec and extension at 72°C for 1 min). The size of the PCR product was determined on an ethidium bromide stained 1% low melting agarose gel. The DNA band of correct size was cut from the gel and purified by MagicPCR Preps (Promega).

The amplified and purified reverse transcriptase gene fragment was used directly as template in a sequence reaction performed with a second PCR. The protocol from the Taq DyeDeoxy™ Terminatior sequencing kit (Applied Biosystems) was used for the sequence reactions. In this cycle sequencing protocol, dye-labeled dideoxynucleotides are used as DNA chain terminators in a fluorescence-based sequence reaction. The same primers as described above for the first PCR amplification was used for the sequencing. Excess dye-labeled dideoxynucleotides were removed by phenol/chloroform extraction prior to analysis on a Model 373A DNA Sequencing System (Applied Biosystems). Fig. 2 represents a schematic overview of the different analysis steps.

The reverse transcriptase of the TSAO-resistant HIV-1 strains HIV-1/TSAO-T, HIV-1/TSAO-m^3T, HIV-1/TSAO-e^3T and HIV-1/TSAO-m^1Hx, contain one amino acid change at position 138. In all these cases glutamic acid-138 has been replaced by lysine. This substitution must have occurred by a single transition mutation of the first base (guanine B adenine) of the 138th codon (GAG B AAG) (Table 5). No amino acid changes at positions 100, 103, 106, 108, 181, 188 or 236 were observed. The latter amino acids were reported to be subjected to mutation in the RT of HIV-1 strains resistant to nevirapine, pyridinone, TIBO or BHAP (see also Table 5) [24-30].

Recently, the structure of the RT enzyme (p66) has been visualized as a right hand, containing a finger, palm, thumb and connection domain [51]. Most, if not all, of the amino acid mutations that have been reported so far to be responsible for HIV-1 resistance to the non-nucleoside HIV-1-specific RT inhibitors are clustered in the palm domain of RT. The amino acid-236 change characteristic for the BHAP-resistant HIV-1 strains lays in the corner between the palm and the thumb domain. In contrast, amino acid-138 that is altered in the TSAO-resistant HIV-1 strains is located on the top of the finger domain, that is at a marked distance of the other amino acids involved in the emergence of resistant

HIV-1 strains. This peculiar characteristic of the TSAO-resistant HIV-1 strains may explain why they retain high sensitivity to the HIV-1-specific non-nucleoside analogues [31-33].

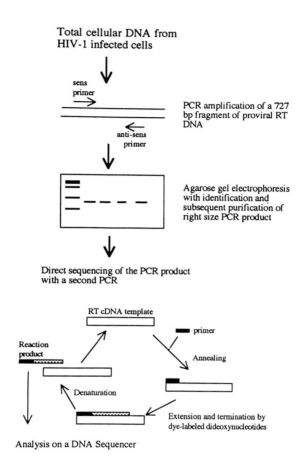

Fig. 2. Procedure to amplify a well-defined nucleotide sequence domain in the HIV-1 reverse transcriptase gene.

Molecular interaction of TSAO derivatives with the HIV-1 reverse transcriptase

The stringent structural requirements the TSAO derivatives have to fulfil to be antivirally active [i.e. obligatory presence of the 3'-spiro substituent in the (R)-(ribo) configuration] must be indicative for direct interaction of the -COOH group of glutamic acid-138 residue of HIV-1 RT with the 4"-amino group of the 3'-spiro moiety of the TSAO derivatives. The nature of this inter-

Table 5. Mutations of HIV-1 RT amino acid residues leading to resistance to AZT, DDI, DDC and non-nucleoside HIV-1-specific RT inhibitors

Compound	Amino acid number[b] in HIV-1 RT	Amino acid change[c]	Nature of amino acids changed[d]
DDI/DDC	69	Thr B Asp	P B –
	74	Leu B Val	H B H
AZT	41	Met B Leu	H B H
	67	Asp B Asn	– B P
	70	Lys B Arg	+ B +
	215	Thr B Phe	P B H
		Thr B Tyr	P B P
	219	Lys B Gln	+ B P
HIV-1-specific	100	Leu B Ile	H B H
RT inhibitors[a]	103	Lys B Asn	+ B P
	106	Val B Ala	H B N
	108	Val B Ile	H B H
	138	Glu B Lys	– B +
	181	Tyr B Cys	P B N
	188	Tyr B His	P B P
		Tyr B Cys	P B N
	236	Pro B Leu	N B H

[a]Non-nucleoside HIV-1-specific RT inhibitors including TIBO, HEPT, nevirapine, pyridinone L697,661, BHAP, α-APA, and TSAO.

[b]Data taken from references 24-33.

[c]Thr, threonine; Leu, leucine; Val, valine; Met, methionine; Asp, aspartic acid; Asn, asparagine; Lys, lysine; Arg, arginine; Phe, phenylalanine; Tyr, tyrosine; Gln, glutamine; Ile, isoleucine; Ala, alanine; Glu, glutamic acid; Cys, cysteine; His, histidine; Pro, proline.

[d]N, neutral; H, hydrophobic; P, polar; +, positively charged; -, negatively charged.

action may be electrostatic or involve hydrogen bonding. In addition to the 4"-amino group of the 3'-spiro moiety other groups of the TSAO derivatives must be involved in the interaction with the HIV-1 RT. The lipophilic (silyl) moieties at C-2' and C-5' and the thiole-2",2"-dioxide group of the 3'-spiro substituent, may represent such pharmacophores involved in the interaction of the TSAO derivatives with functional groups in the RT enzyme. According to an hypothetical model (shown in Fig. 3), the 4"-amino group of the 3'-spiro moiety interacts with the -COOH function of glutamic acid-138, the thiole dioxide function of the 3'-spiro moiety interacts (through hydrogen bonding) with the carboxamide function of asparagine-137 and/or asparagine-136, and the lipophilic silyl group at C-2' binds hydrophobically to isoleucine-135. Whether the TSAO derivatives indeed interact with HIV-1 RT as depicted in the model (Fig. 3) remains to be verified by further investigations (i.e. site-directed mutagenesis).

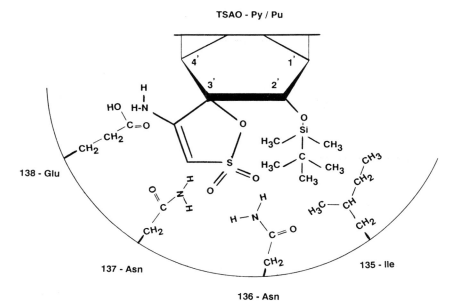

Fig. 3. Hypothetical model of interaction of HIV-1 reverse transcriptase with TSAO nucleosides.

In addition, attempts should be undertaken to strengthen the potential interaction of the TSAO molecules with the asparagine functional groups at position 137 and 136 by chemically modifying the TSAO structures nearby the 4"-amino group. Strengthening of this interaction may also result in a concomitant increase of the antiviral activity of the TSAO molecules. Moreover, since asparagine-137 and asparagine-136, as well as the lipophilic residue at position 135 (either isoleucine, valine or phenylalanine) are highly conserved in the reverse transcriptases of all HIV-1, HIV-2 and SIV strains that have so far been characterized, it may be possible to modify the TSAO molecule such that they acquire activity against a broader range of immunodeficiency viruses.

In conclusion, the TSAO derivatives represent the first HIV-1-specific RT inhibitors for which a well-defined part of the molecule (i.e. the 4"-amino group of the 3'-spiro component) has been identified as an essential pharmacophore interacting with a well-defined moiety (the .-COOH group of Glu-138) of the RT target enzyme. The knowledge gained on the molecular features involved in the interaction of the TSAO derivatives with their target site at the HIV RT should help in the rational design of new TSAO derivatives endowed with more potent anti-HIV-1 activity and/or broader spectrum of anti-HIV activity.

References

1.   M. Baba, H. Tanaka, E. De Clercq, R. Pauwels, J. Balzarini, D. Schols, H. Nakashima, C.-F. Perno, R. T. Walker, T. Miyasaka: Highly specific inhibition of human immunodeficiency virus type 1 by a novel 6-substituted acyclouridine derivative. *Biochem. Biophys. Res. Commun.* **165** (1989) 1375-1381.

2.   T. Miyasaka, H. Tanaka, M. Baba, H. Hayakawa, R. T. Walker, J. Balzarini, E. De Clercq: A novel lead for specific anti-HIV-1 agents: 1-[(2-hydroxyethoxy)methyl]-6-(phenylthio)thymine. *J. Med. Chem.* **32** (1989) 2507-2509.

3. H. Tanaka, H. Takashima, M. Ubasawa, K. Sekiya, I. Nitta, M. Baba, S. Shigeta, R. T. Walker, E. De Clercq, T. Miyasaka: Structure-activity relationships of 1-[(2-hydroxyethoxy)methyl]-6-(phenylthio)thymine analogues: effect of substitutions at the C-6 phenyl ring and at the C-5 position on anti-HIV-1 activity. *J. Med. Chem.* **35** (1992) 337-345.

4. R. Pauwels, K. Andries, J. Desmyter, D. Schols, M. Kukla, H. Breslin, A. Raeymaekers, J. Van Gelder, R. Woestenborghs, J. Heykants, K. Schellekens, M.A.C. Janssen, E. De Clercq, P. Janssen: Potent and selective inhibition of HIV-1 replication in vitro by a novel series of TIBO derivatives. *Nature* **343** (1990) 470-474.

5. M. J. Kukla, H. J. Breslin, R. Pauwels, C. L. Fedde, M. Miranda, M. K. Scott, R. G. Sherrill, A. Raeymaekers, J. Van Gelder, K. Andries, M. A. C. Janssen, E. De Clercq, P. A. J. Janssen: Synthesis and anti-HIV-1 activity of 4,5,6,7-tetrahydro-5-methylimidazo[4,5,1-*jk*][1,4]benzodiazepin-2(1*H*)-one (TIBO) derivatives. *J. Med. Chem.* **34** (1991) 746-751.

6. V. J. Merluzzi, K. D. Hargrave, M. Labadia, K. Grozinger, M. Skoog, J. C. Wu, C.-K. Shih, K. Eckner, S. Hattox, J. Adams, A. S. Rosenthal, R. Faanes, R. J. Ecker, R. A. Koup, J. L. Sullivan: Inhibition of HIV-1 replication by a nonnucleoside reverse transcriptase inhibitor. *Science* **250** (1990) 1411-1413.

7. R. A. Koup, V. J. Merluzzi, K. D. Hargrave, J. Adams, K. Grozinger, R. J. Eckner, J. L. Sullivan: Inhibition of human immunodeficiency virus type 1 (HIV-1) replication by the dipyridodiazepinone BI-RG-587. *J. Infect. Dis.* **163** (1991) 966-970.

8. M. E. Goldman, J. H. Nunberg, J. A. O'Brien, J. C. Quintero, W. A. Schleif, K. F. Freund, S. L. Gaul, W. S. Saari, J. S. Wai, J. M. Hoffman, P. S. Anderson, D. J. Hupe, E. A. Emini, A. M. Stern: Pyridinone derivatives: specific human immunodeficiency virus type 1 reverse transcriptase inhibitors with antiviral activity. *Proc. Natl. Acad. Sci. USA* **88** (1991) 6863-6867.

9. D. L. Romero, M. Busso, C.-K. Tan, F. Reusser, J. R. Palmer, S. M. Poppe, P. A. Aristoff, K. M. Downey, A. G. So, L. Resnick, W. G. Tarpley: Nonnucleoside reverse transcriptase inhibitors that potently and specifically block human immunodeficiency virus type 1 replication. *Proc. Natl. Acad. Sci. USA* **88** (1991) 8806-8810.

10. J. Balzarini, M.-J. Pérez-Pérez, A. San-Félix, D. Schols, C.-F. Perno, A.-M. Vandamme, M.-J. Camarasa, E. De Clercq: 2',5'-Bis-*O*-(*tert*-butyldimethylsilyl)-3'-spiro-5"-(4"-amino-1",2"-oxathiole-2",2"-dioxide)pyrimidine (TSAO) nucleoside analogues: highly selective inhibitors of human immunodeficiency virus type 1 that are targeted at the viral reverse transcriptase. *Proc. Natl. Acad. Sci. USA* **89** (1992) 4392-9396.

11. J. Balzarini, M.-J. Pérez-Pérez, A. San-Félix, S. Velazquez, M.-J. Camarasa, E. De Clercq: [2',5'-Bis-*O*-(*tert*-butyldimethylsilyl)]-3'-spiro-5"-(4"-amino-1",2"-oxathiole-2",2"-dioxide) (TSAO) derivatives of purine and pyrimidine nucleosides as potent and selective inhibitors of human immunodeficiency virus type 1. *Antimicrob. Agents Chemother.* **36** (1992) 1073-1080.

12. M.-J. Camarasa, M.-J. Pérez-Pérez, A. San-Félix, J. Balzarini, E. De Clercq: 3'-Spiro nucleosides, a new class of specific human immunodeficiency virus type 1 inhibitors: synthesis and antiviral activity of [2',5'-bis-*O*-(*tert*-butyldimethylsilyl)-ß-D-xylo- and -ribofuranose]-3'-spiro-5"-[4"-amino-1",2"-oxathiole-2",2"-dioxide] (TSAO) pyrimidine nucleosides. *J. Med. Chem.* **35** (1992) 2721-2727.

13. M.-J. Pérez-Pérez, A. San-Félix, J. Balzarini, E. De Clercq, M.-J. Camarasa: TSAO Analogues. Stereospecific synthesis and anti-HIV-1 activity of 1-[2',5'-bis-*O*-(*tert*-butyldimethylsilyl)-ß-D-ribofuranosyl]-3'-spiro-5"-(4"-amino-1",2"-oxathiole-2",2"-dioxide) pyrimidine and pyrimidine-modified nucleosides. *J. Med. Chem.* **35** (1992) 2988-2995.

14. M.-J. Pérez-Pérez, A. San-Félix, M.-J. Camarasa, J. Balzarini, E. De Clercq: Synthesis of [1-[2',5'-bis-*O*-(*t*-butyldimethylsilyl)-ß-D-xylo- and ß-D-ribofuranosyl]thymine]-3'-spiro-5"-[4"-amino-1",2"-oxathiole-2",2"-dioxide] (TSAO). A novel type of specific anti-HIV agents. *Tetrahedron Lett.* **33** (1992) 3029-3032.

15. R. Pauwels, K. Andries, Z. Debyser, P. Van Daele, D. Schols, A.-M. Vandamme, P. Stoffels, K. De Vreese, R. Woestenborghs, C. G. M. Janssen, J. Anné, G. Cauwenbergh, J. Desmyter, J. Heykants, M. A. C. Janssen, E. De Clercq, P. A. J. Janssen: Potent and highly selective HIV-1 inhibition by a new series of α-anilino phenyl acetamide (α-APA) derivatives targeted at HIV-1 reverse transcriptase. *Proc. Natl. Acad. Sci. USA* (1992) in press.

16. Z. Debyser, R. Pauwels, K. Andries, J. Desmyter, M. Kukla, P. A. J. Janssen, E. De Clercq: An antiviral target on reverse transcriptase of human immunodeficiency virus type 1 revealed by tetrahydroimidazo[4,5,1-*jk*][1,4]benzodiazepin-2(1*H*)-one and -thione derivatives. *Proc. Natl. Acad. Sci. USA* **88** (1991) 1451-1455.

17. E. L. White, R. W. Buckheit Jr., L. J. Ross, J. M. Germany, K. Andries, R. Pauwels, P. A. J. Janssen, W. M. Shannon, M. A. Chirigos: A TIBO derivative, R82913, is a potent inhibitor of HIV-1 reverse transcriptase with heteropolymer templates. *Antiviral Res.* **16** (1991) 257-266.

18. K. B. Frank, G. J. Noll, E. V. Connell, I. S. Sim: Kinetic interaction of human immunodeficiency virus type 1 reverse transcriptase with the antiviral tetrahydroimidazo[4,5,1-*jk*]-[1,4]-benzodiazepine-2-(1*H*)-thione compound, R82150. *J. Biol. Chem.* **266** (1991) 14232-14236.

19. K. A. Cohen, J. Hopkins, R. H. Ingraham, C. Pargellis, J. C. Wu, D. E. H. Palladino, P. Kinkade, T. C. Warren, S. Rogers, J. Adams, P. R. Farina, P. M. Grob: Characterization of the binding site for nevirapine (BI-RG-587), a nonnucleoside inhibitor of human immunodeficiency virus type-1 reverse transcriptase. *J. Biol. Chem.* **266** (1991) 14670-14674.

20. T. J. Dueweke, F. J. Kézdy, G. A. Waszak, M. R. Deibel Jr., W. G. Tarpley: The binding of a novel bis(heteroaryl)piperazine mediates inhibition of human immunodeficiency virus type 1 reverse transcriptase. *J. Biol. Chem.* **267** (1992) 27-30.

21. J. Balzarini, M.-J. Pérez-Pérez, A. San-Félix, M.-J. Camarasa, P. J. Barr, E. De Clercq: Interaction of human immunodeficiency virus type 1 (HIV-1) reverse transcriptase with the novel HIV-1 specific nucleoside analogue 1-[2',5'-bis-*O*-(*tert*-butyldimethylsilyl)-ß-D-ribofuranosyl]-3'-spiro-5"-[4"-amino-1",2"-oxathiole-2",2"-dioxide]thymine (TSAO-T). *Arch. Int. Phys. Biochem. Biophys.* **100** (1992) B28.

22. J. Balzarini, M.-J. Pérez-Pérez, A. San-Félix, M.-J. Camarasa, I. C. Bathurst, P. J. Barr, E. De Clercq: Kinetics of inhibition of human immunodeficiency virus type 1 (HIV-1) reverse transcriptase by the novel HIV-1-specific nucleoside analogue 1-[2',5'-bis-*O*-(*tert*-butyldimethylsilyl)-ß-D-ribofuranosyl]-3'-spiro-5"-[4"-amino-1",2"-oxathiole-2",2"-dioxide]-thymine (TSAO-T). *J. Biol. Chem.* **267** (1992) 11831-11838.

23. M. Baba, E. De Clercq, H. Tanaka, M. Ubasawa, H. Takashima, K. Sekiya, I. Nitta, K. Umezu, H. Nakashima, S. Mori, S. Shigeta, R. T. Walker, T. Miyasaka: Potent and selective inhibition of human immunodeficiency virus type 1 (HIV-1) by 5-ethyl-6-phenylthiouracil derivatives through their interaction with the HIV-1 reverse transcriptase. *Proc. Natl. Acad. Sci. USA* **88** (1991) 2356-2360.

24. J. H. Nunberg, W. A. Schleif, E. J. Boots, J. A. O'Brien, J. C. Quintero, J. M. Hoffman Jr., E. A. Emini, M. E. Goldman: Viral resistance to human immunodeficiency virus type 1-specific pyridinone reverse transcriptase inhibitors. *J. Virol.* **65** (1991) 4887-4892.

25. D. Richman, C.-K. Shih, I. Lowy, J. Rose, P. Prodanovich, S. Goff, J. Griffin: Human immunodeficiency virus type 1 mutants resistant to nonnucleoside inhibitors of reverse transcriptase arise in tissue culture. *Proc. Natl. Acad. Sci. USA* **88** (1991) 11241-11245.

26. J. W. Mellors, G. E. Dutschman, G.-J. Im, E. Tramontano, S. R. Winkler, Y.-C. Cheng: *In vitro* selection and molecular characterization of human immunodeficiency virus-1 resistant to non-nucleoside inhibitors of reverse transcriptase. *Mol. Pharmacol.* **41** (1992) 446-451.

27. E. A. Emini, W. A. Schleif, V. W. Byrnes, J. H. Condra, V. V. Sardanal, J. C. Kappes, M. Saag, G. Shaw: *In vitro* and *in vivo* derivation of HIV-1 variants resistant to inhibition by the nonnucleoside reverse transcriptase inhibitor L-697,661. Abstracts of the HIV Drug-Resistance Workshop, Noordwijk, The Netherlands, 16-18 July 1992, p. 15.

28. G. Maass, U. Immendoerfer, U. Lese, E. Pfaff, R. Goody, B. Koenig: Viral resistance to the thiazolo-iso-indolinones, a new class of non-nucleoside inhibitors of HIV-1 reverse transcriptase. Abstracts of the HIV Drug-Resistance Workshop, Noordwijk, The Netherlands, 16-18 July 1992, p. 29.

29. S. M. Poppe, T. J. Dueweke, I. S. Y. Chen, W. G. Tarpley: Development of resistance to the BHAP non-nucleoside HIV-1 RT inhibitors. Abstracts of the HIV Drug-Resistance Workshop, Noordwijk, The Netherlands, 16-18 July 1992, p. 9.

30. J. Balzarini, A. Karlsson, A.-M. Vandamme, L. Vrang, B. Öberg, M.-J. Pérez-Pérez, A. San-Félix, S. Velázquez, M.-J. Camarasa, E. De Clercq: Human immunodeficiency virus type 1 (HIV-1) strains resistant against the novel class of HIV-1-specific TSAO nucleoside analogues are sensitive to TIBO and nevirapine and the nucleoside analogues AZT, DDI and DDC. Abstracts of the HIV Drug-Resistance Workshop, Noordwijk, The Netherlands, 16-18 July 1992, p. 11.

31. J. Balzarini, A. Karlsson, A.-M. Vandamme, M.-J. Pérez-Pérez, L. Vrang, B. Öberg, A. San-Félix, S. Velázquez, M.-J. Camarasa, E. De Clercq: Human immunodeficiency virus type 1 (HIV-1) strains selected for resistance against the novel class of HIV-1-specific TSAO nucleoside analogues retain sensitivity to HIV-1-specific non-nucleoside inhibitors. Submitted (1992).

32. J. Balzarini, S. Velázquez, A. San-Félix, A. Karlsson, M.-J. Pérez-Pérez, M.-J. Camarasa, E. De Clercq: The HIV-1-specific TSAO-purine analogues show a resistance spectrum that is different from the HIV-1-specific non-nucleoside analogues. *Mol. Pharmacol.* (1992) submitted.

33. J. Balzarini, A. Karlsson, M.-J. Pérez-Pérez, L. Vrang, J. Walbers, B. Öberg, A.-M. Vandamme, M.-J. Camarasa, E. De Clercq: HIV-1-specific reverse transcriptase (RT) inhibitors show differential activity against HIV-1 mutant strains containing different amino acid substitutions in the reverse transcriptase. *Virology* (1992) in press.

34. A. Calvo-Mateo, M.-J. Camarasa, A. Diaz-Ortiz, F. G. de las Heras: Synthesis of 3'-*C*-cyano-3'-deoxy-pentofuranosylthymine nucleosides. *Tetrahedron* **44** (1988) 4895-4903.

35. H. Hayakawa, H. Tanaka, N. Itoh, M. Nakajima, T. Miyasaka, K. Yamaguchi, Y. Iitaka: Reaction of organometallic reagents with 2'- and 3'-ketouridine derivatives: synthesis of uracil nucleosides branched at the 2'- and 3'-positions. *Chem. Pharm. Bull.* **35** (1987) 2605-2608.

36. A. Calvo-Mateo, M.-J. Camarasa, A. Diaz-Ortiz, F. G. de las Heras: Novel aldol-type cyclocondensation of *O*-mesyl(methylsulphonyl)cyanohydrins. Application to the stereospecific synthesis of branched-chain sugars. *J. Chem. Soc. Chem. Commun.* (1988) 1114-1115.

37. M.-J. Pérez-Pérez, M.-J. Camarasa, A. Diaz-Ortiz, A. San-Félix, F. G. de las Heras: Stereospecific synthesis of branched-chain sugars by a novel aldol-type cyclocondensation. *Carbohydr. Res.* **216** (1991) 399-411.

38. G. L. Tong, W. W. Lee, L. Goodman: Synthesis of some 3'-*O*-methylpurineribonucleosides. *J. Org. Chem.* **32** (1967) 1984-1986.

39. J. Yoshimura: Synthesis of branched-ched-chain sugars. *Adv. Carbohydr. Chem. Biochem.* **42** (1984) 69-134.

40. H. Vorbrüggen, K. Kolikiewicz, B. Bennua: Nucleotide synthesis with trimethylsilyl triflate and perchlorate as catalysts. *Chem. Ber.* **114** (1981) 1234-1256.

41. S. Velázquez, A. San-Félix, M. J. Pérez-Pérez, J. Balzarini, E. De Clercq, M.J. Camarasa: *Unpublished results*.

42. L. N. Beigelman, G. U. Gurskaya, E. N. Tsakina, S. N. Mikhailov: Epimerization during the acetolysis of 3-*O*-acetyl-5-*O*-benzoyl-1,2-*O*-isopropylidene-3-*C*-methyl-α-D-ribofuranose. Synthesis of 3'-*C*-methylnucleosides with the ß-D-ribo- and α-D-arabino configurations. *Carbohydr. Res.* **181** (1988) 77-88.

43. E. Walton, S. R. Jenkins, R. F. Nutt, F. W. Holly, M. Nemes: Branched-chain sugar nucleosides. V. Synthesis and antiviral properties of several branched-chain sugar nucleosides. *J. Med. Chem.* **12** (1969) 306-309.

44. T. Sasaki, K. Minamoto, H. Suzuki: Elimination reactions on the di- and trimesylated derivatives of N^3-benzyluridine. *J. Org. Chem.* **38** (1973) 598-607.

45. I. Yamamoto, T. Kimura, Y. Tateoka, K. Watanabe, I. Ho: N-Substituted oxopyrimidines and nucleosides: structure-activity relationship for hypnotic activity as central nervous system depressant. *J. Med. Chem.* **30** (1987) 2227-2231.

46. C. K. Chu, G. V. Ullas, L. S. Jeong, S. K. Ahn, B. Doboszewski, Z. X. Lin, J.W. Beach, R.F. Schinazi: Synthesis and structure-activity relationships of 6-substituted 2',3'-dideoxypurine nucleosides as potential anti-human immunodeficiency virus agents. *J. Med. Chem.* **33** (1990) 1553-1561.

47. B. A. Larder, G. Darby, D. D. Richman: HIV with reduced sensitivity to zidovudine (AZT) isolated during prolonged therapy. *Science* **243** (1989) 1731-1734.

48. B. A. Larder, S. D. Kemp: Multiple mutations in HIV-1 reverse transcriptase confer high-level resistance to zidovudine (AZT). *Science* **246** (1989) 1155-1158.

49. M. H. St. Clair, J. L. Martin, G. Tudor-Williams, M. C. Bach, C. L. Vavro, D. M. King, P. Kellam, S .D. Kemp, B. A. Larder: Resistance to ddI and sensitivity to AZT induced by a mutation in HIV-1 reverse transcriptase. *Science* **253** (1991) 1557-1559.

50. P. Kellam, C. A. B. Boucher, B. A. Larder: Fifth mutation in human immunodeficiency virus type 1 reverse transcriptase contributes to the development of high-level resistance to zidovudine. *Proc. Natl. Acad. Sci. USA* **89** (1992) 1934-1938.

51. L. A. Kohlstaedt, J. Wang, J. M. Friedman, P. A. Rice, T. A. Steitz: Crystal structure at 3.5 °Aresolution of HIV-1 reverse transcriptase complexed with an inhibitor.   *Science* **256** (1992) 1783-1790.

4'-Substituted Nucleosides as Antiviral Agents [1]

Hans Maag*, Arnold J. Gutierrez [2,3], Ernest J. Prisbe [3], Robert M. Rydzewski [4] and
Julien P.H. Verheyden

Institute of Bio-Organic Chemistry
Syntex Discovery Research, Palo Alto, California 94304, USA

Summary. Extending the lead presented by the 4'-substituted nucleoside 4'-azidothymidine (ADRT), a potent and selective inhibitor of HIV *in vitro*, 4'-substituted analogs of other lead structures, such as carbocyclic nucleosides, 3'-thianucleosides (3TC), Oxetanocins and cyclopropylthymines were explored. The synthetic efforts towards the preparation of 4'-azido carbocyclic thymidine derivatives and of 1-(2,2-difluorocycloprop-1-yl)thymines are described in detail.

Introduction

The search for effective antiretroviral drugs for the treatment of human immunodeficiency virus (HIV) infections has dominated the research in the field of antiviral agents in recent years. A very large part of these efforts has been directed at the preparation of inhibitors of reverse transcriptase (RT) of HIV [5]. Among the nucleoside group of RT inhibitors, chain terminating nucleoside analogs of the 2',3'-dideoxy class have taken the center stage with the current clinical use of 3'-azido-3'-deoxythymidine (AZT), 2',3'-dideoxyinosine (ddI) and 2',3'-dideoxycytidine (ddC), in addition to the advanced clinical trials of agents such as (-)-2'-deoxy-3'-thiacytidine (3TC).

AZT   ddC   ddI   3TC

In an effort to identify novel nucleoside analogs with potent and selective activity against HIV we embarked on an extensive program for the synthesis of 4'-substituted nucleosides. Modifications at this position of the ribose ring have rarely been explored, with much of the earlier work having been carried out at Syntex about two decades ago [6,7]. Building on this experience and taking into account the potent activity of AZT, 4'-azidonucleosides became our first targets. In this initial series, 4'-azidothymidine (ADRT, **1**) emerged as a potent and highly selective member of this class and was investigated in great detail in the preclinical setting [8]. Parallel to these investigations, the synthesis of different 4'-modifications was explored and extended to other lead anti-HIV inhibitors. A schematic overview of the efforts in this area is presented in Scheme 1. 4'-Cyanothymidine (**2**) [9], 4'-azidomethylthymidine (**3**) [9], 4'-methoxynucleosides such as **4** [8], 4'-substituted analogs of 3TC **5** [10] and of carba-Oxetanocin A **6** [11] were synthesized at Syntex. In this presentation, carbocyclic analogs of ADRT (**7**) and 1-(2,2-difluoro-3-hydroxymethyl-3-substituted-cycloprop-1-yl)thymines (**8**) will be discussed following a brief introductory overview of the synthesis and biological activity of ADRT (**1**).

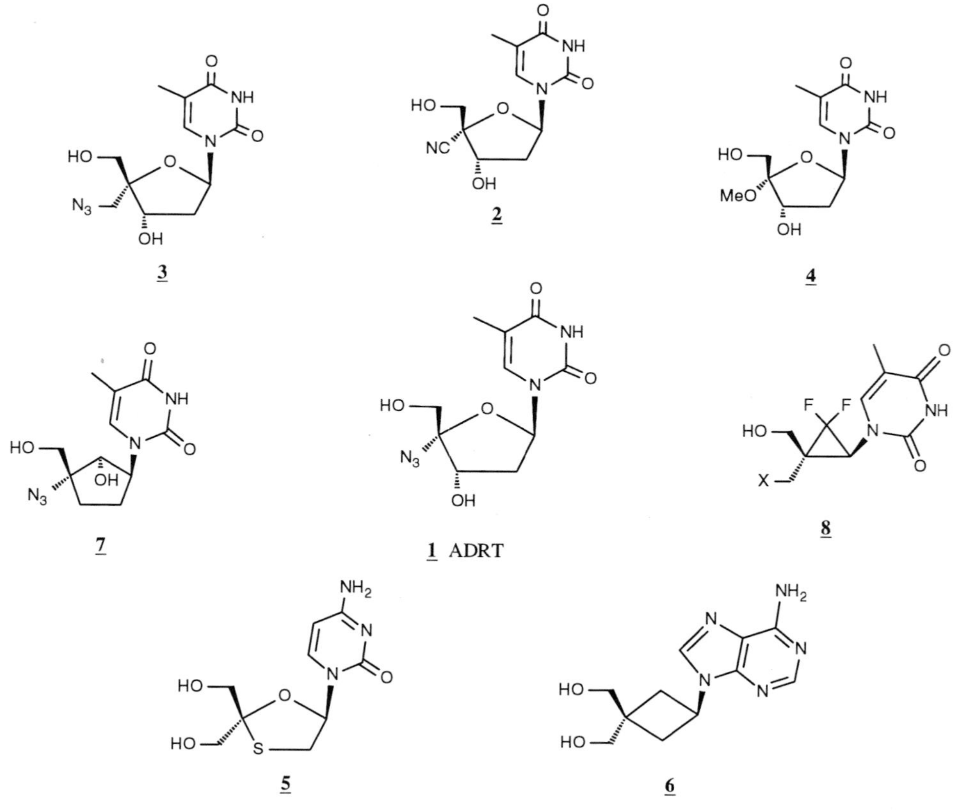

Scheme 1

1. 4'-Azidothymidine (ADRT, 1)

1.1. Synthesis

The synthesis of this initial member of the 4'-substituted family of potential anti-HIV nucleosides was attempted along the route described for the synthesis of 4'-azidocytidine [7] and involved the addition of iodine azide to the exocyclic double bond of 5'-deoxy-4',5'-didehydrothymidine (**11**) (Scheme 2). The olefin **11** was prepared by selective iodination of thymidine **9** with triphenylphosphine, iodine, and either pyridine or imidazole followed by dehydrohalogenation with either DBN or sodium methoxide. Better overall yields and significantly easier purifications were achieved using sodium methoxide as the base. The addition of iodine azide, generated in situ from iodine monochloride and sodium azide in DMF, occurred regioselectively and with high stereoselectivity giving rise to **12** as a 10:1 mixture of epimers at 4'. Since both 4'-epimers were available, assignments could be made in a straightforward manner utilizing NOE data from ^{1}H-NMR experiments. These assignments were later confirmed through a single crystal X-ray structure determination of the final compound. Initial attempts at the displacement of the 5'-iodine of **12** with oxygen nucleophiles were unsuccessful. Lithium, cesium and silver carboxylates in a variety of solvents led only to degradation. A common by-product at higher temperatures was the olefin **11**. Silver nitrate, silver perchlorate and bis(tributyltin) oxide with silver nitrate also did not achieve displacement of the 5'-iodine of **12** or of its acyl derivative **13**.

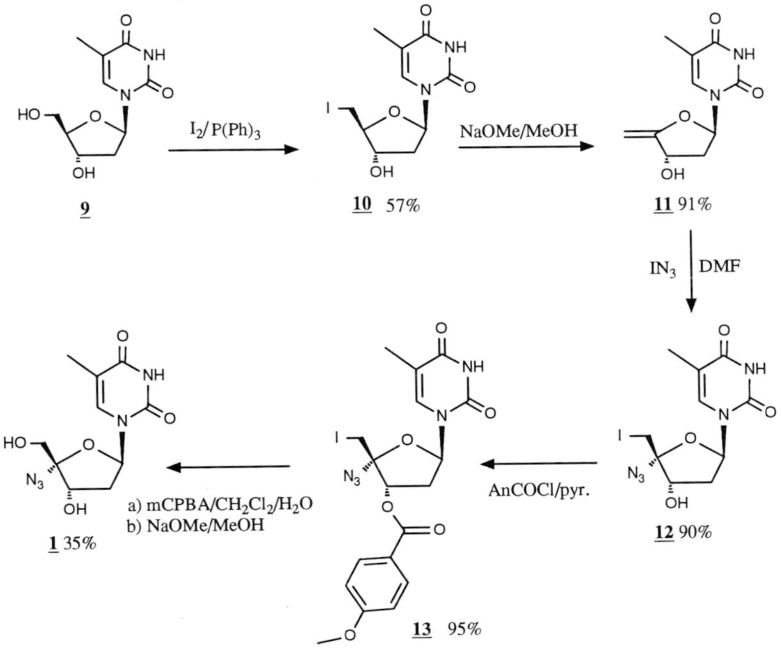

Scheme 2

Finally, oxidation to the hypervalent iodine species in the presence of water [12] allowed for a successful displacement (Scheme 3). It was found, that the activation to the hypervalent iodine species in the presence of water alone was not sufficient and that the participation of the 3'-O-acyl group was essential. In addition higher yields were achieved with the 3'-O-(p-methoxybenzoyl) derivative versus the 3'-O-benzoyl analogue (not shown). This result, as well as the product mixture, can best be explained by the intermediacy of the 3',5'-cyclic benzoxonium ion **14**. Aqueous hydrolysis of the product mixture (**15-18**) under basic conditions provided the target compound ADRT in good overall yield.

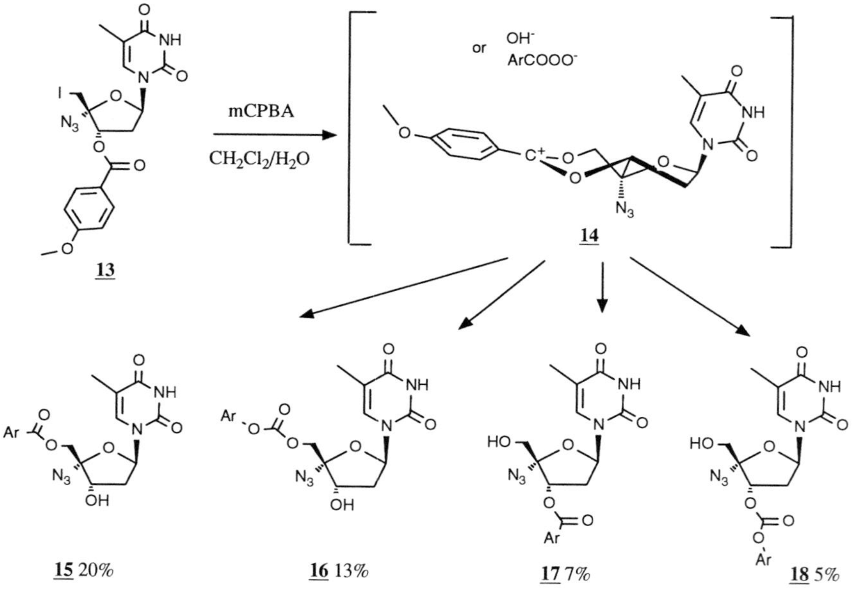

Scheme 3

1.2. Structural Aspects and Biochemistry

Theoretical considerations as well as extensive NMR and X-ray structure determinations clearly indicate, that ADRT preferentially (if not exclusively) exists in a 3'-endo conformation, both in solution and in the solid state [13].

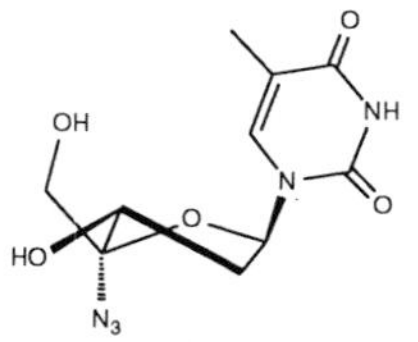

Preferred Conformation of ADRT: 3'-endo

It is assumed that this conformation is retained throughout the metabolic activation to the triphosphate form and will severely retard or abolish the formation of the crucial phosphodiester bond in the DNA polymerase reaction catalyzed by HIV RT (see Figure 1). Detailed biochemical studies by Ming Chen and his colleagues have validated this assumption [14]. Thus, the incorporation of ADRT into the primer strand will retard primer extension and incorporation of two successive ADRT units will completely abolish primer extension. As depicted in the figure, the attachment of the next nucleotide unit is unfavorable in the ADRT terminated primer, as a result of steric and electronic influences resulting from the strict 3'-endo conformation of the ribose ring.

Polymerase Reaction

With ADRT   **Without ADRT**

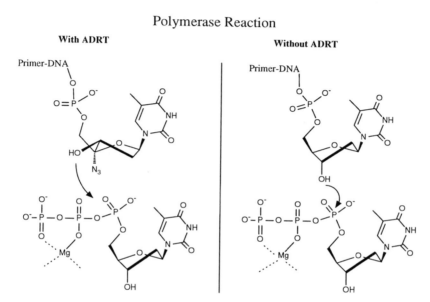

Figure 1

1.3. Antiviral Activity

ADRT, as well as the cytidine, adenosine and guanosine analogs prepared in a similar manner, displayed potent and quite selective activity against HIV *in vitro* [8]. The results achieved against the HIV lab strain LAV (III$_B$) in the CD4$^+$ T-cell line A3.01 are shown in Table 1. As a preliminary measure of selectivity, the cytotoxic effects of the compounds on A3.01 cells were measured and a value for partial toxicity (CC$_{25}$), in which about 25 % of the cells appeared affected by the drug, and complete toxicity (CC$_{100}$), indicating complete destruction of the cell layer, were determined. The selectivity index (SI) for the compound was calculated from the ratio of the partial toxicity (CC$_{25}$) and the antiviral activity (IC$_{50}$). Based on this measure, the thymidine analog (ADRT) is the most selective member of this class and is also one of the most potent analogues. ADRT retains its activity against HIV strains resistant to AZT as shown in Table 2, supporting the argument that the two structures are distinctively different. Furthermore, ADRT analogs lacking a 3'-hydroxy function (3'= H, F, OMe) are devoid of antiviral activity or only weakly active [8]. ADRT was extensively studied in the preclinical setting until toxic effects in monkeys at lower than anticipated doses deferred further development efforts.

Table 1. *In vitro* antiviral activity of 4'-azido-2'-deoxynucleosides

	B	anti-HIVactivity (A3.01 cells) IC_{50} (μM)	Toxicity in A301 cells partial complete CC_{25} (μM) CC_{100}	Selectivity Index SI CC_{25}/IC_{50}
	T	0.01	8 200	800
	A	0.03	<22 67	<733
	I	0.07	<22 200	<314
	C	0.004	0.21 1.9	53
	G	0.003	0.21 1.9	70
	U	0.8	200 >200	250

Table 2. *In vitro* antiviral activity of 4'-azidothymidine (ADRT) and AZT

Virus Strain	Cell Line	ADRT				AZT			
		IC_{50} (μM)	CC_{25} (μM)	CC_{100} (μM)	SI	IC_{50} (μM)	CC_{25} (μM)	CC_{100} (μM)	SI
HIV-1 LAV	A3.01	0.01	8	200	800	0.01	825	3300	82500
	H9	0.04	25	667	625	0.03	44	1200	1467
	PBL	0.02	2.5	>200	125	0.08	4	> 500	50
	MT-2	0.40	56	>500	140	0.16	56	> 500	350
G 762-3 (pre AZT)	MT-2	0.43				0.22			
G 691-2 (post AZT)	MT-2	0.34				10			
H 112-2 (pre AZT)	MT-2	0.40				0.20			
G 910-6p2 (post AZT)	MT-2	0.52				31			

2. 4'-Azido Carbocyclic Thymidines [15]

The potent anti-HIV compound 4'-azidothymidine (ADRT) described above possesses a nitrogen substituent at both glycosidic positions (1' and 4'), giving rise to two possible degradative glycosidic bond cleavage pathways. Thus it is not surprising that ADRT is unstable under both strongly acidic and strongly basic conditions, although the compound has more than adequate stability for clinical use (less than 10% degradation in 2 years in aqueous solutions at neutral pH [16]). Nevertheless, we reasoned that the carbocyclic analogues would be even more stable and could potentially be desirable clinical entities. We intended to combine the features of ADRT with those of carbocyclic nucleosides such as Carbovir **19** [17], which resulted in the carbocyclic thymidine derivative **7** as our synthetic target.

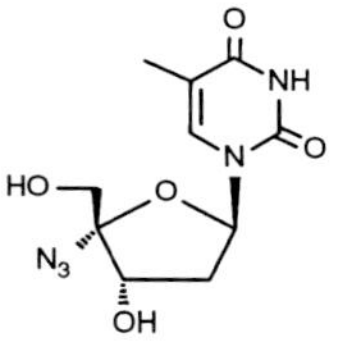

ADRT **1** 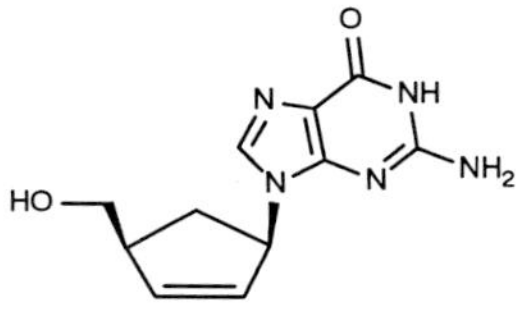   Carbovir **19** 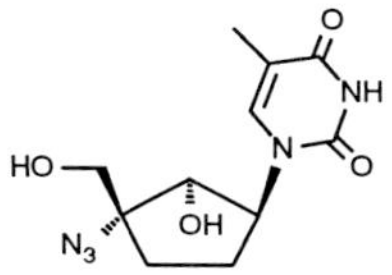   **7**

2.1. Synthesis

Our retrosynthetic analysis is shown in Scheme 4. Based on the extensive literature describing the synthesis of carbocyclic nucleosides, stereo- and regioselective ring opening of an epoxide such as **20** with an appropriate pyrimidine or purine nucleophile was deemed a short and versatile approach to the 4'-azido-2',3'-dideoxy carbocyclic class of compounds, e.g. **7**.

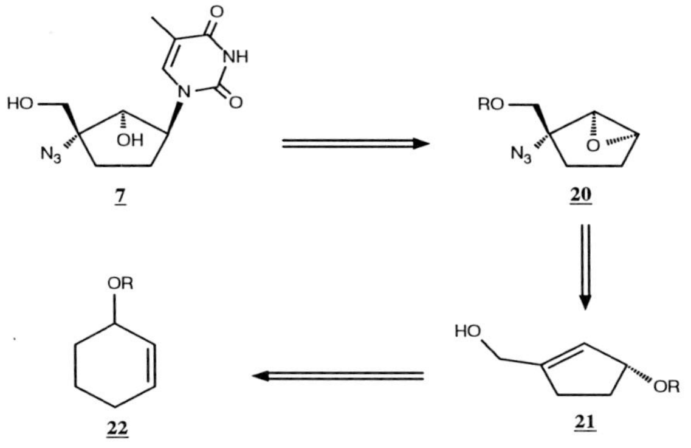

7   **20**

22   **21**

Scheme 4

The cyclopentene **25** (Scheme 5) was prepared according to the method of Jommi [18] starting with the silylation of 2-cyclohexen-1-ol followed by a low temperature ozonolysis. Reductive workup of the ozonide and intramolecular aldol reaction with in situ dehydration catalyzed by dibenzylammonium trifluoroacetate afforded the unstable unsaturated aldehyde 24. Immediate reduction with sodium borohydride gave the desired alcohol 25 in an overall yield of 51%. We assumed that the bulky (triisopropylsilyl)oxy group would direct the addition of iodine azide from the side opposite to the silyl ether. This was not the case. The major product 26 from this reaction had the opposite stereochemistry as determined by X-ray crystallography of the 3,5-dinitrobenzoyl derivative. While it might have been feasible to invert the secondary alcohol group prior to epoxide formation, a more efficient alternative route based on the Lewis acid catalyzed epoxide opening with azide ion was investigated next.

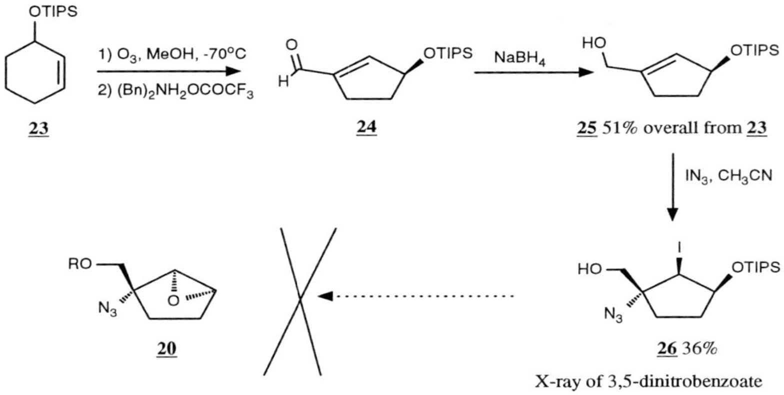

Scheme 5

Epoxidation (Scheme 6) of the allylic alcohol 25 gave a 2:1 mixture of epoxides 27 and 28 with the desired trans epoxide 27 predominating. Assignment of the structures was based on the 0 Hz coupling observed in the trans case (27) between the epoxide proton and the proton at the carbon bearing the secondary silyloxy group in the ^{1}H NMR spectrum. The primary alcohol was protected as the triisopropylsilyl derivative (29) in near-quantitative yield. Attempted Lewis acid (1.0 equiv. of TMS triflate, -78°C) catalyzed azide attack (TMSN$_3$, 2.2 equiv.) on the epoxide 29 proceeded quite cleanly by TLC, but resulted in the gem-diazido alcohol 30 and not in the desired alcohol 31. The structure of this compound was confirmed by a single crystal X-ray analysis of the 3,5-dinitrobenzoyl derivative. The proposed reaction mechanism accounting for the product is an epoxide ring opening accompanied by a 1,2 hydride shift to produce a silyloxy stabilized carbocation (ii). The cation is quenched by azide addition resulting in product iii, which can be isolated from the reaction if less than 2.2 equiv. of TMS azide are employed. Finally 30 is produced through azide displacement of the silyloxy group and hydrolysis of the TMS group upon workup.

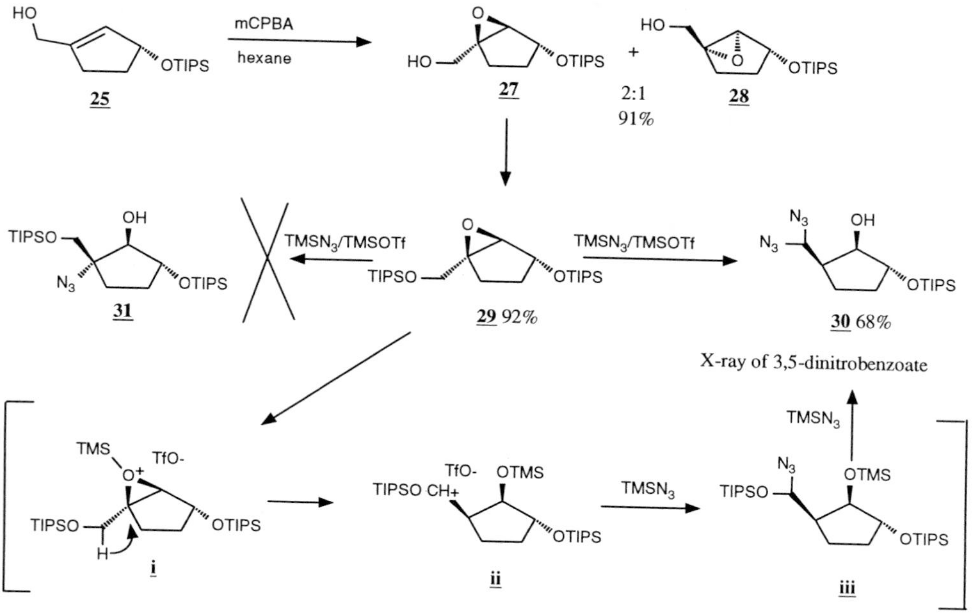

Scheme 6

Allylic halides and sulfonates can be converted to allylic azides under nucleophilic conditions, while allylic carbonates and acetates have been used for palladium-catalyzed azidations [19,20]. Our successful approach to epoxide **35** involved treatment of the allylic methyl ether **32** with an excess of azidotrimethylsilane and a catalytic amount of TMS triflate at -78°C resulting in an equilibrium mixture of the allylic azides **33** and **34** (1:3) (Scheme 7). Rapid equilibration at room temperature did not allow for the separation of the two regioisomers and the ratio was determined by ^{1}H NMR spectroscopy. The formation of the desired epoxide **35** in high yields would require selective epoxidation of the minor and less highly substituted olefin under equilibration conditions, but this could not be achieved. Indeed, in preliminary experiments employing mCPBA in dichloromethane, only a very low yield of **35** (2.4%) was realized. A practical conversion was finally found using the Payne reagent which is normally less selective and less influenced by the degree of alkene substitution [21]. Slow addition of aqueous hydrogen peroxide to a solution of the allylic azide mixture (**33** and **34**), benzonitrile and KHCO$_3$ in methanol gave the mixture of epoxides shown in Scheme 7. Thus the desired epoxide **35** was obtained in a yield of 18% (or 72% based on the abundance of the olefin **33** in the equilibrium mixture) and could be prepared in multigram quantities. Structural assignments were quite difficult and were based on NMR data and supported by the preparation of bromohydrins from the alkenes **33** and **34** and their conversion to epoxides. In this way, consistent assignments were achievable.

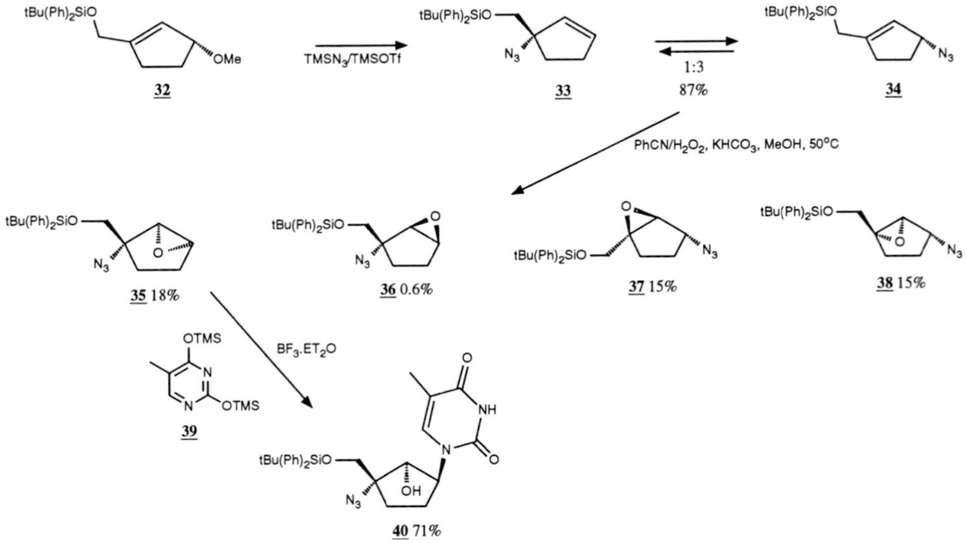

Scheme 7

Condensation of persilylthymine **39** with the epoxide **35** under Vorbrüggen conditions [22] gave the desired carbocyclic nucleoside **40** in a yield of 71%. Deprotection under standard conditions gave (±)-(1'α,2'α,3'β)-1-[1-azido-2-hydroxy-1-hydroxymethyl-3-cyclopentyl]thymine **7**, the desired carbocyclic thymidine derivative (Scheme 8). Inversion of the secondary alcohol group was achieved with the classical sequence of mesylation, internal anhydride formation and hydrolysis providing the isomeric compound **42**.

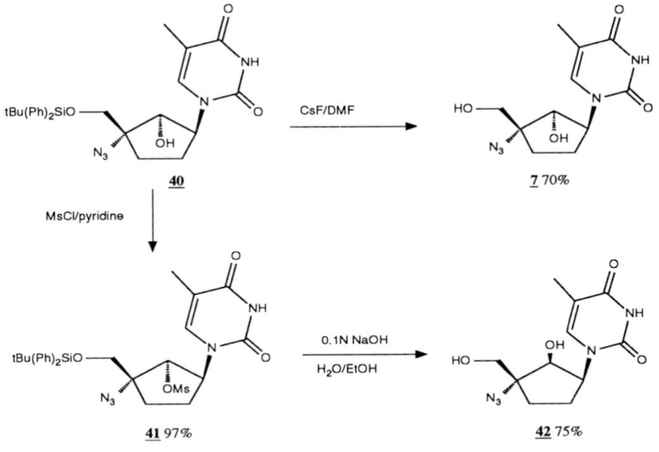

Scheme 8

2.2. Antiviral Activity

Both 4'-azido carbocyclic thymidines (**7** and **42**) were devoid of antiviral activity. This result might be explained by the fact that all strongly active ADRT analogues examined thus far, possess a 3'-hydroxy group which was absent in these carbocyclic compounds.

3. 1-(2,2-Difluoro-3,3-bis(hydroxymethyl)cycloprop-1-yl)thymine and Related Compounds [23]

Cyclopropyl nucleosides present hybrid molecules between acyclic nucleosides and their natural ribose counterparts. Several examples have been prepared, most notably the guanine derivatives **43** [24] and **44** [25]. These analogues are devoid of antiviral activity and we reasoned that this may be due to the rather drastic change in electronic character from the replacement of the oxygen function of the ribose (or the acyclic equivalent) by a methylene group. A considerably better analogue may be one in which the oxygen atom is replaced by the difluoromethylene isostere, resulting in the target structure **45**. A review of the literature revealed that no such compound had been prepared previously and we could not find any example of a 2,2-difluoro-cyclopropylamine, a possible synthetic intermediate for this class of compounds.

43   **44**   **45**

Surveying possible approaches to this compound class, we carefully analyzed the results by Kobayashi et al. who had investigated the difluorocarbene addition to allylic acetates and enol acetates. (Scheme 9) [26,27]. While both additions proceeded in very good yields, the difluorocyclopropanes derived from enol acetates could not be hydrolyzed without concomitant ring opening as shown in the Scheme. The difluorocyclopropanes derived from the allylic acetates on the other hand behaved much better and conversion to the corresponding methyl esters was readily achieved. In contrast to these results, Schlosser and his group reported that the aldehyde **46** was very sensitive and would rearrange rapidly under slightly basic conditions [28].

3.1. Synthesis

Before embarking on a synthesis, we briefly explored the stability of a difluorocyclopropylamine, such as **49**. Addition of difluorocarbene, generated by the thermal decomposition of sodium chlorodifluoro-

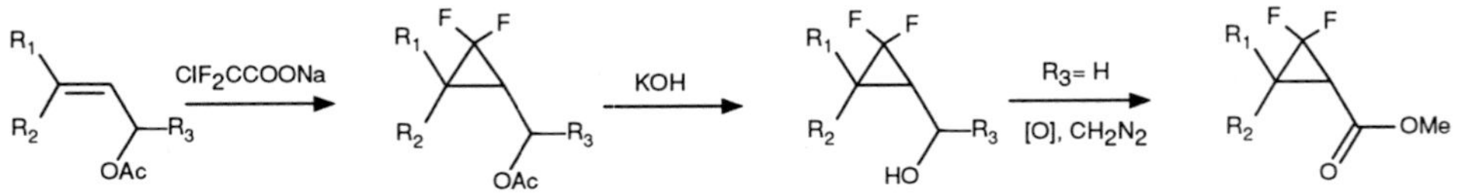

Y. Kobayashi et. al. *Chem. Pharm. Bull.* **31** (1983) 2616.

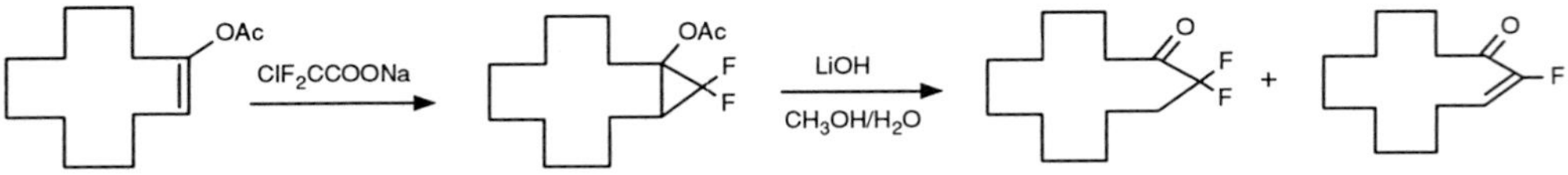

quantitative   48%   29%

Y. Kobayashi et. al. *Chem. Pharm. Bull.* **27** (1979) 3123.

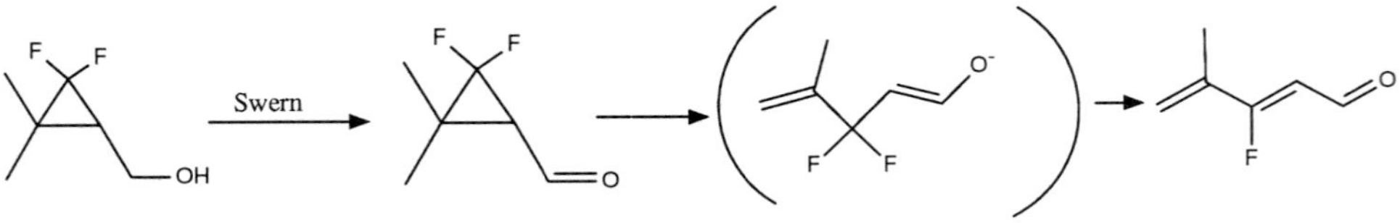

46 18%

M. Schlosser et. al. *Tetrahedron* **46** (1990) 5230.

Scheme 9. Literature results

acetate in diglyme, to the ene phthalimide **47** [29] gave the desired adduct **48** in a yield of only 18%. The removal of the protecting group by hydrazine was followed by ^{1}H NMR spectroscopy and it was concluded that deprotection was successful but that the product (**49**) was both unstable and very volatile. Combining this experience with the Kobayashi results made it quite clear that a free difluorocyclopropylamine should be avoided in the synthetic approach.

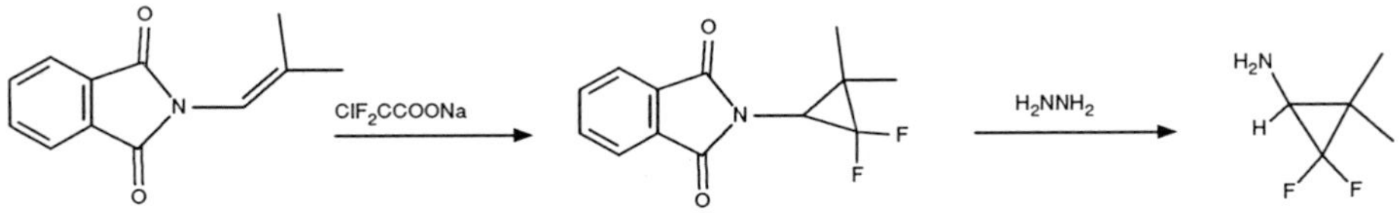

47   **48** 18%   **49** NMR only
                                                       volatile and unstable

The retrosynthetic plan resulting from this analysis is shown in Scheme 10. The difluorocyclopropylcarboxylic acid **52** would be prepared according to Kobayashi [26], and Curtius rearrangement of the derived acyl azide would result in a suitably stable urea **53** which could be elaborated to the thymine nucleus **54** under standard conditions.

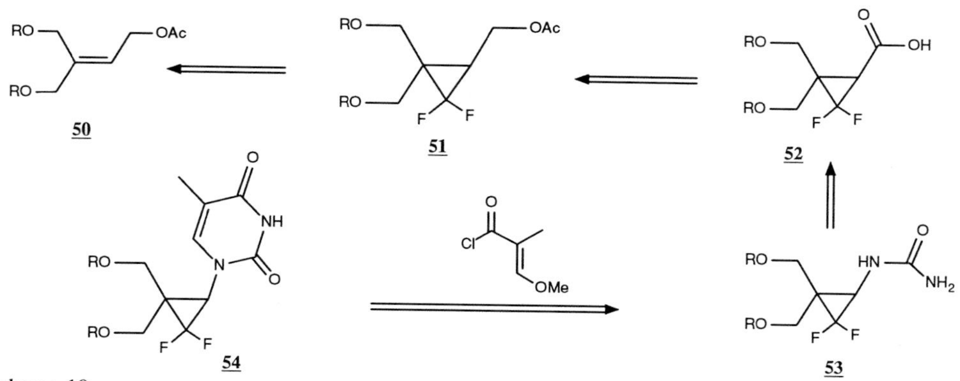

Scheme 10

The first part of the synthesis proceeded according to plan as shown in Scheme 11. Preparation of the required allylic acetate started with the oxidation of 1,3-di-O-benzylglycerol to provide ketone **55**. Horner-Emmons olefination followed by DIBAL-H reduction and acetylation gave the allylic acetate **57** in good overall yield. The key difluorocarbene addition was accomplished using the conditions of Kobayashi [26] (20 equiv. of sodium chlorodifluoroacetate, diglyme, reflux) and gave the desired difluorocyclopropane **58** in good yield (80%). Hydrolysis of the acetate was achieved quantitatively and oxidation to the acid **61** was accomplished with pyridinium dichromate in dimethylformamide, again in very good yield. The intermediate aldehyde **60** was isolated in one experiment and it did not display the instability of the related aldehyde **46** prepared by Schlosser and co-workers (see Scheme 9) [28].

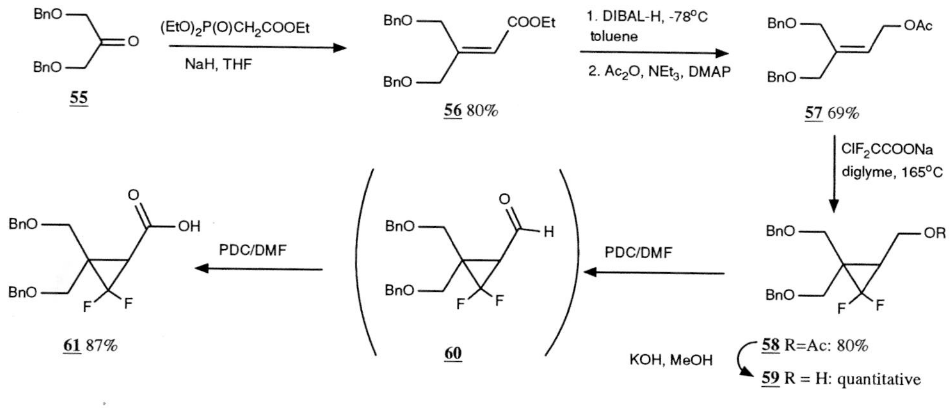

Scheme 11

The acid **61** (Scheme 12) was converted to the acyl azide **62** which was rearranged thermally to the isocyanate **63** in toluene at 85°C. Following a quench with gaseous ammonia at 0°C, the urea **64** was isolated in an overall yield of 79% from the acid **61**. The urea also was quite stable. Its reaction with 1.5 equiv. of 3-methoxy-methacryloyl chloride in dichloromethane in the presence of triethylamine and DMAP gave two products (22 and 55% yield resp.), neither of which proved to be the desired acylated material. In both cases, initial acylation had occurred at the secondary urea nitrogen giving rise to **65** which was partially dehydrated to the nitrile **66** with excess acyl chloride. We explain this unexpected result by invoking the tautomer shown at the bottom of Scheme 12. The electronic influence of the two fluorine atoms is expected to favor the imino tautomer involving the secondary nitrogen of the urea, making this nitrogen more nucleophilic and more prone to electrophilic attack by the acyl chloride. It is not known whether the hydrogen bond shown in the Scheme is actually present and thus contributing to this imino form. More recently, Katagiri and co-workers [24a] described the same acylation reaction in a related system without the fluorine substituents in which acylation proceeded predominantly at the primary nitrogen, supporting our hypothesis that the influence of the gem-difluoro moiety results in the acylation at the secondary position. As part of the structural proof, **65** was cyclized to the thymine derivative **67** and the benzyl protecting groups were removed to provide the N-3 substituted thymine **68** (Scheme 13). Here, the noticeable coupling in the ¹H NMR spectra between the C-6 Proton and the N-1 hydrogen confirmed the structural assignments.

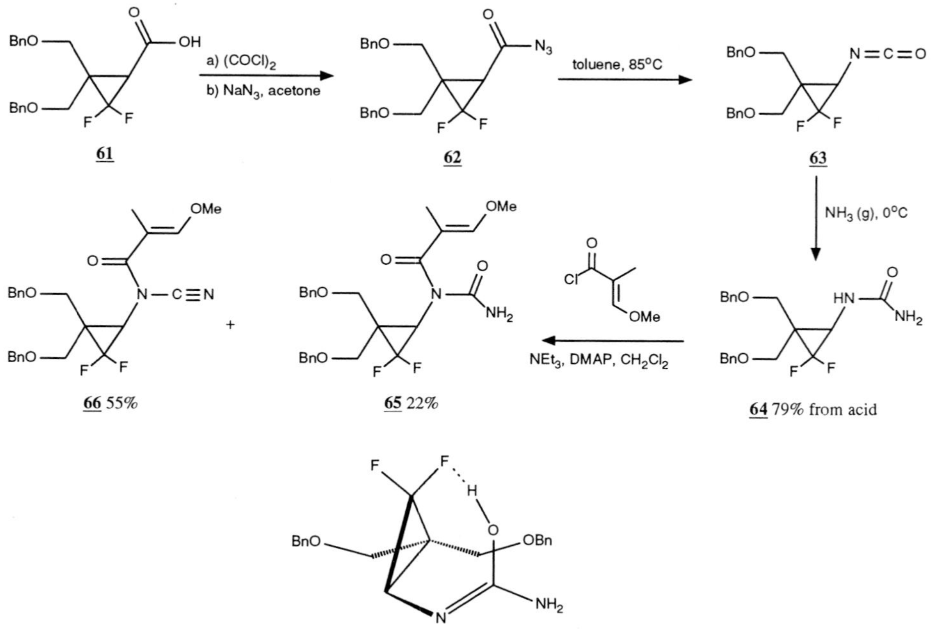

Scheme 12

The synthetic equivalent of the desired acylation was achieved by reaction of the intermediate isocyanate **63** with 3-methoxy-methacryloyl amide in toluene at 80°C (Scheme 13). Cyclization under acidic conditions (2N H₂SO₄, nPrOH, 85°C) of the product **69** gave the desired thymine derivative **70** in 77% yield. Finally, deprotection provided 1-(2,2-difluoro-3,3-bis(hydroxymethyl)cycloprop-1-yl)thymine **71** in a quantitative yield.

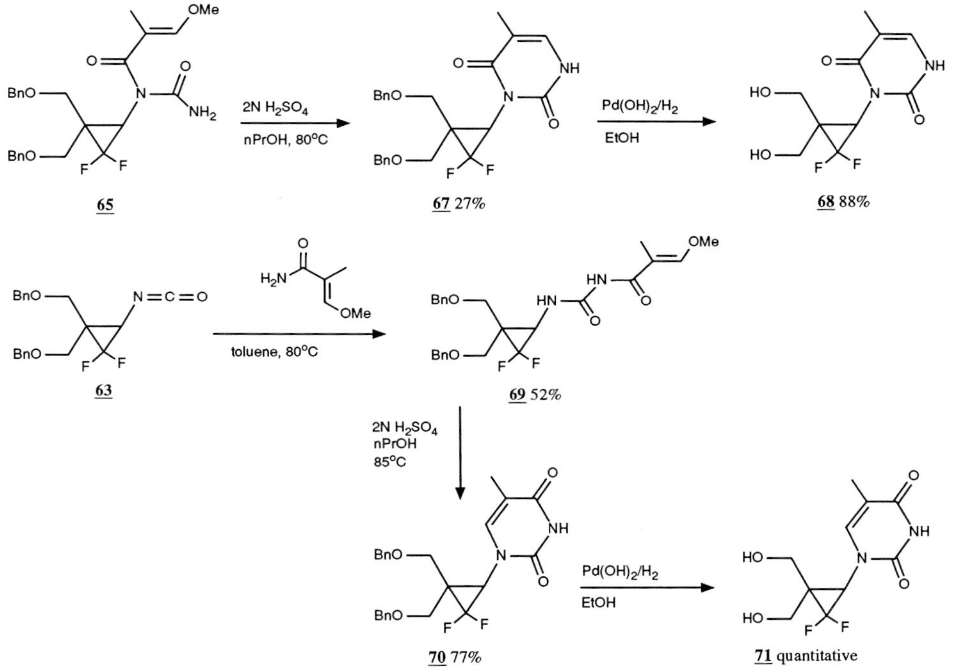

Scheme 13

After several unsuccessful approaches to the selective functionalization of the trans (in respect to the thymine ring) hydroxymethyl group, the sequence shown in Scheme 14 provided a high yielding solution to the problem in some cases. Bis mesylation of the diol **71** in pyridine gave a good yield of **72**, which underwent internal cyclization with DBU to give **73** (78%). Selective hydrolysis to the desired alcohol-mesylate **74** was achieved with dilute aqueous methanesulfonic acid in very good yield. Nucleophilic displacement of the remaining mesylate was only possible with very powerful nucleophiles and these experiments are summarized in Scheme 14. Reaction with sodium azide or lithium chloride in DMF at 50-60°C gave good yields of azide **75** and chloride **76**, respectively. Similarly, the displacement with iodide in acetone gave an excellent yield of the iodide **77**. In contrast, only a moderate yield of the methyl derivative **78** could be achieved with sodium borohydride in HMPA, and sodium cyanide in DMSO gave none of the cyano derivative **79**; instead the ring-opened, allylic alcohol **80** was isolated. This result is consistent with attack by cyanide at the carbon bearing the thymine ring with subsequent ring opening and elimination of methanesulfonic acid.

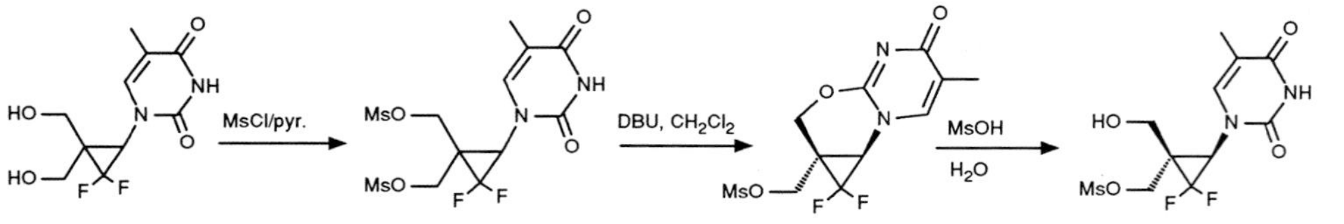

71 | 72 61% | 73 78% | 74 99%

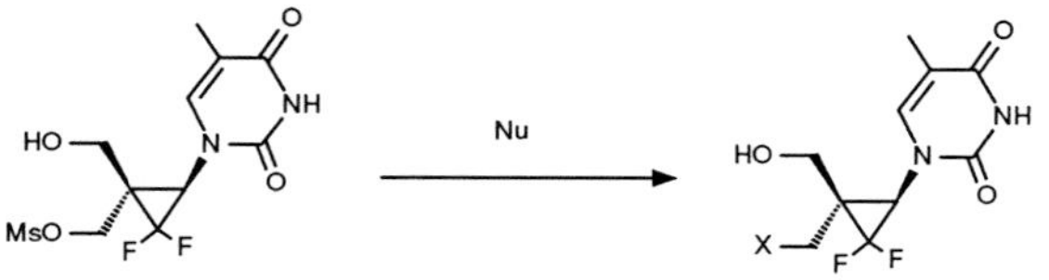

74

Nucleophile	Solvent	X	Yield	Compound
NaN_3[a)]	DMF	N_3	99%	75
LiCl[a)]	DMF	Cl	89%	76
NaI[a)]	Acetone	I	99%	77
$NaBH_4$[b)]	HMPA	H	50%	78

a) 50-60°C, 1-4 h. b) 20°C, 1 h.

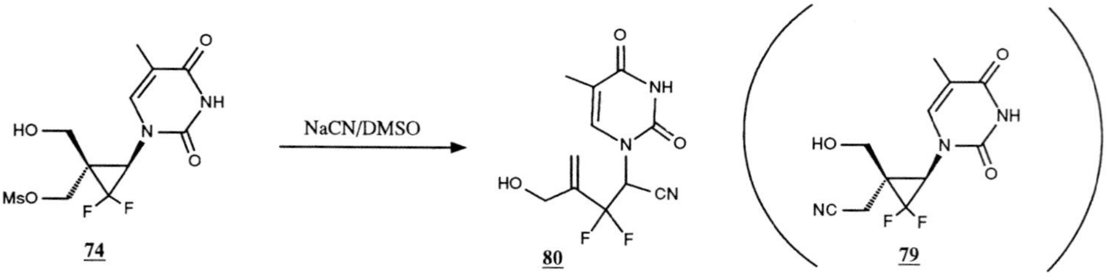

Scheme 14

3.2. Antiviral Activity

Unfortunately, none of the difluorocyclopropanes displayed antiviral activity, nor did they display cytotoxic properties. This is most likely due to a lack of metabolic activation to the corresponding mono-di- and triphosphates.

Conclusions

The potent and selective antiretroviral activity of many 4'substituted-2'-deoxyucleosides such as ADRT (1) appears to be centered around analogs possessing an unmodified 3'-hydroxy group. Thus, while modifications at the 4'-position are well tolerated as indicated by the activity of 4'-cyano- (2) [9], 4'-azidomethyl- (3) [9] and 4'-methoxythymidine (4) [8] in addition to ADRT (1), alterations at the 3'-position leading to 4'-azido-3'-O-methyl- or to 4'-azido-3'-deoxythymidine either significantly diminish or even abolish the antiviral activity [8]. Transposing the 4'-modifications from the natural tetrahydrofuran ring to substitutes such as cyclopentanes (7, 42), oxathiolanes (5) [10], cyclobutanes (6) [11] and difluorocyclopropanes (71, 75-78) leads only to inactive compounds. These modified systems lack the equivalent of the crucial 3'-hydroxy group of the original tetrahydrofuran series. Therefore it is unclear, if the ring replacement or the lack of the 3'-hydroxy group is responsible for the inactivity of these compounds and additional experiments are required to clarify this point.

Acknowledgement. The authors wish to thank Mary Jane McRoberts and Diane Crawford-Ruth for the biological screening of the compounds and the members of Syntex Analytical and Environmental Research for the analytical data. Thanks are also due to Dr. Gregory VanDuyne (Cornell University) for the X-ray structural elucidations.

References

1.   Contribution No. 368 from the Institute of Bio-organic Chemistry.

2.   Syntex Research Postdoctoral Fellow, 1990-1992.

3.   Present Address: Gilead Sciences Inc., 344/346 Lakeside Dr., Foster City, CA 94404, USA.

4.   Present Address: Gensia Pharmaceuticals, 4575 Eastgate Mall, San Diego, CA 92121, USA.

5.   H. Mitsuya, R. Yarchoan, S. Broder, *Science* **249** (1990) 1533-1544.

6.   J. P. H. Verheyden, I. D. Jenkins, G. R. Owen, S. D. Dimitrijevich, C. M. Richards, P. C. Srivastava, N. Le-Hong, J. G. Moffatt, *Ann. N.Y. Acad. Sci.* **255** (1975) 151-165.

7.   J. G. Moffatt in R. T. Walker, E. DeClercq, F. Eckstein (Eds.). Nucleoside Analogues, Plenum Publishing Corp. New York, 1979. pp. 71-164.

8.   H. Maag, R. M. Rydzewski, M. J. McRoberts, D. Crawford-Ruth, J. P. H. Verheyden, E. J. Prisbe, *J.Med. Chem.* **35** (1992) 1440-1451.

9.   C. O'Yang, H. Wu, E. B. Fraser-Smith, K. A. M. Walker, *Tetrahedron Lett.* **33** (1992) 37-40.

10.   J. Ensign Bemis, E. J. Prisbe, H. Maag, 10th International Roundtable on Nucleoside and Nucleotides, Park City, Utah, September 16-20, 1992, Poster Presentation.

11.   R. M. Otoski, Syntex Research Postdoctoral Fellow 1989-1990, unpublished results. Identical or related compounds were prepared by several other groups:

      a) Y. Sato, T. Horii, T. Maruyama, M. Honjo, *Nucleic Acids Res. Symposium Series* **21** (1989)

438   H. Maag et al.

125-126; T. Maruyama, Y. Sato, T. Horii, H. Shiota, K. Nitta, T. Shirasaka, H. Mitsuya, M. Honjo, *Chem. Pharm. Bull.* **38** (1990) 2719-2725.

b) H. Boumchita, M. Legraverend, C. Huel, E. Bisagni, *J. Heterocyclic Chem.* **27** (1990) 1815-1819; H. Boumchita, M. Legraverend, A. Ziel, M. Lemaitre, C. Huel, E. Bisagni, *Eur. J. Med. Chem.* **26** (1991) 613-617.

c) J. M. Henlin, H. Rink, E. Spieser, G. Baschang, *Helv. Chim. Acta* **75** (1992) 589-603.

12.   a) R. C. Cambie, D. Chambers, B. G. Lindsay, P. S. Rutledge, P. D. Woodgate, *J. Chem. Soc., Perkin Trans. 1* (1980) 822-827.

b) T. L. Macdonald, N. Narasimhan, L. T. Burka, *J. Amer. Chem. Soc.* **102** (1980) 7760-7765.

13.   H. Maag, J. E. Nelson, E. J. Prisbe, manuscript in preparation.

14.   M. S. Chen, R. Suttmann, C. Bach, J. C. Wu, E. J. Prisbe, M. J. McRoberts, D. Crawford-Ruth, Fourth International Conference on Antiviral Research, New Orleans, Louisiana, USA, 21-26 April, 1991; Poster 87. Abstract published in *Antiv. Res.* **Suppl. 1** (1991) 91.

15.   H. Maag, R. M. Rydzewski, *J. Org. Chem.* **57** (1992) 5823-5831.

16.   M. Brandl, R. Strickley, T. Bregante, L. Gu, T.W. Chan, *Int. J. Pharmac.* in press.

17.   R. Vince, M. Hua, *J. Med. Chem.* **33** (1990) 17-21.

18.   S. Canonica, M. Ferrari, G. Jommi, M. Sisti, *Synthesis* (1988) 697-699.

19.   a) D. Askin, C. Angst, S. Danishefsky, *J. Org. Chem.* **50** (1985) 5005-5007.

b) T. Rosen, K. J. Guarino, *Tetrahedron* **47** (1991) 5391-5400.

20.   a) S.-I. Murahashi, Y. Tanigawa, Y. Imada, Y. Taniguchi, *Tetrahedron Lett.* **27** (1986) 227-230.

b) S.-I. Murahashi, Y. Taniguchi, Y. Imada, Y. Tanigawa, *J. Org. Chem.* **54** (1989) 3292-3303.

21.   R. G. Carlson, N. S. Behn, C. Cowles, *J. Org. Chem.* **36** (1971) 3832-3833.

22.   U. Niedballa, H. Vorbrüggen, *J. Org. Chem.* **41** (1976) 2084-2086.

23.   H. Maag, A. J. Gutierrez, manuscript in preparation.

24.   a) N. Katagiri, H. Sato, C. Kaneko, *Chem. Pharm. Bull.* **38** (1990) 3184-3186.

b) D. W. Norbeck, H. L. Sham, T. Herrin, W. Rosenbrook, J. P. Plattner, *J. Chem. Soc., Chem. Comm.* (1992) 128-129.

25.   a) S. Nishiyama, S. Ueki, T. Watanabe, S. Yamamura, K. Kato, T. Takita, *Tetrahedron Lett.* **32** (1991) 2141-2142.

b) G. R. Green, M. R. Harnden, M. J. Pratt, *Bioorg. Med. Chem. Lett.* **1** (1991) 347-348.

26.   Y. Kobayashi, T. Tsutomu, T. Taguchi, *Chem. Pharm. Bull.* **31** (1983) 2616-2622.

27.   Y. Kobayashi, T. Taguchi, M. Mamada, H. Shimizu, H. Murohashi, *Chem. Pharm. Bull.* **27** (1979) 3123-3129.

28.   Y. Bessard, L. Kuhlmann, M. Schlosser, *Tetrahedron* **46** (1990) 5230-5236.

29.   J. K. Stille, Y. Becker, *J. Org. Chem.* **45** (1980) 2139-2145.

Synthesis and Antiretroviral Activity of Carbocyclic Phosphonate Nucleoside Analogues

Gerhard Jähne*, Armin Müller, Herbert Kroha, Manfred Rösner, Otto Holzhäuser, Christoph Meichsner, Matthias Helsberg, Irvin Winkler, and Günther Rieß

HOECHST AG, SBU Antiinfectives - Research, G 838, P.O.Box 800320,
D-6230 Frankfurt am Main 80, Germany

Summary: A versatile and high-yielding synthesis of racemic carbocyclic phosphonate nucleosides of adenine, hypoxanthine, guanine, cytosine, uracil, and thymine has been developed using the Pd(0)-catalyzed addition of a nucleobase to 3,4-epoxycyclopentene. These newly prepared compounds are isosteric (and isoelectronic) with (carbo)cyclic 2',3'-dideoxy- and 2',3'-dideoxy-2',3'-didehydronucleoside monophosphates. Some of these carbocyclic phosphonate nucleoside analogues, especially the saturated thymine derivative (c-ddT-iso-MP) show good *in vitro* activity against HIV-1.

1. Introduction

Most therapeutically useful antiviral agents selectively inhibit virally encoded target enzymes produced in virus-infected cells. A number of molecular targets have been identified for strategies to inhibit the replication of HIV (human immunodeficiency virus) [1], the causative agent of AIDS (acquired immune deficiency syndrome): There is, for example, a virally encoded **protease** which is responsible for the processing of the precursor polyproteins into the appropriate structural proteins and replicative enzymes; there is an **integrase**, a protein that is required for the integration of the DNA-provirus into the genome of the host cell; there is a **ribonuclease H (RNase H)** that catalyses the hydrolysis of the RNA-strand of the initially formed RNA-DNA hybrid; and there is a **reverse transcriptase (RT)** that soon after infection of the host cell transcribes the single-stranded RNA genome of HIV into the double-stranded DNA provirus. It is this virally encoded **reverse transcriptase** which is - thus far - the most successfully exploited molecular target for the chemotherapy of HIV-infection.

2. Nucleoside Analogues as Inhibitors of HIV-1 Reverse Transcriptase

Nucleoside analogues like 3'-azido-3'-deoxythymidine (AZT) [2] or 2',3'-dideoxynucleosides (2',3'-dideoxyadenosine, ddA; 2',3'-dideoxyinosine, ddI [3]; 2',3'-dideoxycytidine, ddC [3]) or 2',3'-dideoxy-2',3'-didehydronucleosides (2',3'-dideoxy-2',3'-didehydrothymidine, d4T [4]) block HIV replication *in vitro* by inhibition of the reverse transcriptase after intracellular conversion to their respective triphosphate esters. Members of the dideoxy- or dideoxy-didehydro-class of compounds are perceived to have a number of disadvantages, including lability of the glycosidic bond to acidic or enzymic (hydrolase or phosphorylase) hydrolysis and clinical toxicity [5].

In an attempt to overcome some of these drawbacks, dideoxynucleoside analogues in which the tetrahydrofuran ring has been replaced by other five-membered ring systems have been described. Compounds with a heteroatom in the 3'-position (BCH-189) [6] and analogues having the oxygen atom of the dihydrofuran ring replaced by a methylene group ("Carbovir") show potent *in vitro*-activity [7] (Scheme 1).

Scheme 1: Nucleoside Analogues Active Against HIV

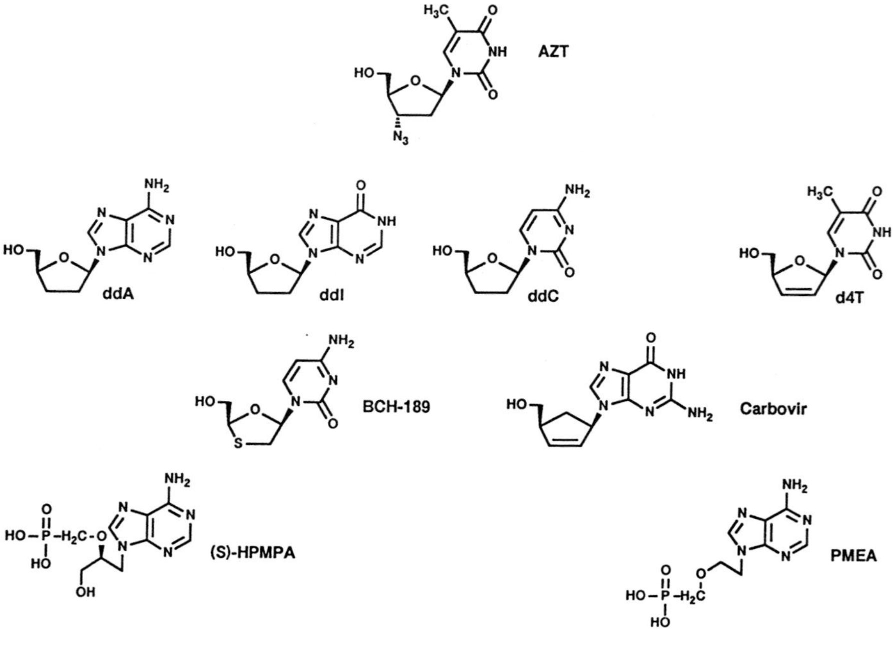

However, not all of these nucleoside analogues are efficiently transformed to their respective and ultimately active triphosphate esters. Above all, it is often the first step, i. e. the generation by cellular enzymes of the monophosphate ester that proceeds very slowly or inefficiently [8].

3. Carbocyclic Phosphonate Nucleoside Analogues

We were thinking of the possibility of overcoming this problem by delivering the nucleotide (i.e. the monophosphate ester) as a hydrolytically stable, isosteric, and isoelectronic phosphonic acid. Phosphonic acids and phosphonate esters of acyclic nucleoside analogues are known, and syntheses and antiviral activity of these compounds have been published by Holy and DeClercq in 1986 [9]. The two adenine derivatives (S)-HPMPA and PMEA represent the prototypes of this class of anti-DNA virus and anti-retrovirus agents. However, their selectivity index (50% cytotoxic dose/50% effective dose = ratio of CD_{50} to ED_{50}) is low.

We [10] and - as we do know now - others [11] set out to synthesize carbocyclic phosphonate nucleosides as isosteric and isoelectronic nucleoside monophosphates (Scheme 2).

Scheme 2: Carbocyclic Phosphonate Nucleosides as Isosteric Nucleoside Monophosphates

Carbocyclic Phosphonate Nucleoside   Nucleoside Monophosphate

In these compounds the carbocyclic moiety should guarantee stability towards hydrolysis or phosphorolysis and the phosphonate group should serve as an analogue of the monophosphate function of natural nucleotides.

The retrosynthetic analysis looked quite straightforward: A 1,4-cis-disubstituted cyclopentene ring with the nitrogen atom of the heterocycle attached to C1 and a hydroxy function attached to C4 was to be etherified with an di-alkyl(hydroxymethyl)phosphonate to afford - after cleavage of the phosphonate ester - the racemic phosphonic acid derivative (Scheme 3).

Scheme 3: Retrosynthetic Analysis

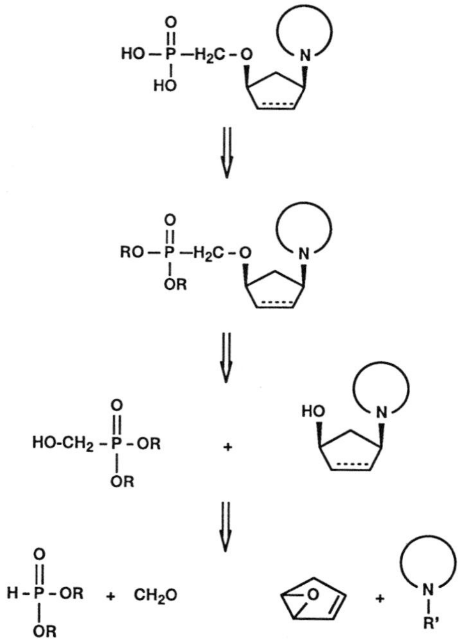

Scheme 4: Pd(O)-catalyzed Addition of Nucleophiles to 3,4-Epoxycyclopentene

B.M. Trost, G.H. Kuo, T. Benneche: J. Am. Chem. Soc. **1988**, *110*, 621

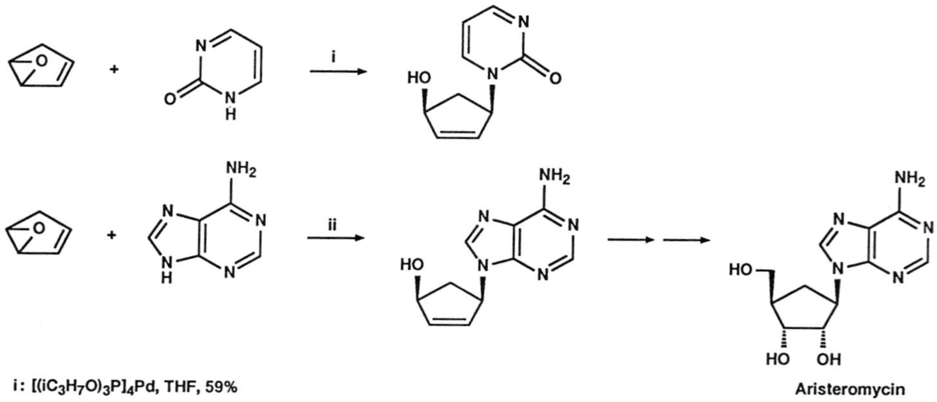

i : [(iC$_3$H$_7$O)$_3$P]$_4$Pd, THF, 59%

ii : 0.6 mol% Pd(OAc)$_2$, 6 mol% (i-C$_3$H$_7$O)$_3$P, 1.2 mol% nC$_4$H$_9$Li, 1 eq adenine, 1:1 THF/DMSO,
    0°C→r.t., 67% (based on recovered adenine)

Aristeromycin

It looked even more straightforward on closer inspection of a publication by Trost et al. [12], who in 1988 were able to add pyrimidine-2(1H)-one and adenine to 3,4-epoxycylopentene in a regio- and stereocontrolled manner using a Pd(0)-based catalyst to give the 1,4-cis-product (Scheme 4).

3.1. Synthesis of Carbocyclic Purine Phosphonate Nucleoside Analogues

3.1.1. The Adenine Congeners: c-d4A-iso-MP and ddA-iso-MP

We hoped to be able to expand this method by using other nucleobases, and to this end we first repeated the published reaction using the nucleobase adenine as the nucleophile. To our surprise the treatment of 3,4-epoxycyclopentene with adenine and tetrakis-triisopropylphosphite-palladium(0) (Pd[P(OiPr)$_3$]$_4$) in a mixture of THF/DMSO first at 0°C and then at r.t. afforded two isomeric products. As with the normal course of alkylation of adenine, we isolated 69% of the desired N9-alkylation product and 20% of the regioisomeric N3-alkylated product (Scheme 5).

Scheme 5: Carbocyclic Phosphonate Nucleosides: Adenine

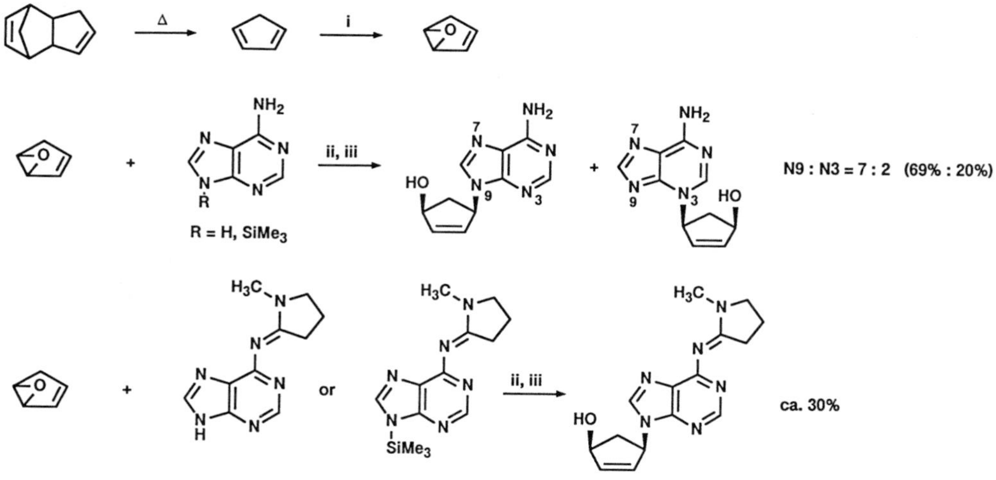

i: CH$_3$C(O)OOH, CH$_2$Cl$_2$, Na$_2$CO$_3$, 0°C → 5°C
ii: Ar, THF, Pd(OAc)$_2$-P(OiPr)$_3$-nBuLi, DMSO/THF, 0°C → RT, 8h
iii: EtOH, r.t., 30 min; chromatography

When we used persilylated adenine, the reaction proceeded very smoothly in solution, but again two regioisomers in a 7:2 mixture were isolated after chromatography in a 89% combined yield (Scheme 5). When N^6-(N-methyl-2-pyrrolidine-ylidene)adenine or its silylated derivative was reacted, only one isomer was isolated (N9-isomer, ca. 30% yield; Scheme 5).

To make sure that the by-product of the above-mentioned reactions was indeed the N3-regioisomer and not the N7-isomer, we treated 6-chloropurine with 3,4-epoxycyclopentene using the standard reaction conditions (Scheme 6). This treatment, however, afforded a single regioisomer in 74% yield (N9-isomer). By switching to silylated 6-chloropurine a 1:2 mixture of the N9- and N7-regioisomers was obtained (Scheme 6).

Reaction of the presumed N9-isomer with ammonia in n-pentanol gave an adenine derivative identical to the major reaction product of the above-mentioned reaction with adenine. Treatment of the presumed N7-isomer with ammonia gave an adenine derivative which was indeed different from the N9- and N3-isomers (Scheme 6).

Scheme 6: Carbocyclic Phosphonate Nucleosides: Adenine

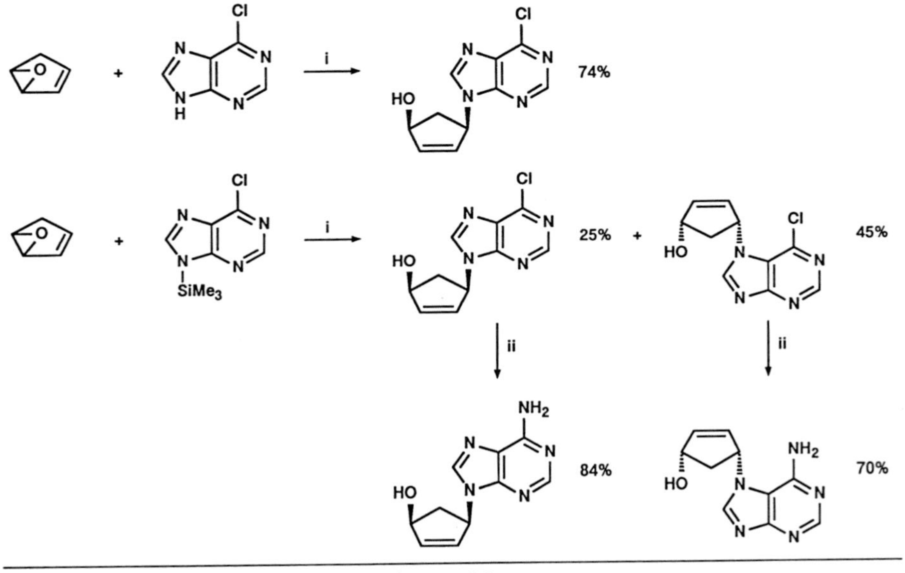

i: Ar, THF, Pd(OAc)$_2$-P(OiPr)$_3$-nBuLi (1-5 mol%), 0°C → r.t., 24h
ii: NH$_3$/n-pentanol, 50 bar N$_2$, 100°C, 10h

By comparison of the ^{1}H- and ^{13}C-NMR- and the UV-VIS-spectra of the pure regioisomers we could prove that in the reaction with adenine the N9- **and** N3-regioisomers were being formed concomitantly (Scheme 7).

Scheme 7: Characterization of the Adenine Regioisomers

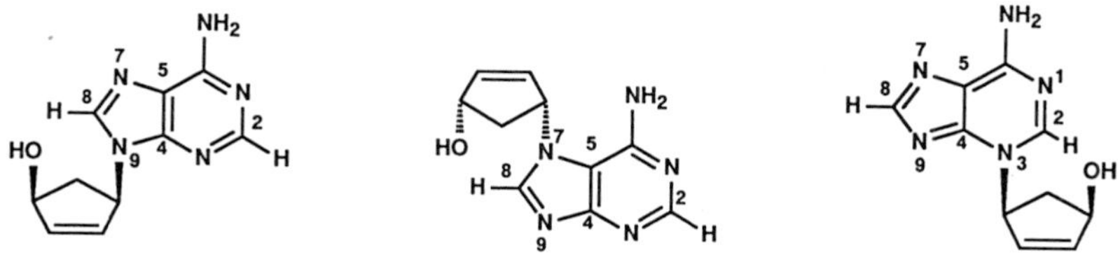

^{1}H-NMR: [ppm]	H-2	8.14	8.18	8.31
	H-8	8.07	8.20	7.74
^{13}C-NMR: [ppm]	C-4	148.83	151.24	148.58
	C-5	119.01	110.56	120.56
UV-VIS: (H$_2$O, pH 7) [λ_{max}, log ε]		262 nm, 4.18	269 nm, 4.15	274 nm, 4.14
		260 nm, 4.14	272 nm, 3.99	273 nm, 4.14

Leonard and Deyrup, JACS 1962, 84, 2148

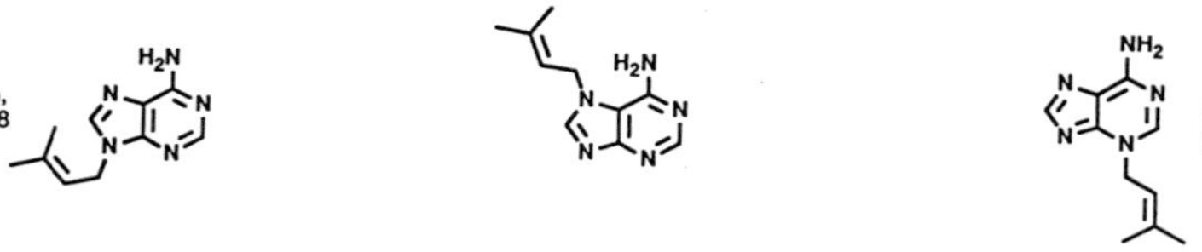

Denayer, Bull.Soc. chim.France, 1962, 1358

Scheme 8: Carbocyclic Phosphonate Nucleosides: Adenine

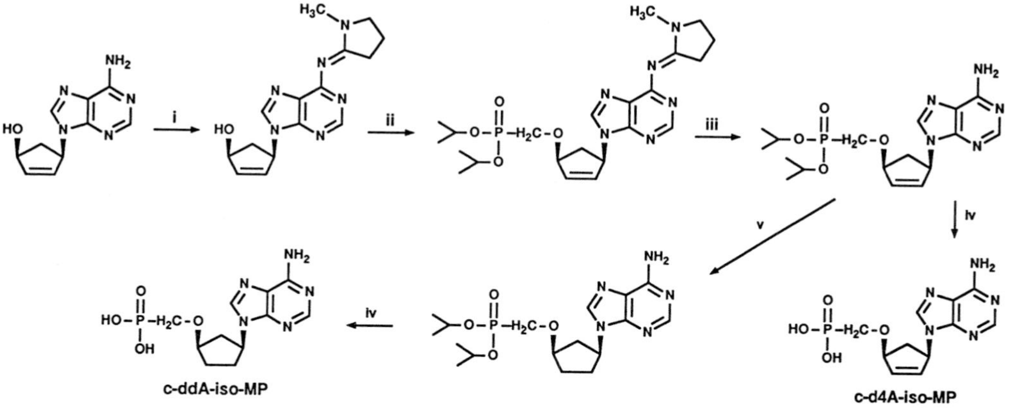

c-ddA-iso-MP

c-d4A-iso-MP

i: OEt, CH$_2$Cl$_2$ or Py, r.t.; 98% or 73%

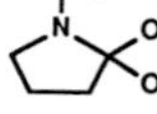

ii: Ar, DMF, NaH,Tos-O-CH$_2$-P(O)(OiPr)$_2$ [prepared via PCl$_3$ + iPrOH → H-P(O)(OiPr)$_2$ (81%) → +(CH$_2$O)$_3$/NEt$_3$/Δ → HO-CH$_2$-P(O)(OiPr)$_2$ (80%) → + Tos-Cl/NEt$_3$/Et$_2$O/-10°C → Tos-O-CH$_2$-P(O)(OiPr)$_2$ (61%)], r.t., 24h; 59%

iii: NH$_4$OH, Δ, 8h; 98%

iv: DMF, TMS-Br, r.t., 3h, then: acetone/H$_2$O (EtOH); 58-66%

v: Pd/C (10%), H$_2$, iPrOH; 74-98%

Having purified the N9-regioisomer of the adenine derivative we protected the amino function by transforming it into the exocyclic amidine derivative. This product, which was obtained by reaction with N-methyl-2,2-diethoxy-pyrrolidine, proved to be more stable than the acyclic N,N-dimethyl-methylidene derivative that is obtained by reaction with DMF-diethylacetal. The protected phosphonate nucleoside analogue of 2',3'-dideoxy-2',3'-didehydro-adenosine was obtained by reaction of the allylic alcohol with sodium hydride and diisopropyl [(p-tolylsulfonyl)oxy]methanephosphonate in DMF at r.t. Ammonolysis of the exocyclic amidine, followed by phosphonate ester cleavage with trimethylsilyl bromide (TMS-Br) [13] in DMF at r.t. and aqueous work-up yielded the carbocyclic phosphonate nucleoside analogue (c-d4A-iso-MP) of d4A monophosphate (Scheme 8). Thus, the first carbocyclic phosphonate nucleoside analogue was obtained via a 6 step synthesis starting from cyclopentadiene and adenine.

Hydrogenation of the olefinic phosphonate ester on 10% Pd/C followed by TMS-Br mediated ester cleavage produced the corresponding saturated phosponic acid (c-ddA-iso-MP), which is the analogue of ddA monophosphate (Scheme 8).

3.1.2. The Hypoxanthine Congeners: c-d4I-iso-MP and c-ddI-iso-MP

The 2',3'-dideoxy- and 2',3'-dideoxy-2',3'-didehydroinosine derivatives were easily prepared by treatment of the respective adenine phosphonates with sodium nitrite/acetic acid/hydrochloric acid. The respective phosphonic acids - the carbocyclic phosphonate nucleotide isosteres of d4I (c-d4I-iso-MP) and

Scheme 9: Carbocyclic Phosphonate Nucleosides: <u>Hypoxanthine</u>

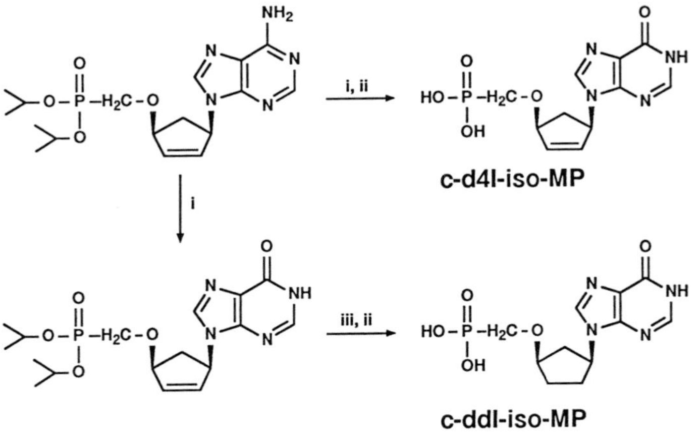

c-d4I-iso-MP

c-ddI-iso-MP

i: NaNO$_2$ / HOAc / 1 N HCl / H$_2$O, r.t., 5h; 86%

ii: DMF / TMS-Br, r.t. 3h; then: acetone / H$_2$O; 69-80%

iii: Pd / C (10%) / H$_2$; MeOH; 92%

ddI **(c-ddI-iso-MP)** were produced, again by reaction with TMS-Br in DMF and aqueous work-up (Scheme 9).

3.1.3. The Guanine Congeners: c-d4G-iso-MP and c-ddG-iso-MP

The guanine derivatives could not be prepared directly, because guanine itself did not react with 3,4-epoxycyclopentene and the Pd(0)-based catalyst; presumably because of its extremely low solubility in the THF/DMSO system.

A homogenous reaction mixture was obtained with persilylated guanine, but a mixture of the N9- and N7-regioisomers was formed in 59% combined yield (Scheme 10). Persilylated N^2-acetyl-guanine, which in alkylation reactions normally shows a strong preference for the thermodynamically more stable N9-isomer, gave a 27% yield of one single regioisomer, which was identified to be the N7-isomer (Scheme 10).

Another guanine derivative designed by Morris Robins et al. [14] to yield the N9-isomer exclusively in ribosylation reactions is 2-acetamido-6-diphenylcarbamoyloxypurine. Neither that one, nor its persilylated derivative, reacted to give the desired N9-regioisomer. Only the N7-isomer was formed in low yield (Scheme 10).

Scheme 10: Carbocyclic Phosphonate Nucleosides: Guanine

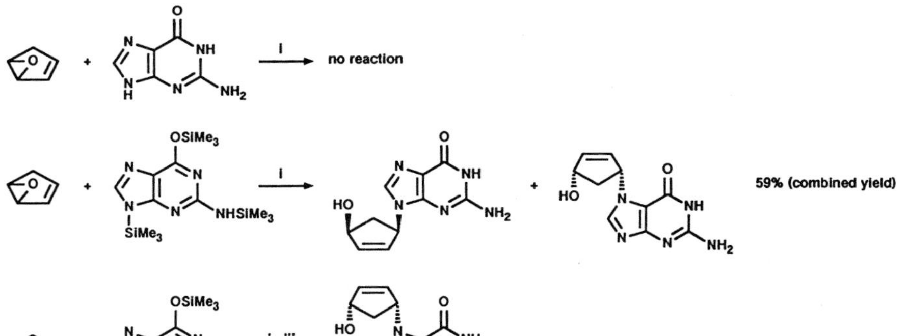

I: Ar, THF, Pd(OAc)$_2$ - P(OiPr)$_3$ - nBuLi (5 mol%), THF/DMSO, 0°C → r.t., 3d
II: MeOH, Δ, 3h, then: Ac$_2$O, DMAP, CH$_2$Cl$_2$, r.t.
III: H$_2$NCH$_3$/H$_2$O (40%), MeOH, Δ

Another chemical equivalent of guanine, however, 2-amino-6-chloropurine reacted smoothly with 3,4-epoxycyclopentene under Pd(0)-catalysis to give the N9-regioisomer exclusively in 82% yield. Treatment of this compound with 2N HCl at 70°C for 2 hours yielded the guanine derivative which, after protection of the amino functionality, was reacted with diisopropyl[(p-tolylsulfonyl)oxy]methanephosphonate to give the protected phosphonate ester in 58% yield. Deblocking of the exocyclic amidine was achieved with aqueous ammonia, and the phosphonate ester was removed by treatment with TMS-Br in DMF and aqueous work-up. Thus, the carbocyclic phosphonate nucleoside analogue (c-d4G-iso-MP) of d4G monophosphate was obtained in good yield (Scheme 11). The saturated phosphonic acid (c-ddG-iso-MP) was prepared by catalytic hydrogenation of the respective olefinic phosphonate ester (Scheme 11).

Scheme 11: Carbocyclic Phosphonate Nucleosides: <u>Guanine</u>

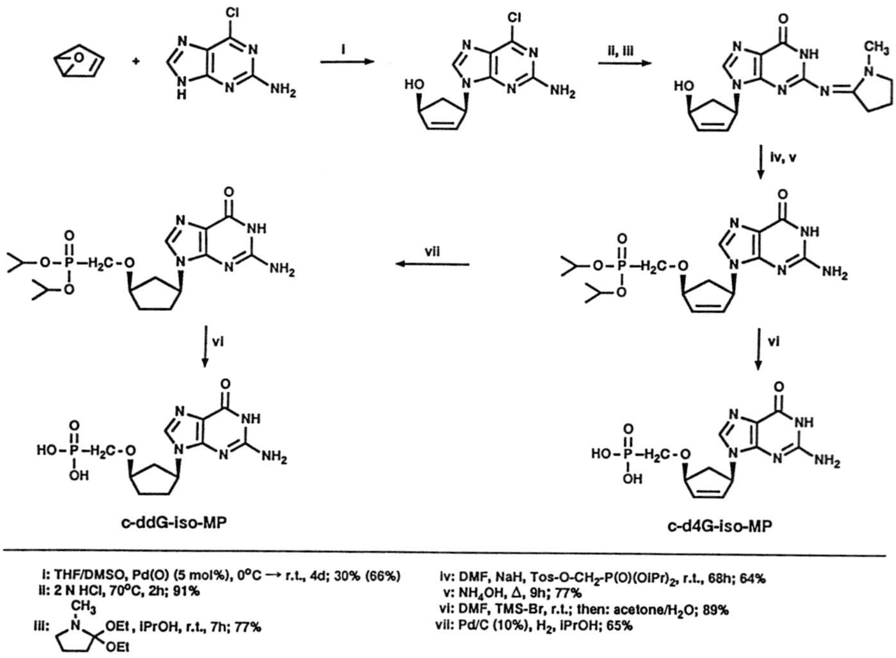

I: THF/DMSO, Pd(0) (5 mol%), 0°C → r.t., 4d; 30% (66%)
II: 2 N HCl, 70°C, 2h; 91%

III: (structure) OEt, IPrOH, r.t., 7h; 77%

iv: DMF, NaH, Tos-O-CH₂-P(O)(OIPr)₂, r.t., 68h; 64%
v: NH₄OH, Δ, 9h; 77%
vi: DMF, TMS-Br, r.t.; then: acetone/H₂O; 89%
vii: Pd/C (10%), H₂, IPrOH; 65%

3.2. Synthesis of Carbocyclic Pyrimidine Phosphonate Nucleoside Analogues

3.2.1. The Cytosine Congeners: c-d4C-iso-MP and c-ddC-iso-MP

The chemistry now developed was next applied to the pyrimidine series. The simplest derivative to prepare was the adduct with cytosine. The Pd(0)-catalyzed reaction of 3,4-epoxycyclopentene with cytosine gave the expected N1-addition product in 79% isolated yield in a regio- and stereospecific manner. As with the adenine derivative, the amino function was protected as the exocyclic amidine. Williamson ether synthesis lead to the phosphonate ester which, after deprotection and TMS-Br-induced cleavage, yielded the carbocyclic phosphonate nucleoside analogue (c-d4C-iso-MP) of d4C-monophosphate (Scheme 12). The cyclopentane derivative (c-ddC-iso-MP) was prepared as shown with the purine series (Scheme 12).

Scheme 12: Carbocyclic Phosphonate Nucleosides: Cytosine

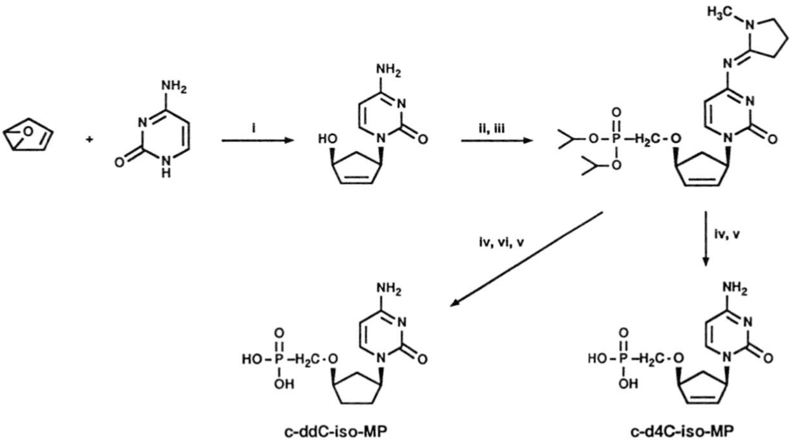

c-ddC-iso-MP

c-d4C-iso-MP

i: THF/DMSO, Pd(0) (5 mol%), 0°C → r.t., 2d; 79%

ii: OEt, Py, r.t., 4h; 90%

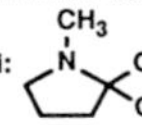

iii: DMF, NaH, Tos-O-CH₂-P(O)(OiPr)₂, r.t., 16h; 70%

iv: NH₄OH, Δ, 4h; 75%

v: DMF, TMS-Br, r.t., 3h; then: acetone/H₂O; 24% (sat: 72%)

vi: Pd/C (10%), MeOH; 96%

3.2.2. The Uracil Congeners: c-d4U-iso-MP and c-ddU-iso-MP

The uracil congeners (c-d4U-iso-MP and c-ddU-iso-MP) were synthesized by reaction of the respective cytosines with sodium nitrite/acetic acid/hydrochloric acid (Scheme 13).

Scheme13: Carbocyclic Phosphonate Nucleosides: Uracil

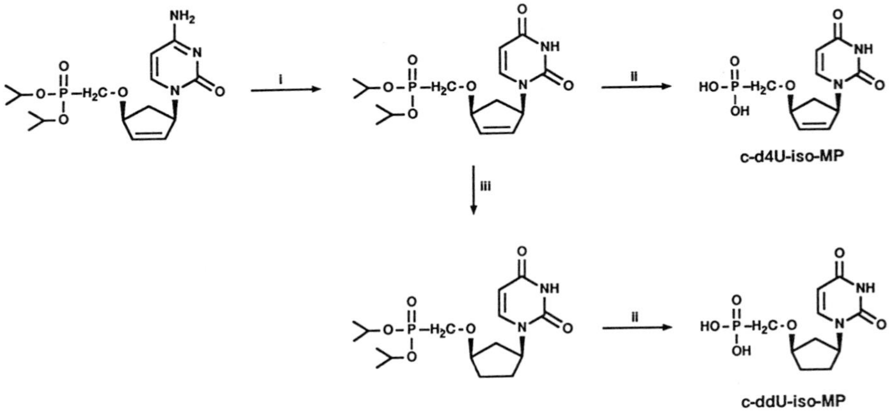

c-d4U-iso-MP

c-ddU-iso-MP

i: NaNO$_2$, HOAc/1N HCl/H$_2$O, r.t., 6h; 72%

ii: DMF/TMS-Br, r.t., 3h; then: acetone/H$_2$O; 46% (sat: 91%)

iii: Pd/C (10%), H$_2$, iPrOH; 70%

Scheme 14: Carbocyclic Phosphonate Nucleosides: Thymine

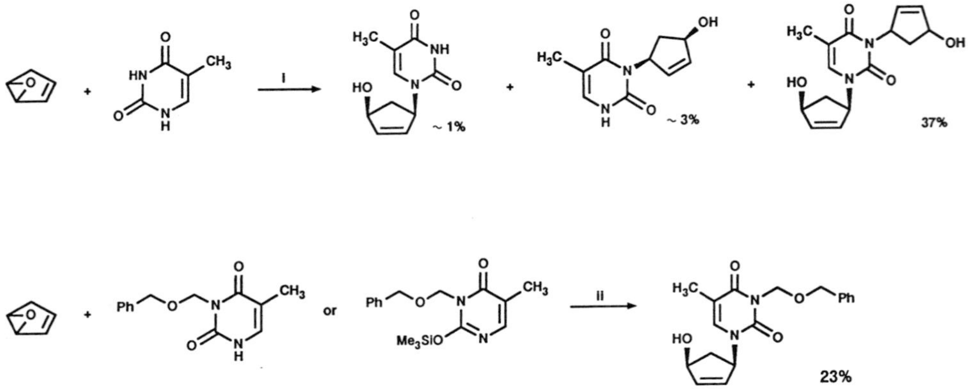

~ 1%   ~ 3%   37%

23%

i: Ar, THF/DMSO, Pd(O) (1 mol%), 0°C → r.t., 20h

ii: Ar, THF, Pd(O) (5 mol%), 0°C → r.t., 48h

3.2.3. The Thymine Congeners: c-d4T-iso-MP and c-ddC-iso-MP

Several trials were needed to find the best way to prepare the thymine derivatives: When thymine was used directly, three products were isolated in low yield: About 1% of the desired N1-adduct besides 3% of the N3-adduct and 37% of a diasteroisomeric mixture of the N1,N3-bis-adduct. A similar product distribution was obtained when persilylated thymine was used (Scheme 14).

We then tried an N3-protected thymine derivative. To this end N3-benzyloxymethylthymine was prepared by reaction of N1-acetylthymine with benzyloxymethyl chloride and subsequent hydrolysis of the N1-acetyl group. Neither N3-benzyloxymethylthymine nor its silylated derivative gave satisfactory yields of the N1-substituted thymine derivative in the Pd(0)-catalyzed reaction with 3,4-epoxycyclopentene (Scheme 14).

Finally, similar to the preparation of the uracil/cytosine pair of compounds, we tried the detour via the 5-methylcytosine derivative. Since the known methods for the preparation of 5-methylcytosine were considered to be too lengthy to perform (four steps starting from thymine via 2,4-dithiothymine [15], or four steps starting from thymine via the labile 2,4-dimethoxy-5-methylpyrimidine [16]) we tried a method that up to now has been applied only to N1-substituted uracils or thymines.

Scheme 15: Carbocyclic Phosphonate Nucleosides: Thymine

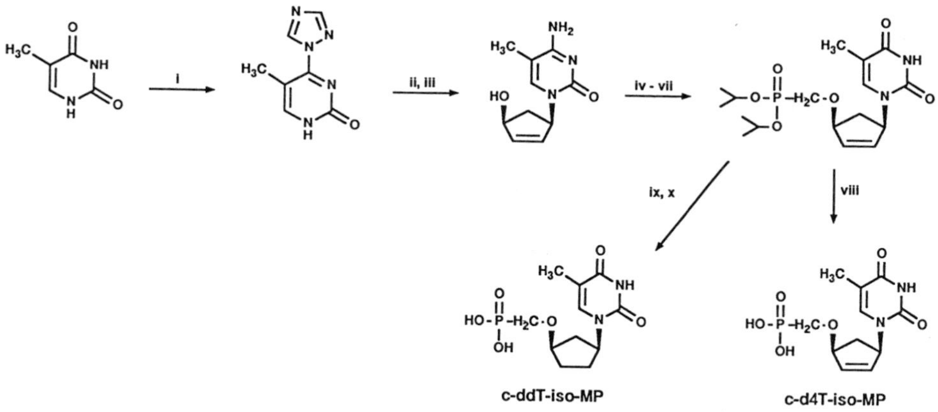

i: a. CH₃CN; 1,2,4-triazole, 0°C b. POCl₃ (2.5h), NEt₃ (0.5h) c. thymine, 0°C, 3h → r.t., 15h d. H₂O/HOAc e. 1 N NaOH, 40°C, 1h; 49%
ii: NH₄OH, Δ, 2h; 79%
iii: Ar, THF/DMSO, 3,4-epoxycyclopentene, Pd(O), 0°C → r.t., 3d; 48%
iv: [structure], Py, r.t., 3h; 98%
v: DMF, NaH, Tos-O-CH₂-P(O)(OiPr)₂, 0°C → r.t., 4h; 71%
vi: DMF, NH₃(g), Δ, 3h; 74%
vii: NaNO₂/H₂O/HOAc/ 1 N HCl, r.t., 4h; 69%
viii: DMF, TMS-Br, r.t., 4h; then: acetone/H₂O; 84%
ix: Pd/C (10%), H₂, iPrOH; 99%
x: DMF, TMS-Br, r.t., 3h; then: acetone/H₂O; 49%

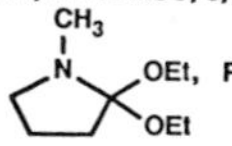

That is, we reacted thymine with phosphorus oxychloride or better with diphenyl phosphorochloridate (POCl(OPh)₂) in the presence of 1,2,4-triazole and N(Et)₃ in dry acetonitrile and isolated (after mild alkaline hydrolysis of the bis-adducts that were also formed) the corresponding triazole derivative in 49% yield [17]. The triazole group at C4 in 5-methyl-4-(1,2,4-triazole-1-yl)-pyrimidin-(1H)2-one is susceptible to nucleophilic displacement reactions and gave 5-methylcytosine on reaction with concentrated aqueous ammonia in 79% yield. This compound added to 3,4-epoxycyclopentene in the usual way giving the 1,4-cis-disubstituted cyclopentene derivative in 48% yield. Protection of the amino function, reaction with the tosylate of diisopropyl(hydroxymethyl)phosphonate, ammonolysis of the exocyclic amidine group followed by reaction with sodium nitrite in an acetic acid/hydrochloric acid mixture and subsequent phosphonate ester cleavage gave the carbocyclic phosphonate nucleoside analogue (c-d4T-iso-MP) of d4T monophosphate (Scheme 15). Catalytic hydrogenation of the respective precursor yielded the saturated derivative (c-ddT-iso-MP), that is the analogue of ddT monophosphate (Scheme 15).

4. Antiviral Testing

Scheme 16:
Carbocyclic Phosphonate Nucleosides
in vitro anti-HIV Activity

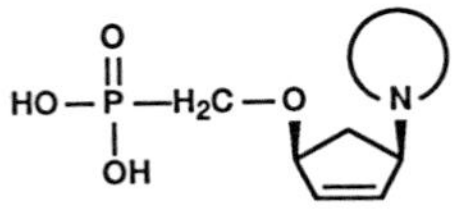

c-d4N-iso-MP

N	IC50 [µg/ml] (p24 antigen)	IC50 [µg/ml] (CPE on PBL)
A	> 20	> 4
H	> 100	> 100
G	20	20
C	> 20	> 100
U	> 100	> 100
T	> 100	> 100

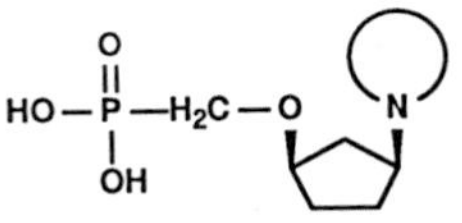

c-ddN-iso-MP

N	IC50 [µg/ml] (p24 antigen)	IC50 [µg/ml] (CPE on PBL)
A	100	> 4
H	> 100	> 100
G	> 100	> 100
C	> 100	> 100
U	> 100	> 100
T	> 4 < 20	20
AZT	< 0.2	< 0.2
ddI	0.8	0.8
ddA	4	4

The antiretoviral activity of these twelve carbocyclic nucleotide isosteres was determined using HIV-1-infected human peripheral blood lymphocytes (PBL's) by comparing the amount of viral protein p24 formed in treated cells versus untreated controls employing a commercial antigen assay kit. In addition, the 50% inhibitory concentration (IC_{50}) of the virally induced cytopathic effect (CPE) was determined in the same assay system.

Whereas the unsaturated hypoxanthine (c-d4I-iso-MP), cytosine (c-d4C-iso-MP), uracil (c-d4U-iso-MP) and thymine (c-d4T-iso-MP) derivatives and the saturated adenine (c-ddA-iso-MP), hypoxanthine (c-ddI-iso-MP), guanine (c-ddG-iso-MP), cytosine (c-ddC-iso-MP), and uracil (c-ddU-iso-MP) derivatives were devoid of activity when compared with the standards AZT, ddI or ddA, the unsaturated guanine analogue (c-d4G-iso-MP) showed some activity, and the **unsaturated adenine derivative (c-d4A-iso-MP)** and notably the **saturated thymine analogue (c-ddT-iso-MP)** exerted antiretroviral *in vitro*-activities comparable to the action of ddA (Scheme 16).

All the carbocyclic phosphonate nucleoside analogues prepared showed hardly any cytotoxic or cytostatic effect when tested in a cell proliferation assay (human lymphocytes).

None of the newly prepared compounds had antiviral activity when tested against DNA or other RNA viruses.

5. Conclusions

In conclusion we can say that a method was devised to prepare carbocyclic phosphonate nucleosides of adenine, hypoxanthine, guanine, cytosine, uracil, and thymine as hydrolytically stable compounds. These analogues are isosteric (and isoelectronic) with (carbo)cyclic 2',3'-dideoxy- and 2',3'-dideoxy-2',3'-didehydronucleosides. The easy to perform syntheses proceed in high yield and require only 6 - 9 steps starting with cyclopentadiene and the respective purine or pyrimidine base.

Some of these newly prepared analogues, especially the saturated carbocyclic thymine phosphonic acid isostere, show anti-HIV-1 activity in *in vitro* systems. The antiretroviral activity is a specific one as there is no activity against other RNA- or DNA-viruses *in vitro*.

The carbocyclic phosphonate nucleoside analogues prepared are non-toxic in a cell-proliferation assay using human lymphocytes.

Acknowledgment. We gratefully acknowledge Dr. H.-W. Fehlhaber, Dr. H. Kogler, and M. Weber of the Analytical Laboratory of Pharma Research, HOECHST AG, for the uv, mass, and nmr analyses of the newly prepared compounds.

References

1.   H. Mitsuya, R. Yarchoan, S. Kageyama, S. Broder: The FASEB Journal **5** (1991) 2369-2381

2.   H. Mitsuya, K. J. Weinhold, P. A. Furman, M. H. St. Clair, S. Nusinoff-Lehrman, R. C. Gallo, D. Bolognesi, D. W. Barry, S. Broder: Proc. Natl. Acad. Sci. USA **82** (1985) 7096-7100; D. D. Richman, M. A. Fischl, M. H. Grieco, M. S. Gottlieb, P. A. Volberding, O. L. Laskin, J. M.

454   G. Jähne et al.

Leedom, J. E. Groopman, D. Mildvan, M. S. Hirsch, G. G. Jackson, D. T. Durack, S. Nusinoff-Lehrman, and the AZT Collaborative Working Group: N. Engl. J. Med. **317** (1987) 192-197

3.   H. Mitsuya, S. Broder: Proc. Natl. Acad. Sci. USA **83** (1986) 1911-1915

4.   T.-S. Lin, R. F. Schinazi, W. H. Prusoff: Biochem. Pharmacol. **36** (1987) 2713-2718

5.   V. E. Marquez, C. K.-H. Tseng, H. Mitsuya, S. Aoki, J. A. Kelley, H. Ford, Jr., J. S. Roth, S. Broder, D. G. Johns, J. S. Driscoll: J. Med. Chem. **33** (1990) 978-985; J. D. Stoeckler, C. Cambor, R. E. Parks, Jr.: Biochemistry **19** (1980) 102-107

6.   B. Belleau, D. Dixit, N. Nguyen-Ga, J. L. Kraus: Abstracts of Papers, Fifth International Conference On AIDS, Montreal; Abstract No. T.C.O.1, Ottawa, Ontario, 1989

7.   L. L. Bondoc, Jr., W. M. Shannon, J. A. Secrist III, R. Vince, A. Fridland: Biochemistry **29** (1990) 9839-9843

8.   E. DeClercq, A. Van Aerschot, P. Herdewijn, M. Baba, R. Pauwels, J. Balzarini: Nucleosides & Nucleotides **8** (1989) 659-671; H. Souydeyns, X.-J. Yao, Q. Gao, B. Belleau, J.-L-Kraus, N. Nguyen-Ba, B. Spira, M. A. Wainberg: Antimicrob. Agents Chemother. **35** (1991) 1386-1390; Y. H. Yeom, R. P. Remmel, S. H. Huang, M. Hua, R. Vince, C. L. Zimmermann: Antimicrob. Agents Chemother. **33** (1989) 171-175

9.   E. DeClercq, A. Holy, I. Rosenberg, T. Sakuma, J. Balzarini, P. C. Maudgal: Nature **323** (1986) 464-467

10.   G. Jähne, A. Müller, H. Kroha, M. Rösner, O. Holzhäuser, C. Meichsner, M. Helsberg, I. Winkler, G. Rieß: Tetrahedron Lett., **33** (1992) 5335-5338

11.   Bristol-Myers Squibb, European Patent Application EP 369409 (priority 14.11.1988); M. Coe, H. Hilpert, A. Noble, M. R. Peel, S. M. Roberts, R. Storer: J. Chem. Soc., Chem. Commun. (1991) 312-314; Merrell Dow Pharmaceuticals Inc., European Patent Application EP 468 119 (priority 24.07.1990); C. U. Kim, J. J. Bronson, L. M. Ferrara, J. C. Martin: Bioorganic & Medicinal Chem. Lett. **2** (1992) 367-370

12.   B. M. Trost, G. H. Kuo, T. Benneche: J. Am. Chem. Soc. **110** (1988) 621-622

13.   C. E. McKenna, M. T. Higa, N. H. Cheung, M.-C. McKenna: Tetrahedron Lett. (1977) 155-158

14.   M. J. Robins, R. Zou, F. Hansske, D. Madej, D. L. J.Tyrrell: Nucleosides & Nucleotides **8** (1989) 725-741

15.   G. H. Hitchings, G. B. Elion, E. A. Falco, P. B. Russell: J. Biol. Chem. **177** (1949) 357-360

16.   H. Ballweg: Tetrahedron Lett. (1969) 2171-2173

Antiviral Acyclic Nucleotide Analogues

Antonín Holý, Hana Dvořáková and Jindřich Jindřich

Institute of Organic Chemistry and Biochemistry

Czechoslovak Academy of Sciences, 166 10 Praha 6 (Czechoslovakia)

Summary: Phosphonomethyl ethers of acyclic nucleosides exhibit potent activity against DNA viruses and retroviruses: N-(3-hydroxy-2-phosphonomethoxypropyl) (HPMP) derivatives act against DNA viruses, N-(2-phosphonomethoxyethyl) (PME) compounds exhibit mixed anti-DNA viral and antiretroviral effect, and N-(3-fluoro-2-phosphonomethoxypropyl) (FPMP) derivatives inhibit exclusively retroviruses. Our present study concerns the principles which govern the specificity of the antiviral effect. We have discovered that related N-(2-phosphonomethoxypropyl) derivatives of purine bases (PMP-derivatives) also suppress exclusively retroviruses. Their congeners containing other alkyl or cycloalkyl substituents at the position C2 of the side-chain are devoid of antiviral activity.

Biological activity of adenine, 2,6-diaminopurine (3-deazaadenine and cytosine) derivatives in the HPMP, FPMP and PMP series is enantiospecific (limited to the (S)-HPMP, (S)-FPMP and (R)-PMP-enantiomers); guanine derivatives in all three series are antivirally active irrespective of the absolute configuration of the molecule.

1. Introduction

Our discovery of antiviral activity in the series of acyclic nucleotide analogs (phosphonomethyl ethers of acyclic nucleosides, I) [1] initiated a detailed investigation of the structure-activity relationship in this structural group [2-5]. This study focused on alterations in three basic components of the molecule: heterocyclic base, side chain and phosphorus acid linkage. The antiviral activity is manifested by three closely related groups of compounds: N-(3-hydroxy-2-phosphonomethoxypropyl) (HPMP-compounds, Ib) [6,7], N-(2-phosphonomethoxyethyl) (PME-compounds, Ia) [6-8] and N-(3-fluoro-2-phosphonomethoxypropyl) derivatives of heterocyclic bases (FPMP-compounds, Ic) [9]; in all three series, adenine, 2,6-diaminopurine and guanine derivatives exhibit the most potent activity. In some cases, also their 2-aminopurine [6,7] and 3-deazaadenine [10] congeners inhibit viral multiplication; the cytosine derivative HPMPC is the only pyrimidine representative in these series which has an antiviral activity. Yet, it is in many respects the most active compound among these analogs [11,12].

The narrow margin for structural variations within these groups excludes any additional substitution of the base, introduction of any branching substituents or enlargement of the side-chain, or any additional changes at the O-C-P linkage [2,13]. In the present work we have focused our attention on the following problems:

(A) Structural principles of antiviral specificity. - None of the above compounds has any effect upon the RNA viruses tested. The HPMP-derivatives inhibit all DNA viruses tested so far. PME-compounds act against both the DNA viruses [1,6-8] and retroviruses [14,15], while FPMP-derivatives suppress solely retroviruses [9]. Therefore, we decided to search for structural parameters which determine the specificity of the antiviral effect.

(B) Enantiospecifity of antiviral action. HPMP- and FPMP-derivatives bear a chirality center at C2. Originally, it was believed that the (S)-enantiomers are the active forms of the two [1,3]; later, there were some indications which directed us to examine a possible involvement of the base in such a distinction. The discrimination of the antipodes might be due to the differences in transport of the compounds through the membrane, activation of the analogue (phosphorylation) [16] or to different response of target enzyme to such metabolites [9,17,18]. To substantiate a biochemical investigation of the principles involved in this selectivity it was necessary to obtain additional data on the differences in biological response between base-modified enantiomers in the HPMP- and FPMP-series.

B-CH$_2$CH-OCH$_2$P(O)(OH)$_2$
                $\|$
                R
	A-CH$_2$CH-OCH$_2$P(O)(OH)$_2$
	                $\|$
	                R

I a R=H I b R=(S)-CH$_2$OH I c R=(S)-CH$_2$F I d R=CH$_3$	II a R=CH$_2$OCH$_3$, CH$_2$OCH(CH$_3$)$_2$, CH$_2$OC$_8$H$_{17}$ II b R=CH$_3$ II c R=CF$_3$ II d R=N$_3$ II e R=NH$_2$ II f R=CH$_2$OCH$_2$P(O)(OH)$_2$ II g R=C$_2$H$_5$, C$_3$H$_7$, i-C$_4$H$_9$ II h R=cyclopropyl, cyclohexyl II i R= C$_6$H$_5$

2. Effect of side-chain substitution on the specificity of antiviral action.

In the first phase of our study we have prepared a series of congeners of HPMPA and FPMPA (II) by substitution of 9-(2-phosphonomethoxyethyl)adenine at the ß-position of the side chain.

Some of the compounds of this series have been prepared earlier: Methyl [2], 2-propyl [19] or n-octyl [2] ethers of HPMPA (IIa), were obtained from the corresponding 9-(3-alkoxy-2-hydroxypropyl)adenines and di(2-propyl) p-toluenesulfonyloxymethylphosphonate (IV). This reaction consists in treatment of protected acyclic nucleosides bearing an isolated hydroxyl group with the synthon IV in the presence of sodium hydride. The resulting diester resists any cleavage of the ester linkage except for the transsilylation with bromo(iodo)trimethylsilane and subsequent hydrolysis:

$$A^*\text{-}CH_2CHCH_2OR \quad + \quad TsOCH_2P(O)(OiPr)_2 \quad \longrightarrow \quad A^*\text{-}CH_2CHOCH_2P(O)(OiPr)_2$$

$$\underset{\text{OH}}{|} \qquad\qquad\qquad\qquad\qquad\qquad\qquad\qquad\qquad \underset{CH_2OR}{|}$$

$$\textbf{III} \qquad\qquad\qquad\qquad \textbf{IV} \qquad\qquad\qquad\qquad\qquad\qquad \textbf{V}$$

$$\downarrow$$

A ...adenin-9-yl , A*...N6-protected adenin-9-yl ,
iPr ...2-propyl , Ts ...p-toluenesulfonyl residue

$$ACH_2CHOCH_2P(O)(OH)_2$$
$$\underset{CH_2OR}{|}$$
$$\textbf{II}$$

This reaction can be performed also with other alkyl esters (methyl, ethyl [20] or benzyl [21]) analogues of compound IV. With the 2-propyl ester, the alkylation reaction at the heterocyclic base by the ester-bound alkyl residue is largely limited. The synthon is easily available by the reaction of di(2-propyl)phosphite with 1,3,5-trioxane or paraformaldehyde followed by tosylation of the intermediary hydroxymethylphosphonate [22].

The azidomethyl derivative IId was prepared in a similar manner from the corresponding alcohol [23]. Its hydrogenation afforded the aminomethyl congener of HPMPA (IIe). None of these compounds was active against either DNA viruses or retroviruses.

The starting alcohols are generally obtainable by the alkylation of adenine with a suitably substituted oxirane. This reaction proceeds largely regiospecifically at the N9 position. The intermediate is then selectively protected at the exocyclic amino function by the formation of its N-benzoyl or N-dimethylaminomethylene derivative.

The presence of a hydroxyl group in the HPMP-compounds is obviously essential for their activity against DNA viruses. Electron pairs at the oxygen atom do not suffice for the interaction with the enzyme in which the hydroxyl group takes part in a proton donating capacity.

Notwithstanding, the replacement of this hydroxyl group by a hydrogen atom, i.e. introduction of the methyl group at the ß-position of PME-substructure as in the 2-phosphonomethoxypropyl derivative IIb resulted in a compound with a clean antiretroviral activity (vide infra), similar to that described for the fluoromethyl congener FPMPA (Ic).

$$CF_3CH(OH)CH_2Br \quad + \quad Ade \quad \longrightarrow \quad A\text{-}CH_2\text{-}CH\text{-}CF_3$$

$$\textbf{VI} \qquad\qquad\qquad\qquad\qquad\qquad\qquad\qquad\qquad \underset{OH}{|}$$

$$\textbf{VII}$$

$$\downarrow$$

$$A\text{-}CH_2CH\text{-}OCH_2P(O)(OH)_2$$
$$\underset{CF_3}{|} \qquad \textbf{IIc}$$

We have synthesized also the trifluoromethyl derivative IIc by the condensation of 9-(2-hydroxy-3,3,3-trifluoropropyl)adenine with the synthon IV. The starting compound VII was obtained by the alkylation of adenine with the compound VI: (It is interesting to note that neither the fluoromethyl nor the trifluoromethyl group undergoes any substantial degradation under the conditions of sodium hydride promoted condensation with the synthon IV.)

However, replacement of the fluoromethyl group by a trifluoromethyl substituent as in compound IIIc causes loss of antiviral activity.

To ascertain whether the anti-retroviral activity could be ascribed solely to the absence of any substituent at the ß-position or is limited to a sterically undemanding group, additional compounds bearing linear or branched alkyl groups were prepared from the appropriate 9-hydroxyalkyladenines. The starting compounds were obtained by the alkylation of adenine with 1-O-tosylalkane-1,2-diols (VIII):

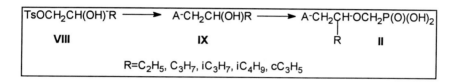

The reaction is best promoted by cesium carbonate. The required synthons were synthesized from alkane-1,2-diols, some of which were commercially available. Additional compounds were obtained by Grignard reaction of benzyloxyacetaldehyde with alkylmagnesium halides followed by hydrogenolysis and tosylation. This procedure was used for the preparation of 2-(2-propyl), 2-(3-methylpropyl) and 2-cyclopropylPMEA (IIg).

$$\text{BnOCH}_2\text{CH}=\text{O} \; + \; \text{RMgX} \longrightarrow \text{BnOCH}_2\text{CH(OH)R} \longrightarrow \longrightarrow \text{VIII}$$

Adenine was alkylated also by substituted oxiranes to afford compounds IX which gave the required phosphonate derivatives II after protection, condensation with the synthon IV and deprotection. This access was chosen for the preparation of the 2-cyclohexyl and 2-phenyl derivative of PMEA (IIh, IIi).

None of these alkyl or cycloalkyl derivatives II displayed any substantial antiviral activity. We have not yet collected conclusive evidence to explain the nature of the structural principles which determine antiviral activity and/or specificity in this series. Investigation of the effects of substituents with distinct electronegative or electropositive character at the ß-carbon might afford the missing data.

3. Effect of absolute configuration on antiviral activity

This study was performed in three series of phosphonate analogues. (R)-and (S)-HPMP-derivatives were synthesized by the reaction sequence which involves the introduction of the phosphonomethyl residue to a specifically protected acyclic nucleoside bearing a 2'-OH function, i.e. base protected 3'-O-trityl derivatives of N-(2,3-dihydroxypropyl) compounds (XII) (vide supra) [24]. These materials were in turn obtained by alkylation of heterocyclic bases with (R)-or (S)-glycidol derivatives (X), N-protection and tritylation. Condensation with the synthon IV in the presence of sodium hydride followed by deprotection gave the enantiomeric pure phosphonates Ib [22,25].

Retroviral inhibitors of the FPMP class (Ic), were obtained with the use of synthon XVI which possess all the structural feature of the side-chain in addition to a suitable leaving group on the α-carbon: such compounds XVI were synthesized by an enantiospecific multi-step procedure [26] which starts from 1-fluoro-2,3-propanediols (XIII). They were transformed into the 3-O-tosyl derivatives XIV and subsequently to the 2-O-methoxymethyl ethers XV.

On treatment with acetic anhydride in the presence of boron trifluoride etherate - diethyl ether complex followed by reaction with tri(2-propyl)phosphite [27], compounds XV were converted to the target molecules XVI. On alkylation with these synthons and following deprotection with bromotrimethylsilane, heterocyclic bases were transformed into the (R)- and (S)-enantiomers of FPMP-compounds (Ic) respectively.

Guanine, 2-aminopurine and 2,6-diaminopurine derivatives XVIII were obtained from a common 2-amino-6-chloropurine intermediate XVII by replacement of the chlorine atom with hydrogen, amino or hydroxy function using appropriate transformations [26].

The third group of phosphonates which was examined in this context are the N-(2-phosphonomethoxypropyl) (PMP-) derivatives Id. In this case, the reaction sequence starts from 1-O-benzyl-(2R)- or -(2S)-propane-2,3-diol (XIX) respectively, which afford upon chloromethylation and subsequent reaction with tri(2-propyl)phosphite the benzyl derivatives XX. These intermediates are subjected to hydrogenolysis on Pd/C and tosylation to give the final synthon XXI. Similarly, as in the FPMP-series, these synthons afford required the PMP-compounds (Id) by condensation with purine or pyrimidine bases in the presence of sodium hydride or cesium carbonate, followed by deprotection and/or cleavage of the ester bonds. In this case, the regiospecificity as well as the reaction rate of the alkylation are favorably influenced by cesium carbonate. Transformation of the heterocyclic base in the intermediates XXII led to guanine, 2-aminopurine and 2,6-diaminopurine derivatives in analogy to XVIII.

$$C_6H_5CH_2OCH_2\overset{*}{C}H\text{-}CH_3 \longrightarrow C_6H_5CH_2OCH_2\overset{*}{C}H\text{-}CH_3 \longrightarrow TsOCH_2\overset{*}{C}H\text{-}CH_3$$

OH	OCH$_2$P(O)(OiPr)$_2$	OCH$_2$P(O)(OiPr)$_2$
XIX	**XX**	**XXI**

$$B\text{-}CH_2\overset{*}{C}H\text{-}CH_3 \longleftarrow B\text{-}CH_2\overset{*}{C}H\text{-}CH_3$$

OCH$_2$P(O)(OH)$_2$	OCH$_2$P(O)(OiPr)$_2$
Id	**XXII**

An alternative access to the PMP-compounds is making use of the attachment of the phosphonomethyl residue to the hydroxyl group of the enantiomeric N-(2-hydroxypropyl) derivatives XXVII. These materials are obtained from 2-O-tetrahydropyranyl lactates XXIII by lithium alanate reduction followed by tosylation of the monoprotected diols XXIV. The tosyl esters XXV are used as alkylating agents for the substitution of the heterocyclic base; the resulting compounds XXVI afford the enantiomers of XXVII by acid hydrolysis.

$$CH_3\overset{*}{C}H\text{-}COOR \longrightarrow CH_3\overset{*}{C}H\text{-}CH_2OH \longrightarrow CH_3\overset{*}{C}H\text{-}CH_2OTs \longrightarrow CH_3\overset{*}{C}H\text{-}CH_2B$$

OHTP	OHTP	OHTP	OHTP
XXIII	**XXIV**	**XXV**	**XXVI**

$$CH_3\overset{*}{C}H\text{-}CH_2B \longleftarrow CH_3\overset{*}{C}H\text{-}CH_2B \longleftarrow CH_3\overset{*}{C}H\text{-}CH_2B$$

OCH$_2$P(O)(OH)$_2$	OCH$_2$P(O)(OiPr)$_2$	OH
Id	**XVII**	**XXVI**

THP ... tetrahydropyran-2-yl residue

In all three series examined, adenine, 2,6-diaminopurine, 2-aminopurine (and in the HPMP-series cytosine and 3-deazaadenine) derivatives display enantiospecific antiviral activity. The activity is restricted to the (S)-HPMP- and (S)-FPMP-compounds. In the series of the PMP-derivatives, solely the (R)-enantiomers inhibit multiplication of retroviruses; indeed, they have the same absolute configuration as (S)-HPMP-derivatives.

It remains to be established whether this enantiospecific antiviral activity can be interpreted in terms of discrimination between the two enantiomers during transport through the cellular membrane, during the metabolic transformation or is due to the selective interaction of the active antimetabolites with the target enzymes. Recently, it has been demonstrated that adenylate kinase from mouse leukemic cells L-1210 distinguishes between the enantiomers of HPMPA and FPMPA, catalyzing the phosphorylation of the (S)-enantiomers only [28].

Contrary to this enantiospecificity, all three types of (R)- and (S)-enantiomers derived from guanine are comparably active. However, the activity of (R)-HPMPG, (R)-FPMPG or (S)-PMPG is always accompanied by a substantial increase in cytotoxicity. It has been demonstrated that guanine phosphonate analogues are potent inhibitors of purine nucleoside phosphorylase [29]. It is possible that the cytotoxicity might be explained by such an interference.

Acknowledgement. The authors express their gratitude to the Grant Agency of the Czechoslovak Academy of Sciences and to Gilead Sciences (Foster City, USA) for supporting this research. The antiviral studies which were performed by Drs. E. De Clercq, J. Balzarini and others in the Rega Institute, Catholic University Leuven (Belgium) are gratefully acknowledged.

References

1. E. De Clercq, A. Holý, I. Rosenberg, T. Sakuma, J. Balzarini, P. C. Maudgal, *Nature* **323** (1986) 464-467.

2. I. Rosenberg, A. Holý, M. Masojídková, *Collect. Czech. Chem. Commun.* **53** (1988), 2753.

3. A. Holý, E. De Clercq, I. Votruba, in Nucleotide Analogues as Antiviral Agents (J. C. Martin, Ed.);ACS Symposium Series No. 401, American Chemical Society, Los Angeles, pp. 51-71 (1989).

4. J. J. Bronson, I. Ghazzouli, M. J. M. Hitchcock, R. R. Webb, J. C. Martin, *J. Med. Chem.* **32** (1989) 1457-1463.

5. R. R. Webb , J. C. Martin, *Tetrahedron Lett.* **28** (1987) 4963-4965.

6. E. De Clercq, T. Sakuma, M. Baba, R. Pauwels, J. Balzarini, I. Rosenberg, A. Holý, *Antiviral Res.* **8** (1987) 261-272.

7. A. Holý, I. Votruba, A. Merta, J. Černý, J. Veselý, J. Vlach, K. Šedivá, I. Rosenberg, M. Otmar, H. Hřebabecký, M. Trávníček, V. Vonka, R. Snoeck, E. De Clercq, *Antiviral Res.* **13** (1990) 295-312 .

8. R. Pauwels, J. Balzarini, D. Schols, M. Baba, J. Desmyter, I. Rosenberg, A. Holý, E. De Clercq, *Antimicrob. Ag. Chemother.* **32** (1988) 1025-1030.

9.   J. Balzarini, A. Holý, J. Jindřich, H. Dvořáková, Z. Hao, R. Snoeck, P. Herdewijn, D. G. Johns, E. De Clercq, *Proc. Natl. Acad. Sci. USA* **88** (1991) 4961-4965.

10.   H. Dvořáková, A. Holý, R. Snoeck, J. Balzarini, E. De Clercq, *Coll. Czech. Chem. Commun. Special Issue No. 1* **55** (1990) 113-116.

11.   R. Snoeck, T. Sakuma, E. De Clercq, I. Rosenberg, A. Holý, *Antimicrob. Ag. Chemother.* **32** (1988) 1839-1844.

12.   E. De Clercq, A. Holý, *Antimicrob. Ag. Chemother.* **35** (1991) 701-706.

13.   C. U. Kim, B. Y. Luh, P. F. Misco, J. J. Bronson, M. J. M. Hitchcock, I. Ghazzouli, C. Martin, *Nucleosides & Nucleotides* **8** (1989) 927-931.

14.   J. Balzarini, L. Naesens, P. Herdewijn, I. Rosenberg, A. Holý, R. Pauwels, M. Baba, D. G. Johns, E. De Clercq, *Proc. Natl. Acad. Sci. U.S.A.* **86** (1989) 332-336.

15.   J. Balzarini, L. Naesens, J. Slachmuylders, H. Niphuis, I. Rosenberg, A. Holý, H. Schellekens, E. De Clercq, *AIDS* **5** (1991) 21-28.

16.   I. Votruba, R. Bernaerts, T. Sakuma, E. De Clercq, A. Merta, I. Rosenberg, A. Holý, *Mol. Pharmacol.* **32** (1987) 524-529.

17.   A. Merta, I. Votruba, I. Rosenberg, M. Otmar, H. Hřebabecký, R. Bernaerts, A. Holý, Antiviral Res. 13 (1990) 209-218.

18.   J. Černý, I. Votruba, V. Vonka, I. Rosenberg, M. Otmar, A. Holý, *Antiviral Res.* **13** (1990) 253-264.

19.   A. Holý, J. Balzarini, E. De Clercq, *Int. J. Purine Pyrimidine Res.* **2** (1991) 61-66.

20.   A. Holý, I. Rosenberg, *Collect. Czech. Chem. Commun.* **47** (1982) 3447-3463.

21.   M. Krečmerová, H. Hřebabecký, A. Holý, *Collect. Czech. Chem. Commun.* **55** (1990) 2521-2536.

22.   A. Holý, *Collect. Czech. Chem. Commun.* , in press.

23.   A. Holý, *Collect. Czech. Chem. Commun.* **54** (1989) 446-454.

24.   A. Holý, I. Rosenberg, H. Dvořáková, *Collect. Czech. Chem. Commun.* **54** (1989) 2470-2501.

25.   J. J. Bronson, L. M. Ferrara, H. G. Howell, P. R. Brodfuehrer, J. C. Martin, *Nucleosides & Nucleotides* **9** (1990) 745-769.

26.   J. Jindřich, H. Dvořáková, A. Holý, *Collect. Czech. Chem. Commun.*, submitted.

27.   P. Alexander, A. Holý, H. Dvořáková, *Collect. Czech. Chem. Commun.*, submitted.

28.   A. Merta, I. Votruba, J. Jindřich, A. Holý, T. Cihlář, I. Rosenberg, M. Otmar, H. Y. Tchoon, *Biochem. Pharmacol.*, in press.

29.   K. Šedivá, A. V. Ananiev, I. Votruba, A. Holý, I. Rosenberg, *Int. J. Purine Pyrim. Res* 2, (1991) 35-40.

Antisense and Triple Helix Forming Oligonucleotides as Potential Antiviral Therapeutics

Eugen Uhlmann

Hoechst AG, Pharma Forschung G 838, D-6230 Frankfurt a. M. 80, Germany

Summary: Antisense and triple helix forming oligonucleotides can be designed and synthesized which recognize and bind to specific nucleic acid target sequences, thus selectively inhibiting gene expression. Although this new approach is still in a very early stage it has the potential to lead to major advances in antiviral and general drug therapy. The scope of this paper is to present a brief overview on the application of synthetic oligonucleotides in drug discovery, to shed some light on the problems associated with this approach, and to summarize recent progress regarding antiviral activity of antisense and triple helix forming oligonucleotides.

1. Introduction

Since Zamecnik and Stephenson in 1978 [1] proposed the use of oligonucleotides directed against complementary viral nucleic acid sequences for inhibition of virus replication, a great deal of work has been devoted to this new therapeutic principle [2 - 9]. Although many of the drugs known today act by directly inhibiting proteins such as enzymes, receptors or ion channels, therapeutic intervention at the mRNA or DNA level appears to be of advantage. On active transcription every gene is transcribed into 10^2 to 10^4 copies of mRNA which in turn are translated into 10^4 to 10^6 copies of protein molecules (Figure 1). *Antisense oligonucleotides* are synthetic oligonucleotides which bind via Watson-Crick base-pairing to complementary regions on the mRNA thereby inhibiting protein biosynthesis. In contrast, *triple helix forming oligonucleotides* bind via Hoogsteen base-pairing to the major groove of double-stranded DNA resulting in inhibition of transcription.

The attributes of these two classes of potential nucleic acids therapeutics, collectively called *oligonucleotide antagonists,* can be summarized as follows. First, due to the general base-pairing rules, therapeutic intervention using oligonucleotide antagonists is a universal approach applicable to many diseases that have been characterized on the DNA level. Potential targets are viruses, oncogenes, receptors, ion channels, and immunomodulators. Second, antisense and triple helix technology may facilitate rational drug design of lead compounds. The design of oligonucleotide antagonists aiming at the inhibition of gene expression is based upon the target nucleic acid base sequence. Thirdly, theoretical considerations as well as *in vitro* hybridization studies suggest that oligonucleotide antagonists should act highly specific with their complementary target sequences. It has been calculated that a 17-mer oligonucleotide sequence appears statistically just once in the human genome ($4 \cdot 10^9$ base pairs). Finally, it ought to be mentioned that the antisense principle is also used in nature to regulate gene expression.

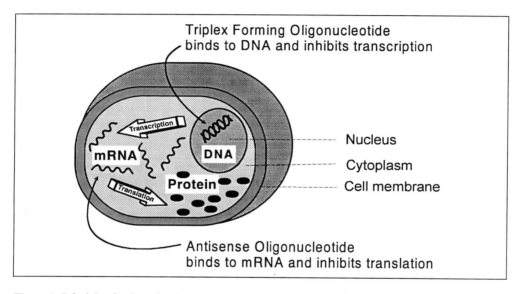

Figure 1: Principle of action of antisense oligonucleotides and triple helix forming oligonucleotides

2. Development Hurdles of Oligonucleotide Antagonists

From a pharmaceutical point of view development of antisense oligonucleotides and triplex forming oligonucleotides as potential drugs is still in its infancy. Some of the development hurdles, like the limited stability of unmodified oligonucleotides in biological systems, as well as their restricted uptake into mammalian cells is subject of numerous experiments. Only recently, the first results on pharmacokinetic properties, organ distribution, toxicity and *in vivo* activity in animal models using antisense oligonucleotides were reported. Up-scaling of oligonucleotide synthesis is still an issue of intense investigations.

2.1 Stability of Oligonucleotides Against Nucleases

Unmodified oligonucleotides with a natural phosphodiester back-bone are degraded in serum within a few hours. Cleavage is mainly brought about by 3'-exonucleases which use the 3'-terminal hydroxy group to attack the neighboring phosphodiester internucleotide linkage of the oligonucleotide. There is a whole range of possibilities for modifying oligonucleotides as summarized in Figure 2. Blocking or removal of the hydroxy group at the 3'-terminus and/or modification of the internucleotide linkage (e.g. as phosphorothioate, phosphoramidate, methylphosphonate or replacement by formacetal) are possible strategies to improve stability against nucleolytic degradation. Replacement of one oxygen atom in the phosphate moiety by sulfur, methyl or alkylamino substituents creates a center of chirality at phosphorus. Therefore, standard synthesis of oligonucleotides having n internucleotidic phosphate groups replaced by methylphosphonate moieties will result in a mixture of 2^n diastereoisomeres. Alteration of the oligonucleotide back-bone often changes binding affinity to complementary sequences as well as the lipophilicity of the oligonucleotides. Modification of the last two phosphate residues at the 3'-end of oligonucleotides as phosphorothioates or methylphosphonate enhances their nuclease stability by at least an order of magnitude while retaining their sequence specificity.

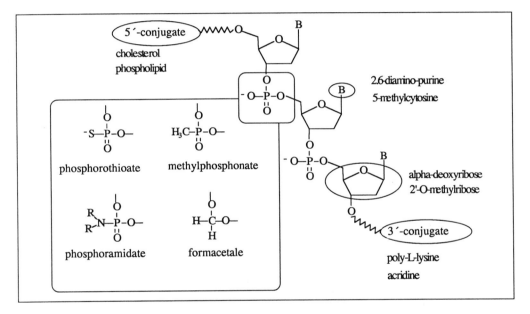

Figure 2: Possibilities for the modification of oligonucleotides (see ref. 6 and 7 for more examples)

2.2 Cellular Uptake of Oligonucleotides

Oligonucleotides are taken up by living cells by an active transport mechanism which most likely is a receptor-mediated endocytosis process [10]. This finding may explain why penetration of oligonucleotides across the cell membrane is much better than one would expect for a polyanionic compound. However, the concentration of free oligonucleotide in the cytoplasm is probably considerably lower due to packaging of the oligomer into lysosomes or endosomes. Interestingly, once the oligonucleotide is set free from this reservoir or when the oligonucleotide is microinjected into the cytoplasm, it rapidly penetrates to the cell nucleus [11]. Improvement of cell-uptake may be effected by lipophilic derivatization of the oligonucleotides, e.g. as lipophilic conjugates at the 5'-terminus [12], by conjugation to poly-L-lysine [13], or by packaging into antibody-targeted liposomes [14]. It is worthwhile to mention that chemical derivatization of oligonucleotides sometimes alters their sequence specificity in hybridizations.

3. Mechanism of Action and Activity in Cell Culture Experiments

Although antisense oligonucleotides are desigend for translational inactivation and triplex forming oligonucleotides for transcriptional inactivation, respectively, other mechanisms of action like blockade of virus adsorption to cells during infection or direct inhibition of viral enzymes (e.g. DNA polymerase or reverse transcriptase) may dominate the antiviral effects seen in cell-culture experiments. At present, some controversy exists regarding the question to what extend the antisense effect is supported through cleavage of the mRNA by cellular RNase H. RNase H recognizes hybrid molecules were one strand is RNA and the complementary strand is DNA and cleaves the RNA strand. It has been suggested that DNA/mRNA hybrids spanning only a few base-pairs are substrates of RNase H and facilitate unspecific breakdown of mRNA. This is why RNase H reaction could be deleterious to the antisense approach. However, many oligonucleotide derivatives such as methylphosphonates, α-anomeric oligonucleotides, 2'-O-alkylribose derivatives and dephospho oligonucleotide analogs do not induce RNase H cleavage.

Viruses	Type of Cells	Type of Oligomer	Effective Dose	Reference
Rous Sarcoma	Fibroblasts	P-O, P-S	10 μM	[1]
HIV	H T-Cells	P-S	0.5 μM	[15]
HSV	Vero	P-Me	50-100 μM	[16]
HSV	Vero	P-Me-(psoralen)	5 μM	[16]
VSV	L 929	P-O-(poly-Lys)	0.1 μM	[13]
Influenza	MDK	P-S	1.25 μM	[18]
Hepatitis B	PLC/PRF/5	P-O, P-S	0.3-17 μM	[19]
SV 40	CV-1	P-O-acridine (tfo)	15-30 μM	[20]

P-O: phosphodiester, P-S: phosphorothioate, P-Me: methylphosphonate, -(psoralen): psoralen-conjugate, -(poly-L-Lys): poly-L-lysine conjugate, -(acridine): acridine-conjugate, tfo: triplex forming oligonucleotide

Table 1: In vitro antiviral activity of antisense and triplex forming oligonucleotides (selected examples).

A number of papers have been published in recent years describing inhibition of virus replication in cell culture experiments. Table 1 summarizes some representative examples of *in vitro* activity of antisense and triplex forming oligonucleotides against different viruses. In most instances, the effective dose for inhibition of virus replication is in the range of 0.1 to 50 μM of oligonucleotide. Unmodified oligonucleotides are active only in absence of serum, phosphorothioate oligonucleotides seem to be more active than methylphosphonate oligonucleotides. But conjugation of psoralen to the latter can enhance their activity significantly when irradiated [16]. Phosphorothioate oligonucleotides are inhibitors of DNA polymerase [17]. Furthermore, it has been suggested that they interact with receptor proteins which are used by viruses [21]. In antiviral assay systems induction of the interferon system may also account for the biological effect. On the other side, oligonucleotide antagonists have demonstrated a broad palette of sequence specific activities against targets other than viruses such as receptors, enzymes, oncogenes, growth factors and other host gene products.

4. Synthesis of Oligonucleotides

Small scale synthesis of oligonucleotides (for reviews see ref. 22-24) is preferably performed using solid-phase chemistry employing the phosphoramidite method as introduced by Caruthers [25]. Commercially available DNA synthesizers allow syntheses at 0.2 μM to 2 mM scale, thus providing gram amounts of oligonucleotides for limited animal testing. Repetitive synthesis cycles are used starting from solid-supports (e.g. Controlled Pore Glass; CPG) to which the first nucleoside has been attached via its 3'-hydroxy group and a base-labile linker (Figure 3).

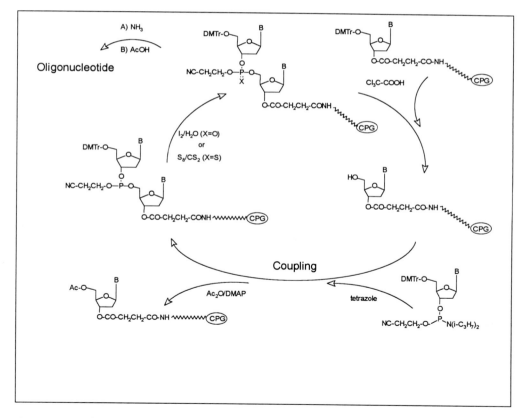

Figure 3: Synthesis of oligonucleotides by the phosphoramidite method

After removal of the acid-sensitive dimethoxytrityl group (DMTr) with trichloroacetic acid the 5'-hydroxy group is reacted with an excess of 5'-DMTr-protected nucleoside-3'-ß-cyanoethyl-N,N-diisopropyl-phosphoramidite using tetrazole as catalyst. Unreacted 5'-hydroxy compound is capped with acetic anhydride/dimethylaminopyridine to prevent failure sequences from further growing. If the resulting phosphite triester is oxidized by iodine/water/pyridine, phosphodiester oligonucleotides are obtained after deprotection. Oxidation with sulfur in CS_2 results in phosphorothioate oligoncleotides. Similarly, methylphosphonate oligonucleotides are obtained starting from phosphoramidite building blocks having a methyl group in place of the cyanoethyloxy function. Yields per cycle are in the range of 95 to 99 % depending on various parameters such as water content of solvents, quality of phosphoramidites, reaction time, type of solid-support, but predominantly on the excess of phosphoramidite used in the coupling step. Crude products are purified by polyacrylamide gel electrophoresis or by HPLC using reversed phase or ion exchange columns. Characerization of the oligomers is done by electrospray mass spectrometry and/or NMR (^{1}H or ^{31}P).

5. Acute Toxicity, Organ Distribution and *In Vivo* Activity of Oligonucleotide Antagonists

No serious acute toxicity has been detected in the few animal studies reported so far. In continuous infusion experiments on adult male rats applying 5 - 150 mg of a 27-mer phosphorothioate oligonucleotide via osmotic pumps no significant decreases in body weight or organ weights were observed [26]. In another study where phosphorothioate oligonucleotides were administered in a single dose, either intravenously or intraperitoneally, the oligomer could be detected for to 48 hours in most of the tissues [27]. Daily administration of 100 mg oligonucleotide/kg of body weight for 14 days did not cause observable toxicity in mice. Organ distribution of a 3'-end modified oligonucleotide four hours after intravenous or intraperitoneal injection has been reported to be in the following order: kidney > liver > spleen > heart, lung > muscle, ear >> brain [28]. In the non-viral field two animal studies suggest that sequence specific inhibition of gene expression by antisense oligonucleotides, administered locally by continuous perfusion or by application as a pluronic solution resulting in decreased tumor growth or inhibition of intimal arterial smooth muscle cell accumulation, respectively [29, 30]. Regarding *in vivo* activity of antisense oligonucleotides against viruses only preliminary studies have been reported thus far. Topical application of methylphosphonate oligonucleotides to mice was effective against a HSV-1 infection [31]. Oligonucleotides with alkylating groups provided protection against tick-borne encephalitis virus infection in mice [32]. Topical application of phosphorothioate antisense oligonucleotides to the cornea of mice infected with HSV-1 inhibited viral growth and cured the infection [33]. In this model the activity of the antisense oligonucleotide was equivalent to that of trifluorothymidine without the concomitant occurence of local or systemic toxicity.

6. Conclusions

It has been demonstrated in several studies that oligonucleotides are taken up by living cells and migrate even to the nucleus. Modified oligonucleotides are stable enough to survive a reasonable period of time in serum. In tissue culture experiments, antiviral activity is usually in the range of 0.1 μM to 50 μM depending mainly on the target sequence and the chemical modification of the oligonucleotide. However, chemical derivatization in some cases results in decreased sequence specificity. New oligonucleotide analogues are being investigated containing dephospho internucleoside linkages possessing enhanced properties [34, 35]. Commercially available DNA synthesizers allow for the synthesis of gram amounts of oligonucleotides thus providing enough material for initial animal studies. Preliminary animal data indicate no acute toxicity with up to 250 mg of oligonucleotide per kg in the rat. Several investigations suggest that oligonucleotide antagonists may be used for the downregulation of gene expression *in vivo*.

7. References

1.   P. Zamecnic and M. Stephenson, *Proc. Natl. Acad. Sci. USA* **75** (1978) 280-284.

2.   C. A. Stein and J. S. Cohen, *Cancer Res.* **48** (1988) 2659-2668.

3.   G. Zon, *Pharm. Res.* 5 (1988) 539-549.

4.   P. S. Miller and P. O. P. Ts'o, *Annu. Rep. Med. Chem.* **23** (1988) 295-304.

5.   C. Hélène and J. J. Toulmé, *Biochem. Biophys. Acta* **1049** (1990) 99-125.

6.   J. Goodchild, *Bioconjugate Chem.* 1 (1990) 165-197.

7.   E. Uhlmann and Peyman, *Chem. Rev.* **90** (1990) 543-584.

8.   S. T. Crooke, *Annu. Rev. Pharmacol. Toxicol.* **32** (1992) 329-376.

9.   L. Neckers, L. Whitesell, A. Rosolen and D. A. Geselowitz, *Crit. Rev. Oncogenesis* **3** (1992) 175-231.

10.   S. L. Loke, C. A. Stein, X. H. Zhang, K. Mori, M. Nakanishi, C. Subasnghe, J. Cohen and L. M. Neckers, *Proc. Natl. Acad. Sci. USA* **86** (1989) 3474-3478.

11.   J. P. Leonetti, N. Mechti, G. Degols, C. Gagnor and B. Lebleu, *Proc. Natl. Acad. Sci. USA* **88** (1991) 2702-2706.

12.   S. Regan, J. C. Marsters and N. Bischofberger, *Nucleic Acids Res.* **18** (1990) 3777-3783.

13.   L. Lemaitre, B. Bayard and B. Lebleu, *Proc. Natl. Acad. Sci. USA* **84** (1987) 648-652.

14.   J. P. Leonetti, P. Machy, G. Geglos, B. Lebleu and L. Leserman, *Proc. Natl. Acad. Sci. USA* **87** (1990) 2448-2451.

15.   M. Matsukura, K. Shinozuka, G. Zon, H. Mitsuya, J. S. Cohen and S. Broder, *Proc. Natl. Acad. Sci. USA* **84** (1987) 7706-7710.

16.   C. C. Smith, L. Aurelian, M. Reddy, P. Miller and P. O. P. Ts'o, *Proc. Natl. Acad. Sci. USA* **83** (1986) 2787-2791.

17.   W. Gao, F. Han, C. Storm, W. Egan and Y. Cheng, *Molec. Pharmacol.* **41** (1992) 223-229.

18.   J. M. Leiter, S. Agrawal, P. Palese and P. Zamecnik, *Proc. Natl. Acad. Sci. USA* **87** (1990) 3430-3434.

19.   G. Goodarzi, S. C. Gross, A. Tewari and K. Watabe, *J. Gen. Virol.* **71** (1990) 3021-3025.

20.   F. Birg, D. Praseuth, A. Zerial, N. T. Thuong, U. Asseline, T. Le Doan and C. Hélène, *Nucleic Acids Res.* **18** (1990) 2901-2908.

21.   C. Stein, M. Neckers, B. Nair, S. Mumbauer, G. Hoke and R. Pal, J. *AIDS* 4 **1991**, 686-693.

22.   E. Sonveaux, *Bioorg. Chem.* **14** (1986) 274-325.

23.   J. Engels and E. Uhlmann, Angew. Chem. **101** (1989) 733-752; *Angew. Chem. Int. Ed. Engl.* **28** (1989) 716-734.

24.   S. Beaucage and R. Iyer, *Tetrahedron* **48** (1992) 2223-2311.

25.   S. Beaucage and M. Caruthers, *Tetrahedron Lett.* **22** (1981) 1859-1862.

26.   P. Iversen, *Anti-cancer Drug Des.* **6** (1991) 531-538.

27.   S. Agrawal, J. Temsamani and J. Y. Tang, *Proc. Natl. Acad. Sci. USA* **88** (1991) 7595-7599.

28.   J. G. Zendegui, K. M. Vasquez, J. Tinsley, D. J. Kessler and M. E. Hogan, *Nucleic Acids Res.* **20** (1992) 307-314.

29.   L. Whitesell, A. Rosolen and L. M. Neckers, *Antisense Res. Developm.* **1** (1991) 343-350.

30.   M. Simons, E. Edelman, J. DeKeyser, R. Langer and R. Rosenberg, *Nature* **359** (1992) 67-70.

31.   P. Miller and P. O. P. Ts'o, *Anti-Cancer Drug Des.* **2** (1987) 117

32.   V. Vlassov, Meeting on "Oligonucleotides as antisense inhibitors of gene expression: Therapeutic implications" June 18-21 (1989), Rockville, MD.

33.   S. T. Crooke, *Bio/Technology* **10** (1992) 882-885.

34.   P.E. Nielsen, M. Egholm, R. Berg and O. Buchardt, *Science* **254** (1991) 1497-1500.

35.   E. Uhlmann and A. Peyman in S. Agrawal (ed.). Methods in Molecular Biology: Synthetic chemistry of oligonucleotide analogs. The Humana Press Inc. New Jersey. In press.

Subject Index

abietic acid 35

acarbose 3

2-acetamido-6-diphenylcarbamoyloxypurine 447

N-acetylallosamine 6

acetyl-CoA carboxylase 133

2-acetylthiazole 69

acosamine 3

acyl ketene 135

acyl radical 91

acyliminium 280

acylpyridinium radical 88, 91

1,6-addition 216

adenine 406, 408, 410, 412, 413, 439, 441, 443, 444, 449, 453, 455, 458 − 461
− alkylation of 443

adraimycin 353

AIDS 2, 4, 10, 63, 360, 439, 454

alcalase 394

aldol condensation 89, 96

aldol cyclisation 336

alkaloids 379, 387

alkaloids, oxoaporphine group 37, 39

alkylating inhibitors 177, 178

allene 319

allosamizoline 1, 6 − 8, 10

allyl tributyltin 270

allylic 1,3-strain 105

allylic anchoring principle 256

allylic azides 429

allylmetal-addition 104

allyloxycarbonyl 257

allylsilane 315 − 318, 321, 323

Alpine-borane 121

Amaryllidaceae 379, 387

amastatin 59

α-amino aldehyde 343, 344, 354
− N,N-diprotected 344, 355
− N,N-monoprotected 60, 348

amino acids, unusual 58, 61

α-amino aldehyde 56, 60

β-amino aldehyde 60

3-aminobenzoic acid 75, 79, 81-85

4-aminobenzoic acid (pABA) 82

2-amino-6-chloropurine 12, 448, 459

6-amino-6-deoxy sugars 343

β-amino-α-hydroxy acids 59

3β-amino-4α-methyl-1-sulfo-2-azetidinone 170

aminoglycoside antibiotics 343

aminopeptidase inhibitor 59

amino sugars 1, 343

3β-amino-1-sulfo-2-azetidinone 169

angiotensin converting enzyme inhibitors 64

angiotensin I convertimg enzyme 263

anhydrogalantinic acid 349 − 351

anomeric organosamarium species 330

anomeric radical 328, 330

anomeric stabilization 96

antamanide 246

anthelmintic activity 345

anthracycline antibiotics 352

anti-AIDS agents 11

antibacterial and antiphage properties 206

antibiotics
− SS-56C 343, 345
− A-396-I 343, 345

anticapsin 215, 220 − 227

antidiabetic 3

antifungal activity 92, 140, 145, 153 − 159

antineoplastic agents 353

antiretoviral activity 453, 456, 458